A Comprehensive Approach to Molecular Biology

A Comprehensive Approach to Molecular Biology

Edited by Dean Watson

SYRAWOOD
PUBLISHING HOUSE

New York

Published by Syrawood Publishing House,
750 Third Avenue, 9th Floor,
New York, NY 10017, USA
www.syrawoodpublishinghouse.com

A Comprehensive Approach to Molecular Biology
Edited by Dean Watson

International Standard Book Number: 978-1-68286-590-3 (Hardback)

Cataloging-in-Publication Data

A comprehensive approach to molecular biology / edited by Dean Watson.
    p. cm.
Includes bibliographical references and index.
ISBN 978-1-68286-590-3
1. Molecular biology. 2. Biomolecules. I. Watson, Dean.
QH506 .C66 2018
572.8--dc23

TABLE OF CONTENTS

PREFACE

The branch of science which studies the biological activities at a molecular level is known as molecular biology. There are a number of techniques used in molecular biology, some of these are molecular cloning, gel electrophoresis, polymerase chain reaction, microarrays, etc. Different approaches, evaluations, methodologies and advanced studies on molecular biology have been included in this book. It includes some of the vital pieces of work being conducted across the world, on various topics related to molecular biology. Through it, we attempt to further enlighten the readers about the new concepts in this field.

After months of intensive research and writing, this book is the end result of all who devoted their time and efforts in the initiation and progress of this book. It will surely be a source of reference in enhancing the required knowledge of the new developments in the area. During the course of developing this book, certain measures such as accuracy, authenticity and research focused analytical studies were given preference in order to produce a comprehensive book in the area of study.

This book would not have been possible without the efforts of the authors and the publisher. I extend my sincere thanks to them. Secondly, I express my gratitude to my family and well-wishers. And most importantly, I thank my students for constantly expressing their willingness and curiosity in enhancing their knowledge in the field, which encourages me to take up further research projects for the advancement of the area.

Editor

Secondary Ion Mass Spectrometry Imaging of *Dictyostelium discoideum* Aggregation Streams

John Daniel DeBord[1], Donald F. Smith[2], Christopher R. Anderton[3], Ron M. A. Heeren[2], Ljiljana Paša-Tolić[3], Richard H. Gomer[4], Francisco A. Fernandez-Lima[1]*

1 Department of Chemistry and Biochemistry, Florida International University, Miami, Florida, United States of America, **2** FOM Institute AMOLF, Science Park 104, Amsterdam, The Netherlands, **3** Environmental Molecular Sciences Laboratory, Pacific Northwest National Laboratory, Richland, Washington, United States of America, **4** Department of Biology, Texas A&M University, College Station, Texas, United States of America

Abstract

High resolution imaging mass spectrometry could become a valuable tool for cell and developmental biology, but both, high spatial and mass spectral resolution are needed to enable this. In this report, we employed Bi_3 bombardment time-of-flight (Bi_3 ToF-SIMS) and C_{60} bombardment Fourier transform ion cyclotron resonance secondary ion mass spectrometry (C_{60} FTICR-SIMS) to image *Dictyostelium discoideum* aggregation streams. Nearly 300 lipid species were identified from the aggregation streams. High resolution mass spectrometry imaging (FTICR-SIMS) enabled the generation of multiple molecular ion maps at the nominal mass level and provided good coverage for fatty acyls, prenol lipids, and sterol lipids. The comparison of Bi_3 ToF-SIMS and C_{60} FTICR-SIMS suggested that while the first provides fast, high spatial resolution molecular ion images, the chemical complexity of biological samples warrants the use of high resolution analyzers for accurate ion identification.

Editor: Jeffrey Graham Williams, University of Dundee, United Kingdom

Funding: The funders had no role in study design, data collection and analysis, decision to publish, or preparation of the manuscript. This work was supported by NIGMS grants R00GM106414 to FF-L and GM102280 to RHG. A portion of this research was performed at the W. R. Wiley Environmental Molecular Sciences Laboratory, a national scientific user facility sponsored by the Department of Energy's Office of Biological and Environmental Research (DOE-BER) and located at Pacific Northwest National Laboratory (PNNL). This work is part of the research program of the Foundation for Fundamental Research on Matter (FOM), which is part of The Netherlands Organization for Scientific Research (NWO).

Competing Interests: The authors have declared that no competing interests exist.

* E-mail: fernandf@fiu.edu

Introduction

The interrogation of biological systems with secondary ion mass spectrometry (SIMS) has seen significant growth over the last decade. [1,2,3,4,5,6,7] This relatively newfound application of a surface technique traditionally limited to the study of inorganic and small molecule analytes is largely derived from the advent of larger, cluster primary ion probes (e.g., C_{60}, [8,9,10,11] Ar clusters, [12,13] and Au nanoparticles [14,15,16,17,18,19]) which provide enhanced secondary ion yields of molecular and fragment ions from biological samples. While the use of traditional time of flight (TOF-SIMS) and magnetic sector based methodologies have intrinsic advantages for the *in situ* analysis of surfaces (e.g., speed, sensitivity, dynamic range, depth profiling), the complexity and number of components usually encountered in the analysis of biological systems warrant the coupling of these new sources to high mass accuracy and resolution analytical devices for direct identification of the molecules of interest. [20,21,22,23,24] In particular, this requirement grows out of the need for improved identification certainty for molecular ions generated from biological samples, which are substantially more complex relative to semiconductor and polymer-based applications, where the number of sample components is limited and the analyte of interest is typically predetermined.

Previous mass spectrometry imaging studies have shown the advantages of correlating spatial information with molecular composition for the study of a variety of biological systems. [25] A common drive has been the search for biological models and better interrogation probes with higher spatial resolution and improved molecular identification. To this end, we used *Dictyostelium discoideum* as a biological model for evaluating the performance of two different mass spectrometry imaging approaches. *D. discoideum* cells are eukaryotic cells that normally live on soil surfaces and eat bacteria. [26,27] An interesting feature of their biological cycle is that when the cells overgrow their food supply and starve, they aggregate together in dendritic streams to form groups of ~20,000 cells. The aggregated cells eventually form a fruiting body consisting of a 1–2 mm tall stalk supporting a mass of spore cells which can then be dispersed by the wind to start new colonies. Because soil surfaces are exposed to rain water, the cells can survive and undergo development in water. This feature makes *D. discoideum* a good model for *in situ* mass spectrometry imaging since it does not require the use of cleaning protocols that can potentially compromise the spatial information (e.g., removal of buffer salts and/or media components). In addition, this cell averages 10 μm in size, which is at the frontier of various surface interrogation techniques (e.g., SIMS, DESI and MALDI). [5,25,28] Although the lipid composition of *D. discoideum* has been studied at different developmental stages using traditional chromatographic techniques and mass spectrometry, [26,29,30,31,32] nothing is known about their distribution during chemotaxis and the aggregation process. In this article, we explore the potential for

SIMS imaging of unknown biological samples by employing traditional TOF-SIMS and accurate mass determination via FTICR-SIMS for direct molecular ion identification of biological components in *D. discoideum* during aggregation.

Experimental Method

Sample Preparation

D. discoideum Ax2 cells were grown in shaking culture at 21°C in Formedium HL-5 as previously described. [33] Mid-log cells (1–2×10^6 cells/ml) were collected by centrifugation at 1,500 x g for 4 minutes, resuspended in PBM (20 mM KH_2PO_4, 10 μM $CaCl_2$, 1 mM $MgCl_2$, pH 6.1), and collected by centrifugation. The resuspension and centrifugation were repeated two more times. The cells were resuspended in PBM to 5×10^6 cells/ml, and 10 ml of cells was placed in a 125 ml Erlenmeyer flask and shaken at room temperature for 4 hours. The cells were then diluted 1:6 with PBM, collected by centrifugation, and resuspended in deionized water. The collection and resuspension in deionized water were repeated twice, and the cells were diluted to 9×10^5 cells/ml. 80 μl droplets of the cells were then spotted onto gold-coated silicon chips (Sigma Aldrich). After allowing cells to settle for 30 minutes, 40 μl of the overlaying water was removed and the chips were placed in a humid box at 21°C. 17 hours later, the chips with aggregating cells were gently drained by touching to a kimwipe, and placed cell-side down on a piece of dry ice. This was covered by a piece of aluminum foil, inverted, and placed in a vacuum chamber. After 12 hours, the dry ice had evaporated and the sample was dessicated. The chips with cells were then stored over a $CaCl_2$ desiccant at room temperature.

Instrumentation

Duplicate *D. discoideum* samples were analyzed in positive ion mode using a ToF SIMS[5] instrument (ION-TOF, Münster, Germany) and a custom C_{60} FTICR-SIMS. The custom C_{60} FTICR-SIMS instrument (more details in refs [21,23]) utilizes a 40 keV C_{60} primary ion gun (Ionoptika Ltd., Hampshire, England) that is coupled to a SolariX 9.4T FTICR mass spectrometer (Bruker Daltonics Inc, Billerica, MA). The vacuum pumping scheme of the SolariX cart was modified so that the pressure in the source chamber was reduced to 3×10^{-5} mbar instead of the ~3 mbar at which it typically operates. The C_{60} FTICR-SIMS images were acquired using 40 keV C_{60}^+ projectiles over a field of view of approximately 4 mm×6 mm with a pixel size of 125 μm and a total primary ion dose of 2.78×10^{13} ions/cm^2. Spectra were acquired using a broadband excitation over the 100<m/z<1,500 range, with 1.0 s transients collected for each pixel. Transients were zero-filled and Sine-Bell apodized prior to fast Fourier transformation. An ion accumulation time of 0.40 s was used to obtain sufficient S/N in the resulting spectra. In the case of the ToF-SIMS analysis, no modifications were made to the instrument. The analysis was performed by rastering the 25 keV Bi_3^+ beam over a 500 μm^2 field of view with a pixel size of 3.9 μm and a total primary ion dose of 8.16×10^{12} ions/cm^2.

Data Analysis

Spectra and images from the Bi_3 ToF-SIMS analysis were processed using SurfaceLab 6 software (ION-TOF, Münster, Germany). C_{60} FTICR-SIMS images were visualized using FlexImaging software (Bruker Daltonics Inc., Billerica, MA). Peak signals were identified using mMass software [34,35] from the summed spectrum of all pixels within the region of interest. A signal to noise threshold of 10 was used to generate a peak list containing 2,595 peaks. This peak list was then searched against the LIPID MAPS database (www.lipimaps.org), which contains ~37,000 entries, using the mMass compound search tool.[34,35] In addition to the typical protonated ions, sodium and potassium adducts as well as dehydration rearrangement products (-H_2O) were considered. These assignment criteria returned 293 peaks which could be matched to a lipid ion with better than 5 ppm mass accuracy. All reported ion masses were measured from the total spectrum summed over all image pixels.

Results and Discussion

We observed that when *D. discoideum* cells are starved in water, aggregation stream formation begins at about 16 hours, compared to the ~8 hours when cells are starved in buffer. However, stream formation was not further delayed when cells were starved on a gold surface instead of the usual glass or plastic surfaces used for most work with this organism. Typical mass spectra of aggregating *D. discoideum* cells from Bi_3 ToF-SIMS and C_{60} FTICR-SIMS are shown in Figure 1. The mass range and ion relative abundances are similar for each instrument and both are characteristic of SIMS analyses of biological targets. That is, the SIMS spectra are dominated by singly charged ions in the 0<m/z<500 range with some larger ion species (500< m/z <1200) present at lower abundance. A common feature between the spectra is the fact that the most intense peaks correspond to gold cluster and gold cluster hydrocarbon adduct ions derived from the gold-coated silicon wafer substrate (as expected, since this constitutes the majority of the surface area within the analyzed region). The gold cluster species were used to internally calibrate the FTICR-SIMS spectrum summed over all pixels to a mass accuracy below 5 ppm. As a figure of merit, a mass resolving power of ~150,000 ($m/\Delta m_{50\%}$) was measured at m/z = 393.9326 (Au_2^+ peak), where $\Delta m_{50\%}$ is the magnitude mode spectral peak width at half-maximum peak height. The C_{60} FTICR-SIMS spectrum also shows numerous lipid-specific fragments, with the most abundant being the phosphatidylcholine head group ($C_5H_{15}NPO_4^+$) at m/z 184. A total of 293 peaks in the C_{60} FTICR-SIMS spectrum can be attributed to lipid species. When comparing the Bi_3 ToF-SIMS and C_{60} FTICR-SIMS spectra, there are some key differences that become apparent. (1) The radio-frequency ion guides and quadrupole (set to transmit m/z 160 and above) used to transfer ions from the source to the ICR cell induce a low mass cutoff as seen by the significant reduction in ion signal below m/z = 200 (relative to the ToF-SIMS spectrum). [36] (2) The greater number of lipid signals detected in the 650<m/z<900 range for C_{60} show that this large cluster projectile is more efficient for generating intact lipid molecular ions than smaller primary ions such as Bi_3 (as previously noted in ref [37]). It is important to note that the ion fluences used were 2.78×10^{13} ions/cm^2 and 8.16×10^{12} ions/cm^2 for the C_{60} and Bi_3 analysis, respectively. These values are at or slightly above the static limit, meaning that erosion of the sample is expected. According to the reported sputter yields for these projectiles in organic matrices at the similar fluences and kinetic energies, [38,39,40] the sampled depths are estimated to be approximately 50 nm and 15 nm for the C_{60}^+ and Bi_3^+ analyses, respectively.

Figure 2 shows optical and selected ion images from the Bi_3 ToF-SIMS (Figure 2: B–F) and C_{60} FTICR-SIMS (Figure 2: H–L) spectra. The optical microscopy images (Figure 2: A, G) clearly show the cellular aggregation streams which form branched structures <200 μm in width and a few millimeters in length. The ToF-SIMS total secondary ion image (Figure 2B) shows higher intensity for ions originating from the aggregation streams and lower overall intensity from the gold substrate. $C_5H_{13}NPO_3^+$ (m/z

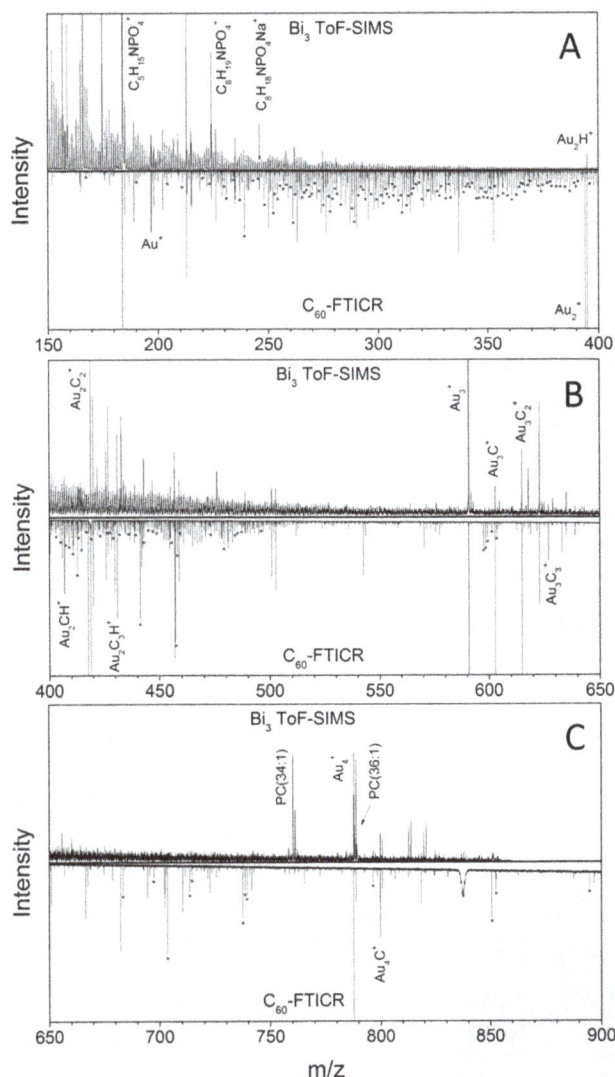

Figure 1. Comparison of Bi₃ ToF-SIMS and C₆₀ FTICR-SIMS spectra. Bi₃ ToF-SIMS (top of each panel) and C₆₀ FTICR-SIMS (inverted in each panel) spectra of aggregating *D. discoideum* cells. Lipid fragments, gold clusters, and gold cluster adduct peaks are labeled with their corresponding molecular formulas while lipid species identified from the LIPID MAPS database are denoted with asterisks (*).

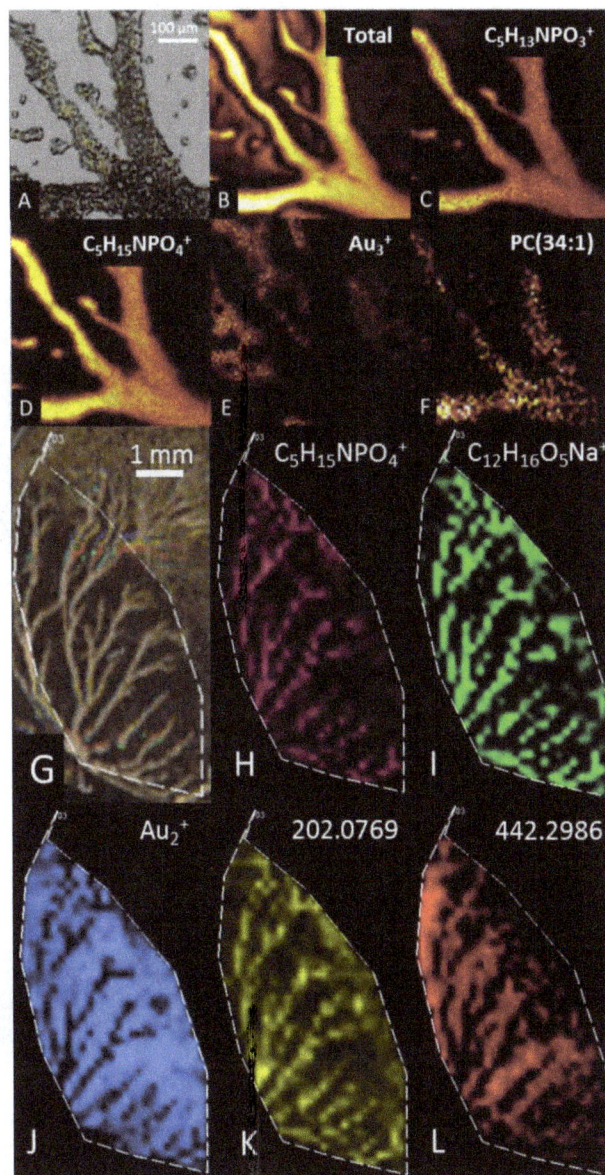

Figure 2. Optical and secondary ion images of *D. discoideum* aggregation streams. (A) Optical and (B–F) ion images of aggregation streams generated by 25 keV Bi₃ TOF-SIMS analysis. (G) Optical and (H–L) ion images of aggregation streams generated by 40 keV C₆₀ FTICR-SIMS analysis.

= 166.1), $C_5H_{15}NPO_4^+$ (m/z = 184.1), and m/z = 760.6 give spatial distributions corresponding to the aggregation streams. The $C_5H_{13}NPO_3^+$ and $C_5H_{15}NPO_4^+$ species are head group fragment ions from glycerophosphatidylcholines, which make up ~25% of all lipids present in *D. discoideum*. [31] The signal at m/z = 760.6 appears to be a lipid molecular ion due to its co-localization with the aggregations streams, the observed isotopic pattern which contains significant ^{13}C contributions, and the presence of another peak at m/z = 788.6 corresponding to the same molecule with a fatty acyl chain two carbons longer. [41,42] However, due to the limited mass accuracy afforded by ToF analysis and their absence from the C₆₀ FTICR-SIMS spectrum, the precise identities of these supposed lipids was not determined. Viewing the sample as a binary system containing signals from the cellular aggregations and from the substrate, we are also able to show that the Au_3^+ ion image represents only the substrate as this signal is not observed from the aggregations streams.

Analogously, molecular ion images can also be obtained from the C₆₀ FTICR-SIMS spectra. Two of the mass spectral features (Figure 2: H, I) which display spatial distributions corresponding to the aggregation streams are $C_5H_{15}NPO_4^+$ (m/z = 184.0737, δ = 2.1 ppm) and $C_{12}H_{16}O_5Na^+$ (m/z = 263.0889, δ = −0.2 ppm). As mentioned above, $C_5H_{15}NPO_4^+$ corresponds to the phosphatidylcholine head group, while according to the LIPID MAPS database, the $C_{12}H_{16}O_5Na^+$ species corresponds to the heterocyclic fatty acyl 3-carboxy-4-methyl-propyl-2-furanpropanoic acid (LIPID MAPS ID: LMFA01150004), which has previously been detected from human uremic serum as a sodiated ion using SIMS. [43] Ion images for two unidentified peaks from the FTICR spectrum are shown in panels K and L. The image of m/z 202.0769 shows a distribution consistent with the aggregation

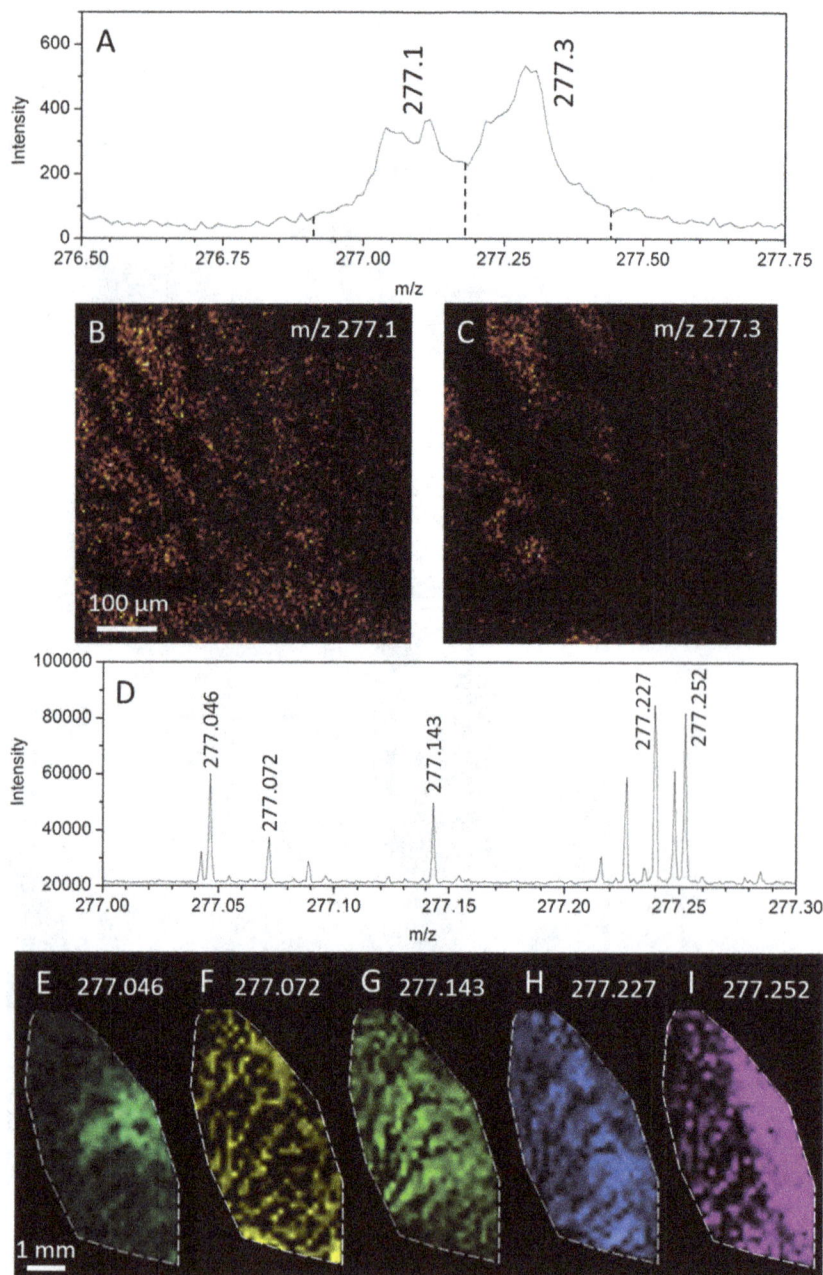

Figure 3. Secondary ion images from within the m/z = 277 nominal mass. (A) Bi₃ ToF-SIMS and (D) C₆₀ FTICR-SIMS spectra excerpts showing multiple peaks within the 277 nominal mass. (B,C) Bi₃ ToF-SIMS ion images obtained from the first "peak" and second "peak" within the 277 nominal mass. (E–I) C₆₀ FTICR-SIMS ion images generated for the corresponding peaks in D with a m/z bin size of +/− 0.001.

streams with lower level concentrations between the aggregation streams. The m/z 442.2986 ion is located on the surface in proximity to, but not within the aggregation streams. Such an arrangement may mean this ion corresponds to a metabolite which is secreted from the *D. discoideum* cells. The m/z 202.0769 and 442.2986 ions did not return lipid matches within the 5 ppm mass accuracy threshold, suggesting these lipids are not contained in the database, these compounds are not lipids, or the mass errors for these peaks fall outside the applied threshold range. As such, identities for these ions can not be determined from this analysis. As in the ToF-SIMS analysis, a Au-related ion, Au₂⁺ (m/z

393.9326, δ = −0.1 ppm), can be used to visualize the substrate and not the aggregation streams.

The molecular ion images shown in Figure 2 demonstrate that ions throughout the mass range can be used to display meaningful spatial distributions. Moreover, the mass resolving power of the FTICR-SIMS instrument is most apparent when the true complexity of the sample is revealed. The excerpted mass spectrum (from the sum of all spectra) shown in Figure 3D shows that within the spectrum, there can be upwards of 10 ions within a given nominal mass, and that each of these ions may arise from different regions within the sample. Assuming a composition of carbon, hydrogen, nitrogen, oxygen, and phosphorus and a 5 ppm

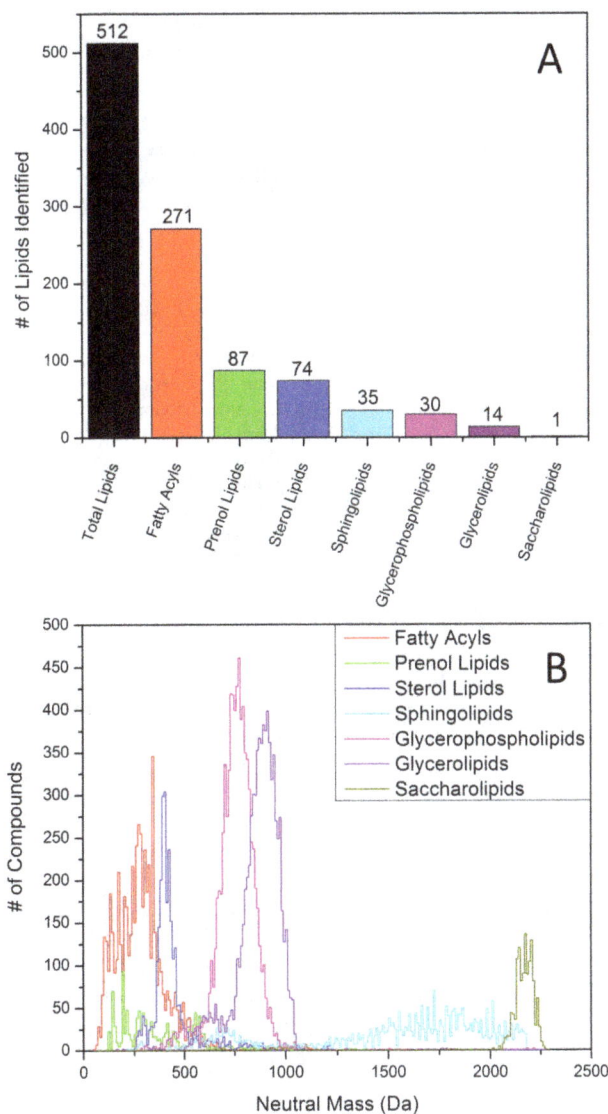

Figure 4. Overlap between the detected lipid classes and the LIPID MAPS database. (A) The number of peaks identified from the C_{60}-FTICR-SIMS spectrum by lipid class. (B) Mass distributions for compounds from the LIPID MAPS database organized by lipid class. Each data point represents the number of lipids for a given class binned every 10 mass units.

spatial distribution, but the unique distributions of the summed peaks are lost due to insufficient mass resolving power. Further attempts to segment the peaks resulted in insufficient counts per window to generate ion images (see Figure S1).

The search for lipid IDs from the high resolution FTICR-SIMS spectrum against the LIPID MAPS database resulted in 293 hits throughout the spectrum within 5 ppm mass measurement accuracy. Depending on the uniqueness of each detected m/z, peaks can be assigned to a single lipid, any of multiple isomers within a given class, or to any of multiple isomers from multiple lipid classes. A summary plot showing the 512 lipid class assignments for the 293 peaks is provided in Figure 4A, with the detailed list of peak assignments included as Table S1. Figure 4B shows the number of compounds from each lipid class in the LIPID MAPS database binned every 10 Da. The three most commonly detected classes of lipids, the fatty acyls, prenol lipids, and sterol lipids, feature mass distributions which reside almost entirely within the 200–500 Da range. This happens to be the range over which most of the C_{60} FTICR-SIMS signal is observed. As a general trend, as the mass of the compound increases, the probability of ion formation/survival decreases in SIMS analyses. Only one saccharolipid (LMSL05000001, $C_{18}H_{32}O_8$) was detected due to the fact that nearly all saccharolipids in the database reside at >2,000 Da. Most of the detected species, especially the nonpolar and electronegative compounds (like fatty acyls), were detected as sodiated or potassiated ions, with most undergoing dehydration reactions in order to generate positive ions.

Previous reports of the *D. discoideum* lipid profile are almost exclusively limited thin-layer chromatographic measurements of the types and relative abundances of the general lipid classes without regard to the specific lipids present. As an example, Paquet et al. recently reported that neutral, phosphoethanolamine, and phosphocholine lipids constitute over 80% of a total lipid extract from *D. discoideum*. However, variations in ionization probability between lipid classes and a mass dependent detection probability preclude quantitative comparisons of this type from mass spectrometric data. There have also been many reports of the fatty acid profile of *D. discoideum* obtained from hydrolyzed lipid extracts, but these fatty acyls have not been linked back to their parent lipid class. In order to obtain a more detailed lipid profile, the fatty acyls should be detected along with their corresponding head groups. This could be done either by using a solvent prefractionation method to isolate the various lipid class prior to hydrolysis and subsequent GC analysis [45] or by analysis of the original intact molecular ions [24] as was done here. This has been done for the most abundant sphingolipids from *D. discoideum* using liquid chromatography mass spectrometry; [32] however, this analysis was performed in negative ion mode while our MS analyses were acquired in positive ion mode. The author did propose identities for the four most abundant lipids observed in positive ion mode to be PC(36:4), PC(34:4), PC(32:2), and PS(32:1), but these ions were not observed in the C_{60} FTICR-SIMS spectrum.

Despite the fact that lipid profiling using this approach is biased by the mass range and ionization probability of the desorbed molecules, it does offer a rapid tool for molecular differentiation and cell state classification. A current limitation of this approach (e.g., compared to LC-MS lipid profiling) lies in the inability to differentiate isobaric species. The identity of structural isomers is often important in lipid analysis and efforts have been made to incorporate MS/MS capabilities into SIMS analysis. [46,47]. The current FTICR-SIMS instrument is also capable of MS/MS measurements, [21] though none were performed during the course of this study. Another limitation of the current prototype

threshold, we can suggest molecular formulas for the peaks at m/z = 277.046, 277.072, 277.143, 277.227, and 277.252 m/z to be $C_{12}H_9N_2O_6^+$ or $C_{10}H_{14}O_7P^+$, $C_{14}H_{13}O_6^+$, $C_{16}H_{21}O_4^+$, $C_{17}H_{29}N_2O^+$, and $C_{19}H_{33}O^+$ respectively. From the ion images it is apparent that two of the ions (277.072 and 277.252) originate from the aggregation streams. As expected, these two ions are the only two which match entries from the LIPID MAPS database. The m/z = 277.072 ion can be identified as Thysanone (LIPID MAPS ID: LMPK13030001, $[C_{14}H_{12}O_6+H]^+$, $\delta = -4.9$ ppm). [44] The peak at 277.252 has the formula $[C_{19}H_{32}O+H]^+$ ($\delta = -0.2$ ppm) and has 13 possible lipid matches with the same stoichiometry. This list of matches includes a sterol lipid, 4 prenol lipids, and 8 sphingolipids. Analysis with ToF-SIMS (Figure 3A) reveals at least two unresolved peaks within the m/z 277 nominal mass. The selected ion images generated by integrating the left and right halves of the peak cluster (3B,3C) show some differences in lies in the sub-optimal focusing of the C_{60}^{+} primary ion beam which has a diameter of ~ 75 μm and the lack of ion raster optics which means mechanical stage movement must be used to generate ion images rather than the more precise method of beam rastering. Other groups have shown that C_{60} beams can be focused down to 200 nm and rastered to create images with sub-micron spatial resolution. [48] Such improvements would be necessary for the current instrument to resolve smaller surface features such as lipid distributions within *Dictyostelium* aggregation streams or individual *Dictyostelium* cells.

Conclusions

Bi_3 ToF-SIMS and C_{60} FTICR-MS offer complementary information, where the first analysis provides short analysis times and high spatial resolution while the second demonstrates the need for higher mass resolving power when interrogating biological samples. In particular, the use of high mass resolving power in SIMS (e.g., FTICR-SIMS) was shown to be effective for the analysis of a variety of chemical classes with molecular ion masses $<1,000$ Da (e.g., fatty acyls, prenol lipids, and sterol lipids). Further incorporation of high resolution mass analyzers with high spatial resolution surface probes will permit a better identification of molecular components in biological matrices, a necessary step in the progression towards single cell mass spectrometry imaging.

Supporting Information

Figure S1 Bi_3 ToF-SIMS secondary ion images from within the m/z = 277 nominal mass. (A–D) Bi_3 ToF-SIMS selected ion images produced from the (E) segmented peaks within the m/z 277 nominal mass. (F) Optical image and (G) summed image of the full m/z 277 peak.

Table S1 List of lipids with $<$5 ppm mass error identified from the LIPID MAPS database.

Acknowledgments

The authors would like to thank Chris Thompson (Bruker Daltonics, Inc.) for advice on data processing and calibration of the FTICR-SIMS images. This work was supported by NIGMS grants R00GM106414 to FF-L and GM102280 to RHG. A portion of this research was performed at the W. R. Wiley Environmental Molecular Sciences Laboratory, a national scientific user facility sponsored by the Department of Energy's Office of Biological and Environmental Research (DOE-BER) and located at Pacific Northwest National Laboratory (PNNL). This work is part of the research program of the Foundation for Fundamental Research on Matter (FOM), which is part of The Netherlands Organization for Scientific Research (NWO).

Author Contributions

Conceived and designed the experiments: FFL RHG. Performed the experiments: FFL DFS JDD. Analyzed the data: FFL JDD CRA. Contributed reagents/materials/analysis tools: RMAH LPT FFL RHG. Wrote the paper: JDD FFL DFS RMAH LPT RHG.

References

1. Gormanns P, Reckow S, Poczatek JC, Turck CW, Lechene C (2012) Segmentation of multi-isotope imaging mass spectrometry data for semi-automatic detection of regions of interest. Plos One 7: e30576.
2. Touboul D, Brunelle A, Laprévote O (2011) Mass spectrometry imaging: Towards a lipid microscope? Biochimie 93: 113–119.
3. Fletcher JS, Lockyer NP, Vickerman JC (2011) Developments in Molecular SIMS Depth Profiling and 3D Imaging of Biological Systems using Polyatomic Primary Ions. Mass Spectrometry Reviews 30: 142–174.
4. Benabdellah F, Seyer A, Quinton L, Touboul D, Brunelle A, et al. (2010) Mass spectrometry imaging of rat brain sections: nanomolar sensitivity with MALDI versus nanometer resolution by TOF–SIMS. Analytical & Bioanalytical Chemistry 396: 151–162.
5. Chughtai K, Heeren RMA (2010) Mass Spectrometric Imaging for Biomedical Tissue Analysis. Chemical Reviews 110: 3237–3277.
6. Frisz JF, Lou K, Klitzing HA, Hanafin WP, Lizunov V, et al. (2013) Direct chemical evidence for sphingolipid domains in the plasma membranes of fibroblasts. Proceedings of the National Academy of Sciences 110: E613–E622.
7. Kraft ML, Weber PK, Longo ML, Hutcheon ID, Boxer SG (2006) Phase Separation of Lipid Membranes Analyzed with High-Resolution Secondary Ion Mass Spectrometry. Science 313: 1948–1951.
8. van Stipdonk MJ, Harris RD, Schweikert EA (1996) A Comparison of Desorption Yields from C+60 to Atomic and Polyatomic Projectiles at keV Energies. Rapid Communications in Mass Spectrometry 10: 1987–1991.
9. Cheng J, Winograd N (2005) Depth Profiling of Peptide Films with TOF-SIMS and a C60 Probe. Analytical Chemistry 77: 3651–3659.
10. Fletcher JS, Conlan XA, Jones EA, Biddulph G, Lockyer NP, et al. (2006) TOF-SIMS Analysis Using C60. Effect of Impact Energy on Yield and Damage. Analytical Chemistry 78: 1827–1831.
11. Fletcher JS, Lockyer NP, Vaidyanathan S, Vickerman JC (2007) TOF-SIMS 3D Biomolecular Imaging of Xenopus laevis Oocytes Using Buckminsterfullerene (C60) Primary Ions. Analytical Chemistry 79: 2199–2206.
12. Rabbani S, Barber AM, Fletcher JS, Lockyer NP, Vickerman JC (2011) TOF-SIMS with argon gas cluster ion beams: a comparison with C60+. Analytical Chemistry 83: 3793–3800.
13. Bich C, Havelund R, Moellers R, Touboul D, Kollmer F, et al. (2013) Argon cluster ion source evaluation on lipid standards and rat brain tissue samples. Analytical Chemistry 85: 7745–7752.
14. Fernandez-Lima FA, DeBord JD, Schweikert EA, Della-Negra S, Kellersberger KA, et al. (2013) Surface characterization of biological nanodomains using NP-ToF-SIMS. Surface and Interface Analysis 45: 294–297.
15. Fernandez-Lima FA, Post J, DeBord JD, Eller MJ, Verkhoturov SV, et al. (2011) Analysis of Native Biological Surfaces Using a 100 kV Massive Gold Cluster Source. Analytical Chemistry 83: 8448–8453.
16. Della-Negra S, Depauw J, Guillermier C, Schweikert EA (2011) Massive clusters: Secondary emission from qkeV to qMeV. New emission processes? New SIMS probe? Surface and Interface Analysis 43: 62–65.
17. Novikov A, Caroff M, Della-Negra S, Depauw J, Fallavier M, et al. (2005) The Aun cluster probe in secondary ion mass spectrometry: Influence of the projectile size and energy on the desorption/ionization rate from biomolecular solids. Rapid Communications in Mass Spectrometry 19: 1851–1857.
18. Bouneau S, Della-Negra S, Depauw J, Jacquet D, Le Beyec Y, et al. (2004) Heavy gold cluster beams production and identification. Nuclear Instruments and Methods in Physics Research Section B: Beam Interactions with Materials and Atoms 225: 579–589.
19. Brunelle A, Della-Negra S, Depauw J, Jacquet D, Le Beyec Y, et al. (2001) Enhanced secondary-ion emission under gold-cluster bombardment with energies from keV to MeV per atom. Physical Review A 63: 022902.
20. Gilmore IS (2013) SIMS of organics—Advances in 2D and 3D imaging and future outlook. Journal of Vacuum Science & Technology A 31: 050819.
21. Smith DF, Robinson EW, Tolmachev AV, Heeren RMA, Pasa-Tolic L (2011) C60 Secondary Ion Fourier Transform Ion Cyclotron Resonance Mass Spectrometry. Analytical Chemistry 83: 9552–9556.
22. Green FM, Gilmore IS, Seah MP (2011) Mass Spectrometry and Informatics: Distribution of Molecules in the PubChem Database and General Requirements for Mass Accuracy in Surface Analysis. Analytical Chemistry 83: 3239–3243.
23. Smith D, Kiss A, Leach F III, Robinson E, Paša-Tolić L, et al. (2013) High mass accuracy and high mass resolving power FT-ICR secondary ion mass spectrometry for biological tissue imaging. Analytical and Bioanalytical Chemistry 405: 6069–6076.
24. Fhaner CJ, Liu S, Ji H, Simpson RJ, Reid GE (2012) Comprehensive Lipidome Profiling of Isogenic Primary and Metastatic Colon Adenocarcinoma Cell Lines. Analytical Chemistry 84: 8917–8926.
25. Ellis S, Bruinen A, Heeren RA (2013) A critical evaluation of the current state-of-the-art in quantitative imaging mass spectrometry. Analytical and Bioanalytical Chemistry: 1–15.
26. Paquet VE, Lessire R, Domergue F, Fouillen L, Filion G, et al. (2013) Lipid Composition of Multilamellar Bodies Secreted by Dictyostelium discoideum Reveals Their Amoebal Origin. Eukaryotic Cell 12: 1326–1334.
27. Jang W, Gomer RH (2011) Initial Cell Type Choice in Dictyostelium. Eukaryotic Cell 10: 150–155.
28. Wu C, Dill AL, Eberlin LS, Cooks RG, Ifa DR (2013) Mass spectrometry imaging under ambient conditions. Mass Spectrometry Reviews 32: 218–243.

29. Long BH, Coe EL (1974) Changes in Neutral Lipid Constituents during Differentiation of the Cellular Slime Mold, Dictyostelium discoideum. Journal of Biological Chemistry 249: 521–529.

30. Birch GL (2011) Lipidomic profiling of Dictyostelium Discoideum. Purdue University.

31. Weeks G, Herring FG (1980) The lipid composition and membrane fluidity of Dictyostelium discoideum plasma membranes at various stages during differentiation. Journal of Lipid Research 21: 681–686.

32. Birch GL (2011) Lipidomic Profiling of Dictyostelium Discoideum. Purdue University.

33. Brock DA, Gomer RH (2005) A secreted factor represses cell proliferation in Dictyostelium. Development 132: 4553–4562.

34. Strohalm M, Kavan D, Novák P, Volný M, Havlíček VR (2010) mMass 3: A Cross-Platform Software Environment for Precise Analysis of Mass Spectrometric Data. Analytical Chemistry 82: 4648–4651.

35. Strohalm M, Hassman M, Košata B, Kodíček M (2008) mMass data miner: an open source alternative for mass spectrometric data analysis. Rapid Communications in Mass Spectrometry 22: 905–908.

36. Beu S, Hendrickson C, Marshall A (2011) Excitation of Radial Ion Motion in an rf-Only Multipole Ion Guide Immersed in a Strong Magnetic Field Gradient. Journal of the American Society for Mass Spectrometry 22: 591–601.

37. Ostrowski SG, Szakal C, Kozole J, Roddy TP, Xu J, et al. (2005) Secondary Ion MS Imaging of Lipids in Picoliter Vials with a Buckminsterfullerene Ion Source. Analytical Chemistry 77: 6190–6196.

38. Brison J, Muramoto S, Castner DG (2010) ToF-SIMS Depth Profiling of Organic Films: A Comparison between Single-Beam and Dual-Beam Analysis. The Journal of Physical Chemistry C 114: 5565–5573.

39. Muramoto S, Brison J, Castner DG (2011) Exploring the Surface Sensitivity of TOF-Secondary Ion Mass Spectrometry by Measuring the Implantation and Sampling Depths of Bin and C60 Ions in Organic Films. Analytical Chemistry 84: 365–372.

40. Delcorte A, Leblanc C, Poleunis C, Hamraoui K (2013) Computer Simulations of the Sputtering of Metallic, Organic, and Metal–Organic Surfaces with Bin and C60 Projectiles. The Journal of Physical Chemistry C 117: 2740–2752.

41. Petković M, Schiller J, Müller M, Benard S, Reichl S, et al. (2001) Detection of Individual Phospholipids in Lipid Mixtures by Matrix-Assisted Laser Desorption/Ionization Time-of-Flight Mass Spectrometry: Phosphatidylcholine Prevents the Detection of Further Species. Analytical Biochemistry 289: 202–216.

42. Hase A (1981) Fatty acid composition and sterol content of Dictyostelium discoideum cells at various stages of development. Journal of the Faculty of Science, Hokkaido University 12: 183–194.

43. Takeda N, Niwa T, Tatematsu A, Suzuki M (1987) Identification and quantification of a protein-bound ligand in uremic serum. Clinical chemistry 33: 682–685.

44. Singh SB, Cordingley MG, Ball RG, Smith JL, Dombrowski AW, et al. (1991) Structure of stereochemistry of thysanone: a novel human rhinovirus 3C-protease inhibitor from Thysanophora penicilloides. Tetrahedron Letters 32: 5279–5282.

45. Yoshioka S, Nakashima S, Okano Y, Hasegawa H, Ichiyama A, et al. (1985) Phospholipid (diacyl, alkylacyl, alkenylacyl) and fatty acyl chain composition in murine mastocytoma cells. Journal of Lipid Research 26: 1134–1141.

46. Ferreri C, Chatgilialoglu C (2005) Geometrical trans Lipid Isomers: A New Target for Lipidomics. ChemBioChem 6: 1722–1734.

47. Piehowski PD, Carado AJ, Kurczy ME, Ostrowski SG, Heien ML, et al. (2008) MS/MS Methodology To Improve Subcellular Mapping of Cholesterol Using TOF-SIMS. Analytical Chemistry 80: 8662–8667.

48. Fletcher JS, Rabbani S, Henderson A, Blenkinsopp P, Thompson SP, et al. (2008) A New Dynamic in Mass Spectral Imaging of Single Biological Cells. Analytical Chemistry 80: 9058–9064.

2

Genetic Diversity of Bacterial Communities and Gene Transfer Agents in Northern South China Sea

Fu-Lin Sun[1,2], You-Shao Wang[1,2]*, Mei-Lin Wu[1], Zhao-Yu Jiang[1], Cui-Ci Sun[1,2], Hao Cheng[1]

1 State Key Laboratory of Tropical Oceanography, South China Sea Institute of Oceanology, Chinese Academy of Sciences, Guangzhou, China, 2 Daya Bay Marine Biology Research Station, South China Sea Institute of Oceanology, Chinese Academy of Sciences, Shenzhen, China

Abstract

Pyrosequencing of the 16S ribosomal RNA gene (rDNA) amplicons was performed to investigate the unique distribution of bacterial communities in northern South China Sea (nSCS) and evaluate community structure and spatial differences of bacterial diversity. Cyanobacteria, Proteobacteria, Actinobacteria, and Bacteroidetes constitute the majority of bacteria. The taxonomic description of bacterial communities revealed that more Chroococcales, SAR11 clade, Acidimicrobiales, Rhodobacterales, and Flavobacteriales are present in the nSCS waters than other bacterial groups. Rhodobacterales were less abundant in tropical water (nSCS) than in temperate and cold waters. Furthermore, the diversity of Rhodobacterales based on the gene transfer agent (GTA) major capsid gene (g5) was investigated. Four g5 gene clone libraries were constructed from samples representing different regions and yielded diverse sequences. Fourteen g5 clusters could be identified among 197 nSCS clones. These clusters were also related to known g5 sequences derived from genome-sequenced Rhodobacterales. The composition of g5 sequences in surface water varied with the g5 sequences in the sampling sites; this result indicated that the Rhodobacterales population could be highly diverse in nSCS. Phylogenetic tree analysis result indicated distinguishable diversity patterns among tropical (nSCS), temperate, and cold waters, thereby supporting the niche adaptation of specific Rhodobacterales members in unique environments.

Editor: Bas E. Dutilh, Universiteit Utrecht, Netherlands

Funding: This research was supported by the National Natural Science Foundation of China (41406130, 41176101, 41430966, 31270528 and 41206082), the Strategic Priority Research Program of the Chinese Academy of Sciences (XDA10020225), and the key projects in the National Science & Technology Pillar Program in the Eleventh Five-year Plan Period (2012BAC07B0402). The funders had no role in study design, data collection and analysis, decision to publish, or preparation of the manuscript.

Competing Interests: The authors have declared that no competing interests exist.

* Email: yswang@scsio.ac.cn

Introduction

The bacterioplankton phylotypes of α-Proteobacteria are among the largest heterotrophic marine bacteria and often detected in various marine regions on Earth [1,2]. Studies on marine microbial populations have suggested that Order Rhodobacterales (α-Proteobacteria) members are ubiquitous in marine environments and can account for >25% of total marine bacterioplankton [2–4]. Although Rhodobacterales has also been found as most abundant members in temperate and cold waters [5,6], Rhodobacterales in tropical waters have been rarely investigated.

The complete genome sequences of Rhodobacterales contain gene transfer agent (GTA) gene clusters [7,8]; these genes are not found in other major bacterioplankton groups. GTA is a small phage-like particle released by bacteria; each particle contains a random ca. 4.5 kb fragment of bacterial genomic DNA [9] that can be transferred between cells [10]. GTAs are present in phylogenetically diverse prokaryotes, indicating that this mode of DNA transfer may be important in shaping microbial genomes and communities [6]. GTA-related gene transfer has also been considered as a potential adaptive mechanism of these bacteria to maintain metabolic flexibility in changing marine environments [10,11]. A capsid protein-encoding gene (g5) of GTA has been used as a marker to estimate the diversity of Rhodobacterales in temperate and cold waters because GTA genes are conserved in Rhodobacterales [5,6].

Northern South China Sea (nSCS) is a marginal sea encompassing the Pearl River Estuary and a broad continental shelf. nSCS is characterized by tropical and subtropical climate and represents typical oligotrophic characteristics with significant environmental gradients from the discharge of the Pearl River; physical forces, such as mesoscale eddies, monsoon, upwelling, Kuroshio Current, and so on, influence nSCS [12]. All of these physical disturbances can influence water-column stability in different temporal and spatial scales [13]. Furthermore, nSCS consists of various ecosystems (such as mangrove forests, seagrass beds, coral reefs) marked with high biodiversities. However, the roles of heterotrophic bacterioplankton in these waters have not been explicitly characterized.

Although the distribution of the Rhodobacterales community in cold waters and temperate coast has been reported [2,5,6,14,15], the members of Rhodobacterales in tropical waters have not been described in detail. This study aimed to (i) determine the bacterial community and relative abundance of Rhodobacterales in nSCS, (ii) analyze the diversity and spatial genetic variations of the g5 gene in nSCS, and (iii) compare g5 structure of the nSCS with those from other areas.

Methods

Study stations and water sampling

E701, E703, E709, E403, SCS15, SCS17, and SCS19 are sampling stations in the South China Sea. Water samples (E701, E703, E709, and E403) were collected in September 2011. Samples of SCS15, SCS17, and SCS19 were collected in May 2013 (Figure 1). Water samples at each station were collected and 1000 mL of seawater was filtered with 0.22 μm pore size filters (47 mm in diameter, Millipore Corp., Bedford, USA) at low vacuum pressure to collect prokaryotic cells. Each sample was prepared in three replicates. After filtration was performed, the membranes were immediately frozen in liquid nitrogen and then stored at $-20°C$ until DNA extraction was conducted in our laboratory.

Ethics statement

No specific permits were required for the described field studies. Our study area is not privately owned or protected in any way. Our field studies did not involve endangered or protected species. The South China Sea Institute of Oceanology and Chinese Academy of Sciences issued the permissions to investigate each location.

DNA extraction, PCR amplification, and pyrosequencing

For each sample, triplicate DNA aliquots were extracted according to the special DNA protocol for marine bacterial communities [16]. A region of 444 bp in the 16S rRNA gene covering the V1–V3 region was selected to construct a community library by tag pyrosequencing. The broadly conserved bar-coded primers 27F and 533R containing A and B sequencing adaptors (454 Life Sciences) were used to amplify this region. The forward primer (B-27F) sequence was 5′-*CCTA-TCCCCTGTGTGCCTTGGCAGTCTCAG*AGAGTTTGATCCT-GGCTCAG-3′, in which the B adaptor sequence is italicized and underlined. The reverse primer (A-533R) sequence was 5′-*CCATCT-CATCCCTGCGTGTCTCCGACTCAG*NNNNNNNNTTACCGC-GGCTGCTGGCAC-3′, in which the sequence of the A adaptor is italicized and underlined. Ns represent an eight-base sample-specific barcode sequence. Amplicon pyrosequencing was performed from the A-end by using a 454/Roche A sequencing primer kit on a Roche Genome Sequencer GS FLX Titanium platform at Majorbio Bio Tech Co. Ltd (Shanghai, China). We eliminated sequences that contained more than one ambiguous nucleotide and a primer at one end or sequences that were shorter than 200 bp after barcode and primer sequences were removed. Pyrosequencing reads were simplified using the 'unique.seqs' command to generate a unique set of sequences. These pyrosequencing sequences were aligned using the 'align.seqs' command and compared with the Bacterial SILVA database (SILVA version115, http://www.arb-silva.de). Aligned sequences were trimmed further and redundant reads were eliminated using the 'screen.seqs', 'filter.seqs', and 'unique.seqs' commands in that order. The 'chimera.slayer' command was used to determine chimeric sequences. The 'dist.seqs' command was used, and unique sequences were assigned to operational taxonomic units (OTUs, 97% similarity). In the present study, data preprocessing and OTU-based analysis were performed on Mothur [17]. Taxonomic assignments with <80% confidence were marked as unknown. All of the sequences can be downloaded from the NCBI Sequence Read Archive database under the accession numbers SRX547142-SRX547144.

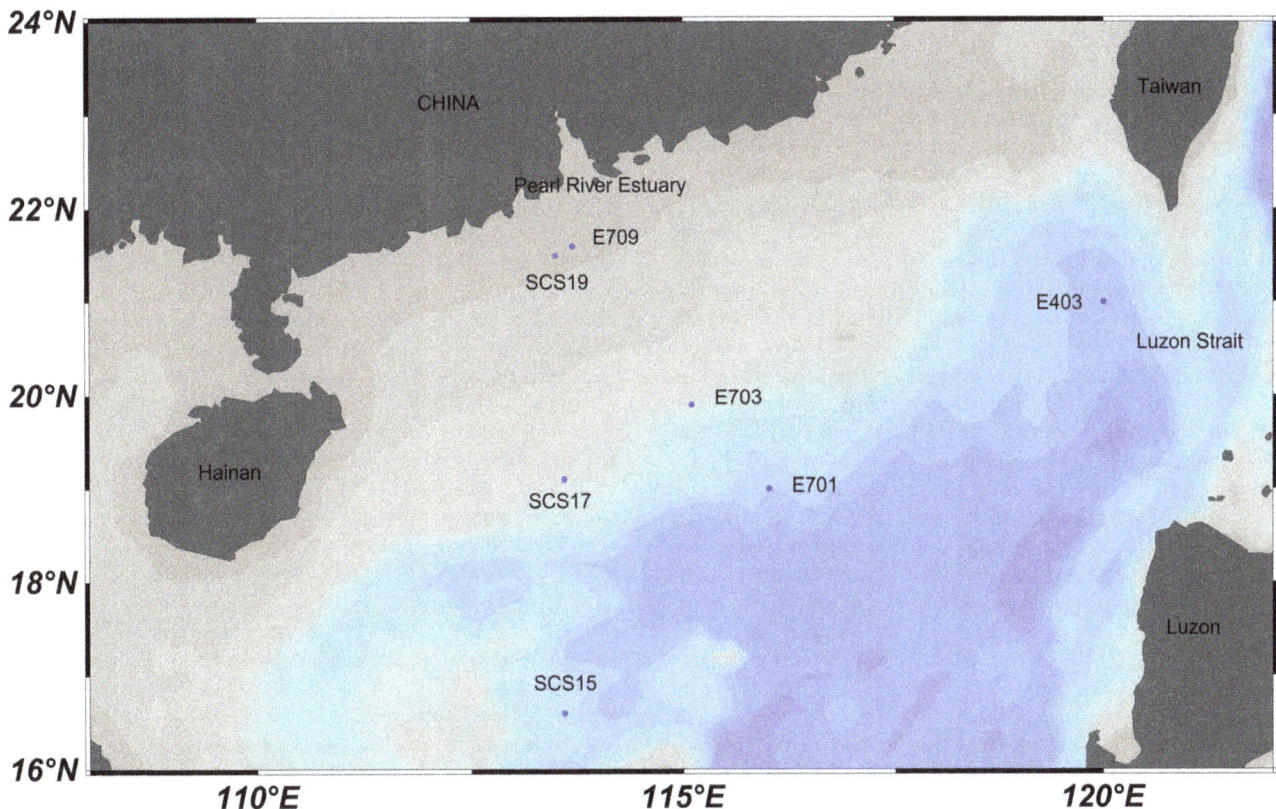

Figure 1. Map of sampling stations in the northern South China Sea.

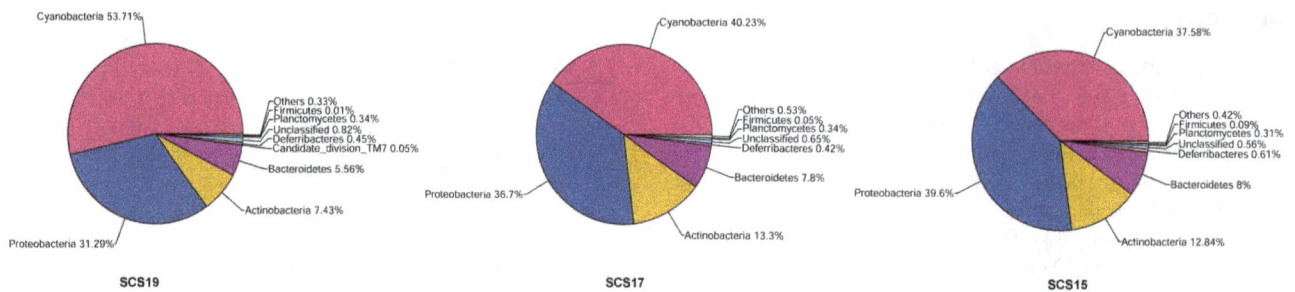

Figure 2. Bacterial compositions of the different samples in the nSCS.

PCR amplification and clone library analyses of GTA *g5* genes

The primers used to amplify GTA *g5* genes described in a previous study [5] were used in the present study. These primers include MCP-109F, 5′-GGC TAY CTG GTS GAT CCS CAR AC-3′ and MCP-368R, and 5′-TAG AAC AGS ACR TGS GGY TTK GC-3′. Target DNA was amplified in a single round of PCR in reaction volumes of 50 µl containing 10 pmol of each primer, 4 µL of 5 mM dNTPs, 1.25 U of *Taq* DNA polymerase (Takara, Japan), and 3% DMSO (v/v). The thermocycling conditions used in this study were listed as follows: 5 min at 95°C; 35 cycles at 95°C for 30 s, 60°C for 30 s, and 72°C for 30 s; and a final extension step at 72°C for 7 min.

The purified PCR products of *g5* genes were inserted into the pMD-18T vector (Takara, Japan) to construct clone libraries. Positive clones were selected to sequence and analyze using an ABI3730 DNA sequencer. All of the *g5* sequences were edited using CROSS-MATCH to remove vector and primer sequences [18]. The DNA sequences were subsequently translated into an amino acid sequence. The resulting capsid protein sequences obtained in this study were aligned and compared with the reference sequences in the GenBank database. Neighbor-joining phylogenetic trees were constructed using the MEGA 5.0 software [19]. Evolution distances were calculated using Jones-Taylor-Thornton model with a rate variation among sites and complete gap deletions to translate the *g5* gene sequence into its corresponding amino acid sequence [5]. The sequences obtained from the four clone libraries were deposited in the GenBank database with the accession numbers of KC422732 to KC422774.

The aligned sequences in each clone library were analyzed using Mothur software [17] to determine operational taxonomic units (OTUs) at a 3% dissimilarity cut-off. Simultaneously, Mothur was used to estimate the richness indices (Chao1 and Shannon), diversity index (Simpson), and coverage [17]. The structure of g5 genes was analyzed using Euclidean distance by multi-dimensional scaling (MDS) analysis in SPSS 18.0 for Windows. MDS is an ordination technique that represents the samples as points in a multi-dimensional space. Sample communities with the highest similarity in the data set are shown as the closest plotted points, and the communities with the lowest similarity are indicated by the points that are the farthest apart.

To directly assess the relationship between the structure of the *g5* gene and water environment of the nSCS, a canonical correspondence analysis (CCA) was carried out using the CANOCO 4.5 for Windows [20]. Statistical significance (at the 5% level) of relationships between *g5* gene data and environmental variables were assessed using the Monte Carlo permutation test (499 permutations).

Results

Taxonomic composition analysis

A total of 31,831 valid sequences and 3,392 OTUs (1331, 1321, and 1340) were obtained from the three samples (SCS19, SCS17, and SCS15) by 454 pyrosequencing analyses; among these sequences, two reads corresponded to eukaryotes and were excluded in the subsequent analyses. The remaining sequences were then assigned to 15 different phyla or groups.

The three samples showed similar bacterial community distributions in phylum level (Figure 2). Overall, the most abundant groups in surface water were affiliated to the phylum Cyanobacteria, which represented 43.58% of the pyrosequencing tags. The second most abundant group was Proteobacteria (35.97%), which were mainly Alphaproteobacteria (32.33%), followed with Actinobacteria (11.29%) and Bacteroidetes (7.16%).

From the area near the shore station to the far sea area, the Cyanobacteria group decreased sharply from 53.71% in SCS19 to 37.58% in SCS15. By contrast, Proteobacteria, Actinobacteria, and Bacteroidetes content increased gradually from SCS19 to SCS15 (Figure 2). In the order level (Figure 3), the Chroococcales group had the highest percentage of specific taxonomy in SCS19, SCS17, and SCS15 with 51.75%, 39.23%, and 35.94%, respectively. The Chroococcales group in SCS19, SCS17, and SCS15 consisted mainly of *Synechococcus* (30.67%, 5.90%, and 9.23%) and *Prochlorococcus* (20.22%, 32.76%, and 26.01%). Other bacterial orders that dominated the samples included SAR11 clade (17.48%, 22.23%, and 23.17%), Acidimicrobiales (7.14%, 12.83%, and 12.46%), Flavobacteriales (5.28%, 7.38% and 7.10%), and Rhodobacterales (6.67%, 4.75%, and 4.89%). Other groups were rare and many sequences were present at abundance <1% of the total population.

Among all of the Rhodobacterales sequences, only 9% sequences could be determined to identify genus. In the bacterial genus level, bacterial groups related to Rhodobacterales from the nSCS waters were identified by phylogenetic tree analysis. These groups were diverse: *Donghicola*, *Labrenzia*, *Loktanella*, *Maritimibacter*, *Paracoccus*, *Pelagibaca*, *Rhodovulum*, *Roseobacter*, *Ruegeria*, *Thalassococcus*, and *Citreicella* (Figure S1).

Diverse and unique *Rhodobacterales* in the nSCS

GTA diversity was assessed in the four samples representing different regions (E709, E703, E701, and E403). A total of 197 sequences were recovered from these four clone libraries. The phylogenetic analysis of the g5 clone sequences fell within the Rhodobacterales and corresponded to 14 phylogenetic clusters (designated as A–N; Figure 4).

The coverage of four clone libraries ranged from 85.7% to 98.1% at the 3% distance cut-off, indicating that clone libraries adequately covered the diversity of *g5* genes (Table S1). Shannon–

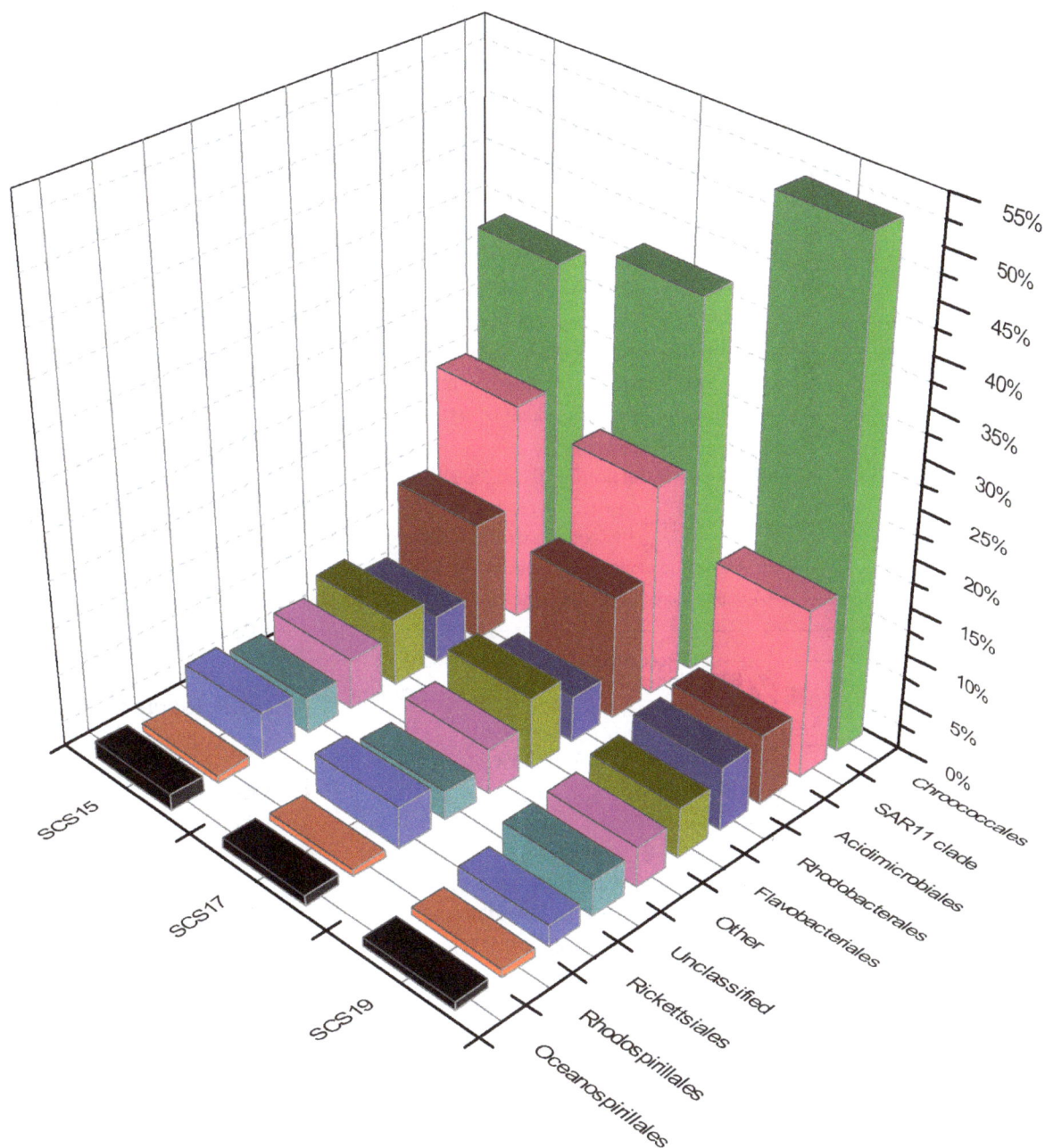

Figure 3. Bacterial compositions of the different communities in the nSCS.

Weaver and Simpson indices revealed that $g5$ gene diversity was higher in site E701 than in sites E709, E703, and E403. Chao 1 demonstrated that the richness at sample E701 was greater than that at other samples. However, low Simpson index was observed in all samples (Table S1).

Two-dimensional plots of MDS for samples showed a spatial diversity in the $g5$ gene structure (Figure 5). The results revealed that four plots that represented $g5$ structure from samples E701, E703, E709, and E403 had large distances with one another and had an MDS stress value of 0.02. Stress values below 0.2 indicate that an MDS ordination plot is a good spatial representation of differences between data. Overall, MDS ordination plots indicated that the composition of the $g5$ structure varied with the sampling sites (ANOVA, p<0.01).

The CCA of the $g5$ gene data explained 75.3% of the variation in the first two axes (Figure S2). According to Monte Carlo analysis, only latitude (F = 1.62, P = 0.038) showed a significant correlation to the $g5$ gene structure. By contrast, other environmental factors (temperature, salinity, and Chla) had no significant correlation to $g5$ gene structure (p>0.05).

Variation of GTA capsid genotypes in the nSCS

A spatial variation of $g5$ composition in the nSCS was evident (Figure 4, Table S1, Table S2). $g5$ sequence data was grouped into 14 clusters, labeled A–N. Clusters D, H, I, and L were unique to E709 and closely related to *Roseobacter*, *Ruegeria*, and *Citreicella*. Cluster F was unique to E703 and closely related to uncultured Rhodobacteraceae bacterium. Clusters E, J, and K only appeared

69
99

E709 clones (30)
E703 clones (49)
E403 clones (25)
E701 clones (27)
— E403 clones (2)

Cluster A

Cluste B

Celeribacter

Celeribacter baekdonensis (ZP_11132954)
Oceanicola batsensis (ZP_01001458)
Oceanicola sp. (ZP_09514753)

97 **E403 clones (2)** | Cluster C

Oceanicola

99
Rhodobacter capsulatus (YP_003577839)
Uncultured Rhodobacteraceae bacterium (ACK77246)
Dinoroseobacter shibae (YP_001533511)
Rhodobacterales bacterium (ZP_01741236)

78 — E709 clone (1) | Cluster D
Roseobacter sp. (ZP_01902399)

Roseobacter

— E701 clone (1) | Cluster E
99 └ Maritimibacter alkaliphilus (ZP_01015311)

Maritimibacter

88 — uncultured Rhodobacteraceae bacterium (ACK77260)
52 — uncultured Rhodobacteraceae bacterium (ACK77261)
— E703 clones (2) | Cluster F

Roseobacter litoralis (YP_004690365)
81 ◀ **E709 clones (5)**
E703 clone (1) | Cluster G
99 └ Silicibacter sp. (ZP_05742458)

Silicibacter

97 — Ruegeria sp. (YP_613056)
94 └ E709 clones (2) | Cluster H
— E709 clone (1) | Cluster I

Ruegeria

98 — E701 clone (1) | Cluster J
Rhodobacterales bacterium (ZP_05078457)
82 — Roseobacter sp. (ZP_01755050)
Roseobacter sp. (ZP_01057324)
52 — Silicibacter lacuscaerulensis (ZP_05787326)
Ruegeria sp. (ZP_08862092)
Ruegeria pomeroyi (YP_167486)
◀ **E701 clones (4)** | Cluster K

Ruegeria

74 99 — E709 clone (1) | Cluster L
Citreicella sp. (ZP_05782348)

Citreicella

94 **E701 clones (7)**
E403 clones (13) | Cluster M
99 **E403 clones (4)**
98 **E701 clones (17)** | Cluster N

0.05

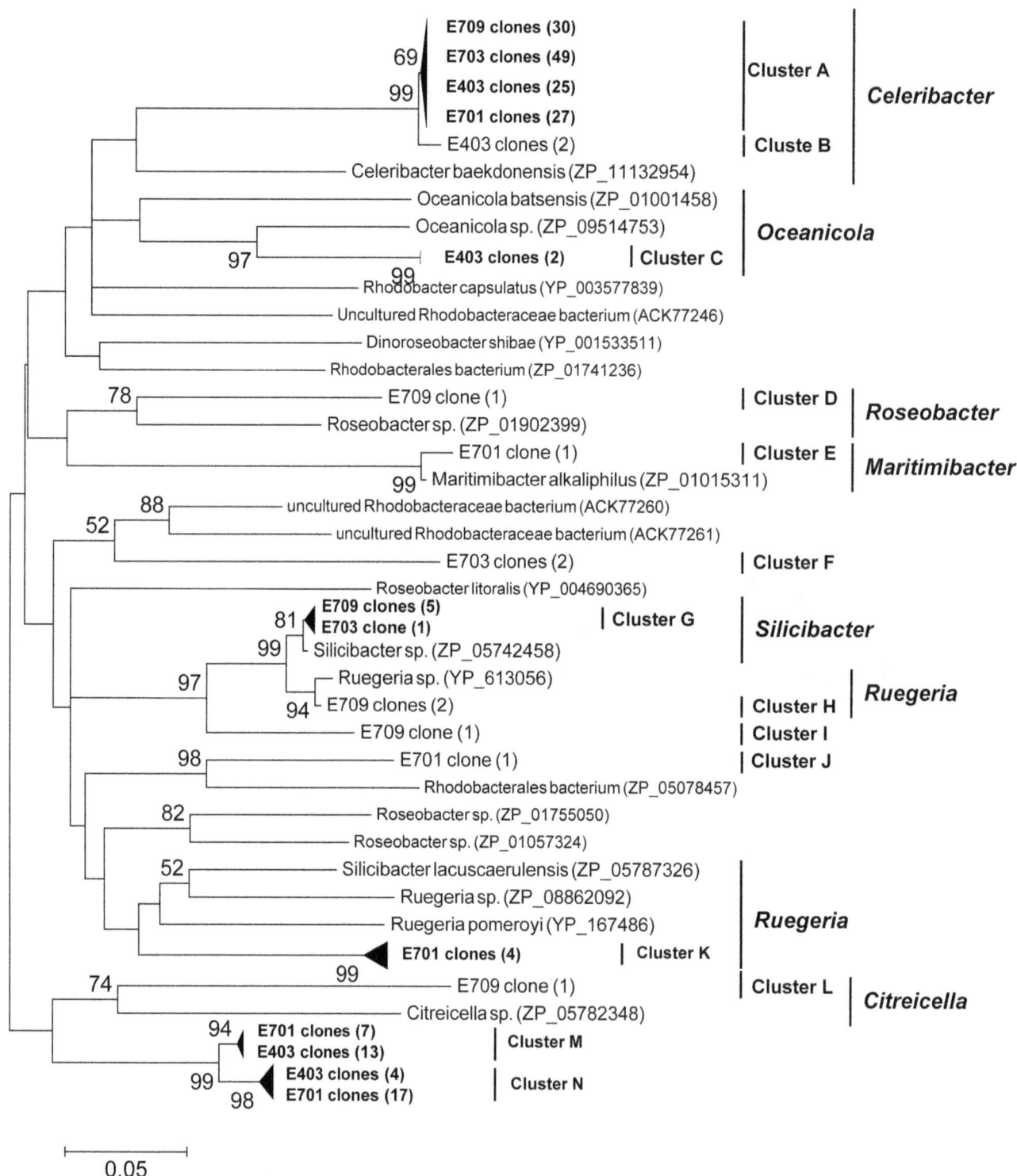

Figure 4. Neighbor-joining phylogenetic tree based on partial *g5* amino acid sequences (ca. 250 aa) showing the phylogenetic diversity of *g5* in the northern South China Sea.

in site E709, and clusters E and K were closely related to *Maritimibacter* and *Ruegeria*. Clusters B and C only appeared in site E403 and were related to *Celeribacter and Oceanicola*. Cluster A constitutes more than 47% of the *g5* clones in four clone libraries, especially in E709 and E703, which was related to *Celeribacter*, achieved 76.19% and 94.23%. Cluster G was present

in sites E709 and E703 and related to *Silicibacter*. Clusters M and N were present in sites E701 and E403; however, we did not find high matching sequences in the GenBank database.

Clusters D, H, I, and L accounted for 11.9% of the E709 clone library and were not detected in other libraries (Figure 4, Table S3). Station E703 had the lowest *g5* diversity among the

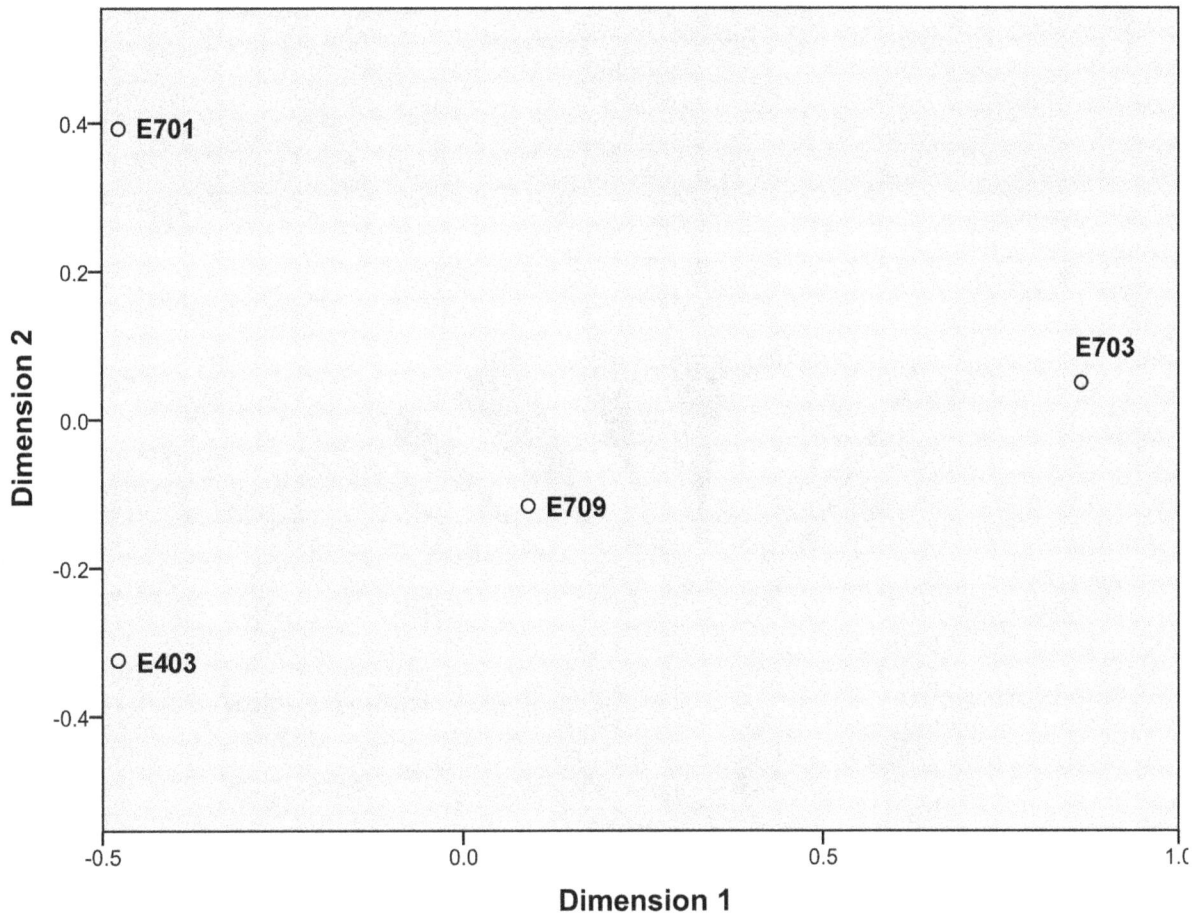

Figure 5. Two-dimensional plots of MDS analysis from *g5* gene clone library to compare broad-scale differences between Rhodobacterales communities.

investigation stations. The highest number of clones in cluster A was observed in station E703 among all of the other libraries; this result suggested that *Celeribacter* was dominant in Rhodobacterales bacteria (Figure 4). Clusters B and C in site E403 were more representative of a typical Rhodobacterales cluster and may have been derived from the external area of the Pacific Ocean. Station E701 had three common clusters (cluster A, M, and N) with E403 and only one common cluster (cluster A) with E709.

Comparison of *g5* structure of the nSCS with those from other areas

The *g5* gene sequences that belonged to the uncultured environmental samples from the Subartic North Atlantic Ocean (cold water), the Arctic Ocean (cold water), and Chesapeake Bay (temperate water) were retrieved from the GenBank database. These sequences were aligned and analyzed with Mothur to determine the OTUs at a 3% dissimilarity cut-off. The resulting capsid protein sequences obtained in this study were aligned and compared with these 134 OTUs. Homology analysis was conducted to align the nSCS gene sequence (36 OTUs) with these *g5* gene OTU sequences. A phylogenetic tree was constructed using MEGA5.0 for the translated amino acid sequence of the *g5* gene (Figure 6). Our results showed that the majority of g5 sequences from temperate water were most similar to sequences obtained from Subartic North Atlantic and Atlantic Ocean waters. The *g5* genes in the nSCS had unique sequences,

and the majority of the *g5* gene OTUs had no similarity to the *g5* gene OTUs from other regions. Furthermore, a few OTUs were similar to OTUs in temperate and cold ocean waters.

Discussion

In many studies, Rhodobacterales abundance could reach above 25% through sequence analysis from the Atlantic Ocean to the Pacific Ocean [21–24]. These findings indicated that Rhodobacterales is the primary bacteria group in the cold and temperate water marine ecosystems. Thus far, studies focusing exclusively on Rhodobacterales in tropical water have not been reported.

In the current study, Rhodobacterales and other bacterial communities in the nSCS were assessed for the first time. Our results showed that the average abundance of Rhodobacterales was 5.44% for all the sequences. Although Rhodobacterales had a relatively higher abundance than other groups in nSCS waters, the Rhodobacterales content was lower than that from temperate and cold waters. Nevertheless, Rhodobacterales was one of the dominant orders of bacterial communities in nSCS.

Pyrosequencing analysis indicated that Cyanobacteria and Proteobacteria dominated the nSCS. The overwhelming majority of the identified Cyanobacteria sequences were related to *Synechococcus* and *Prochlorococcus*. These bacteria dominated the cyanobacterial communities in coastal and offshore station of

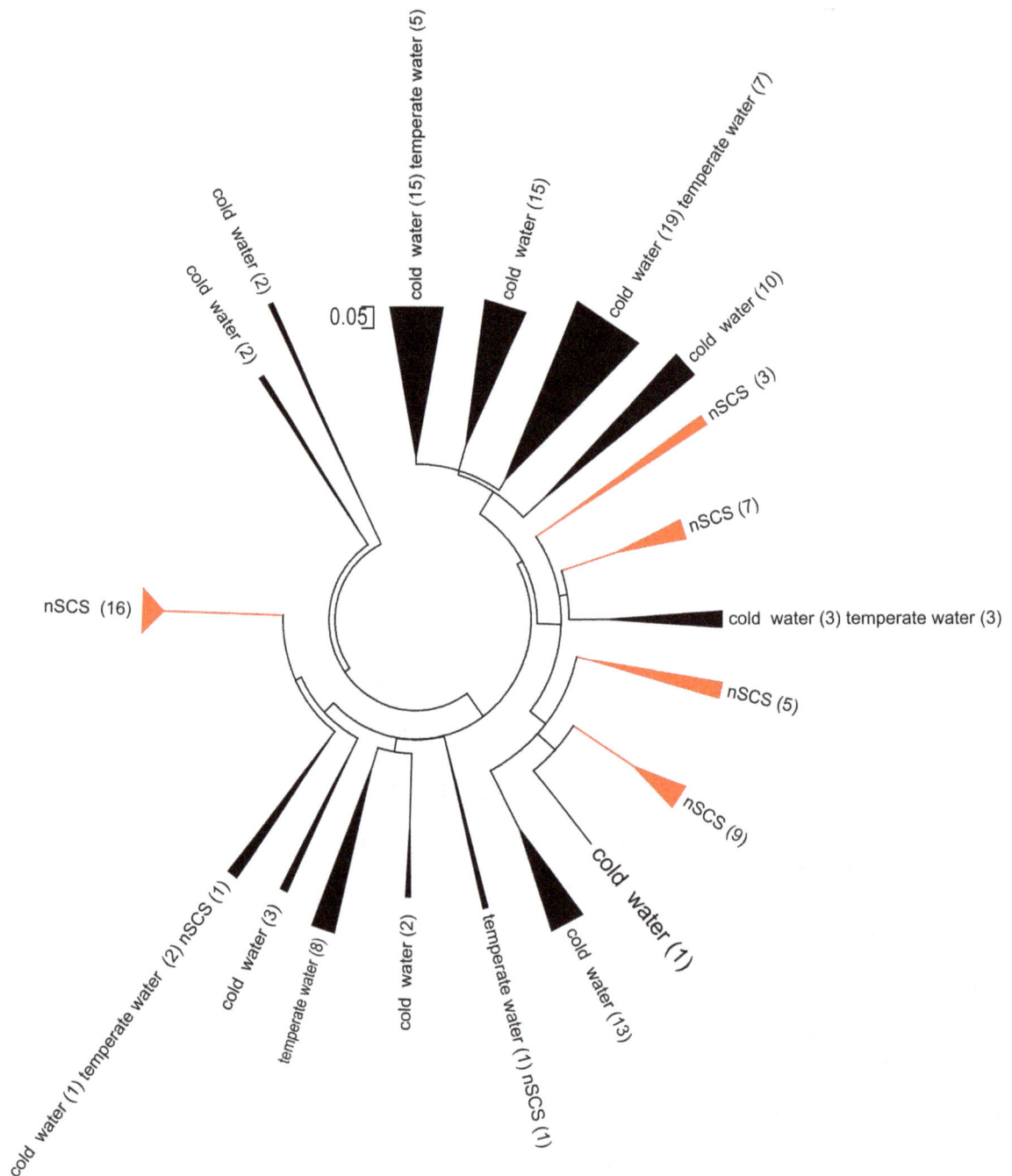

Figure 6. Biogeographic distribution of *g5* gene OTUs in the nSCS (red) compared with those from Subartic North Atlantic Ocean (cold water), the Arctic Ocean (cold water) and Chesapeake Bay (temperate water).

the nSCS and were considered dominant groups in tropical ocean ecosystems. *Prochlorococcus* and *Synechococcus*, the most abundant photosynthetic microorganisms in oceans, contribute significantly to primary production [25,26]. *Prochlorococcus* and *Synechococcus* are likely to contribute significantly to the primary production in the SCS because most of the nSCS exhibits oligotrophic characteristics.

Although *Synechococcus* and *Prochlorococcus* often occur simultaneously, they have different adaptation types depending on biogeochemical conditions. *Synechococcus* has also been reported to be abundant in environments with low salinities and/or low temperatures. *Synechococcus* is more abundant in nutrient-rich areas than in oligotrophic areas. Our results indicated that the abundance of *Synechococcus* decreased from 30.67% (SCS19) to 5.90% (SCS15). In the nSCS, a lower temperature (24.42°C) was detected in SCS19 than in SCS17 (28.97°C) and SCS15 (29.89°C). Temperature may have regulated the abundances of *Synechococcus* in nSCS waters. In contrast to

Synechococcus, *Prochlorococcus* is generally absent in brackish or well-mixed waters and more abundant in warm oligotrophic areas, which correspond to a major part of the oceans on Earth [27]. The northern part of SCS has typical oligotrophic characteristics with significant environmental gradients from the discharge of the Pearl River. *Prochlorococcus* could adapt to oligotrophic environments and was more abundant in SCS17 (32.76%) and SCS15 (26.01%) than in SCS19 (20.22%).

SAR11 clade accounted for 17.48%–23.17% of rRNA genes that have been identified in the nSCS by pyrosequencing methods in our study. Bacteria belonging to the SAR11 clade frequently constituted 25% or more of the cloned 16S rRNA gene sequences retrieved from seawater samples around the world [28]. SAR11 bacteria were responsible for about 50% of the amino acid assimilation and 30% of the DMSP assimilation in surface waters because these bacteria are highly abundant and active [29]. The high abundance of SAR11 suggested that members of this clade could play an important role in C, N, and S cycling in nSCS. Significant correlations were observed between the abundance of SAR11 and the abundance of *Prochlorococcus* [30]. Both *Prochlorococcus* and SAR11 have maximized their ability to consume nutrients efficiently at very low nutrient concentrations [30,31].

The pyrosequencing analysis results showed that the Rhodobacterales bacteria exhibited highly diverse sequences despite a relatively low abundance. In the bacterial genus level, Rhodobacterales from nSCS waters included 11 identified genera, which displayed high diversity. In the $g5$ gene cluster, only six genera of Rhodobacterales were found. Most of these genera were included in the results of the 16S rRNA phylogenetic analysis. $g5$ was highly conserved among all of the Rhodobacterales bacteria, and the phylogeny based on $g5$ was consistent with that based on 16S rRNA genes [32]. Inconsistency phylogeny between $g5$ and 16S rRNA gene was observed in this study possibly because of sampling site differences. This study also indicated that Rhodobacterales bacteria in the nSCS showed an evident spatial heterogeneity because of complex hydrographic conditions in nSCS (Table S2; Table S3). Most of the Rhodobacterales genera obtained in this study, such as *Roseobacter*, *Silicibacter*, and *Ruegeria*, could undergo aerobic anoxygenic photosynthesis, sulfur oxidation, carbon monoxide oxidation, and DMSP demethylation [3,33–35]. These traits are important in the nSCS ecosystem because most parts belong to oligotrophic waters. The roles of Rhodobacterales in tropical waters should be investigated in future studies.

Most $g5$ clones in nSCS had low amino acid identities (<90%) compared with known g5 sequences that were derived from genome-sequenced Rhodobacterales. Although Rhodobacterales bacteria had high levels of 16S rRNA sequence similarities with known GenBank sequences, a lower similarity match was found when comparing $g5$ gene sequences with GenBank database. This result also suggested that different Rhodobacterales bacteria contain highly diverse GTA genes in unique environments.

The diversity of the $g5$ gene of Rhodobacterales was higher in the offshore stations (E403 and E701) than near shore stations (E703 and E709). A significant correlation between geographic distance (latitude) and $g5$ compositions (p<0.05) was found when the relationship between location, temperature, salinity, and $g5$ compositions was analyzed; by contrast, other environmental factors (temperature, salinity, and Chla) had no apparent effect on $g5$ structure (p>0.05). Zhao et al. [5] also found that the composition of g5 sequences varies remarkably in different locations along the Chesapeake Bay. Furthermore, distinguishable diversity patterns are found between temperate and subarctic waters [6]. Geographic distance could be accounted for $g5$ gene diversity differences in nSCS.

The g5 gene was observed to have different clusters among the sampling sites in the nSCS. Clusters D, H, I, and L of the E709 clone library were not detected in other libraries, suggesting that a unique Rhodobacterales may be present near shore water (Figure 4). Station E709 is located in the Pearl River Estuary, and inshore area input affects this station in wet season. The Rhodobacterales community of this station had a distinct structure compared with other sites. Site E403 is near the Luzon Straits, and the Pacific Ocean largely influences this station [36]. Clusters B and C may represent typical Rhodobacterales cluster better than the other clusters recovered in this study. This cluster may have been derived from the external Pacific Ocean water. Offshore stations represented E701, which is influenced by near shore water and Pacific Ocean water in geographically. The site E701 sample had one common $g5$ gene cluster with site E709, and had three common $g5$ clusters with E403, which indicated that the Pacific Ocean water affected the Rhodobacterales composition in E701 to a great extent.

Studies on $g5$ genetic diversity have mainly focused on cold and temperate waters, such as the Subartic North Atlantic Ocean [6], Arctic marine [15], and Chesapeake Bay [5]. The overwhelming majority of $g5$ gene OTUs in the South China Sea was different from these areas. Only a few $g5$ gene OTUs sequences in the South China Sea had a close genetic distance to the sequences from the cold and temperate waters. The results also indicated that the $g5$ gene in the nSCS had a characteristic regional distribution. Furthermore, Shannon index in the nSCS ($H' = 2.35$) was similar to temperature water ($H' = 2.39$) [5] and lower than cold water ($H' = 3.71$) [6]. Overall diversity of the Rhodobacterales community, as inferred from g5 gene sequences, in the subarctic and arctic water appears higher than that in the temperate and tropical waters.

Conclusions

The present study demonstrated the spatial distribution of bacterial communities in nSCS environments. The South China Sea had more Chroococcales, SAR11 clade, Acidimicrobiales, Flavobacteriales, and Rhodobacterales than other bacterial groups. Rhodobacterales exhibited high diversity in nSCS despite relatively low abundance. Differences of $g5$ gene composition from tropical, temperate, and cold regions also suggested the specific adaptations of Rhodobacterales to different environments. Further research on the isolation and characterization of indigenous Rhodobacterales in nSCS may improve our understanding of the ecological roles of Rhodobacterales in tropical waters.

Supporting Information

Figure S1 Phylogenetic tree analysis based on partial 16S rRNA gene of Rhodobacterales in the northern South China Sea.

Figure S2 Canonical correspondence analysis (CCA) ordination diagram of $g5$ gene composition in northern South China Sea, with environmental factors as arrow.

Table S1 Characteristics of environmental parameters and clone information for each sampling station.

Table S2 Comparison of *g*5 gene OTUs composition and distribution in four clone libraries.

Table S3 Comparison of g5 gene cluster and distribution in the nSCS.

Acknowledgments

This research was supported by the National Natural Science Foundation of China (41406130, 41176101, 41430966, 31270528 and 41206082), the Strategic Priority Research Program of the Chinese Academy of Sciences (XDA10020225), and the key projects in the National Science & Technology Pillar Program in the Eleventh Five-year Plan Period (2012BAC07B0402). We also thank the open cruises of the South China Sea Institute of Oceanology and the National Natural Science Foundation of China in 2011 and 2013.

Author Contributions

Conceived and designed the experiments: FLS. Performed the experiments: FLS ZYJ. Analyzed the data: FLS MLW YSW CCS HC. Contributed reagents/materials/analysis tools: YSW. Wrote the paper: FLS. Conducted field work: FLS MLW ZYJ.

References

1. Hagström Å, Pommier T, Rohwer F, Simu K, Stolte W, et al. (2002) Use of 16S ribosomal DNA for delineation of marine bacterioplankton species. Appl Environ Microbiol 68: 3628–3633.
2. Buchan A, González JM, Moran MA (2005) Overview of the Marine *Roseobacter* Lineage. Appl Environ Microbiol 71: 5665–5677.
3. Moran MA, González JM, Kiene RP (2003) Linking a Bacterial Taxon to Sulfur Cycling in the Sea: Studies of the Marine Roseobacter Group. Geomicrobiol J 20: 375–388.
4. Newton RJ, Griffin LE, Bowles KM, Meile C, Gifford S, et al. (2010) Genome characteristics of a generalist marine bacterial lineage. ISME J 4: 784–798.
5. Zhao Y, Wang K, Budinoff C, Buchan A, Lang A, et al. (2008) Gene transfer agent (GTA) genes reveal diverse and dynamic *Roseobacter* and *Rhodobacter* populations in the Chesapeake Bay. ISME J 3: 364–373.
6. Fu Y, MacLeod DM, Rivkin RB, Chen F, Buchan A, et al. (2010) High diversity of Rhodobacterales in the subarctic North Atlantic Ocean and gene transfer agent protein expression in isolated strains. Aquat Microb Ecol 59: 283.
7. Lang AS, Beatty J (2000) Genetic analysis of a bacterial genetic exchange element: the gene transfer agent of *Rhodobacter capsulatus*. P Natl Acad Sci USA 97: 859–864.
8. Biers EJ, Sun S, Howard EC (2009) Prokaryotic genomes and diversity in surface ocean waters: interrogating the global ocean sampling metagenome. Appl Environ Microbiol 75: 2221–2229.
9. Solioz M, Marrs B (1977) The gene transfer agent of *Rhodopseudomonas capsulata*: Purification and characterization of its nucleic acid. Arch Biochem Biophys 181: 300–307.
10. Biers EJ, Wang K, Pennington C, Belas R, Chen F, et al. (2008) Occurrence and expression of gene transfer agent genes in marine bacterioplankton. Appl Environ Microbiol 74: 2933–2939.
11. McDaniel LD, Young E, Delaney J, Ruhnau F, Ritchie KB, et al. (2010) High frequency of horizontal gene transfer in the oceans. Science 330: 50–50.
12. Han WY (1998) Marine chemistry of the South China Sea. China Science Press, Beijing.
13. Lu ZM, Gan JP, Dai MH, Cheung AY (2010) The influence of coastal upwelling and a river plume on the subsurface chlorophyll maximum over the shelf of the northeastern South China Sea. J Mar Syst 82: 35–46.
14. Giebel H-A, Brinkhoff T, Zwisler W, Selje N, Simon M (2009) Distribution of *Roseobacter* RCA and SAR11 lineages and distinct bacterial communities from the subtropics to the Southern Ocean. Environ Microbiol 11: 2164–2178.
15. Fu Y, Keats KF, Rivkin RB, Lang AS (2013) Water mass and depth determine the distribution and diversity of Rhodobacterales in an Arctic marine system. FEMS Microbiol Ecol 84: 564–576.
16. Boström KH, Simu K, Hagström Å, Riemann L (2004) Optimization of DNA extraction for quantitative marine bacterioplankton community analysis. Limnol Oceanogr 2: 365–373.
17. Schloss PD, Westcott SL, Ryabin T, Hall JR, Hartmann M, et al. (2009) Introducing mothur: open-source, platform-independent, community-supported software for describing and comparing microbial communities. Appl Environ Microbiol 75: 7537–7541.
18. Gordon D, Desmarais C, Green P (2001) Automated finishing with autofinish. Genome Res 11: 614–625.
19. Tamura K, Dudley J, Nei M, Kumar S (2007) MEGA4: molecular evolutionary genetics analysis (MEGA) software version 4.0. Mol biol evol 24: 1596–1599.
20. Lepš J, Šmilauer P (2003) Multivariate analysis of ecological data using CANOCO: Cambridge Univ Pr.
21. Henriques IS, Almeida A, Cunha Â, Correia A (2004) Molecular sequence analysis of prokaryotic communities in the middle and outer sections of the Portuguese estuary Ria de Aveiro. FEMS microbiol ecol 49: 269–279.
22. Mullins TD, Britschgi TB, Krest RL, Giovannoni SJ (1995) Genetic comparisons reveal the same unknown bacterial lineages in Atlantic and Pacific bacterioplankton communities. Limnol Oceanogr 40: 148–158.
23. Rappe MS, Kemp PF, Giovannoni SJ (1997) Phylogenetic diversity of marine coastal picoplankton 16S rRNA genes cloned from the continental shelf off Cape Hatteras, North Carolina. Limnol Oceanogr 42: 811–826.
24. Suzuki M, Preston C, Beja O, De La Torre J, Steward G, et al. (2004) Phylogenetic screening of ribosomal RNA gene-containing clones in bacterial artificial chromosome (BAC) libraries from different depths in Monterey Bay. Microbial Ecol 48: 473–488.
25. Liu HB, Nolla HA, Campbell L (1997) Prochlorococcus growth rate and contribution to primary production in the equatorial and subtropical North Pacific Ocean. Aquat Microb Ecol 12: 39–47.
26. Jardillier L, Zubkov MV, Pearman J, Scanlan DJ (2010) Significant CO_2 fixation by small prymnesiophytes in the subtropical and tropical northeast Atlantic Ocean. ISME J 4: 1180–1192.
27. Partensky F, Blanchot J, Vaulot D (1999) Differential distribution and ecology of Prochlorococcus and Synechococcus in oceanic waters: a review. In: Marine Cyanobacteria (Charpy, L and Larkum, AWD, Eds) Special No. 19: 457–476.
28. Giovannoni S, Stingl U (2007) The importance of culturing bacterioplankton in the'omics' age. Nat Rev Microbiol 5: 820–826.
29. Malmstrom RR, Kiene RP, Cottrell MT, Kirchman DL (2004) Contribution of SAR11 bacteria to dissolved dimethylsulfoniopropionate and amino acid uptake in the North Atlantic ocean. Appl Environ Microb 70: 4129–4135.
30. Eiler A, Hayakawa DH, Church MJ, Karl DM, Rappé MS (2009) Dynamics of the SAR11 bacterioplankton lineage in relation to environmental conditions in the oligotrophic North Pacific subtropical gyre. Environ Microbiol 11: 2291–2300.
31. Hill PG, Zubkov MV, Purdie DA (2010) Differential responses of Prochlorococcus and SAR11-dominated bacterioplankton groups to atmospheric dust inputs in the tropical Northeast Atlantic Ocean. FEMS Microbiol Lett 306: 82–89.
32. Lang AS, Beatty JT (2007) Importance of widespread gene transfer agent genes in alpha-proteobacteria. Trends Microbiol 15: 54–62.
33. Allgaier M, Uphoff H, Felske A, Wagner-Döbler I (2003) Aerobic Anoxygenic Photosynthesis in *Roseobacter* Clade Bacteria from Diverse Marine Habitats. Appl Environ Microbiol 69: 5051–5059.
34. Brinkhoff T, Giebel H-A, Simon M (2008) Diversity, ecology, and genomics of the Roseobacter clade: a short overview. Arch Microbiol 189: 531–539.
35. Lenk S, Moraru C, Hahnke S, Arnds J, Richter M, et al. (2012) Roseobacter clade bacteria are abundant in coastal sediments and encode a novel combination of sulfur oxidation genes. ISME J 6: 2178–2187.
36. Qu TD, Mitsudera H, Yamagata T (2000) Intrusion of the North Pacific waters into the South China Sea. J Geophys Res 105: 6415–6424.

3

Limited Density of an Antigen Presented by RMA-S Cells Requires B7-1/CD28 Signaling to Enhance T-Cell Immunity at the Effector Phase

Xiao-Lin Li[1], Marjolein Sluijter[2], Elien M. Doorduijn[2], Shubha P. Kale[1], Harris McFerrin[1], Yong-Yu Liu[3], Yan Li[4,1], Madhusoodanan Mottamal[1], Xin Yao[1], Fengkun Du[1], Baihan Gu[1], Kim Hoang[1], Yen H. Nguyen[1], Nichelle Taylor[1], Chelsea R. Stephens[1], Thorbald van Hall[2], Qian-Jin Zhang[1]*

1 Department of Biology, Xavier University of Louisiana, New Orleans, Louisiana, United States of America, 2 Clinical Oncology, K1-P, Leiden University Medical Center, Leiden, the Netherlands, 3 Department of Basic Pharmaceutical Sciences, University of Louisiana at Monroe, Monroe, Louisiana, United States of America, 4 College of Chemistry & Environmental Science, Hebei University, Hebei Province, Baoding, China

Abstract

The association of B7-1/CD28 between antigen presenting cells (APCs) and T-cells provides a second signal to proliferate and activate T-cell immunity at the induction phase. Many reports indicate that tumor cells transfected with B7-1 induced augmented antitumor immunity at the induction phase by mimicking APC function; however, the function of B7-1 on antitumor immunity at the effector phase is unknown. Here, we report direct evidence of enhanced T-cell antitumor immunity at the effector phase by the B7-1 molecule. Our experiments *in vivo* and *in vitro* indicated that reactivity of antigen-specific monoclonal and polyclonal T-cell effectors against a Lass5 epitope presented by RMA-S cells is increased when the cells expressed B7-1. Use of either anti-B7-1 or anti-CD28 antibodies to block the B7-1/CD28 association reduced reactivity of the T effectors against B7-1 positive RMA-S cells. Transfection of Lass5 cDNA into or pulse of Lass5 peptide onto B7-1 positive RMA-S cells overcomes the requirement of the B7-1/CD28 signal for T effector response. To our knowledge, the data offers, for the first time, strong evidence that supports the requirement of B7-1/CD28 secondary signal at the effector phase of antitumor T-cell immunity being dependent on the density of an antigenic peptide.

Editor: Xue-feng Bai, Ohio State University, United States of America

Funding: This study was supported by funding from NIH (RCMI, 8G12MD007595), Louisiana Cancer Research Consortium (LCRC) and Xavier University's Center for Undergraduate Research (CUR) to Dr. Qian-Jin Zhang. Dr. Thorbald van Hallwas supported by Dutch Cancer Society (UL2010-4785). Dr. Harris McFerrin was supported by funding from the NIGMS (P20GM103424). This study was also supported by funding from Louisiana Board of Regents Eminent Alumni Scholars Program, Kellogg Professorship IV in the Arts and Sciences to Dr. Shubha P. Kale. The funders had no role in study design, data collection and analysis, decision to publish, or preparation of the manuscript.

Competing Interests: The authors have declared that no competing interests exist.

* Email: qzhang2@xula.edu

Introduction

It is well established that in the induction phase of CD8[+] T-cell responses, T cells require two signals through cell-cell interactions with antigen presenting cells (APCs) for their activation and proliferation [1,2]. Major Histocompatibility Complex class I (MHC-I) presentation of antigen to the T-Cell Receptor (TCR) serves as the first signal, while association of B7-1 (or CD80) with the CD28 molecule expressed on T cells triggers the second signal. B7-1 is not expressed on most tumor cells; therefore, if tumors express MHC-I and trigger the first signal, they may not fully activate anti-tumor specific T cells [3]; however, transfecting the B7-1 gene into tumor cells can render them capable of effectively stimulating antitumor T-cell activation, leading to cancer eradication *in vivo* [4–8]. The augmented antitumor T-cell responses by B7-1 expressing tumor cells occur in the induction phase of immunity.

Transporter associated with antigen processing (TAP)-deficient tumors represent immune-escape variants [9]. Presentation of MHC-I-restricted antigen in these tumors is insufficient; therefore, the induction of the T-cell responses is either difficult [10] or less

efficient [11]. Introduction of the B7-1 gene into TAP-deficient tumor cells stimulates immune system to generate stronger T-cell mediated immune responses against B7-1 negative parental counterparts [10–12], suggesting that the induction phase of T-cell immunity is augmented by B7-1. Recent evidence indicates that CD8[+] T cells generated by B7-1 expressing tumor cells recognized a panel of the TAP independent antigens [13]. One of the antigens, Lass5, derived from the ceramide synthase Lass5 (or Trh4/CerS5) protein, located in the endoplasmic reticulum (ER) lumen, associates with H-2D[b] and is presented by many TAP-deficient, but not TAP-proficient, mouse cells [11,13]. Although both TAP-proficient and TAP-deficient mouse cells express Lass5 protein, peptide/D[b] complexes are selectively presented on TAP-deficient counterparts, most likely due to competition of TAP-mediated peptide antigens [14].

In this study, we have addressed whether expression of B7-1 on TAP-deficient tumor cells can functionally enhance T-cell immunities at the effector phase. We have confirmed that B7-1/CD28 signaling at the effector phase of immunity is required to enhance T-cell based immune response against Lass5 antigen expressed by TAP-deficient tumor cells, and this requirement can be overcome when the targets express high levels of the Lass5 antigen.

Materials and Methods

Ethics Statement

The Xavier University of Louisiana Institutional Animal Care and Use Committee (IACUC) approved animal protocol (012711-001BI) used in this study. C57BL/6 mice (6-week-old females) were purchased from Charles River Laboratories and were maintained in pathogen-free animal facilities at Xavier University of Louisiana. Each ventilated and sealed cage contained 5 mice with bedding materials of aspen shavings or shreds. All mice were treated in accordance with the Institute of Laboratory Animal Research (NIH, Bethesda, MD) Guide for the Care and Use of Laboratory Animals. In *in vivo* experiments, the tumor size reached a volume 30×10^2 (mm^3) or the mice were sacrificed by CO_2 upon observed distress.

Peptide

H-2D^b restricted peptide Lass5 (MCLRMTAVM) at 98% purification was purchased from GL Biochem Ltd (Shanghai, China) and used for this study. The peptide was dissolved in pure DMSO at a stock concentration of 10 mg/ml and stored at $-20°C$.

Cell Lines and Cell Culture

Mouse TAP2-deficient RMA-S cells were transfected with either pUB6-vector or pUB6-based B7-1 cDNA [11]. The transfectants were designated as RMA-S/pUB and RMA-S/B7-1 cells and were maintained in RPMI 1640 (Mediatech Inc., Manassas, VA., USA) supplemented with 10% FCS, 2 mM L-glutamine, 100 IU/ml penicillin, 100 microgram/ml streptomycin and 20 mM HEPES and supplemented with 10 microgram/ml Blasticidin. In addition, both cell lines were further transfected with Lass5 (Trh4/CerS5) expressing LZRS-retroviral vector [14]. The Lass5-vector transfectants were designated as RMA-S/B7-1.Trh4 and RMA-S/pUB.Trh4 cells respectively.

Hybridoma

Hybridoma producing anti-mouse NK1.1 monoclonal antibody (mAb), clone PK 136 was obtained from ATCC (Manassas, VA). Culture of the hybridoma and purification of the NK1.1 mAb was performed using a published protocol [15] with slight modification. The mAb was concentrated and purified using the ammonium sulfate method and purified mAb was obtained at a concentration of about 100 mg per milliliter and used for *in vivo* depletion of mouse NK cells.

FACS Assays

FACS assays were performed to detect B7-1 on transfected cells and to detect the NK1.1 cell population in mouse splenocytes. B7-1 expressed on RMA-S/pUB and RMA-s/B7-1 transfectants was labeled with a FITC-conjugated anti-mouse CD80 mAb (clone 16-10A1, Biolegend, San Diego, CA, USA). The NK cell population was detected in mouse splenocytes by labeling with anti-mouse CD16/32 (Fc-receptor) mAb (clone 93, Biolegend, San Diego, CA, USA), followed by labeling with FITC-conjugated anti-mouse NK1.1 mAb (clone PK136, Biolegend, San Diego, CA, USA). After extensively washing, the cell pellets were suspended in PBS at 1×10^6 cells/ml concentration. Expression of cell surface B7-1 molecule and NK1.1 protein was determined by using a BD FACScalibur.

Quantitative PCR analysis of Lass5 expressing transfectants

Total RNA isolation and cDNA preparation from RMA-S/B7-1.Trh4 and RMA-S Trh4/pUB cells were performed using an RNeasy Mini Kit (Qiagen, MD, USA). Five hundred nanograms of purified total RNA were used to synthesize cDNA using a High Capacity RNA-to-cDNA Kit (Applied Biosystems, Foster City, USA). Quantitative PCR on short and long transcripts of Trh4 was done as described previously [13]. SensiMix SYBR No-ROX kit from GC Biotech Bioline (Alphen aan den Rijn, NL) was used in a C1000 Thermal Cycler (Bio-Rad, Hercules, CA, USA) and results were analyzed using Bio-Rad CFX manager software. Long Trh4 (Lass5) transcripts were amplified with Power SYBR Green Master Mix (Applied Biosystems) on a GeneAmp 7300 System (Applied Biosystems).

Generation of Cytolytic T Lymphocytes (CTL) and ^{51}Cr-release Assays

Antigens used for CTL generation were prepared using the following procedures: RMA-S/B7-1 or RMA-S/pUB cells were incubated at 26°C overnight with 100 micromole D^b-restricted and TAP-independent Lass5 peptide [13]. Afterwards, the cells were treated with 30 microgram/ml mitomycin-c for 3-hours at 26°C and washed extensively. The peptide-pulsed RMA-S/B7-1 or RMA-S/pUB cells were then injected i.p. into C57BL/6 mice (5×10^6 cells/mouse). After a 9-day immunization, the RMA-S/pUB- or RMA-S/B7-1-immunized mice were killed by CO_2. The immunized spleens were re-stimulated with mitomycin-c treated, 100 micromole Lass5-pulsed RMA-S/pUB or RMA-S/B7-1 cells (1×10^7 cells/1×10^8 splenocytes). ^{51}Cr-release assays were conducted by using target cells indicated in each figures. Percentage data were converted to logarithmic data before statistical analysis. Two-way ANOVA followed by Dunnett's Multiple Comparison test or Unpaired Student's t-test were performed. Results were considered significant if P value ≤ 0.05.

T-cell activation assays

Lass5-specific T cell clone LnB5 was generated as previously described [13]. T-cell activities were measured by intracellular IFN-gamma staining of T-cells conducted as previously described [16,17]. In brief, 8×10^3 Lass5-specific LnB5 cells were incubated with indicated amounts of stimulator cells for 4-h in the presence of 1 microgram/ml GolgiPlug (BD Biosciences). After incubation the cells were fixed, permeabilized and stained with PE-conjugated IFN-gamma-specific mAb, using an intracellular cytokine staining starter kit (BD Biosciences). Afterwards, the cells were stained with FITC-conjugated anti-mouse CD8a mAb and washed extensively. The cell samples were then analyzed using a FACS Calibur flow cytometer (BD Biosciences). Percentage data were converted to logarithmic data before statistical analysis. Two-way ANOVA followed by Dunnett's Multiple Comparison test or Student's t-test were performed. Results were considered significant if P value ≤ 0.05.

Reduction of CTL Killing Activity by Blocking of B7-1/CD28 Binding

mAbs against mouse B7-1 (Clone 16-10A1; Armenian Hamster IgG), CD28 (Clone 37.51; Golden Syrian Hamster IgG), and relevant purified Hamster IgG-isotype controls were purchased (eBioscience, San Diego, CA). Both mAbs were reported to functionally block B7-1/CD28 binding [18,19]. Before adding bulk-cultured CTLs or the LnB5 T-cell clone into target cell cultures for ^{51}Cr- release assays or intracellular IFN-gamma secretion assays, either T cells or target RMA-S/B7-1-culture was added with 10 microgram/ml relevant mAbs against either mouse CD28 (for CTL-culture) or mouse B7-1 (for RMA-S/B7-1-culture) for 1 hour at room temperature. The relevant purified Hamster IgG-isotype control antibody was used as an experimental control. The antibody-containing cultures were then used for ^{51}Cr-release assays (for bulk-cultured CTLs) or intracellular IFN-gamma secretion assays (for LnB5 T-cells).

In Vivo Tumor Growth

C57BL/6 mice were treated with three alternate procedures before tumor cell challenge. 1) The mice were immunized i.p with PBS; 2) The mice were immunized i.p. with Lass5-peptide-pulsed and mitomycin-c-treated RMA-S/pUB cells or RMA-S/B7-1 cells at 5×10^6 cells/mouse; and 3) After one week of immunization with 5×10^6 cells/mouse Lass5-peptide-pulsed and mitomycin-c-treated RMA-S/pUB cells, the mice were depleted of NK effectors by using concentrated NK1.1 mAb (clone 16-10A1, 0.5 mg/mouse injection). The mAb treatment was performed every other day for the first one and half weeks and once a week for the following weeks. Twenty three days post-immunization, the mice were challenged s.c. with 5×10^6 live RMA-S/pUB or RMA-S/B7-1 cells per mouse. Tumor growth was initially detected by palpation daily, and once tumor were palpable, tumor volume was measured by a caliper and calculated by the formula $V = \pi$ x abc/6 (where a, b, and c are the orthogonal diameters). The experimental mice were terminated at animal facility by CO2 inhalation when the tumor size reached a volume 30×10^2 (mm^3). Each experimental group contained 4 to 5 mice described in table 1.

Results

Inhibition of RMA-S/B7-1 cell growth in immunized syngeneic mice

B7-1 molecule expression on tumor cells can elicit anti-tumor immunity at the induction phase [11,12,20,21]; however, there has been no direct evidence to support the enhancement of anti-tumor immunity at the effector phase by B7-1. To test this possibility, RMA-S cells were transfected with the B7-1 gene (designated as RMA-S/B7-1) or a relevant vector (designated as RMA-S/pUB). B7-1 expression on RMA-S/B7-1 but not RMA-S/pUB cells was confirmed by FACS assay (Fig. 1A-a).

To test if B7-1 enhanced T-cell based antitumor immunity at the effector phase, we conducted an *in vivo* tumor-growth inhibition experiment. Since RMA–S cells present a well-known H-2D^b-restricted Lass5 peptide, we immunized mice with Lass5-peptide-pulsed and mitomycin-c-treated RMA-S/pUB and RMA-S/B7-1 cells, respectively. PBS-immunization was used as control.

Twenty-three-days after immunization, each group was divided into two sub-groups that were challenged with 5×10^6 cells/mouse of live RMA-S/B7-1 or RMA-S/pUB cells, respectively. Tumor sizes were measured twice a week after challenge with live tumor cells. The tumors appeared in all mice during the initial week in control PBS-immunized groups while the tumors appeared in most mice at 1.5 weeks in tumor-immunized groups (table 2, Fig. 1B-e insert), suggesting that antitumor immunity was established in tumor-immunized groups. This established immunity dramatically inhibited the growth of B7-1 expressing tumors at 1.5 weeks (table 2). During this time point, both RMA-S/pUB- or RMA-S/B7-1-immunized mice challenged with RMA-S/B7-1 cells had tumors that were much smaller in size, and tumors were found in only two out of nine mice, compared to those challenged with the RMA-S/pUB cells in which larger tumors grew quickly in all mice. The difference in tumor sizes between RMA-S/pUB- and RMA-S/B7-1-cell challenged groups at 1.5 week time point was statistically significant (P<0.05). Results suggested that anti-tumor immunity at the effector phase played an important role in inhibiting B7-1 expressing tumor growth. After the initial two weeks of tumor growth, the RMA-S/pUB tumors continued to grow quickly in both RMA-S/pUB and RMA-S/B7-1 immunized mice while no tumors could be detected in the immunized mice challenged with RMA-S/B7-1 cells (Fig. 1B-e and 1C). In PBS-immunized mice, RMA-S/pUB and RMA-S/B7-1 tumors continued to grow dramatically except in one mouse in which the RMA-S/B7-1 tumor had regressed during initial 1.5 weeks (data not shown). Our results suggested that a major component of the anti-B7-1 expressing tumor immunity is T effectors but not NK effectors because: 1) the RMA-S/B7-1 tumors grew quickly in PBS-immunized mice while no RMA-S/B7-1 tumors appeared in tumor-immunized mice at initial week and 2) NK activity could only inhibit less than 1×10^6 challenged B7-1 expressing RMA-S cells per mouse [22]. In our experiment, 5×10^6 tumor cells per mouse were injected. To further confirm T effectors provided anti-RMA-S/B7-1 tumor protective immunity, we treated the peptide-pulsed RMA-S/pUB-immunized mice with anti NK1.1 mAb before live cell challenge. Figure 1A (b, c and d) indicated that anti-NK1.1 mAb treatment depleted NK cells in the mice. These mice challenged with RMA-S/pUB or RMA-S/B7-1 cells displayed tumor growth patterns (Fig. 1B-f) similar to the peptide-pulsed RMA-S/pUB-immunized mice without anti-NK1.1 mAb treatment (see Fig. 1B-e insert). The RMA-S/B7-1 cells in the mAb-treated mice grew and formed small tumors that disappeared at week 2 after tumor cell challenge while the RMA-S/pUB cells continuously grew to form large tumors in the mAb-treated mice (Fig. 1B-f). Statistical analysis of tumor sizes indicated significant differences between the two mouse groups during the initial week and 1.5 week time points (P<0.05 and <0.01

Table 1. C57/BL6 mice used in each different experimental group.

number of mice	RMA-S/pUB -challenge*	RMA-S/B7-1-challenge*
RMA-S/pUB-immunized	4	5
RMA-S/B7-1-immunized	4	5
PBS-immunized	4	4
NK depletion and RMA-S/pUB-immunized	4	4

*indicates the number of mice per group.
Results of statistical analysis for mouse tumor sizes at specific time points were obtained using Paired Student *t* test, and differences were considered significant at P< 0.05.

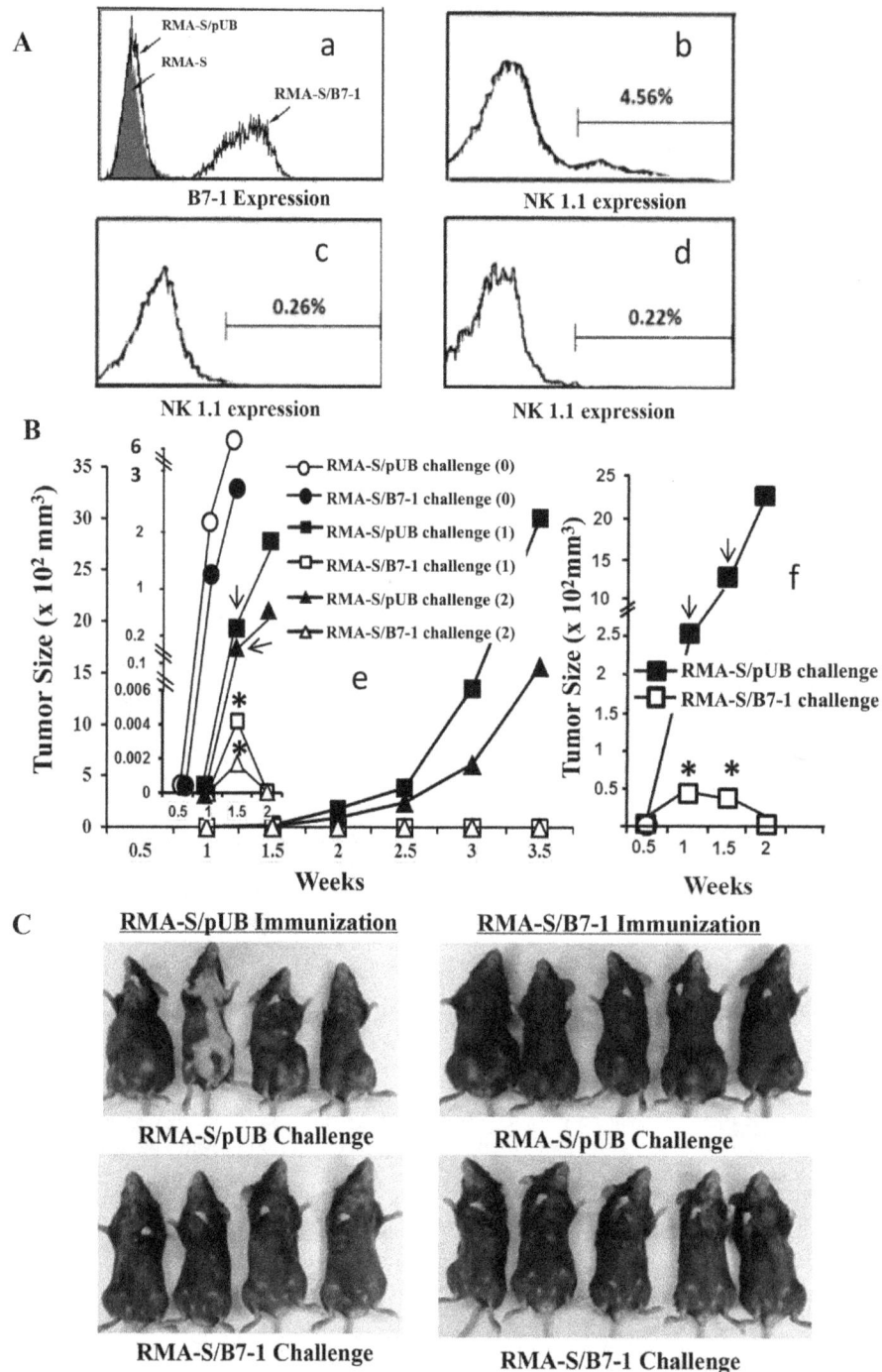

Figure 1. Inhibition of B7-1 expressing RMA-S tumor growth in Lass5-antigen immunized mice. A: a) B7-1 expression in the transfectants. B7-1 expression was determined by FACS assay using FITC-conjugated anti-mouse CD80 mAb; b, c and d) NK1.1 population in mouse splenocytes were detected by anti-NK1.1 mAb. b) Normal mouse splenocytes, c) and d) the splenocytes from tumor-immunized and anti-NK1.1 mAb treated mouse (c: on the tumor cell challenge time and d: end of experiment). B and C: *In vivo* tumor growth assays. B: e) mice immunized with PBS (0), Lass5-peptide-pulsed and mitomycin-c-treated RMA-S/pUB (1) or RMA-S/B7-1 (2) cells. After immunization, the mice were challenged s.c with RMA-S/pUB or RMA-S/B7-1 cells. The insert indicates tumor growth during the time point of the initial tumor cell injection through two weeks. f) Mice immunized with Lass5-peptide-pulsed and mitomycin-c-treated RMA-S/pUB cells and followed by anti-NK1.1 mAb treatment. Afterwards, the mice were challenged s.c with RMA-S/pUB or RMA-S/B7-1 cells. Statistical analysis of tumor sizes indicated significant differences between RMA-S/pUB '↓' and RMA-S/B7-1 '*' cell challenge groups at relevant time points (P value≤0.05 or 0.01). C: Tumor sizes at the endpoint were shown in the mice immunized with Lass5-peptide-pulsed and mitomycin-c-treated RMA-S/pUB or RMA-S/B7-1 cells and followed by challenge with live RMA-S/pUB or RMA-S/B7-1 cells.

Table 2. Tumor formation in the mouse groups during the initial time points.

Mice immunized With or without Tumor cells	Challenge of live tumor cells	
	RMA-S/pUB Number of mice with tumor	RMA-S/B7-1 Number of mice with tumor
RMA-S/pUB-immunized group	4*	1*
RMA-S/B7-1-immunized group	4*	1*
PBS immunized group	4#	4#
RMA-S/pUB- and mAb treated group	4#	4#

#indicates that tumors appear at initial week after the inoculation.
*indicates that tumors appear at initial 1.5 weeks after the inoculation. Total mice per group were shown in the Material and Method Section.

respectively). NK activities could play an auxiliary function in controlling RMA-S/B7-1 tumor growth. In the NK depleted and tumor-immunized mice, RMA-S/B7-1 tumors appeared at initial week and disappeared at week 2 (table 2; Fig. 1A-f), while in the tumor-immunized mice RMA-S/B7-1 tumors appeared at 1.5 weeks and disappeared at week 2 (Fig. 1A-e insert). These results indicated that NK activity could only control early or late appearance of RMA-S/B7-1 tumors and could not inhibit tumor growth.

Bulk-culture T cells more efficiently kill RMA-S/B7-1 cells, and the killing activities require the B7-1/CD28 axis

To confirm *in vivo* experiments, *in vitro* ^{51}Cr-release assays were performed. Two T-cell bulk cultures generated by immunization of mice with Lass5-peptide-pulsed and mitomycin-C-treated RMA-S/pUB or RMA-S/B7-1 cells were used to determine if the B7-1/CD28 axis could enhance T-cell killing activity. Figure 2 showed that two T-cell bulk cultures killed B7-1-expressing RMA-S/B7-1 targets more efficiently than RMA-S/pUB targets (Fig. 2A and B). These results suggested that the role of B7-1 molecule in increasing immune response at the effector phase could occur in Lass5-peptide-stimulated T-cell bulk cultures.

To confirm enhanced T-cell killing activity was associated with the B7-1/CD28 axis, blocking antibodies against B7-1 and CD28 molecules were used. We first performed assays to block the B7-1/CD28 axis using a mAb against mouse B7-1, and an IgG isotype antibody was used as a control. After incubation of RMA-S/B7-1 targets with the mAb or the isotype antibody at room temperature for 1 hour, the targets were mixed with effectors, and the effector killing activities were determined. Results showed that T-cell killing activities against the antibody-incubated RMA-S/B7-1 targets were reduced to a level similar to those observed in RMA-S/pUB cells incubated with isotype-control antibody while isotype-blocked RMA-S/B7-1 cell killing remained at higher levels (Fig. 3A and B). In addition, blocking of the B7-1/CD28 axis by using a mAb against mouse CD28 displayed similar results (Fig. 3C and D). These assays suggested that enhanced killing activities of T effectors required B7-1/CD28 binding.

It has been reported that NK activity can be triggered *in vitro* by B7-1, and this occurred even in the absence of CD28 and could not be blocked by anti-CD28 mAb [23]. Our preparation of T-cell bulk-cultures displayed killing activities for RMA-S/B7-1 targets being reduced by anti-CD28 mAb, suggesting that the role of NK cells was negligible.

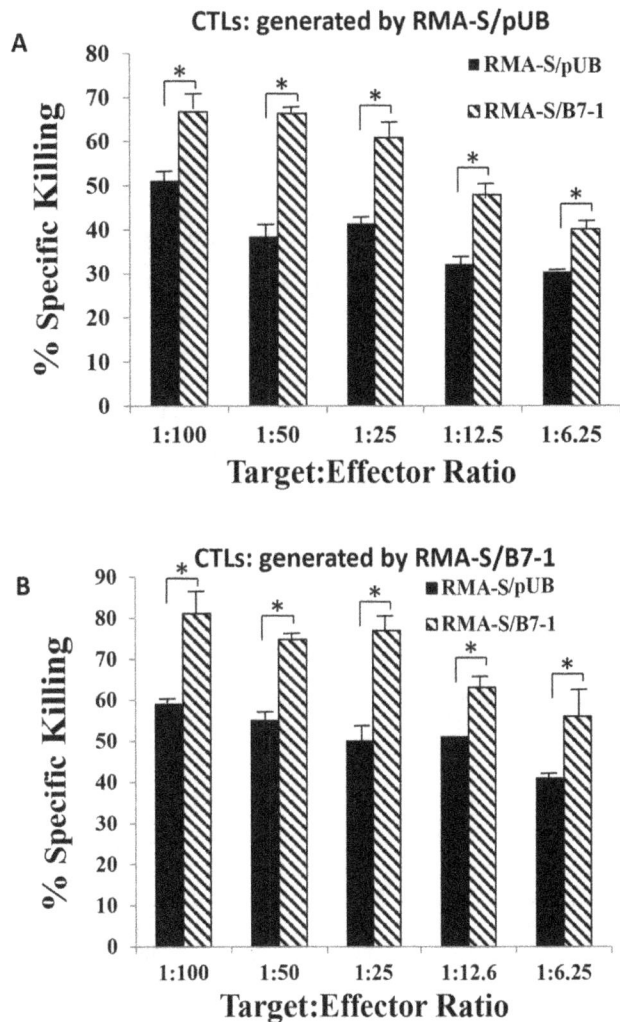

Figure 2. Efficient killing of B7-1 expressing tumor cells by bulk culture T cells. *In vitro* ^{51}Cr-release assays were conducted. (A): Bulk-culture T effectors were generated by immunizing mice with Lass5 peptide-pulsed mitomycin-c-treated RMA-S/pUB cells. (B): Bulk-culture T effectors were generated by immunizing mice with Lass5 peptide-pulsed mitomycin-c-treated RMA-S/B7-1 cells. One out of three experiments with similar results was shown. * indicated that P-values were less than 0.05.

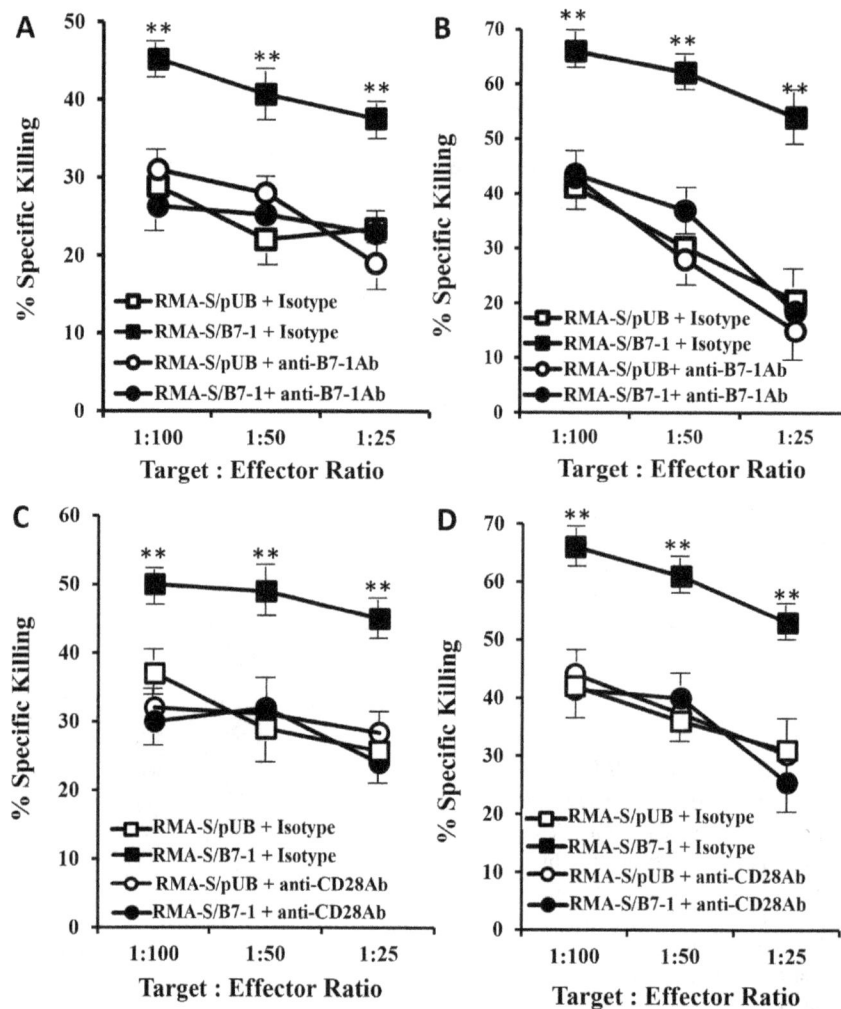

Figure 3. Effects of anti-CD80 and CD28 antibodies on reducing killing activities of bulk culture T effectors against RMA-S/B7-1 cells. Lift-panel (A and C): The cytolytic T effectors were generated by immunization of mice with mitomycin-c-treated RMA-S/pUB cells pulsed with Lass5 peptide. Right-panel (B and D): The cytolytic T effectors were generated by immunization of mice with mitomycin-c-treated RMA-S/B7-1 cells pulsed with Lass5 peptide. Up-panel (A and B): ^{51}Cr-labeled RMA-S/B7-1 and RMA-S/pUB target cells were incubated with either anti-mouse B7-1 mAb or relevant IgG-control. After incubation, the cells were then incubated with antigen-specific bulk culture T effectors for *in vitro* ^{51}Cr-release assays. Bottom-panel (C and D): Cytolytic bulk culture T effectors were incubated with either anti-mouse CD28 mAb or relevant IgG-control. After incubation, the T-cells were then incubated with ^{51}Cr-labeled RMA-S/B7-1 and RMA-S/pUB target cells for *in vitro* ^{51}Cr-release assays. ** indicated that P-values were less than 0.05 among 'RMA-S/B7-1+ Isotype' and other targets at each 'Target: Effector' ratio.

B7-1/CD28 axis plays a major role in increasing LnB5 T-cell activation

To confirm that the role of the B7-1/CD28 axis in delivering a signal into and activating the T-cells at the effector phase was not due simply to binding, the LnB5 T-cell clone specific for the Lass5 peptide [13] was employed. We incubated the LnB5 cells with different amount of either RMA-S/B7-1 or RMA-S/pUB cells and measured the concentration of IFN-gamma secretion by the LnB5 T-cells. Results clearly showed that RMA-S/B7-1 cells stimulated T-cell activation more efficiently than the RMA-S/pUB cells as indicated by more IFN-gamma secretion (Fig. 4A). Enhanced T-cell activation was confirmed to be due to the B7-1/CD28 axis because blocking B7-1/CD28 binding between RMA-S/B7-1 targets and LnB5 effectors by either anti-B7-1 or anti-CD28 antibodies or both reduced IFN-gamma secretion to the levels similar to that of LnB5 T-cells incubated with RMA-S/pUB cells (Fig. 4B, C and D). These results indicate that the B7-1/CD28 axis provides a second signal, triggering enhancement of Lass5 antigen specific T-cell activation at the effector phase.

Requirement of B7-1/CD28 signaling at the effector phase of immunity is overcome by Lass5-overexpressing targets

Why does enhanced response to Lass5 antigen require the secondary signal at the effector phase? The possible reasons are 1) the Lass5 peptide has a low affinity for H-2D^b binding and/or 2) the Lass5 peptide is generated at a limited level. Both of these possibilities would reduce antigenic peptide surface stability or expression. These situations may reduce the strength of the first signal and therefore require help by the secondary signal to efficiently activate function of T effectors. We have previously performed peptide-binding and peptide-stability assays demonstrating binding and stability of the Lass5 peptide to H-2D^b at levels comparable to the levels of high affinity binders such as the

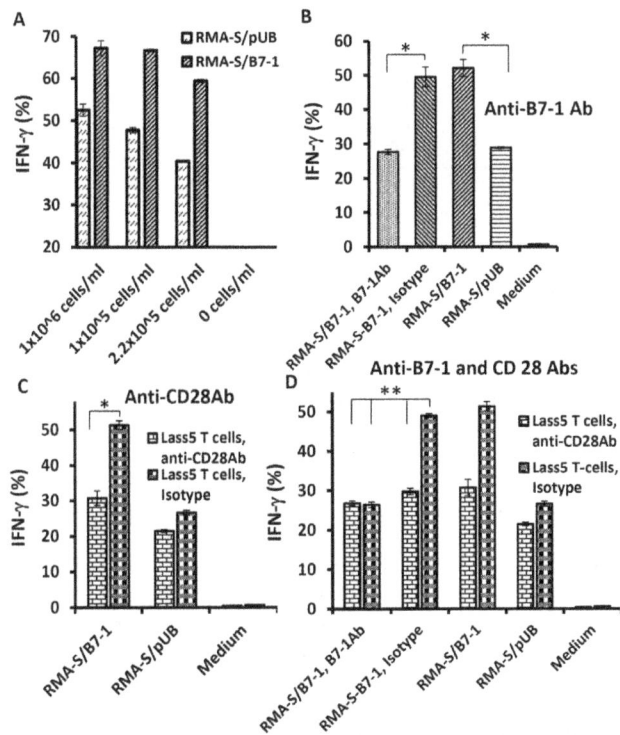

Figure 4. Importance of B7-1:CD28 axis in enhancing a Lass5 specific LnB5 T-cell clone activation. The RMA-S/pUB and RMA-S/B7-1 transfectants were used as targets recognized by a Lass5 specific LnB5 T-cell clone. Lass5 specific T-cell clone activation detected by the intracellular IFN-gamma release assays were conducted with stimulators RMA-S/pUB and RMA-S/B7-1 cells in (A) to (D). (A): 8×10^3 T-cells were incubated with indicated amounts of RMA-S/pUB and RMA-S/B7-1 cells. (B): 8×10^3 T-cells were incubated with 1×10^5 stimulators that previously incubated with either anti-B7-1 mAb or isotype control (for RMA-S/B7-1). (C): 8×10^3 T cells were incubated with either anti-CD28 mAb or isotype control before co-culture with 1×10^5 stimulators (RMA-S/pUB or RMA-S/B7-1). (D): Before co-culture of the T-cells and stimulators, 8×10^3 T-cells were incubated with either anti-CD28 mAb or Isotype control and 1×10^5 RMA-S/B7-1 stimulator cells were incubated with either anti-B7-1 mAb or Isotype control. One out of at least two experiments with similar results was shown. * and ** indicated that P-values were less than 0.05.

viral gp33 epitope (KAVYNFATM) from LCMV [14]. Computer modeling analysis of Lass5 peptide and two immunodominant viral epitopes, ASNENMETM from the influenza-A virus and KAVYNFATM from LCMV virus, demonstrated that the relative binding capacity of the Lass5 peptide is weaker than influenza-A viral peptide but stronger than LCMV viral peptide (data not shown). These results suggested that binding capacity of the Lass5 epitope to the H-2D^b molecule is similar to immunodominant viral epitopes.

To test if increased Lass5 expression could overcome the requirement of the B7-1/CD28 axis for enhancing immune response, RMA-S/B7-1 and RMA-S/puB cells were further transfected with a Lass5 (Trh4) cDNA-carrying LZRS retroviral vector. Lass5 mRNA over-expression in the transfectants was detected by quantitative PCR (no antibody available). Long and short Lass5 transcripts were detected, and only the long transcript contained a Lass5 coding sequence [13]. Table 3 shows that both RMA-S/B7-1.Trh4 and RMA-S/pUB.Trh4 cells expressed higher levels of Lass5 mRNA compared to that detected in RMA-S

cells. The levels of the increased Lass5 transcripts in RMA-S/B7-1.Trh4 and RMA-S/pUB.Trh4 cells were about 822 and 535 respectively.

Overexpression of Lass5 mRNA in transfectants enhanced LnB5 T-cell recognition. Both RMA-S/B7-1.Trh4 and RMA-S/pUB.Trh4 cells stimulated LnB5 effectors to secrete IFN-gamma at levels higher than that found in Trh4-untransfected counterparts (Fig. 5A), suggesting that higher IFN-gamma secretion in the T-effectors was induced by the recognition of increased number of D^b/Lass5 complexes on the surface of the transfectants. In addition, LnB5 T-effectors stimulated by RMA-S/B7-1.Trh4 or RMA-S/pUB.Trh4 cells secreted similar levels of IFN-gamma (Fig. 5A). Apparently, B7-1 expression on the RMA-S/B7-1.Trh4 cells provided a negligible role in serving as a secondary signal for T-cell activation. This was further confirmed by antibody blocking assays in which both anti-B7-1 and/or anti-CD28 antibodies could not reduce T-effector activation (Fig. 5A). The results might indicate that the transfectants expressed an increased number of D^b/Lass5 complexes which provided a stronger first signal for T effector activation and thus overcame the requirement for the B7-1/CD28 signal. To further confirm the increased number of D^b/Lass5 complexes being a critical factor for providing enhanced T-cell killing activity that bypass the requirement of B7-1/CD28 signaling, RMA-S/B7-1 and RMA-S/pUB cells were pulsed with Lass5 peptide as targets in polyclonal T-cell based ^{51}Cr-release assays. The peptide-pulsed targets should express much more surface D^b/Lass5 complexes, and they displayed higher responses for T-cell killing, compared to RMA-S/B7-1 and RMA-S/pUB cells (Fig. 5B). The blockage of the B7-1/CD28 axis by the antibodies did not reduce T-cell killing activities on the peptide-pulsed RNA-S/B7-1 targets (Fig. 5B).

Taken together, the results indicated that naturally expressed Lass5 epitope provides a relatively weak first signal for T-effector response and thus the secondary signal is required. Increasing the number of Lass5 epitopes on the cell surface compensates for the inadequate first signal and bypasses the requirement for the B7-1/CD28 secondary signal for T-effector responses.

Discussion

We have demonstrated that, in comparison with RMA-S/pUB cells, RMA-S/B7-1 cells are more efficiently recognized by Lass5 specific T-cell clones or bulk-cultures of T effectors. The enhanced T-cell based immune response against RMA-S/B7-1 cells occurs at the effector phase of the immunity and requires binding of B7-1 on tumor cells to CD28 on antigen specific T effectors. This requirement can be overcome by an increase in Lass5 expression in tumor cells.

In antitumor immunity, B7-1-transfected tumor cells are potent immunogens which provoke robust T-cell-based antitumor immune reactions [10,11]. The existence of the enhanced immunity may reflect the involvement of tumor-direct priming for antitumor-specific T-cell generation [12]. Although numerous accumulated data support the importance of B7-1 in the induction phase of antiviral and antitumor immunity, the involvement of this molecule in the effector phase has emerged recently. There is a report indicating that in influenza-infected mice, B7-expressing dendritic cells (DCs) trigger both CTL cytotoxicity and release of inflammatory mediators while B7-negative epithelial cells trigger only CTL cytotoxicity [24]. Furthermore, the authors show that inhibiting B7/CD28 interactions significantly decreases the release of inflammatory mediators and that this decrease coincides with a corresponding reduction in mediator-producing CD8$^+$ T cells [24]. Another report indicates that absence of costimulation by

Table 3. Lass5 mRNA expression in RMA-S transfectants.

Lass5	RMA-S		RMA-S/pUB.Trh4		RMA-S/B7-1.Trh4	
mRNA	Mean	StDev	Mean	StDev	Mean	StDev
Long	1.00	0.12	534.84	26.09	821.84	33.01
Short	3.86	0.32	12.21	1.25	8.27	1.19

Note: Lass5 mRNA expression was determined by quantitative PCR using specific primers. Levels of Lass5 mRNA expression of two natural splice variants (long and short) were normalized with mRNA of the GAPDH housekeeping gene. Only long transcript is coding for the Lass5 peptide MCLRMTAVM.

B7/CD28 association at the effector phase leads to reduced survival of influenza virus specific effector cells [25]. Apparently, B7/CD28 association at the effector phase was associated with an increase in the number of virus specific CD8+ T cells. In antitumor

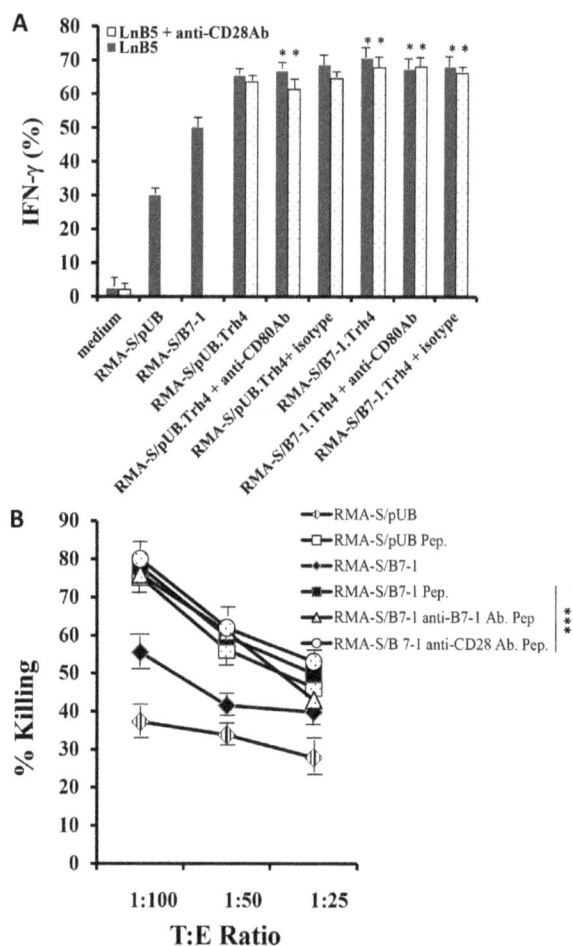

Figure 5. Increase in Lass5 expression Bypasses B7-1/CD28 requirement for T effectors' response. Lass5 specific LnB5 T-cell clone (A) and T-cell bulk culture (B) were used to determine B7-1/CD28 requirement. (A): Lass5 high expressing RMA-S/pUB.Trh4 and RMA-S/B7-1.Trh4 cells were used as targets that were recognized by LnB5 T-cell clone. The antibodies against CD80 (B7-1) or CD28 molecules were used to block B7-1/CD28 axis. The isotype Ig was used as a control. (B): Lass5-peptide (50 micromole) pulsed RMA-S/pUB and RMA-S/B7-1 cells were used as targets that were recognized by T-cell bulk culture for ^{51}Cr-release assays. Pep means Lass5 peptide. One out of two experiments with similar results for each assay was shown. ** and *** indicated no statistical significance.

immunity, one report suggested that B7-1 was involved in enhanced antitumor immunity at the effector phase. Bai et al [26], by determining the sizes of murine B7-1 positive and negative tumors in tumor-carrying RAG−/− mice that were administered tumor-antigen specific CTLs, found that the CTLs inhibited growth of the B7-1 positive tumors more efficiently than the B7-1 negative counterparts. These results are very similar to our *in vivo* results (Fig. 1B-e insert). Our work *in vitro* expands upon these *in vivo* findings by removing confounding factors *in vivo* to further confirm that B7-1/CD28 signaling is involved at the effector phase of antitumor immunity. Specifically, our results of CTL activation and killing assays provide important information that directly indicates the association of B7/CD28 signaling with the effector phase of antitumor immunity because our *in vitro* working system contains only cloned or bulk-cultured CTLs with B7-1 positive or negative targets and thus this system eliminates possible confounding factors. Our results from *in vitro* experiments also indicate that the same number of CTLs provide higher activation/killing activities against B7-1 positive than B7-1 negative tumor cells. This differs from that reported by other research groups [24,25] who demonstrated that the influenza viral specific immune responses at the effector phase with or without B7/CD28 association were influenced by the numbers of the CTLs. Of particular note, the enhanced CTL activities in our experiments cannot be attributed simply to B7-1/CD28 association leading to target/T-cell close binding, because the association activates the T effectors to secrete more IFN-gamma suggesting that a signal is delivered into the T effectors (Fig. 4).

Others have demonstrated that NK activities were involved in B7-1 expressing RMA-S cells *in vitro* and *in vivo* [22,23]. In *in vitro* assays, the report [23] indicated that NK activities were independent of B7-1/CD28 association, since an anti-CD28 mAb was unable to block NK reactivity. In our experiments, the enhanced activity of the polyclonal T effectors can be blocked by an anti-CD28 mAb (Fig. 3C and D), suggesting negligible NK activities in the T-cell bulk-cultures. In *in vivo* assays, NK activities were reported [22] to control B7-1 expressing RMA-S tumor growth, and this control was dependent on initial cell numbers in the inoculate. In the case of inoculation with more than 1×10^6 B7-1 expressing tumor cells per mouse, NK activities only temporally inhibited but did not block tumor formation and growth [22]. Our results support this point of view (Fig. 1B-e insert). In PBS-immunized mice, all RMA-S/B7-1-inoculated mice grew tumors during the first week and the growth rate of the tumors was decreased 2.34-fold, compared to growth rate of the RMA-S/pUB tumors. However, both B7-1 positive and negative tumors grew quickly in the following weeks with one exception in which one RMA-S/B7-1 tumor was regressed.

T-cell-based immunity but not NK activity plays a major role in controlling B7-1 expressing RMA-S tumor growth at the effector phase. Our *in vivo* tumor immunization and NK depletion experiment (Fig. 1B-f) demonstrates this issue. Without NK activity, antigen specific T effectors inhibited growth of B7-1 positive RMA-S tumors more efficiently than growth of B7-1 negative counterparts. At least, the results at the initial week reflect inhibitive function of T effectors at the effector phase. The following weeks may suggest both the induction and effector phase of T cell immunity being activated by challenged B7-1 positive tumor stimulation.

Lass5 peptide is a suitable H-2D^b binder, similar to immunodominant viral epitopes [14] (and unpublished data). Its expression at a limited level on the surface of RMA-S cells was suggested by the evidence indicating that it cannot be presented by TAP-proficient RMA cells [13] (because of other TAP-dependent peptides' competition) and can be presented by Lass5-transfected RMA cells [14]. Transfection of Trh4 (Lass5) gene into or Lass5 peptide-pulse on RMA-S/B7-1 and RMA-S/pUB cells enhances T-cell responsiveness and bypasses the requirement for B7-1/CD28 signaling at the effector phase (Fig. 5A and B). Reports showed that the association between MHC-I/peptide complexes on targets and T-cell receptors (TCRs) on T cells served as first signal for T-cell responsiveness and this signal requires clustering of the TCRs with the MHC-I/peptide complexes at the interface [27–29]. Recent report indicated that the density of the MHC-I/peptide complexes can regulate TCR signaling [30]. Our results indicating enhanced T-cell responsiveness and decreased B7-1/CD28 requirement (Fig. 5A and B) may be ascribed to increased Lass5 peptide densities on target cells associated with relative larger TCR clusters on the effectors that provide a stronger first signal for T-cell responses without requirement of B7-1/CD28 signaling.

Besides B7-1/CD28 signaling, association of B7-1 with cytotoxic T lymphocyte-associated antigen 4 (CTLA-4) provides another signal to T-cells. This B7-1/CTLA-4 signal, unlike the B7-1/CD28 signal, terminates T effector activation [31]. Blocking B7-1/CD28 association by anti-CD28 mAb reduced T effector activation and killing activity (Fig. 3C–D and 4C–D). The reduction in T-effector function cannot be attributed to blockage of B7-1/CD28 positive signal thereby activating the B7-1/CTLA-4 negative signal, because blocking both signals by combinations of anti-CD28 and anti-CTLA-4 (clone: 9H10) mAbs did not recover T-effector killing activity against RMA-S/B7-1 targets (data not shown). In Trh4-transfected or Lass5 peptide-pulsed RMA-S/B7-1 target system (Fig. 5), blockage of B7-1/CD28 association by anti-CD28 mAb did not activate the B7-1/CTLA-4 negative signaling because reduction of T-effector activities was not observed. It is not clear that why CTLA-4 does not promote a negative signal to inhibit T-effector function in our working system. Some reports have provided an opposite evidence in which CTLA-4 played active signal for T-cell activation [5,32]. In our current work, the results of B7-1/CTLA-4 signaling are limited but we are interested in investigating further.

TAP2-deficient RMA-S cells can present many different TAP-independent antigens, as demonstrated by different T-cell clones being generated [13]. In future studies, we will investigate if the results observed with Lass5 antigenic peptide presentation can be expanded to other TAP-independent antigens. If these antigens display similar results, it suggests that 1) T-cell responses to TAP-independent antigens require B7-1/CD28 signaling at the effector phase and 2) a potential mechanism in which the first signal strength regulates the requirement of secondary B7-1/CD28 signaling shown in Lass5 antigen presentation can be confirmed to be an important role for T-cell response to TAP-independent antigens at the effector phase. Since many types of human cancers down-regulate TAP molecules [33,34], understanding how T-cells respond to these types of cancers may provide useful information for cancer immunotherapy.

Acknowledgments

We would like to thank Dr. Ian Davenport (Xavier University) for reviewing the manuscript and to thank Mr. Reginald Starks (Xavier University) for taking care of the animals used in the study. We would also like to thank RCMI and LCRC Core Facility for supporting this study.

Author Contributions

Contributed reagents/materials/analysis tools: QJZ TvH SPK YYL HM. Wrote the paper: QJZ. Designed T cell clone experiments: TvH. Designed all other experiments: QJZ. Conducted most of the experiments: XLL. Conducted the T cell clone experiments: MS ED TvH. Analyzed data and participated in the many discussions on the findings and follow up experiments: SPK YYL HM. Did computer modeling analysis: MM. Performed animal experiments: XY YL FD. Performed animal experiments: BG. Undergraduate students, supported by Xavier's Center for Undergraduate Research, who participated in and assisted with the experiments: KH YHN NT CRS.

References

1. Robey E, Allison JP (1995) T-cell activation: integration of signals from the antigen receptor and costimulatory molecules. Immunol Today 16: 306–310.
2. Van Gool SW, Vandenberghe P, de Boer M, Ceuppens JL (1996) CD80, CD86 and CD40 provide accessory signals in a multiple-step T-cell activation model. Immunol Rev 153: 47–83.
3. Zang X, Allison JP (2007) The B7 family and cancer therapy: costimulation and coinhibition. Clin Cancer Res 13: 5271–5279.
4. Townsend SE, Allison JP (1993) Tumor rejection after direct costimulation of CD8+ T cells by B7-transfected melanoma cells. Science 259: 368–370.
5. Chen L, Ashe S, Brady WA, Hellstrom I, Hellstrom KE, et al. (1992) Costimulation of antitumor immunity by the B7 counterreceptor for the T lymphocyte molecules CD28 and CTLA-4. Cell 71: 1093–1102.
6. Bixby DL, Yannelli JR (1998) CD80 expression in an HLA-A2-positive human non-small cell lung cancer cell line enhances tumor-specific cytotoxicity of HLA-A2-positive T cells derived from a normal donor and a patient with non-small cell lung cancer. Int J Cancer 78: 685–694.
7. Boyerinas B, Park SM, Murmann AE, Gwin K, Montag AG, et al. (2012) Let-7 modulates acquired resistance of ovarian cancer to Taxanes via IMP-1-mediated stabilization of multidrug resistance 1. Int J Cancer 130: 1787–1797.
8. Bueler H, Mulligan RC (1996) Induction of antigen-specific tumor immunity by genetic and cellular vaccines against MAGE: enhanced tumor protection by coexpression of granulocyte-macrophage colony-stimulating factor and B7-1. Mol Med 2: 545–555.
9. Dunn GP, Bruce AT, Ikeda H, Old LJ, Schreiber RD (2002) Cancer immunoediting: from immunosurveillance to tumor escape. Nat Immunol 3: 991–998.
10. Wolpert EZ, Petersson M, Chambers BJ, Sandberg JK, Kiessling R, et al. (1997) Generation of CD8+ T cells specific for transporter associated with antigen processing deficient cells. Proc Natl Acad Sci U S A 94: 11496–11501.
11. Li XL, Liu YY, Knight D, Odaka Y, Mathis JM, et al. (2009) Effect of B7.1 costimulation on T-cell based immunity against TAP-negative cancer can be facilitated by TAP1 expression. PLoS One 4: e6385.
12. Li XL, Zhang D, Knight D, Odaka Y, Glass J, et al. (2009) Priming of immune responses against transporter associated with antigen processing (TAP)-deficient tumours: tumour direct priming. Immunology 128: 420–428.
13. van Hall T, Wolpert EZ, van Veelen P, Laban S, van der Veer M, et al. (2006) Selective cytotoxic T-lymphocyte targeting of tumor immune escape variants. Nat Med 12: 417–424.
14. Oliveira CC, Querido B, Sluijter M, Derbinski J, van der Burg SH, et al. (2011) Peptide transporter TAP mediates between competing antigen sources generating distinct surface MHC class I peptide repertoires. Eur J Immunol 41: 3114–3124.
15. Levitsky HI, Lazenby A, Hayashi RJ, Pardoll DM (1994) In vivo priming of two distinct antitumor effector populations: the role of MHC class I expression. J Exp Med 179: 1215–1224.
16. van Hall T, Sijts A, Camps M, Offringa R, Melief C, et al. (2000) Differential influence on cytotoxic T lymphocyte epitope presentation by controlled

expression of either proteasome immunosubunits or PA28. J Exp Med 192: 483–494.

17. Ly LV, Sluijter M, van der Burg SH, Jager MJ, van Hall T (2013) Effective cooperation of monoclonal antibody and peptide vaccine for the treatment of mouse melanoma. J Immunol 190: 489–496.

18. Razi-Wolf Z, Freeman GJ, Galvin F, Benacerraf B, Nadler L, et al. (1992) Expression and function of the murine B7 antigen, the major costimulatory molecule expressed by peritoneal exudate cells. Proc Natl Acad Sci U S A 89: 4210–4214.

19. Yu XZ, Bidwell SJ, Martin PJ, Anasetti C (2000) CD28-specific antibody prevents graft-versus-host disease in mice. J Immunol 164: 4564–4568.

20. Boussiotis VA, Freeman GJ, Gribben JG, Nadler LM (1996) The role of B7-1/B7-2:CD28/CLTA-4 pathways in the prevention of anergy, induction of productive immunity and down-regulation of the immune response. Immunol Rev 153: 5–26.

21. Kaufmann AM, Gissmann L, Schreckenberger C, Qiao L (1997) Cervical carcinoma cells transfected with the CD80 gene elicit a primary cytotoxic T lymphocyte response specific for HPV 16 E7 antigens. Cancer Gene Ther 4: 377–382.

22. Kelly JM, Takeda K, Darcy PK, Yagita H, Smyth MJ (2002) A role for IFN-gamma in primary and secondary immunity generated by NK cell-sensitive tumor-expressing CD80 in vivo. J Immunol 168: 4472–4479.

23. Chambers BJ, Salcedo M, Ljunggren HG (1996) Triggering of natural killer cells by the costimulatory molecule CD80 (B7-1). Immunity 5: 311–317.

24. Hufford MM, Kim TS, Sun J, Braciale TJ (2011) Antiviral CD8+ T cell effector activities in situ are regulated by target cell type. J Exp Med 208: 167–180.

25. Dolfi DV, Duttagupta PA, Boesteanu AC, Mueller YM, Oliai CH, et al. (2011) Dendritic cells and CD28 costimulation are required to sustain virus-specific CD8+ T cell responses during the effector phase in vivo. J Immunol 186: 4599–4608.

26. Bai XF, Bender J, Liu J, Zhang H, Wang Y, et al. (2001) Local costimulation reinvigorates tumor-specific cytolytic T lymphocytes for experimental therapy in mice with large tumor burdens. J Immunol 167: 3936–3943.

27. Germain RN (1997) T-cell signaling: the importance of receptor clustering. Curr Biol 7: R640–644.

28. Boniface JJ, Rabinowitz JD, Wulfing C, Hampl J, Reich Z, et al. (1998) Initiation of signal transduction through the T cell receptor requires the multivalent engagement of peptide/MHC ligands [corrected]. Immunity 9: 459–466.

29. Cochran JR, Aivazian D, Cameron TO, Stern LJ (2001) Receptor clustering and transmembrane signaling in T cells. Trends Biochem Sci 26: 304–310.

30. Anikeeva N, Gakamsky D, Scholler J, Sykulev Y (2012) Evidence that the density of self peptide-MHC ligands regulates T-cell receptor signaling. PLoS One 7: e41466.

31. Teft WA, Kirchhof MG, Madrenas J (2006) A molecular perspective of CTLA-4 function. Annu Rev Immunol 24: 65–97.

32. Wu Y, Guo Y, Huang A, Zheng P, Liu Y (1997) CTLA-4-B7 interaction is sufficient to costimulate T cell clonal expansion. J Exp Med 185: 1327–1335.

33. Seliger B, Maeurer MJ, Ferrone S (1997) TAP off-tumors on. Immunol Today 18: 292–299.

34. Ritz U, Seliger B (2001) The transporter associated with antigen processing (TAP): structural integrity, expression, function, and its clinical relevance. Mol Med 7: 149–158.

4

A Sialoreceptor Binding Motif in the *Mycoplasma synoviae* Adhesin VlhA

Meghan May[1]*, Dylan W. Dunne[2], Daniel R. Brown[3]

1 Department of Biomedical Sciences, College of Osteopathic Medicine, University of New England, Biddeford, Maine, United States of America, **2** Department of Biological Sciences, Jess and Mildred Fisher College of Science and Mathematics, Towson University, Towson, Maryland, United States of America, **3** Department of Infectious Diseases and Pathology, College of Veterinary Medicine, University of Florida, Gainesville, Florida, United States of America

Abstract

Mycoplasma synoviae depends on its adhesin VlhA to mediate cytadherence to sialylated host cell receptors. Allelic variants of VlhA arise through recombination between an assemblage of promoterless *vlhA* pseudogenes and a single transcription promoter site, creating lineages of *M. synoviae* that each express a different *vlhA* allele. The predicted full-length VlhA sequences adjacent to the promoter of nine lineages of *M. synoviae* varying in avidity of cytadherence were aligned with that of the reference strain MS53 and with a 60-a.a. hemagglutinating VlhA C-terminal fragment from a Tunisian lineage of strain WVU1853[T]. Seven different sequence variants of an imperfectly conserved, single-copy, 12-a.a. candidate cytadherence motif were evident amid the flanking variable residues of the 11 total sequences examined. The motif was predicted to adopt a short hairpin structure in a low-complexity region near the C-terminus of VlhA. Biotinylated synthetic oligopeptides representing four selected variants of the 12-a.a. motif, with the whole synthesized 60-a.a. fragment as a positive control, differed ($P<0.01$) in the extent they bound to chicken erythrocyte membranes. All bound to a greater extent ($P<0.01$) than scrambled or irrelevant VlhA domain negative control peptides did. Experimentally introduced branched-chain amino acid (BCAA) substitutions Val3Ile and Leu7Ile did not significantly alter binding, whereas fold-destabilizing substitutions Thr4Gly and Ala9Gly tended to reduce it ($P<0.05$). Binding was also reduced to background levels ($P<0.01$) when the peptides were exposed to desialylated membranes, or were pre-saturated with free sialic acid before exposure to untreated membranes. From this evidence we conclude that the motif P-X-(BCAA)-X-F-X-(BCAA)-X-A-K-X-G binds sialic acid and likely mediates VlhA-dependent *M. synoviae* attachment to host cells. This conserved mechanism retains the potential for fine-scale rheostasis in binding avidity, which could be a general characteristic of pathogens that depend on analogous systems of antigenically variable adhesins. The motif may be useful to identify previously unrecognized adhesins.

Editor: Mitchell F. Balish, Miami University, United States of America

Funding: This work was supported by the Robert M. Fisher Foundation (MM). The funder had no role in study design, data collection and analysis, decision to publish, or preparation of the manuscript.

Competing Interests: The authors have declared that no competing interests exist.

* Email: mmay3@une.edu

Introduction

The bacterial pathogen *Mycoplasma synoviae* is associated with a broad spectrum of clinical manifestations ranging from inapparent infection to systemic disease of poultry. Infection is most commonly associated with inflammatory lesions of the joints, respiratory and/or reproductive tract and results in reduced feed conversion and poor egg quality. Less commonly, *M. synoviae* can be found infecting additional tissues in galliform birds (*e.g.* spleen, liver, central nervous system, skeletal muscle, and eye) [1–4] and respiratory tissues or synovial membranes of distantly related avian species such as ducks, geese, pigeons, and sparrows [5].

Attachment to sialylated receptors on host cells is mediated by the *M. synoviae* variable lipoprotein hemagglutinin VlhA [6–7]. Previous analyses indicated that the *vlhA* gene family has been laterally transferred between *M. synoviae* and *Mycoplasma gallisepticum* possibly during coinfection of a shared avian host [8–9]. In *M. synoviae*, antigenic variants of this adhesin result from

unidirectional recombination between a single expression site and a large reservoir of *vlhA* pseudogenes [10]. In contrast, altered expression in *M. gallisepticum* stems from the expansion and contraction of a poly-GAA repeat upstream of the promoters of each copy of *vlhA* [11]. The selective pressure of specific host immune responses to these antigens is thought to drive diversity in *vlhA* allele expression [10–13]. Despite the critical importance of cytadherence to the establishment and maintenance of infection, discrete VlhA types were demonstrated to have significantly different avidities for host cell binding, which can be quantified by agglutination of erythrocytes [14]. *M. synoviae*'s capacity for cytadherence maps surprisingly to a hypervariable C-terminal domain of VlhA called MSPA [15–16]. The precise means of attachment and how this capacity is retained despite such extensive sequence polymorphism and allele switching are not known. We sought to identify and characterize the specific motif that mediates adhesion of VlhA proteins to host cells.

PHM Residue	1	2	3	4	5	6	7	8	9	10	11	12
Strains/lineages												
F10-2-AS and K4907	P	K	V	T	F	D	V	A	Q	K	E	G
FMT	P	K	V	T	F	N	L	A	A	K	E	G
K5016	P	K	V	T	F	T	V	T	A	K	N	G
K5395	P	T	V	T	F	N	L	A	A	K	E	G
MS53	P	K	V	T	F	N	L	T	P	K	E	G
MS117, MS173, MS178	P	T	V	T	F	T	V	A	A	K	D	G
WVU1853^T/Florida and Tunisia	P	K	V	T	F	T	V	E	A	K	P	G
Preliminary consensus:	P	X	V	T	F	X	(B)	X	X	K	X	G
Site-directed Mutants												
FMT T4G	P	K	V	G	F	N	L	A	A	K	E	G
FMT A9G	P	K	V	T	F	N	L	A	G	K	E	G
FMT V3I	P	K	I	T	F	N	L	A	A	K	E	G
FMT L7I	P	K	V	T	F	N	I	A	A	K	E	G
FinalPHM:	P	X	(B)	T	F	X	(B)	X	A	K	X	G
FMT – scrambled:	L	A	F	G	A	V	K	K	T	P	E	N
Irrelevant peptide:	P	N	A	V	F	V	Q	Q	M	K	D	N

Figure 1. Aligned PHM and control peptide sequences. The putative hemagglutination motif (PHM) was deduced by aligning the adhesin protein VlhA allele present at the expression site of ten specimens of *M. synoviae* with a 60-a.a. hemagglutinating VlhA C-terminal fragment from the Tunisian lineage of strain WVU1853^T, then inspecting the alignment for contiguous residues inferred to be under stabilizing selection. Peptides representing five variants of the PHM, including strains having a >20-fold range in quantitative hemagglutination phenotypes [14], were synthesized. Directed mutations were introduced at selected residues relative to the PHM from strain FMT, which had only one difference (Thr6Asn) from the most common amino acid at each residue. The mutations Val3Ile and Leu7Ile were predicted to be inconsequential, while Thr4Gly and Ala9Gly were predicted to affect PHM structure and/or function. Negative control peptides used in erythrocyte membrane-binding assays are also shown. Functionally non-synonymous differences relative to the most common amino acid at each residue are shaded in black, synonymous differences are shaded in gray, and identical residues are not shaded. (B) = branched chain amino acid.

Materials and Methods

Identification and Structural Modeling of the Putative Hemagglutination Motif (PHM)

The predicted full-length VlhA sequences adjacent to the single transcription promoter of nine lineages of *M. synoviae* varying in avidity of cytadherence (F10-2AS, FMT, K4907, K5016, K5395, MS117, MS173, MS178, and a >30X-passaged Florida lineage of strain WVU1853^T) [14] were aligned with that of the reference strain MS53 [8] and with a 60-a.a. hemagglutinating VlhA C-terminal fragment from a ca. 12X-passaged Tunisian lineage of strain WVU1853^T [15] by using ClustalΩ [17]. The multiple alignment was manually inspected for conserved motifs, evident as contiguous residues inferred to be under stabilizing selection ($\omega < 1$) by using Bayesian models of sequence evolution in the Selecton v2.4 software suite [18]. The secondary structures of full-length VlhA, MSPA and its C-terminal 60 residues, and of the putative hemagglutination motifs (PHMs) described were modeled using the Phyre2 suite of template-directed and *ab initio* protein structure prediction algorithms (http://www.sbg.bio.ic.ac.uk/ phyre2) [19]. The effects of individual amino acid substitutions on peptide structural stability were predicted by applying the Site Directed Mutator algorithm (http://mordred.bioc.cam.ac.uk/ ~sdm/sdm.php) [20] to the.pdb files generated by Phyre2. Substitutions having stability scores ($\Delta\Delta G$) between −0.5 and 0.5 were predicted to be neutral, whereas those < −2 or >2 were predicted to be highly destabilizing. The potential to bind sialic acid (KEGG Compound C00270; PubChem.sdf 445063) or any other ligand in the KEGG Compound database was predicted by applying the eFindSite ligand binding site prediction algorithm (http://brylinski.cct.lsu.edu/) [21–22] also to the.pdb files generated by Phyre2.

Quantitative Binding of PHM Peptides

Twelve-a.a. peptides representing five variants of the PHM from strains FMT, K5016, K5395, MS53 and WVU1853^T, plus the whole 60-a.a. hemagglutinating fragment of the Tunisian lineage of strain WVU1853^T, were synthesized, biotinylated and lyophilized (Biomatik, Wilmington, DE). Purity of each lyophilized preparation was confirmed by HPLC to be 90–92% full-length peptide. Those strains were chosen because FMT, K5016, K5395 and the Florida lineage of WVU1853^T spanned a >20-fold range in quantitative hemagglutination phenotypes, and the entire *vlhA* locus sequence of the reference strain MS53 has been published. [8,14]. Peptides having single directed mutations introduced at the conserved residues 3 or 4, or non-conserved residues 7 or 9, were also synthesized using the strain FMT motif PKVTFNLAAKEG as a parent. FMT was chosen as the parent motif because it had only one difference (Thr6Asn) from the most commonly observed amino acid at each residue (Figure 1). The functionally synonymous substitutions Val3Ile and Leu7Ile (BLOSUM62 [23] scores >0) were predicted to be inconsequential, while non-synonymous Thr4Gly and Ala9Gly (BLOSUM62 scores ≤0) were predicted to affect PHM structure and/or function.

The capacity of the peptides to bind to native or desialylated chicken erythrocyte membranes was assessed quantitatively in an ELISA format. Microtiter plates were coated with 5% v/v suspensions of chicken erythrocytes (Lampire Biologicals,

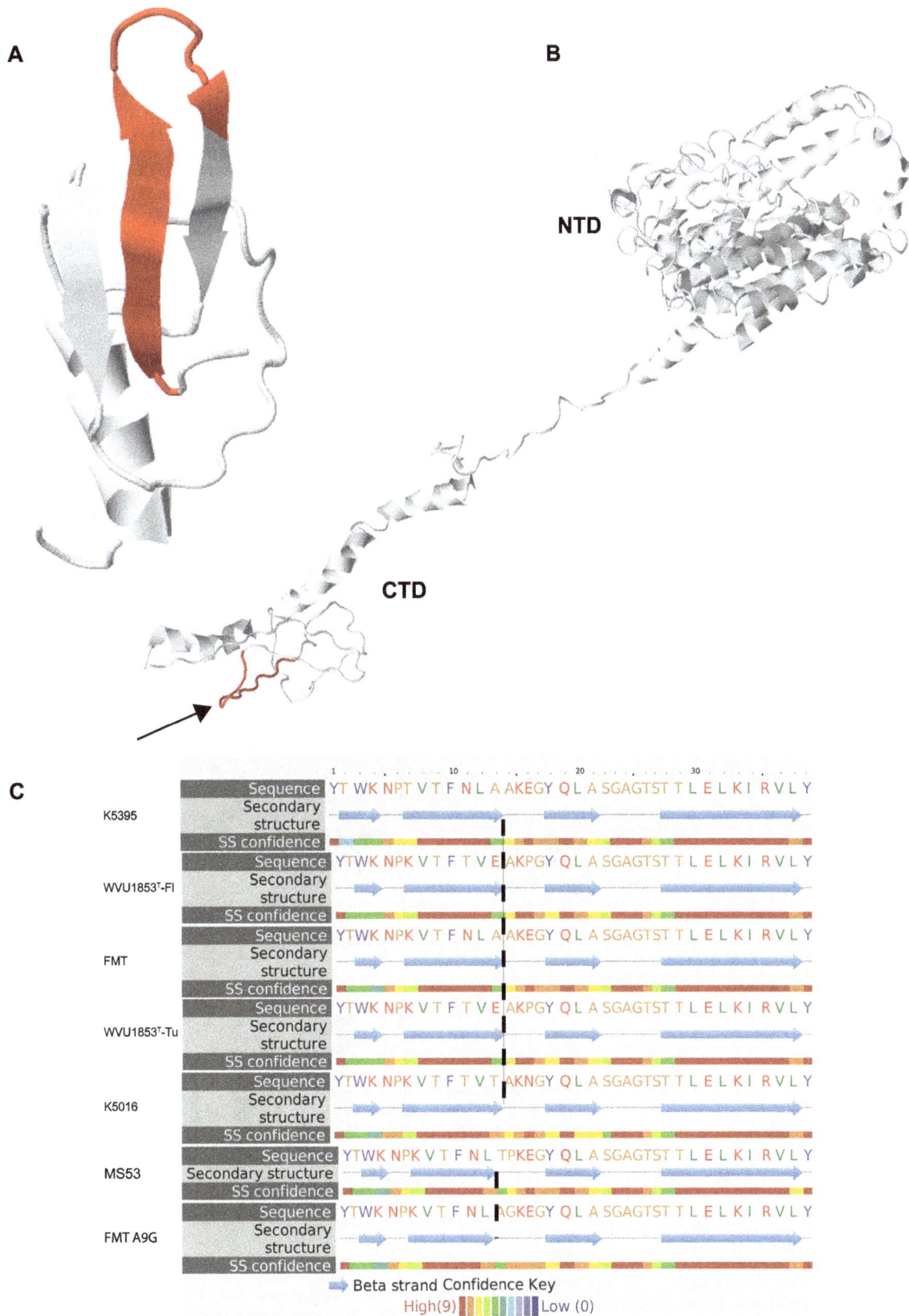

Figure 2. PHM structural predictions. (**A**) The putative hemagglutination motif (PHM; red) was predicted to adopt a hairpin structure of two anti-parallel β strands separated by a short disordered loop. (**B**) The motif (red, indicated by arrow) mapped to a low-complexity region near the carboxyterminal domain (CTD) of the *M. synoviae* adhesin protein VlhA cleavage product MSPA, shown here in the structure predicted for the Tunisian lineage of strain WVU1853[T]. The N-terminal domain (NTD) of MSPA was predicted to have much greater 3-dimensional complexity. (**C**) The length of the disordered loop was predicted to be longer in PHM peptides that bound to avian erythrocyte membranes (representing Florida and

Tunisian lineages of strain WVU1853[T] and strains FMT, K5016 and K5395) than in the reduced-binding peptide mutant FMT-Ala9Gly and the non-binding peptide representing strain MS53.

Pipersville, PA) diluted 1:3 in 0.5 M sodium bicarbonate lysis buffer, pH 10.0, to a total volume of 300 µL per well. Desialylated membranes were prepared by pre-treatment of the erythrocytes with 10 U/ml of sialidase purified from *Clostridium perfringens* (Sigma-Aldrich, St. Louis, MO) for 1 hr at 37°C. Following coating for 12 hr at 4°C, cellular debris including hemoglobin was removed by washing each well 3× with 300 µL of PBS, pH 7.4, and sealed plates were blocked 1 hr at 37°C with 300 µL per well of 5% v/v fetal bovine serum in PBS.

After washing the membrane-coated and blocked wells 3× with 300 µL of PBS, 50 µg of biotinylated peptide solubilized in 50 µL of water was added to each of duplicate wells and allowed to bind for 1 hr at 37°C. After washing each well 3× with 300 µL of PBS, bound peptides were detected using horseradish peroxidase-conjugated streptavidin (2 µg/mL, Sigma-Aldrich, St. Louis, MO) and the chromogenic substrate 3,3',5,5'-tetramethylbenzidine (Thermo Fisher Scientific, Waltham, MA) with an acid stop followed by spectrophotometric analysis ($\lambda = 450$ nm). The hemagglutinating 60-mer of the Tunisian lineage of strain WVU1853[T] served as the positive control peptide, and negative controls were a scrambled version of the PHM from strain FMT (LAFGAVKKTPEN) and an irrelevant peptide (PNAVFVQQMKDD) from a distant site in the expressed VlhA of the Florida lineage of strain WVU1853[T] (GenBank AEA01932.1). The effect of pre-saturation with ligand was tested by first incubating the peptides in 250 mg/ml N-acetylneuraminic acid (Sigma-Aldrich, St. Louis, MO) in water without pH adjustment at a peptide: ligand molar ratio of $1:2\times10^4$ for 1 hr at 37°C.

Statistical Procedures

The effect of peptide sequence on extent of adherence to membranes ($n = 3$ independent replications of each treatment combination, with duplicate measurements of each peptide within replicate) was analyzed by ANOVA, with Tukey-Kramer

Honestly Significant Difference (HSD) post-hoc comparisons used to group the means when the main effect was significant ($P<0.05$ or less). The effects of membrane pre-treatment with sialidase and peptide pre-saturation with sialic acid were analyzed by ANOVA, with HSD or Dunnett's post-hoc comparisons to the corresponding native specimens when the main effect was significant. Statistical analyses were performed using Origin 9 (OriginLab, Northampton, MA) software.

Motif Distribution in *M. synoviae* and *M. gallisepticum*

M. synoviae strain MS53 *vlhA* pseudogene sequences and *M. gallisepticum* strains R, F, WI01, NY01, NC06, CA06, VA94, NC95, NC08, and NC96 were obtained from GenBank (accession numbers NC_007294.1, NC_004829.2, NC_017503.1, NC_018410.1, NC_018409.1, NC_018411.1, NC_018412.1, NC_018406.1, NC_018407.1, NC_018413.1, and NC_018408.1, respectively). Occurrences of PHM-encoding sequences were totaled and normalized to the total length of *vlhA*-encoding sequence in each strain. Each member of the *vlhA* pseudogene reservoir of *M. synoviae* strain MS53 was used to construct a neighbor-joining tree (bootstrap n = 100) using ClustalW2 [24]. The designated outgroup was *vlhA* 4.02 from *M. gallisepticum*.

Results

Identification of the PHM

When the full-length expressed VlhA protein MSPA sequences of nine strains of *M. synoviae* that vary in avidity of cytadherence were aligned with MSPA of the reference strain MS53 [8] and a 60-a.a. hemagglutinating peptide derived from the C-terminus of MSPA expressed by the Tunisian lineage of strain WVU1853[T] [15], an imperfectly conserved 12-a.a. motif was evident in all sequences (Figure 1). A total of seven different PHM sequence variants were evident among the 11 total sequences aligned.

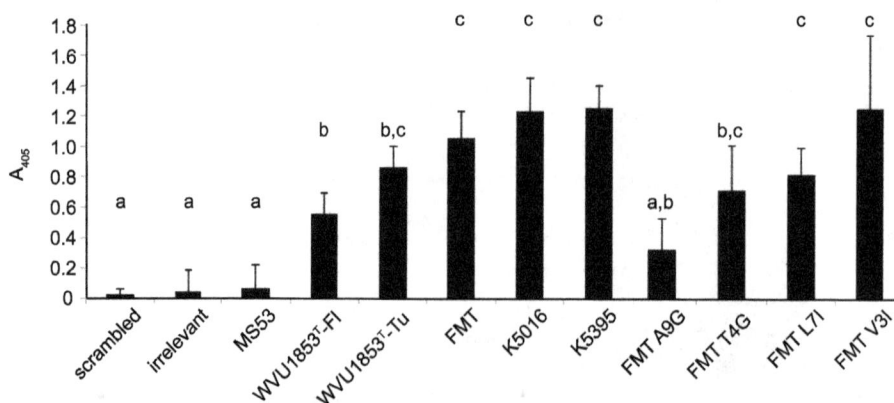

Figure 3. Erythrocyte membrane binding by PHM peptides. Bars depict mean ± standard error of the amount of synthetic peptide bound to avian erythrocyte membranes in an ELISA format (n = 3 independent replicates, with duplicate measurements of each peptide within replicate). The peptides represented variants of the putative hemagglutination motif (PHM) at the VlhA expression site of *M. synoviae* strains MS53, WVU1853[T] (Florida and Tunisian lineages), FMT, K5016 and K5395, which spanned a >20-fold range in quantitative hemagglutination phenotypes [14]. The positive control was the Tunisian lineage of strain WVU1853[T], and negative controls were scrambled strain FMT peptide and an irrelevant peptide from a distant site in VlhA from the Florida lineage of strain WVU1853[T]. Different letters above the bar indicate means that differ ($P<0.05$ or less) by Tukey-Kramer Honestly Significant Difference test. As predicted, the directed substitution Ala9Gly significantly reduced binding versus the parent peptide from strain FMT, and Thr4Gly tended to reduce binding, whereas Val3Ile and Leu7Ile did not significantly alter binding.

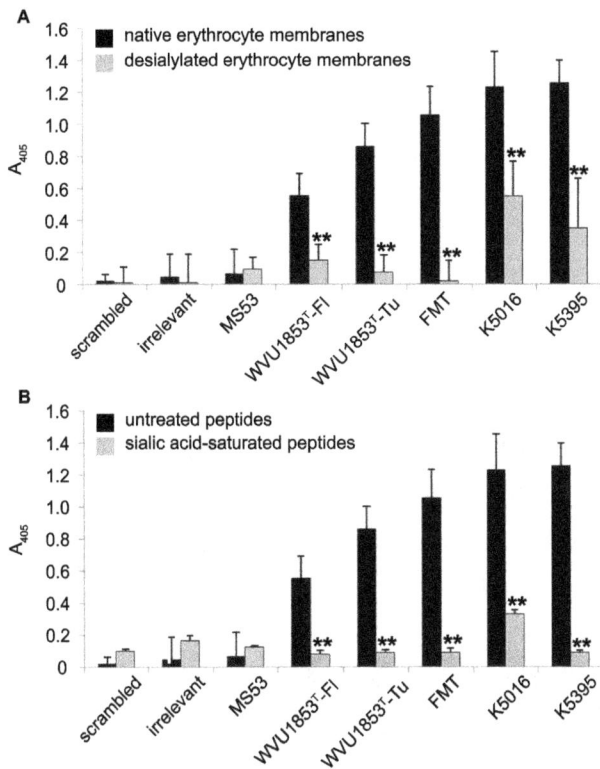

Figure 4. Effects of sialylation and desialylation on PHM peptide binding. Bars depict mean ± standard error of the amount of synthetic peptide bound to avian erythrocyte membranes in an ELISA format (n = 3 independent replicates, with duplicate measurements of each peptide within replicate). **(A)** Desialylation of erythrocyte membranes significantly reduced PHM peptide binding relative to native membranes (** = $P<0.01$) for all strains of *M. synoviae* except MS53, which bound to native or desialylated erythrocyte membranes at background levels. **(B)** Presaturation of PHM peptides with free sialic acid before exposure to native erythrocyte membranes significantly reduced binding relative to untreated peptides (** = $P<0.01$) for all strains except MS53, on which sialic acid had no effect.

Strains FMT, K5016, K5395 and MS53 all had unique PHM sequences; the sequences in strains F10-2-AS and K4907 were identical; the sequences in Florida and Tunisian lineages of WVU1853[T] were identical; and the sequences in Argentine strains MS117, MS173 and MS178 were all identical. Six of twelve residues in the PHM were perfectly conserved across strains, two (residues 6 and 7) were conserved in polarity and hydrophobicity, respectively, and four were variable. Polar Asn_6 or Asp_6 were invariably paired with Leu_7, while Thr_6 was invariably paired with Val_7. The motif was predicted to adopt a short hairpin secondary structure of two anti-parallel beta strands, separated by a disordered loop of four or five residues, in a region of low structural complexity (regional structure prediction confidence < 70%) near the C-terminus of MSPA (Figure 2a, b). Fifty-three percent of residues in the full-length VlhA were modeled at >90% confidence [19], with the regions of greatest confidence being similar to the streptococcal adhesin emb (99.8% confidence) and the staphylococcal extracellular matrix-binding protein ebhA (99.4%). The degree of structural complexity in the C-terminus of MSPA was otherwise too low for the algorithms to predict binding of any specific ligand.

Synthetic biotinylated peptides representing the full-length 60-a.a. hemagglutinating fragment and four strain variants of its

candidate 12-a.a. cytadherence motif (Figure 1) bound to chicken erythrocyte membranes in an ELISA format and could be detected by probing with horseradish peroxidase-conjugated streptavidin. Four of the peptides bound to membranes to a significantly greater extent ($P<0.05$) than scrambled or irrelevant control peptides did, but a peptide representing the corresponding motif from strain MS53 did not bind to membranes to any extent greater than background (Figure 3). Single neutral substitutions (predicted $\Delta\Delta G = -0.25$) experimentally introduced at conserved residue 3 (Val3Ile) or non-conserved residue 7 (Leu7Ile) did not alter binding to membranes with respect to the extent of binding by the parent motif of strain FMT, whereas the experimental destabilizing substitution Thr4Gly (predicted $\Delta\Delta G = -2.31$) tended to reduce binding (Figure 3). The motif of strain MS53 differs naturally from all others by Ala9Pro (BLOSUM62 = -1; predicted $\Delta\Delta G = -2.22$), and the even more destabilizing substitution Ala9Gly (predicted $\Delta\Delta G = -3.88$) nearly abolished binding when introduced into the parent motif of strain FMT ($P<0.05$; Figure 3). These effects correlated with a predicted change in length of the disordered loop in the hairpin secondary structure of the motif (Figure 2c).

Binding of the peptides to desialylated membranes was significantly reduced ($P<0.01$) relative to untreated membranes for all peptides except those representing strain MS53 and the scrambled and irrelevant controls (Figure 4a). When pre-incubated with free sialic acid, all peptides except the one representing strain MS53 and the scrambled and irrelevant controls had significantly diminished ($P<0.01$) capacity for membrane binding (Figure 4b). From this evidence we conclude that the composite amino acid motif P-X-(BCAA)-X-F-X-(BCAA)-X-A-K-X-G binds sialic acid and likely mediates VlhA-dependent *M. synoviae* attachment to sialylated receptors on the surface of avian erythrocytes.

PHM Distribution among *M. synoviae* Strain MS53 *vlhA* Pseudogenes and *Mycoplasma gallisepticum vlhA* Homologs

Candidate PHM sequences occurred in 45 of the 70 putative *vlhA* pseudogenes of *M. synoviae* strain MS53 [8], 39% of the time with no deviation from the consensus among the alleles expressed by the strains examined, 20% with a single deviation, and 17% with two deviations from consensus. Phylogenetic clustering of *vlhA* pseudogenes containing intact copies of the PHM did not correlate with their syntenic order in the strain MS53 genome (Figure S1). The PHM occurred at least 18-fold more frequently in strain MS53 (0.65 motifs/kb of *vlhA* sequence) than in the genomes of any of 10 strains of *M. gallisepticum* (0.014–0.037 motifs/kb of *vlhA* sequence), a species known to employ a different primary cytadherence mechanism [25] (Figure 5a). The rate of occurrence of imperfect PHMs was comparable between the two species (Figure 5b).

Discussion

One of the defining moments of many infections is the attachment of a disease-causing agent to its host. Understanding how the parasitic bacterial species *M. synoviae* colonizes a host cell's surface is paramount to understanding how to prevent infection. It is known that the protein family VlhA is responsible for attachment by *M. synoviae*, but the functional motifs of the adhesin and the molecular basis for rheostasis in binding avidity have not been characterized. Proteins in this family from multiple strains of *M. synoviae* have been identified as having a role in the attachment to host blood cells [7,15]. Khiari *et al.* [15] mapped

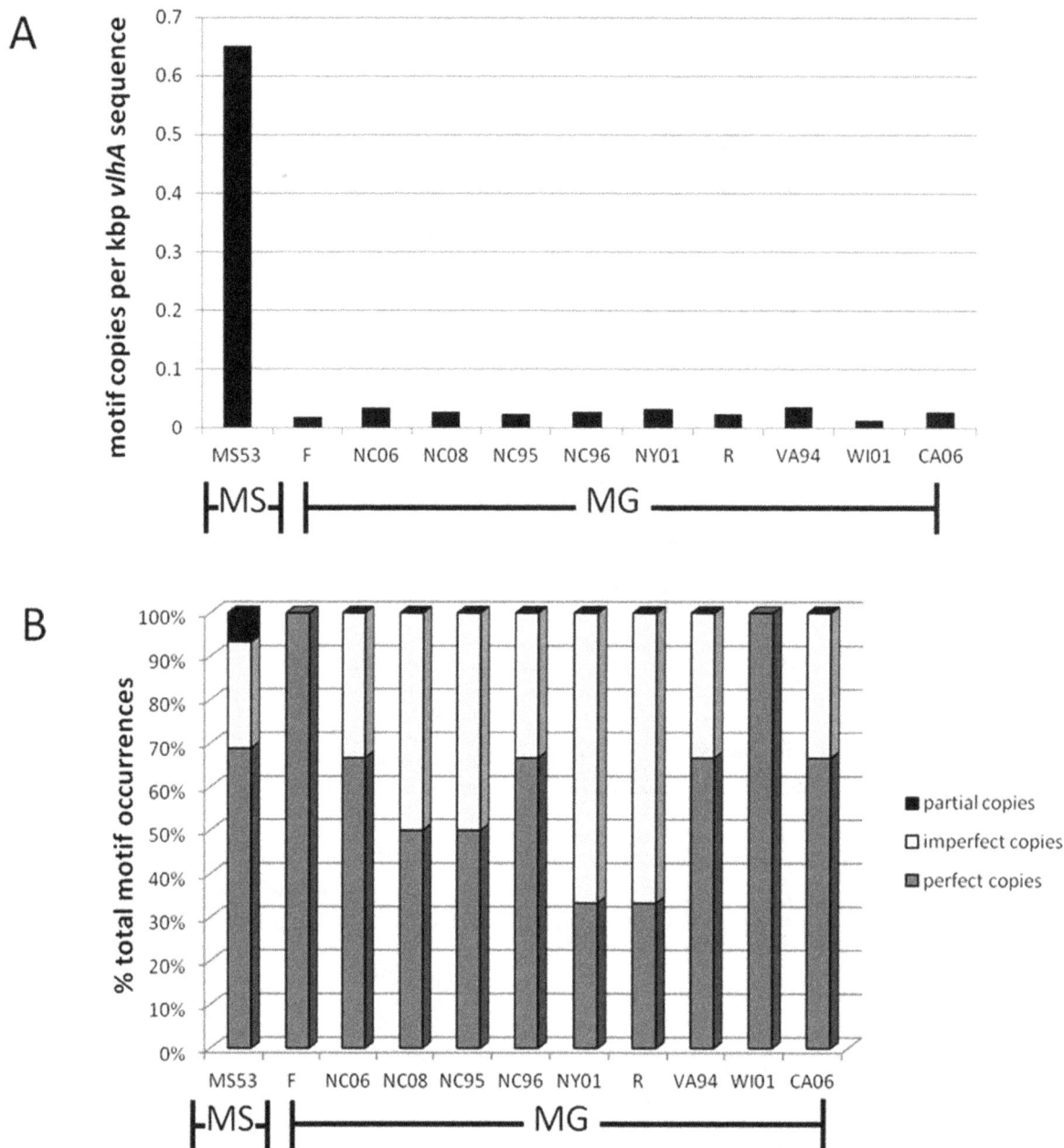

Figure 5. PHM distribution in *vlhA* genes and pseudogenes in *M. synoviae* and *M. gallisepticum*. (A) PHM-encoding sequence as a function of total kbp of *vlhA* sequence is elevated 18-fold in *M. synoviae* (MS) reference strain MS53, the only strain for which the entire *vlhA* locus sequence has been published, relative to 10 fully-sequenced strains of *M. gallisepticum* (MG), evidence that it is far more common in *M. synoviae*. **(B)** The relative proportions of perfect and imperfect PHM copies were comparable between strains of *M. synoviae* and *M. gallisepticum*.

the capacity for attachment to the carboxyterminus of VlhA, and we utilized that finding to identify a specific motif sufficient to mediate VlhA binding to sialylated host cells.

Sequence conservation across adherent strains enabled the identification of a 12-residue putative hemagglutination motif that could be characterized further. This motif was predicted to have remarkably little structural complexity, in contrast to the complex topology of sialic acid ligand-binding domains of other microbes [26–27]. While residues at PHM positions 3 and 7 were conserved, substitution with similar residues having BLOSUM62 scores>0 did not alter function. The conserved Thr residue at position 4 could be changed to the dissimilar residue Gly (BLOSUM62

$= -3$) without loss of function. It is thus likely that the binding mechanism will tolerate synonymous substitutions at positions 3 and 7, and nonsynonymous substitutions at position 4. Residue 9 was a conserved Ala in all adherent strains. Strain MS53, which has an unknown attachment phenotype but is an attenuated strain, had the nonsynonmymous substitution Ala9Pro (BLOSUM62 $= -1$). Changing the strain FMT peptide to Gly_9 (BLOSUM62 $= 0$) significantly diminished binding, and the strain MS53 peptide was non-adherent. Taken together, these results indicate that Ala_9 is critical to PHM domain function. Our results indicate that the composite amino acid motif P-X-(BCAA)-X-F-X-(BCAA)-X-A-K-X-G mediates MSPA binding to avian erythrocytes. The potential to accommodate all amino acids with BLOSUM62 scores>0 at PHM positions 3 and 7 (*i.e.*, Ala, Met, Thr and Met, Phe, respectively) rather than restricting the parameters to branched-chain amino acids (Ile, Leu, Val) merits further analysis.

Previous studies indicated that whole *M. synoviae* cells interact with sialylated host cell receptors in order to facilitate attachment. Extrapolation from PHM peptide-binding to whole cell attachment necessarily requires demonstration of peptide-sialic acid interactions. Desialylation of avian erythrocytes prior to antigen preparation resulted in significant losses of binding capacity for all PHM peptides except the scrambled and irrelevant controls and strain MS53, for which desialylation had no effect on binding. In a reciprocal experiment, pre-adsorption of peptides with free sialic acid prior to exposure to intact erythrocyte antigen similarly diminished binding capacity for all PHM peptides except the scrambled and irrelevant controls and strain MS53. These results indicate a specific interaction between sialic acid and the PHM and support the hypothesis that the PHM domain mediates attachment of whole *M. synoviae* cells to host sialoreceptors.

The occurrence of PHM domains was not uniform among the pseudogenes of *M. synoviae* strain MS53, the only strain for which the entire pseudogene reservoir has been sequenced [8]. A majority (69%) of pseudogenes had perfect or near-perfect PHMs, while 31% had no discernible PHMs. To provide some context for the distribution of PHM domains in the sample of VlhA sequences existing within *M. synoviae* strain MS53, we examined the frequency and distribution in an alternative sample of VlhA sequences that exist distributed across multiple strains of *M. gallisepticum*. In contrast to the 45 copies in *M. synoviae* strain MS53, sequenced *M. gallisepticum* strains ranged from having just a single copy of *vlhA* encoding a PHM domain (strains WI01 and F) up to a maximum of only 3 copies (strains R, NY01, NC06, CA06, VA94, NC95, and NC96). Normalization to the total amount of *vlhA* sequence within species confirmed that *M. synoviae* has a greatly elevated instance of PHM-encoding sequence relative to *M. gallisepticum*, and that the low frequency of PHM is consistent across strains of *M. gallisepticum*. The multiple independent cytadherence mechanisms of *M. gallisepticum* [28–32] may allow the decay of PHM domains within VlhA proteins, while selective pressure to retain the functional motif in the homologous proteins in *M. synoviae* is substantially greater due to the absence of other mechanisms of cytadherence.

This work describes a novel functional motif associated with adherence to sialic acid, and its distribution across *vlhA* pseudogenes. This very specific protein fragment pattern may be a target to design novel drug therapies or vaccines to alleviate or prevent infection due to *M. synoviae* as well as other pathogens that use similar mechanisms to attach to their hosts, and allows for the identification of currently unrecognized microbial adhesins targeting sialoreceptors.

Supporting Information

Figure S1   Distribution and relatedness of PHM-encoding pseudogenes. PHM-encoding pseudogenes (shaded) did not cluster together as a separate group from non-encoding pseudogenes. Relatedness of pseudogenes did not reflect gene synteny.

Acknowledgments

We thank Edan Tulman (University of Connecticut) for helpful discussions regarding *vlhA* loci in *M. gallisepticum*. This work was supported by the Robert M. Fisher Foundation (MM).

Author Contributions

Conceived and designed the experiments: MM DRB. Performed the experiments: DD MM. Analyzed the data: MM DRB. Contributed reagents/materials/analysis tools: MM. Wrote the paper: MM DRB.

References

1. Stipkovits L, Kempf I (1996) Mycoplasmoses in poultry. Rev Sci Tech. 15(4): 1495–525.
2. Senties-Cué G, Shivaprasad HL, Chin RP (2005) Systemic *Mycoplasma synoviae* infection in broiler chickens. Avian Pathol. 34(2): 137–42.
3. Chin RP, Meteyer CU, Yamamoto R, Shivaprasad HL, Klein PN (1991) Isolation of *Mycoplasma synoviae* from the brains of commercial meat turkeys with meningeal vasculitis. Avian Dis. 35(3): 631–7.
4. Lockaby SB, Hoerr FJ, Lauerman LH, Kleven SH (1998) Pathogenicity of *Mycoplasma synoviae* in broiler chickens. Vet Pathol. 35(3): 178–90.
5. Brown DR, May M, Bradbury JM, Balish MF, Calcutt MJ, et al. (2010) Genus I. *Mycoplasma*. In: Krieg NR, Ludwig W, Brown DR, Whitman WB, Hedlund BP, Paster BJ, Staley JT, et al., editors. Bergey's Manual of Systematic Bacteriology Volume 4. Springer, Inc.: New York, NY.
6. Manchee R, Taylor-Robinson D (1969) Utilization of neuraminic acid receptors by mycoplasmas. J Bacteriol. 98(3): 914–9.
7. Noormohammadi A, Markham P, Duffy M, Whithear K, Browning G (1998) Multigene families encoding the major hemagglutinins in phylogenetically distinct mycoplasmas. Infect Immun. 66(7): 3470–5.
8. Vasconcelos A, Ferreira H, Bizarro C, Bonatto S, Carvalho M, et al. (2005) Swine and poultry pathogens: the complete genome sequences of two strains of *Mycoplasma hyopneumoniae* and a strain of *Mycoplasma synoviae*. J Bacteriol. 187(16): 5568–77.
9. Szczepanek SM, Tulman ER, Gorton TS, Liao X, Lu Z, et al. (2010) Comparative genomic analyses of attenuated strains of *Mycoplasma gallisepticum*. Infect Immun. 78(4): 1760–71.
10. Noormohammadi A, Markham P, Kanci A, Whithear K, Browning G (2000) A novel mechanism for control of antigenic variation in the haemagglutinin gene family of mycoplasma synoviae. Mol Microbiol. 35(4): 911–23.
11. Glew MD, Baseggio N, Markham PF, Browning GF, Walker ID (1998) Expression of the pMGA genes of *Mycoplasma gallisepticum* is controlled by variation in the GAA trinucleotide repeat lengths within the 5' noncoding regions. Infect Immun. 66(12): 5833–41.
12. Citti C, Browning GF, Rosengarten R (2005) Phenotypic diversity and cell invasion in host subversion by pathogenic mycoplasmas. In: Blanchard A, Browning GF, editors. Mycoplasmas Molecular Biology Pathogenicity and Strategies for Control. Horizon Bioscience: Norfolk, UK.
13. Zimmerman C-U (2014). Current insights into phase and antigenic variation in mycoplasmas. In: Browning GF, Citti C, editors. Mollicutes Molecular Biology and Pathogenesis. Caister Academic Press: Norfolk, UK.
14. May M, Brown DR. (2011) Diversity of expressed vlhA adhesin sequences and intermediate hemagglutination phenotypes in *Mycoplasma synoviae*. J Bacteriol. 193(9): 2116–21.
15. Khiari AB, Guériri I, Mohammed RB, Mardassi BB (2010) Characterization of a variant vlhA gene of *Mycoplasma synoviae*, strain WVU 1853, with a highly divergent haemagglutinin region. BMC Microbiol. 10: 6. [doi: 1471-2180-10-6 [pii] 10.1186/1471-2180-10-6].
16. Noormohammadi A, Markham P, Whithear K, Walker I, Gurevich V, et al. (1997) *Mycoplasma synoviae* has two distinct phase-variable major membrane antigens, one of which is a putative hemagglutinin. Infect Immun. 65(7): 2542–7.
17. Sievers F, Wilm A, Dineen D, Gibson TJ, Karplus K, et al. (2011) Fast, scalable generation of high-quality protein multiple sequence alignments using Clustal Omega. Mol Syst Biol. 7: 539. [doi: 10.1038/msb.2011.75.]
18. May M, Brown DR (2009) Diversifying and stabilizing selection of sialidase and N-acetylneuraminate catabolism in *Mycoplasma synoviae*. J Bacteriol. 191(11): 3588–93.
19. Kelley LA, Sternberg MJ (2009) Protein structure prediction on the Web: a case study using the Phyre server. Nat Protoc. 4(3): 363–71.
20. Worth CL, Preissner R, Blundell TL (2011). SDM–a server for predicting effects of mutations on protein stability and malfunction. Nucleic Acids Res. 39(Web Server issue): W215–22. [doi: 10.1093/nar/gkr363].
21. Brylinski M, Feinstein WP (2013) eFindSite: improved prediction of ligand binding sites in protein models using meta-threading, machine learning and auxiliary ligands. J Comput Aided Mol Des. 27(6): 551–67.
22. Feinstein W, Brylinski M (2014) eFindSite: Enhanced fingerprint-based virtual screening against predicted ligand binding sites in protein models. Mol Inform. 33(2): 15 [doi: 10.1002/minf.201300143]
23. Henikoff S, Henikoff JG (1992) Amino acid substitution matrices from protein blocks. Proc Natl Acad Sci U S A. 89(22): 10915–9.

24. Larkin MA, Blackshields G, Brown NP, Chenna R, McGettigan PA, et al. (2007) Clustal W and Clustal X version 2.0. Bioinformatics. 23(21): 2947–8.

25. Papazisi L, Frasca S, Gladd M, Liao X, Yogev D, Geary SJ (2002) GapA and CrmA coexpression is essential for *Mycoplasma gallisepticum* cytadherence and virulence. Infect Immun. 70(12): 6839–45.

26. Tharakaraman K, Jayaraman A, Raman R, Viswanathan K, Stebbins NW, et al. (2013) Glycan receptor binding of the influenza A virus H7N9 hemagglutinin. Cell. 153(7): 1486–93.

27. Pang SS, Nguyen ST, Perry AJ, Day CJ, Panjikar S, et al. (2014) The three-dimensional structure of the extracellular adhesion domain of the sialic acid-binding adhesin SabA from *Helicobacter pylori*. J Biol Chem. 289(10): 6332–40.

28. Boguslavsky S, Menaker D, Lysnyansky I, Liu T, Levisohn S, et al. (2000) Molecular characterization of the *Mycoplasma gallisepticum* pvpA gene which encodes a putative variable cytadhesin protein. Infect Immun. 68(7): 3956–64.

29. Forsyth MH, Tourtellotte ME, Geary SJ (1992) Localization of an immuno-dominant 64 kDa lipoprotein (LP 64) in the membrane of *Mycoplasma gallisepticum* and its role in cytadherence. Mol Microbiol. 6(15): 2099–106.

30. Goh MS, Gorton TS, Forsyth MH, Troy KE, Geary SJ (1998) Molecular and biochemical analysis of a 105 kDa *Mycoplasma gallisepticum* cytadhesin (GapA). Microbiology. 144 (11): 2971–8.

31. Jenkins C, Geary SJ, Gladd M, Djordjevic SP (2007) The *Mycoplasma gallisepticum* OsmC-like protein MG1142 resides on the cell surface and binds heparin. Microbiology. 153(5): 1455–63.

32. May M, Papazisi L, Gorton TS, Geary SJ (2006) Identification of fibronectin-binding proteins in *Mycoplasma gallisepticum* strain R. Infect Immun. 74(3): 1777–85.

5

The Generation of Successive Unmarked Mutations and Chromosomal Insertion of Heterologous Genes in *Actinobacillus pleuropneumoniae* Using Natural Transformation

Janine T. Bossé[1]*, Denise M. Soares-Bazzolli[1,2], Yanwen Li[1], Brendan W. Wren[3], Alexander W. Tucker[4], Duncan J. Maskell[4], Andrew N. Rycroft[5], Paul R. Langford[1], on behalf of the BRaDP1T consortium[¶]

1 Section of Paediatrics, Imperial College London, St Mary's Campus, London, United Kingdom, 2 Laboratório de Genética Molecular de Micro-organismos, Departamento de Microbiologia - DMB – BIOAGRO, Universidade Federal de Viçosa – Viçosa, Brazil, 3 Department of Pathogen Molecular Biology, London School of Hygiene and Tropical Medicine, London, United Kingdom, 4 Department of Veterinary Medicine, University of Cambridge, Cambridge, United Kingdom, 5 Department of Pathology and Pathogen Biology, The Royal Veterinary College, North Mymms, Hatfield, United Kingdom

Abstract

We have developed a simple method of generating scarless, unmarked mutations in *Actinobacillus pleuropneumoniae* by exploiting the ability of this bacterium to undergo natural transformation, and with no need to introduce plasmids encoding recombinases or resolvases. This method involves two successive rounds of natural transformation using linear DNA: the first introduces a cassette carrying *cat* (which allows selection by chloramphenicol) and *sacB* (which allows counter-selection using sucrose) flanked by sequences to either side of the target gene; the second transformation utilises the flanking sequences ligated directly to each other in order to remove the *cat-sacB* cassette. In order to ensure efficient uptake of the target DNA during transformation, *A. pleuropneumoniae* uptake sequences are added into the constructs used in both rounds of transformation. This method can be used to generate multiple successive deletions and can also be used to introduce targeted point mutations or insertions of heterologous genes into the *A. pleuropneumoniae* chromosome for development of live attenuated vaccine strains. So far, we have applied this method to highly transformable isolates of serovars 8 (MIDG2331), which is the most prevalent in the UK, and 15 (HS143). By screening clinical isolates of other serovars, it should be possible to identify other amenable strains.

Editor: Glenn Francis Browning, The University of Melbourne, Australia

Funding: This work was supported by a Longer and Larger (LoLa) grant from the Biotechnology and Biological Sciences Research Council (grant numbers BB/G020744/1, BB/G019177/1, BB/G019274/1 and BB/G018553/1), the UK Department for Environment, Food and Rural Affairs, and Zoetis (formerly Pfizer Animal Health) awarded to the Bacterial Respiratory Diseases of Pigs-1 Technology (BRaDP1T) consortium, a grant from Conselho Nacional de Desenvolvimento Científico e Tecnológico (CNPq grant number PDE 201840/2011-1) awarded to DMSB, and a BBSRC Imperial-Brazil partnering award (BB/K021109/1) awarded to PRL. The funders had no role in study design, data collection and analysis, decision to publish, or preparation of the manuscript.

Competing Interests: The authors have declared that no competing interests exist.

* Email: j.bosse@imperial.ac.uk

¶ Membership of the BRaDP1T consortium is provided in the Acknowledgments.

Introduction

Porcine pleuropneumonia, caused by *Actinobacillus pleuropneumoniae*, is an endemic disease that continues to cause considerable economic losses in the swine industry worldwide [1,2]. After good husbandry practices are taken into account, there are two basic methods used to limit endemic infection: vaccines and antibiotics. Increasing resistance to antibiotics limits their efficacy, and there is growing pressure against the use of antibiotics in livestock production. Therefore development of an effective vaccine is required for control of this important disease.

Although bacterin (killed whole cell) and subunit vaccines have been developed for *A. pleuropneumoniae*, none has conferred complete protection against infection with all serovars (for a review, see [3]). There is growing interest in development of live attenuated vaccines (LAVs), as they have the potential to protect against homologous and heterologous serovars [4–6]. For licensing purposes, a LAV should not contain antibiotic resistance markers, and ideally should be easily differentiated from clinical isolates [5–7]. Furthermore, an ideal LAV for *A. pleuropneumoniae* might also be used as a vector for heterologous protection against other pig pathogens.

At present, the only system for introducing unmarked mutations into *A. pleuropneumoniae* is based on the use of suicide vectors (pBMK1 and pEMOC2) carrying the counter-selectable *sacB* gene [8,9]. First developed and most widely used in serovar 7 strain AP76 [8–17], it has been successfully applied to selected strains of serovars 1, 2, and 5 [6,7,18]. However, this system does not work in all strains [7]. Because of the nature of the system, which involves co-integration of the vector and formation of a merodiploid, upon counter-selection, resolution of the integrated plasmid can result either in the strain retaining the mutated copy of the target gene or in a return to the wild-type genotype. Although there should be an equal likelihood of either result, this is not always the case, and detection of the desired mutant strain may require screening of large numbers of colonies.

We have previously reported that some strains of *A. pleuropneumoniae* are capable of natural transformation [19,20]. The reference strains of serovars 1, 3, 4, 5 and 8 all showed low frequencies of transformation (10^{-8}–10^{-9}), whereas the serovar 15 reference strain, HS143 [21], had a transformation frequency of 10^{-4} [20]. Despite the low transformation frequency of the serovar 1 reference strain, Shope 4074, we and others, have used natural transformation for generation of insertion-deletion mutations [19,22–25]. Here we describe a simple two-step transformation system using linear DNA for generation of unmarked mutations in highly transformable isolates of *A. pleuropneumoniae*.

Materials and Methods

Bacterial strains and growth conditions

Escherichia coli XL1-Blue (Stratagene) or Stellar (Clontech), used for plasmid construction, were propagated on Luria-Bertani (LB; Difco) agar or in LB broth supplemented, when necessary, with 20 μg/ml chloramphenicol (Cm) or 100 μg/ml ampicillin (Amp). *A. pleuropneumoniae* serovar 8 (UK clinical isolates, including MIDG2331) and serovar 15 (reference strain, HS143) were grown at 37°C in 5% CO_2 on brain heart infusion agar (BHI; Difco) supplemented with 0.01% β-nicotinamide adenine dinucleotide (BHI-NAD) or in BHI-NAD broth. When required, 1 μg/ml Cm was added for selection of transformants. For sucrose counter-selection, bacteria were plated onto salt-free LB agar consisting of 10 g tryptone, 5 g yeast extract, and 1.5 g agar per L supplemented with 10% filter-sterilised sucrose (LB-S) for *E. coli* clones, or onto salt-free LB agar supplemented with 10% sucrose, 10% horse serum and 0.01% NAD (LB-SSN) for *A. pleuropneumoniae* clones.

DNA manipulations

Genomic DNA was prepared from bacterial strains using a QIAamp mini DNA kit, and plasmid extractions were performed using Qiaprep spin columns (Qiagen), according to the manufacturer's protocols. DNA concentrations were measured using a NanoDrop ND-1000 UV-Vis Spectrophotometer (NanoDrop Technologies). Unless otherwise stated, restriction enzymes were obtained from Roche and used according to the manufacturer's protocol. PCR was performed using either the QIAGEN Fast Cycling PCR Kit (Qiagen) or the CloneAmp HiFi PCR Premix (Clontech), according to the manufacturers' protocols.

Identification of highly transformable serovar 8 isolate(s)

In order to identify more highly transformable isolates of serovar 8 of *A. pleuropneumoniae*, we tested 15 UK clinical isolates (collected between 1992 and 2003 from different parts of the UK) by the plate transformation assay previously described [19]. Briefly, individual isolates were grown in BHI-NAD broth to an OD_{600} of approximately 0.5, and 10 μl were spotted in duplicate onto BHI-NAD agar (8 spots per plate). Strain HS143 (serovar 15 reference strain), previously shown to be highly transformable [20], was used as a positive control. Following 100 min incubation at 37°C in 5% CO_2, 750 ng of marked genomic DNA (serovar 15 *sodC*::Cm) were added to one spot of each strain (10 μl of 75 ng/μl), and cultures incubated for a further 4 h. Using a 1 μl loop, a small amount of culture was removed, bisecting each spot, and streaked for isolated colonies on

BHI-NAD-Cm. The selection plates were incubated overnight at 37°C in 5% CO_2. Strains resulting in good growth on BHI-NAD-Cm plates were tested further to determine transformation frequency, as previously described [19].

Construction of the counter-selectable cassette

A 2.1 kb sequence containing the *omlA* promoter and *sacB* gene was amplified by PCR from pBMK1 [8], a generous gift from Professor Gerald-F. Gerlach, using primers sacB_For and sacB_Rev (see Table 1 for all primers used in this study), which added ApaI sites on both ends of the amplicon. The PCR product was digested with ApaI, cleaned using a Qiaquick spin column (Qiagen), and ligated using T4 DNA ligase (New England Biolabs) into ApaI-digested pUSScat vector (a pGEMT plasmid containing an 842 bp insert comprised of a *cat* gene flanked by 2 copies of the uptake signal sequences (USS) required for natural transformation in *A. pleuropneumoniae* [23,26]), which was dephosphorylated using Shrimp Alkaline Phosphatase (Roche). The ligation mix was transformed into *E. coli* XL1-Blue cells (Stratagene). Transformants were selected on LB-Cm, and screened by colony PCR for the presence of the *sacB* gene. Sucrose sensitivity of selected clones was confirmed by patching onto LB-S. Restriction mapping of the pUSScatsac plasmid confirmed the insertion of the *sacB* gene downstream of, and in the same orientation as, the *cat* gene.

Deletion of *sodC* and/or *ureC*

The primers used in creation of the constructs are shown in Table 1. Where required, 15 bp extensions were added to the 5′ end of primers to allow directional cloning of the PCR fragments using the In-Fusion kit (Clontech) according to the manufacturer's protocol. The 3 kb *cat-sacB* cassette (Figure 1A) was amplified from pUSScatsac using primers catsacB_for and catsacB_rev. As mentioned above, this cassette contains 2 copies of the USS to facilitate natural transformation in *A. pleuropneumoniae*. Flanking sequences for the gene deletions were amplified from MIDG2331 chromosomal DNA using appropriate primer pairs. In cases where the amplified *A. pleuropneumoniae* sequences did not contain native USS, these were engineered into primers so that the deletion constructs would be efficiently taken up in the second transformation step. All fragments for the *cat-sacB* insertion and deletion constructs were amplified using proof-reading CloneAmp HiFi PCR Premix. Initially, PCR amplicons containing the genes to be deleted (Figures 1B and 1C), flanked by at least 600–1000 bp to either side, were cloned into pGEMT (Promega) to create pTsodCF and pTureCF. Inverse PCR was then used to open up the vectors, removing the target sequence and adding 15 bp overhangs to allow insertion of the *cat-sacB* cassette by In-Fusion cloning. The resulting In-Fusion products were transformed into *E. coli* Stellar cells (Clontech) and were selected on LB agar containing 20 μg/ml Cm, as required. PCRs were performed using the QIAGEN Fast Cycling PCR Kit (Qiagen) on selected colonies in order to confirm the presence of inserts. Selected *cat-sacB*-containing clones were confirmed as being sensitive to sucrose by patching onto LB-S plates. The deletion constructs were generated by amplifying the left and right flanking sequences with added 15 bp overhangs designed to allow direct fusion by overlap-extension (OE) PCR. For example, the *sodC* flanking regions were amplified using the primer pairs sodCleft_for/deltasodC_left and deltasodC_right/sodCright_rev. The resulting amplicons were combined, diluted 1/100, and used as template for OE-PCR using the primer pair sodCleft_for/sodCright_rev. The resulting deletion constructs were cloned into pGEMT.

Gene knockouts were achieved by two sequential transformation steps. In the first step, the plasmids containing the *cat-sacB*

Table 1. Primers used in this study.

Name	Sequence
sacB_for	GCGTAATACGACTCACTATAGGGCCCATTG
sacB_rev	TTCCGCTTCCTTTAGGGGCCCTTG
catsacB_for	GATTCGCGGATCCGAGCTCTCTAAC
catsacB_rev	GCGTGAAGCTCGAGGTATGGGATTC
sodCleft_for	GGATTCGCCAATaCCGCTTGtACG
sodCright_rev	CCTTATTAAATGGCGGACCGACTTTCC
sodCcat_left	**TCGGATCCGCGAATC**GATGCGCCGAATAATGTAAAAGCAAGAG
sacBsodC_right	**CCTCGAGCTTCACGC**GGCTTGCGGCGTCATCAAATAGC
deltasodC_left	**ATGACGCCGCAAGCC**GATGCGCCGAATAATGTAAAAGCAAGAG
deltasodC_right	**ATTATTCGGCGCATC**GGCTTGCGGCGTCATCAAATAGC
ureCleft_for	CGGTCATAAaCAAGCGGTCTATTTTCAG
ureCright_rev	GATTGTGCCGATATTGAGTTCTGTACCAAAC
ureCcat_left	**TCGGATCCGCGAATC**CCATTTTCTGCCCCCTATAATTTGC
sacBureC_right	**CCTCGAGCTTCACGC**CGTGTGGACGGCGAGCATATTACTTG
deltaureC_left	**CTCGCCGTCCACACG**CCATTTTCTGCCCCCTATAATTTGC
deltaureC_right	**GGGGGCAGAAAATGG**CGTGTGGACGGCGAGCATATTACTTG
ureCnadVleft	**GGGCTCGGTTACTAG**CCATTTTCTGCCCCCTATAATTTGC
nadVureC_right	**ACTCGTGCGGCCGCC**CGTGTGGACGGCGAGCATATTACTTG
nadV_for	CTAGTAACCGAGCCCGCCTAATGAG
nadV_rev	GGCGGCCGCACTAGTGATTACAAG

ApaI sites in the sacB_for and sacB_rev primers are underlined. The USS present in sodCleft_for and ureCleft_for are indicated in italics with the lower case letters indicating a base change from the native sequence in order to generate a USS. The 15-bp extensions required for In-Fusion cloning are indicated in bold text.

cassette flanked by *A. pleuropneumoniae*-specific sequence were linearised with NotI and transformed into the different *A. pleuropneumoniae* strains by natural transformation on agar plates, as previously described [19]. Cm-resistant transformants were screened for the appropriate insertion-deletion by PCR, and were tested for sensitivity to sucrose on LB-SSN. Subsequently, deletion constructs (either purified OE-PCR products, or linearised pGEMT clones containing the OE-PCR products) were used to transform appropriate insertion-deletion mutants in order to remove the *cat-sacB* cassette. Transformants were plated on LB-SSN, and sucrose-resistant transformants were screened for Cm sensitivity on BHI-NAD-Cm. Selected Cm-sensitive clones were tested by PCR to confirm the appropriate deletion. The double mutant (serovar 8 Δ*sodC*Δ*ureC*) was obtained by transformation of the serovar 8 Δ*sodC* mutant with linearised pTΔ*ureC::catsacB* construct, followed by removal of the *cat-sacB* cassette using the linearised pTΔ*ureC* construct in a second transformation. Loss of urease activity was confirmed by addition of urea base medium (Difco) to overnight broth cultures, as previously described [27]. Loss of SodC was confirmed by dot blot using the monoclonal antibody HD1, as previously described [28].

Replacement of *ureC* with the *Haemophilus ducreyi nadV* gene

A 1.5 kb sequence containing the *nadV* gene was amplified from *H. ducreyi* genomic DNA (using primers listed in Table 1) and was directionally cloned into the appropriate inverse PCR product of pTΔ*ureCF* (amplified using ureCnadV_left and nadVureC_right as primers) using the In-Fusion kit. Following transformation into *E. coli* Stellar cells, clones were screened by PCR to identify the correct insertion. Plasmid prepared from a

selected clone was linearised with NotI prior to use as template DNA to transform the sero8Δ*ureC::catsacB* mutant in order to remove the *cat-sacB* cassette. Sucrose resistant colonies were screened for Cm-sensitivity and the ability to grow on BHI without addition of NAD.

Results

Identification of a highly transformable serovar 8 isolate

Of the 15 UK clinical isolates tested, we identified one that had a transformation frequency of 1.9×10^{-5} (serovar 8 strain MIDG2331). This transformation frequency is at least 3 logs greater than previously shown for the serovar 8 reference strain [20].

The unmarked mutation system

The *cat-sacB* cassette (Figure 1A) facilitated generation of multiple successive mutations in *A. pleuropneumoniae* using the two-step transformation protocol. Cm selection was very stringent, and all Cm-resistant clones tested were confirmed to contain the *cat-sacB* cassette by PCR (data not shown). Following counter-selection after the second round of transformation, spontaneous resistance to sucrose was evident, but the high transformation frequencies for HS143 and MIDG2331 (10^{-4} to 10^{-5} for each transformation) made it possible to isolate transformants and confirm the deletion by PCR.

Deletion of *sodC* and/or *ureC*

As proof of principle, the two-step transformation system was used to generate unmarked mutations of *sodC* and/or *ureC* in serovars 8 and 15 of *A. pleuropneumoniae*. These genes (encoding

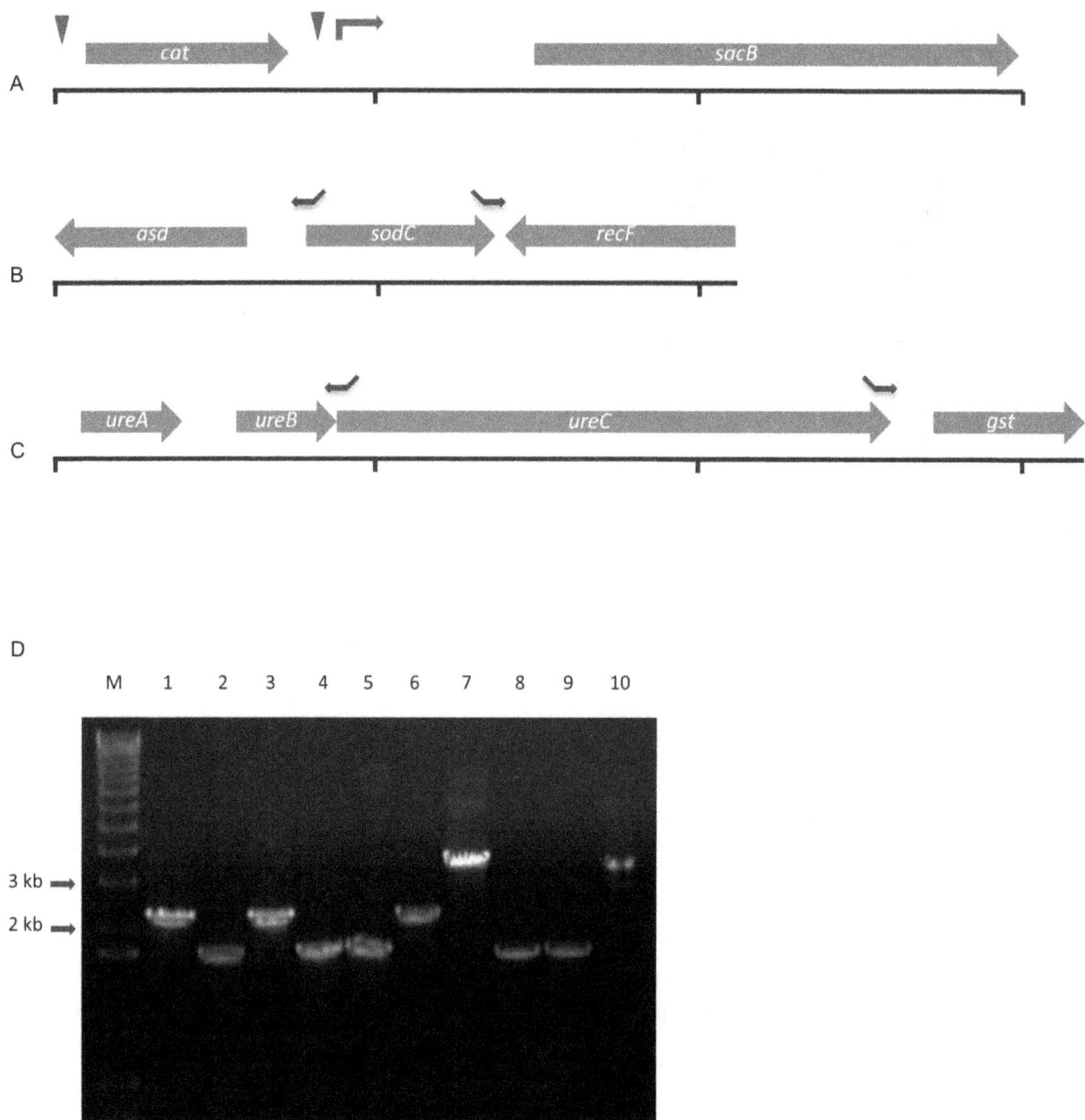

Figure 1. Construction and PCR verification of *sodC* and *ureC* deletions. A) Map showing the 3.0 kb *cat-sacB* cassette amplified from pUSScatsac using using catsacB_for and catsacB_rev. Triangles above the map indicate positions of the 2 USS required for efficient transformation, the bent arrow indicates the position of the *omlA* promoter. B) Map showing 2.1 kb sequence amplified using sodCleft_for and sodCright_rev (cloned into pTsodCF). Arrows above the map indicate positions of primers used in inverse PCR to delete a 504 bp region of *sodC*, and to add 15-bp overhangs required for fusion to the *cat-sacB* cassette (sodCcat_left and sacBsodC_right) or for direct fusion of the left and right flank sequences (deltasodC_left and deltasodC_right). C) Map showing the 3.2 kb sequence amplified using ureCleft_for and ureCright_rev (cloned into pTureCF). Arrows above the map indicate positions of primers used in inverse PCR to delete a 1641 bp region of *ureC*, and to add 15-bp overhangs required for fusion to the *cat-sacB* cassette (ureCcat_left and sacBureC_right), for direct fusion of the left and right flank sequences (deltaureC_left and deltaureC_right), or fusion to a 1914 bp fragment containing the *nadV* gene from *H. ducreyi* (ureCnadVleft and nadVureC_right). D) PCR amplification using primers sodCleft_for and sodCright_rev (lanes 1–6) or ureCleft_for and ureCright_rev (lanes 7–10) with template DNA from: 1) sero 15 WT; 2) sero 15 ΔsodC; 3) sero 8 WT; 4) sero 8 ΔsodC; 5) sero 8 ΔsodCΔureC; 6) sero 8 ΔureC; 7) sero 8 WT; 8) sero 8 ΔureC; 9) sero 8 ΔsodCΔureC; 10) sero 8 ΔureC::nadV. M = 1 kb DNA ladder (Invitrogen).

a [Cu,Zn]-superoxide dismutase and a subunit of the urease enzyme, respectively) were chosen because they are present in all *A. pleuropneumoniae* serovars and have easily detectable phenotypes. Deletion of *sodC* and/or *ureC* was confirmed in the different strains as shown by PCR (Figure 1D), SodC dot blot (Figure 2A), and urease activity assay (Figure 2B), as appropriate.

Replacement of *ureC* with the *H. ducreyi nadV* gene

In order to illustrate the usefulness of this method for introducing foreign genes into targeted locations in the chromo-

some of *A. pleuropneumoniae*, a portion of the *ureC* gene was replaced with the *H. ducreyi nadV* gene. The *nadV* gene was chosen as heterologous expression from a plasmid was previously shown to result in NAD-independence in *A. pleuropneumoniae* [29], making it easy to phenotypically verify the insertion following sucrose counterselection. Expression of the chromosomally inserted *nadV* gene rendered *A. pleuropneumoniae* strains NAD-independent (Figure 2C), while elimination of urease activity in the mutant (Figure 2B), along with PCR verification (Figure 1D), confirmed the targeted location of the insertion.

A

Sero 15　　Sero 8

WT

Δ sodC

Δ sodC Δ ureC

Δ ureC

B

1　　2　　3　　4

C

BHI-NAD　　　　　　　　　BHI

WT | ΔureC

ΔureC::nadV

WT | ΔureC

ΔureC::nadV

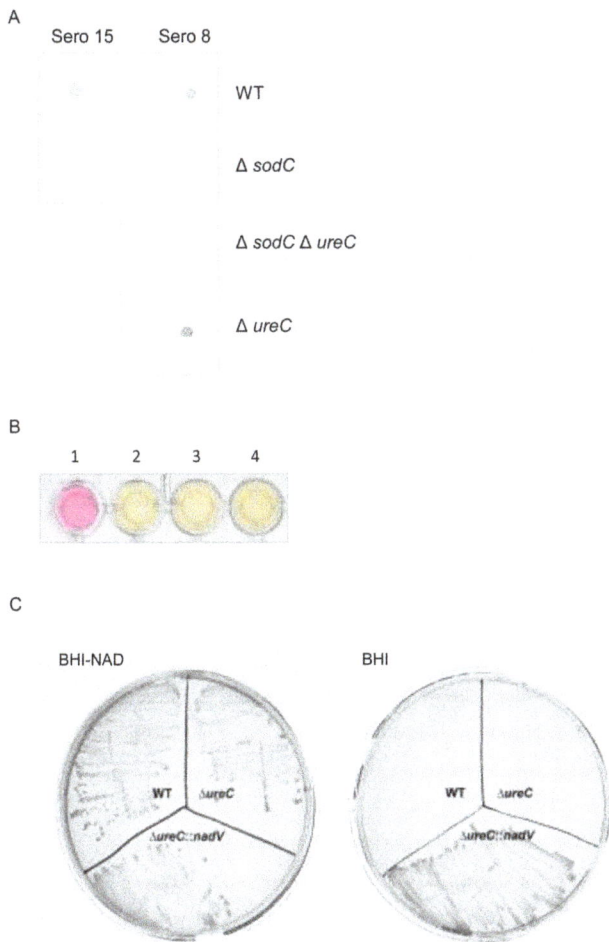

Figure 2. Phenotypic confirmation of mutations. A) Detection of SodC. Whole cell lysates (10 µg protein) were tested for reactivity with mouse monoclonal antibody HD1 by dot blot. B) Detection of urease activity in broth cultures of 1) sero 8 WT; 2) sero 8 ΔureC; 3) sero 8 ΔsodCΔureC; and 4) sero 8 ΔureC::nadV. A change in colour from yellow to pink indicates a positive reaction for urease activity. C) Growth of sero 8 strains on BHI-NAD and BHI (no NAD).

Discussion

The introduction of unmarked mutations into the bacterial chromosome is particularly desirable for generation of multiple mutations and LAVs. In *A. pleuropneumoniae,* suicide vectors (pBMK1 and pEMOC2) have been used by some groups to generate (multiple) unmarked mutations in selected strains [5,6,7,18]. In our experience, however, although co-integrates are readily selected following conjugation of constructs based on these plasmids, counter-selection on sucrose tends to yield high numbers of wild-type revertants, making identification of true deletion mutants extremely laborious and often impossible.

Recently, a markerless mutation system was described for *Actinobacillus succinogenes* [30]. The method used a combination of natural transformation for introduction of insertion/deletion mutations with FRT sites flanking the selective marker, and electroporation with a plasmid expressing the Flp recombinase to drive excision of the marker. This method leaves a residual FRT site (scar) with each deletion, and requires curing of the plasmid expressing the recombinase. Although multiple successive mutations are possible, the build up of FRT scars in the chromosome

could lead to recombination hotspots, which would not be desirable in a LAV strain.

Previously, we have shown the utility of natural transformation for generation of insertion-deletion mutations in *A. pleuropneumoniae* using linear dsDNA constructs [19,23,25]. By adding a counter-selectable gene into the insertion-deletion cassette and a second transformation step to remove the cassette, we have further exploited this simple technique to generate scarless unmarked deletions, and to insert a heterologous gene into a targeted site in the chromosome. This method can also be used for the generation of targeted point mutations by incorporating these into the unmarked sequence used in the second transformation step. The use of linear DNA templates (either linearised plasmid or PCR product) in both rounds of transformation ensures that allele replacement is by double-crossover, avoiding problems associated with merodiploid formation/resolution that can arise with suicide vectors. Furthermore, DNA taken up by natural transformation is not prone to degradation by the abundant restriction systems present in *A. pleuropneumoniae*, which can affect efficiency of electroporation [31].

A similar two-step natural transformation method has been described for creating unmarked mutations in *Helicobacter pylori* [32] and *Acinetobacter* sp. strain ADP1 [33], transformable bacteria that do not require specific USS. When creating mutants in *A. pleuropneumoniae* by this method, it is essential to include the 9 bp USS (ACAAGCGGT) required for efficient uptake of DNA by this bacterium [20,26] in donor DNA used in both steps. To this end, we have generated a *cat-sacB* cassette containing 2 perfect copies of the *A. pleuropneumoniae* USS flanking the *cat* gene. This ensures efficient uptake of the insertion-deletion construct. In the second step, if the unmarked deletion fragment does not contain an endogenous USS, then it can be engineered into primer sequence(s), as we have done.

In this study, we have generated mutations in the highly transformable serovar 15 reference strain HS143 [21], as well as in the serovar 8 clinical isolate MIDG2331. We chose serovar 8 to reflect its high prevalence in the UK [34]. MIDG2331, amenable to this method of creating unmarked deletions, was identified after screening only 15 isolates. In countries where other serovars predominate, we recommend testing a selection of clinical isolates for identification of appropriate transformable strains. Testing must be empirical, as the presence of known competence genes is not sufficient to ensure successful transformation of *A. pleuropneumoniae* strains [20]. Even in *Haemophilus influenzae*, where natural transformation has been extensively studied, the reason for variation in levels of competence of different isolates is not clear [35].

With the availability of whole genome sequences for most serovars of *A. pleuropneumoniae* [36–39], it is now possible, using HS143 and/or MIDG2331, to systematically mutate specific highly conserved core genes in order to determine their contribution to the biology and pathogenesis of this bacterium, with a view to improving diagnostics, therapies and vaccine strategies.

Acknowledgments

The BRaDP1T Consortium comprises: Duncan J. Maskell, Alexander W. (Dan) Tucker, Sarah E. Peters, Lucy A. Weinert, Jinhong (Tracy) Wang, Shi-Lu Luan, Roy R. Chaudhuri (University of Cambridge; present address for R. Chaudhuri is Centre for Genomic Research, University of Liverpool, Crown Street, Liverpool, L69 7ZB, UK.); Andrew N. Rycroft, Gareth A. Maglennon, Dominic Matthews (Royal Veterinary College); Brendan W. Wren, Jon Cuccui, Vanessa Terra (London School of Hygiene and Tropical Medicine); and Paul R. Langford, Janine T. Bossé, Yanwen Li (Imperial College London).

Author Contributions

Conceived and designed the experiments: JTB PRL. Performed the experiments: JTB DMSB YL. Analyzed the data: JTB DMSB PRL. Contributed reagents/materials/analysis tools: JTB PRL. Wrote the paper: JTB DMSB YL ANR BWW AWT DJM PRL.

References

1. Bossé JT, Janson H, Sheehan BJ, Beddek AJ, Rycroft AN, et al. (2002) *Actinobacillus pleuropneumoniae*: pathobiology and pathogenesis of infection. Microbes Infect 4: 225–235.

2. Gottschalk M, Taylor DJ (2006) *Actinobacillus pleuropneumoniae* In: Straw BE, Zimmerman JJ, Dallaire S, Taylor DJ, editors. Diseases of Swine, 9th ed. Ames: Blackwell Publishing Professional. pp. 563–576.

3. Ramjeet M, Deslandes V, Goure J, Jacques M (2008) *Actinobacillus pleuropneumoniae* vaccines: from bacterins to new insights into vaccination strategies. Anim Health Res Rev 9: 25–45.

4. Inzana TJ, Todd J, Veit HP (1993) Safety, stability, and efficacy of noncapsulated mutants of *Actinobacillus pleuropneumoniae* for use in live vaccines. Infect Immun 61: 1682–1686.

5. Maas A, Jacobsen ID, Meens J, Gerlach GF (2006) Use of an *Actinobacillus pleuropneumoniae* multiple mutant as a vaccine that allows differentiation of vaccinated and infected animals. Infect Immun 74: 4124–4132.

6. Maas A, Meens J, Baltes N, Hennig-Pauka I, Gerlach GF (2006) Development of a DIVA subunit vaccine against *Actinobacillus pleuropneumoniae* infection. Vaccine 24: 7226–7237.

7. Tonpitak W, Baltes N, Hennig-Pauka I, Gerlach GF (2002) Construction of an *Actinobacillus pleuropneumoniae* serotype 2 prototype live negative-marker vaccine. Infect Immun 70: 7120–7125.

8. Oswald W, Tonpitak W, Ohrt G, Gerlach G (1999) A single-step transconjugation system for the introduction of unmarked deletions into *Actinobacillus pleuropneumoniae* serotype 7 using a sucrose sensitivity marker. FEMS Microbiol Lett 179: 153–160.

9. Baltes N, Tonpitak W, Hennig-Pauka I, Gruber AD, Gerlach GF (2003) *Actinobacillus pleuropneumoniae* serotype 7 siderophore receptor FhuA is not required for virulence. FEMS Microbiol Lett 220: 41–48.

10. Baltes N, Hennig-Pauka I, Jacobsen I, Gruber AD, Gerlach GF (2003) Identification of dimethyl sulfoxide reductase in *Actinobacillus pleuropneumoniae* and its role in infection. Infect Immun 71: 6784–6792.

11. Baltes N, Kyaw S, Hennig-Pauka I, Gerlach GF (2004) Lack of influence of the anaerobic [NiFe] hydrogenase and L-1,2 propanediol oxidoreductase on the outcome of *Actinobacillus pleuropneumoniae* serotype 7 infection. Vet Microbiol 102: 67–72.

12. Baltes N, N'diaye M, Jacobsen ID, Maas A, Buettner FF, et al. (2005) Deletion of the anaerobic regulator HlyX causes reduced colonization and persistence of *Actinobacillus pleuropneumoniae* in the porcine respiratory tract. Infect Immun 73: 4614–4619.

13. Baltes N, Tonpitak W, Gerlach GF, Hennig-Pauka I, Hoffmann-Moujahid A, et al. (2001) *Actinobacillus pleuropneumoniae* iron transport and urease activity: effects on bacterial virulence and host immune response. Infect Immun 69: 472–478.

14. Buettner FF, Maas A, Gerlach GF (2008) An *Actinobacillus pleuropneumoniae* arcA deletion mutant is attenuated and deficient in biofilm formation. Vet Microbiol 127: 106–115.

15. Jacobsen I, Gerstenberger J, Gruber AD, Bossé JT, Langford PR, et al. (2005) Deletion of the ferric uptake regulator Fur impairs the in vitro growth and virulence of *Actinobacillus pleuropneumoniae*. Infect Immun 73: 3740–3744.

16. Jacobsen I, Hennig-Pauka I, Baltes N, Trost M, Gerlach GF (2005) Enzymes involved in anaerobic respiration appear to play a role in *Actinobacillus pleuropneumoniae* virulence. Infect Immun 73: 226–234.

17. Tonpitak W, Thiede S, Oswald W, Baltes N, Gerlach GF (2000) *Actinobacillus pleuropneumoniae* iron transport: a set of *exbBD* genes is transcriptionally linked to the *tbpB* gene and required for utilization of transferrin-bound iron. Infect Immun 68: 1164–1170.

18. Lin L, Bei W, Sha Y, Liu J, Guo Y, et al. (2007) Construction and immunogencity of a DeltaapxIC/DeltaapxIIC double mutant of *Actinobacillus pleuropneumoniae* serovar 1. FEMS Microbiol Lett 274: 55–62.

19. Bossé JT, Nash JH, Kroll JS, Langford PR (2004) Harnessing natural transformation in *Actinobacillus pleuropneumoniae*: a simple method for allelic replacements. FEMS Microbiol Lett 233: 277–281.

20. Bossé JT, Sinha S, Schippers T, Kroll JS, Redfield RJ, et al. (2009) Natural competence in strains of *Actinobacillus pleuropneumoniae*. FEMS Microbiol Lett 298: 124–130.

21. Blackall PJ, Klaasen HL, van den Bosch H, Kuhnert P, Frey J (2002) Proposal of a new serovar of *Actinobacillus pleuropneumoniae*: serovar 15. Vet Microbiol 3: 47–52.

22. Ali T, Oldfield NJ, Wooldridge KG, Turner DP, Ala'Aldeen DAA (2008) Functional characterization of AasP, a maturation protease autotransporter protein of *Actinobacillus pleuropneumoniae*. Infect Immun 76: 5608–5614.

23. Bossé JT, Sinha S, Li MS, O'Dwyer CA, Nash JH, et al. (2010) Regulation of *pga* operon expression and biofilm formation in *Actinobacillus pleuropneumoniae* by sigmaE and H-NS. J Bacteriol 192: 2414–2423.

24. Izano EA, Sadovskaya I, Vinogradov E, Mulks MH, Velliyagounder K, et al. (2007) Poly-N-acetylglucosamine mediates biofilm formation and antibiotic resistance in *Actinobacillus pleuropneumoniae*. Microb Pathog 43: 1–9.

25. Mullen LM, Bossé JT, Nair SP, Ward JM, Rycroft AN, et al. (2008) *Pasteurellaceae* ComE1 proteins combine the properties of fibronectin adhesins and DNA binding competence proteins. PLoS One. 3:e3991.

26. Redfield RJ, Findlay WA, Bossé J, Kroll JS, Cameron AD, et al. (2006) Evolution of competence and DNA uptake specificity in the *Pasteurellaceae*. BMC Evol Biol 6: 82.

27. Bossé JT, Gilmour HD, MacInnes JI (2001) Novel genes affecting urease activity in *Actinobacillus pleuropneumoniae*. J Bacteriol 183: 1242–1247.

28. Fung WW, O'Dwyer CA, Sinha S, Brauer AL, Murphy TF, et al. (2006) Presence of copper- and zinc-containing superoxide dismutase in commensal *Haemophilus haemolyticus* isolates can be used as a marker to discriminate them from nontypeable *H. influenzae* isolates. J Clin Microbiol 44: 4222–4226.

29. Bossé JT, Durham AL, Rycroft AN, Kroll JS, Langford PR (2009) New plasmid tools for genetic analyses in *Actinobacillus pleuropneumoniae* and other *Pasteurellaceae*. Appl Environ Microbiol 75: 6124–6131.

30. Joshi RV, Schindler BD, McPherson NR, Tiwari K, Vieille C (2014). Development of a markerless knockout method for *Actinobacillus succinogenes*. Appl Environ Microbiol 80: 3053–3061.

31. Jansen R, Briaire J, Smith HE, Dom P, Haesebrouck F, et al. (1995) Knockout mutants of Actinobacillus pleuropneumoniae serotype 1 that are devoid of RTX toxins do not activate or kill porcine neutrophils. Infect Immun 63: 27–37.

32. Copass M, Grandi G, Rappuoli R (1997) Introduction of unmarked mutations in the *Helicobacter pylori vacA* gene with a sucrose sensitivity marker. Infect Immun 65: 1949–1952.

33. Jones RM, Williams PA (2003) Mutational analysis of the critical bases involved in activation of the AreR-regulated sigma54-dependent promoter in *Acinetobacter* sp. strain ADP1 Appl Environ Microbiol 69: 5627–5635.

34. O'Neill C, Jones SC, Bossé JT, Watson CM, Williamson SM, et al. (2010) Prevalence of *Actinobacillus pleuropneumoniae* serovars in England and Wales. Vet Rec 167: 661–662.

35. Maughan H, Redfield RJ (2009) Extensive variation in natural competence in *Haemophilus influenzae*. Evolution 63: 1852–1866.

36. Foote SJ, Bossé JT, Bouevitch AB, Langford PR, Young NM, et al. (2008) The complete genome sequence of *Actinobacillus pleuropneumoniae* L20 (serotype 5b). J Bacteriol 190: 1495–1496.

37. Li G, Xie F, Zhang Y, Wang C (2012) Draft genome sequence of *Actinobacillus pleuropneumoniae* serotype 7 strain S-8. J Bacteriol 194: 6606–6607.

38. Xu Z, Chen X, Li L, Li T, Wang S, et al. (2010) Comparative genomic characterization of *Actinobacillus pleuropneumoniae*. J Bacteriol 192: 5625–5636.

39. Zhan B, Angen Ø, Hedegaard J, Bendixen C, Panitz F (2010) Draft genome sequences of *Actinobacillus pleuropneumoniae* serotypes 2 and 6. J Bacteriol 192: 5846–5847.

6

A Single-Step Method for Rapid Extraction of Total Lipids from Green Microalgae

Martin Axelsson, Francesco Gentili*

Department of Wildlife, Fish, and Environmental Studies, Swedish University of Agricultural Sciences, Umeå, Sweden

Abstract

Microalgae produce a wide range of lipid compounds of potential commercial interest. Total lipid extraction performed by conventional extraction methods, relying on the chloroform-methanol solvent system are too laborious and time consuming for screening large numbers of samples. In this study, three previous extraction methods devised by Folch et al. (1957), Bligh and Dyer (1959) and Selstam and Öquist (1985) were compared and a faster single-step procedure was developed for extraction of total lipids from green microalgae. In the single-step procedure, 8 ml of a 2:1 chloroform-methanol (v/v) mixture was added to fresh or frozen microalgal paste or pulverized dry algal biomass contained in a glass centrifuge tube. The biomass was manually suspended by vigorously shaking the tube for a few seconds and 2 ml of a 0.73% NaCl water solution was added. Phase separation was facilitated by 2 min of centrifugation at 350 g and the lower phase was recovered for analysis. An uncharacterized microalgal polyculture and the green microalgae *Scenedesmus dimorphus*, *Selenastrum minutum*, and *Chlorella protothecoides* were subjected to the different extraction methods and various techniques of biomass homogenization. The less labour intensive single-step procedure presented here allowed simultaneous recovery of total lipid extracts from multiple samples of green microalgae with quantitative yields and fatty acid profiles comparable to those of the previous methods. While the single-step procedure is highly correlated in lipid extractability ($r^2 = 0.985$) to the previous method of Folch et al. (1957), it allowed at least five times higher sample throughput.

Editor: Miyako Kusano, RIKEN PSC, Japan

Funding: The financial support from the Swedish Energy Agency (http://www.energimyndigheten.se/en/) and Processum Biorefinery Initiative AB (http://www.processum.se/sv/processum) was greatly appreciated. The funders had no role in study design, data collection and analysis, decision to publish, or preparation of the manuscript.

Competing Interests: We have the following interests. This study was partly funded by Processum Biorefinary Initiative AB. There are no patents, products in development or marketed products to declare.

* E-mail: francesco.gentili@slu.se

Introduction

Oleaginous microalgae have received considerable attention as a renewable source of oil for production of biodiesel and other fuels [1]. Some microalgal lipids are already valued ingredients in aquaculture feeds [2,3] and as nutritional supplements for humans [4,5]. However, microalgae are known to synthesize a diversity of unusual lipid compounds which may be commercially exploitable [6,7,8] and have been proposed as a suitable biorefinery feedstock for value added co-production of fine chemicals and fuels [9]. While specific analytical methods may exist for selected lipid compounds, total lipid extraction is favourable when screening for a variety of lipids. The extraction methods devised by Folch et al. [10] and by Bligh and Dyer [11] have found general acceptance as standard procedures for recovery of total lipids [12,13]. Both methods rely on chloroform and methanol to form a monophasic solvent system to extract and dissolve the lipids. A biphasic system is then produced in a purification step by the addition of water, leading to the separation of polar and non-polar compounds into an upper and lower phase respectively [10,11].

The method by Folch et al. [10] is regarded as the most reliable method for complete recovery of total lipids but the Bligh and Dyer [11] procedure is more widely known [14] and has been favoured for extraction of lipids from tissues of vascular plants [15]. It is also the more commonly used method but has often been incorrectly applied resulting in incomplete recovery [12]. While fatty acid profiles may remain intact, the method systematically underestimates concentrations in samples containing more than 2% lipids [14]. This limitation is important to consider in microalgal research as oleaginous microalgae contain an average of 25.5% lipid by dry weight during normal circumstances and 45.7% when subjected to stress [16].

The main differences between the protocols of Folch et al. [10] and Bligh and Dyer [11] are the volume of solvent system in relation to the amount of sample, the ratios between solvents within the systems, and the presence or absence of NaCl in the added water fraction. While Folch et al. [10] employed 20 times the sample volume of a 2:1 (v/v) chloroform-methanol mixture (assuming that the tissue had the specific gravity of water), Bligh and Dyer [11] used a chloroform-methanol step wise extraction of 1:2 and 1:1 (v/v) amounting to a final volume of only four times the equivalent sample amount. Other solvent systems have been developed as alternatives to the toxic chloroform systems, but these are generally less efficient for total lipid extraction and are sensitive to the water content of the sample [15,17–20]. The conventional chloroform based methods are considered superior for total lipid extraction, but they are also notorious for being laborious and time consuming, thus limiting sample throughput. The protocols generally involve sequential addition of solvents and several steps of manual sample manipulation, such as homogenization and filtration, making them unsuitable for screening large numbers of samples.

The aim of this study was to a) compare three established total lipid extraction methods relying on the chloroform-methanol solvent system and b) to find a faster and simpler procedure to simultaneously obtain extracts from multiple small samples of green freshwater microalgae (a few hundred mg of wet weight biomass). Presented here is the development of a one-step total lipid extraction procedure to facilitate screening for quantitative total lipid yields and fatty acid compositions among green freshwater microalgae.

Methods

Ethics statement

"N/A".

No specific permissions were required for these locations/ activities because we used local species or species present naturally in our environment. We confirm that the field studies did not involve endangered or protected species.

We had the permission to collect our samples from the private land at the power plant.

1. Cultivation and harvest of microalgae

The green microalgal strains *Scenedesmus dimorphus* (417), *Selenastrum minutum* (326) and *Chlorella protothecoides* (25) were purchased from UTEX The Culture Collection of Algae at the University of Texas at Austin (in parenthesis is the UTEX id); while an uncharacterized polyculture of algae cultured in municipal wastewater, was retrieved from a bioreactor installed at a combined heat and power plant (Umeå Energi, Umeå).

For the lipid extraction method comparison, two independent experiments were carried out. In experiment 1, *S. dimorphus* (417), *S. minutum* (326) and *C. protothecoides* (25) were grown in Proteose medium [21] and the polyculture of endogenous algae was cultured in untreated final municipal wastewater treatment plant effluent. All cultures were grown in bottles with 1 liter of volume sparged with sterile air at a flow of 170 ml/min and subjected to approximately 12 hours of natural light from a window in Umeå, Sweden (March to April, 2012, 63°49′30″N). After four weeks of growth, biomass was harvested by centrifugation at 3584 g for 10 min and pellets were stored at −20°C overnight. Experiment 1 had two replicates.

In experiment 2, the same algae were grown in untreated municipal wastewater treatment plant influent. All cultures were grown in bottles of 1 liter of volume sparged with flue gases at a flow of 170 ml/min containing approximately 10% CO_2, from a combined heat and power plant (Umeå Energi, Umeå). *S. dimorphus* (417), *S. minutum* (326) and the algal polyculture were grown for four days while *C. protothecoides* was grown for 11 days in a greenhouse in Umeå in April 2013 at an average temperature of 19°C, receiving approximately 16 hours of natural light a day at an average PAR (photosynthetic active radiation) intensity of 715 $\mu Em^{-2}s^{-1}$. The PAR was measured and recorded every 5 minutes using a LiCor 1400 datalogger connected to a spherical light sensor LI 193 (LiCor Lincoln, Nebraska USA). The algae were harvested as mentioned in experiment 1 and immediately subjected to lipid extraction. Experiment 2 had four replicates.

For the cell disruption method comparison, the same algae were cultured in autoclaved (121°C, 20 min.) municipal wastewater

treatment plant effluent. The growth conditions were the same as described for experiment 1.

For the oven dried experiment, biomass was harvested from an uncharacterized polyculture of microalgae grown in a 650 l bioreactor with municipal wastewater and treated flue gas, 3 l/min containing approximately 10% CO_2, from a combined heat and power plant (Umeå Energi, Umeå). The bioreactor was placed in a greenhouse on the roof of the combined heat and power plant. The algal polyculture was grown under a batch regime in February 2012 at an average temperature of 19°C receiving approximately 10 hours of natural light a day at an average PAR (photosynthetic active radiation) intensity of 361 $\mu Em^{-2}s^{-1}$. Samples of algae, 12.73±0.31 mg (mean ± SE) dry weight harvested in the morning, were pelletized by centrifugation at 3584 g and stored overnight at −20°C.

2. Lipid extraction: Previous methods

All reagents were of analytical grade, chloroform was purchased from VWR and methanol from Fischer-Scientific.

The protocols of Folch et al. [10], Bligh and Dyer [11], and Selstam and Öquist [22], the latter based on the method of Bligh and Dyer for vascular plant material, were adapted to provide a final solvent system volume of 10 ml for extraction of about 200 mg of wet weight microalgal biomass. All weight measurements were done with a high precision balance (Kern ABT 120-5DM; readout 0.01 mg; Kern, Germany). Microalgal paste was homogenized in a glass Potter-Elvehjelm homogenizer together with solvents. Cell debris was removed by means of vacuum filtration through a Whatman grade GF/C glass microfiber filter (1.2 μm) into a glass centrifuge tube. Phase separation was facilitated by centrifugation at 350 g for 2 min and the organic, lower phase was placed in an aluminium foil cup for overnight solvent evaporation at room temperature followed by gravimetrical determination of the lipid extract.

In the Folch method, microalgal paste was homogenized in a 2:1 chloroform:methanol (v/v) mixture and cell debris was removed by filtration. The homogenizer and collected cell debris were rinsed with fresh solvent mixture and the rinse was pooled with the previous filtrate prior to the addition of a 0.73% NaCl water solution, producing a final solvent system of 2:1:0.8 chloroform:methanol:water (v/v/v).

In the method of Bligh and Dyer, microalgal paste was mixed with deionized water, chloroform, and methanol to reach 1:2:0.8 parts chloroform:methanol:water (v/v/v) and homogenized. One part chloroform was added and the mixture was further homogenized. Then, one part deionized water was added to the homogenate giving a final ratio of 2:2:1.8 chloroform:methanol:-water (v/v/v); the homogenate was re-homogenized and finally filtered to remove cell debris.

In the Selstam and Öquist procedure, microalgal paste was homogenized in chloroform and a 4:1 mixture of methanol and 0.73% NaCl water solution producing a 1:2:0.5 chloroform:-methanol:water (v/v/v) system. The homogenate was filtered and the homogenizer and cell debris rinsed with fresh methanol-water mixture and chloroform resulting in 2:3.6:0.9 parts chloroform:-methanol:water (v/v/v) as the rinse was collected to the previous filtrate. Finally, more chloroform and 0.73% NaCl water solution were added to give a ratio of 1:1:0.8 chloroform:methanol:water (v/v/v).

3. Lipid extraction: Single-step procedure

Microalgal paste was resuspended in 2:1 parts of chloroform:-methanol (v/v) by manually shaking the tube vigorously for a few seconds or until the biomass was dispersed in the solvent system.

Finally a 0.73% NaCl water solution was added to produce a 2:1:0.8 system of chloroform:methanol:water (v/v/v).

4. Extraction of lipids from oven dried microalgae

Two quadruplicate sets of frozen pelletized algae (for cultivation details see 2.1 Cultivation and harvest of microalgae) were placed in tin foil cups and dried at 65°C overnight. Lipid extractions were performed from the dried biomass without cell disruption or using a glass Potter-Elvehjelm homogenizer and from frozen pellets according to the single-step procedure.

5. Cell disruption

Different methods of cell disruption, allowing simultaneous treatment of several samples, were assessed as alternatives to the Potter-Elvehjelm homogenizer in the single-step extraction procedure. Grinding in liquid nitrogen and ultrasonication using a probe were not investigated as they would not allow simultaneous treatment of multiple samples. The microalgal paste was subjected to either 1) no treatment, 2) Potter-Elvehjelm homogenization, 3) microwaves at full effect (557 W) for 1 min followed by low effect (254 W) for 4 min, 4) microwaves at full effect (557 W) for 3 min or 5) ultrasonication for 30 min in a sonicator bath (47 kHz, 60 W, Branson B-2200). The microwave treatments were performed prior to the addition of solvents, the Potter-Elvehjelm homogenization was performed in 4 ml of the solvent mixture following a rinse with the same amount while the samples of the other treatments were directly subjected to 8 ml of the 2:1 chloroform-methanol (v/v) solvent mixture. Post treatment, extracts were washed with 2 ml 0.73% NaCl water solution and recovered as previously described.

6. Chemical analysis of lipid extracts

For a general screening of extracted compounds, solvents were removed by evaporation and the dry extracts were dissolved in dichloromethane prior to a full scan analysis by GC/MS for m/z fragments up to 700 (DB50 column, EI 35 eV, quadrupole).

Qualitative and quantitative FA (fatty acids) profile analyses were performed at the Department of Food Science, Swedish University of Agricultural Sciences (SLU) Uppsala, Sweden. FA were methylated as previously described [23] and analyzed with a gas chromatograph with a FID detector (GC-FID) [24]. The GC-FID analyses were done with two and three replicates for qualitative analyses (experiment 1 and 2) and with three replicates with internal standards (461 standard reference mixture from Nu-Chek Prep Ink. USA) for quantitative analyses (experiment 2).

7. Statistical analysis

Data were analyzed at a 95% confidence level either using a two-sample t-test or one-way analysis of variance (ANOVA) with Tukey's post-hoc test and by regression analysis (Minitab 16.1.0).

Results and discussion

The methods of Folch et al. [10], Bligh and Dyer [11], and Selstam and Öquist [22] were compared, adapted and evaluated prior to selecting a method suitable for a rapid and simple procedure for simultaneous extraction of multiple samples.

1. Comparison of the previous methods

For optimum lipid recovery, the order at which the individual solvents are added is important [25]. Yields were indeed significantly smaller if water was added prior to the methanol and chloroform mixture in the single-step procedure (data not shown). In addition, the endogenous water of the sample must be taken into consideration while performing an extraction as it should mix with the chloroform and methanol to form a monophasic ternary system [11]. The water content of wet microalgal paste is notably high and was accounted for in this work. The increased solvent-to-sample ratios assumed here would however reduce the importance of this factor as the capacity of the solvent system for retaining the monophasic system increases with its volume [14]. Increased solvent-to-sample ratios should make the extraction system more robust and allow more variation in sample content and size. In the present study, twice the volume of final solvent system used by Folch et al. [10] and almost nine times the volume used by Bligh and Dyer [11] were employed. Bligh and Dyer wanted to avoid large solvent volumes but in the present work their solvent-to-sample ratio would result in an inconveniently small volume. A larger solvent-to-sample ratio is also justified in view of the reported limitations of the Bligh and Dyer procedure when employed on lipid rich samples [14,26].

The method of Folch et al. [10] was easier and faster to perform as it involved less sample manipulation compared to the other previous methods. In addition, with the exception of *S. minutum*, the Folch method resulted in significantly higher gravimetrical yields of extracted lipids compared to the other two methods (Fig. 1). Thus, the method of Folch et al. [10] was selected as a basis for further development. While performing extractions using the previous methods, homogenization and subsequent filtration of the solvent-sample system were recognized as particular impediments to simultaneous extraction of many samples.

2. Filtration and multiple step extractions

A filtration step to remove obstructive tissue from the sample was employed in the method by Folch et al. [10] to allow efficient phase separation and recovery of the lipid fraction. However, in the present study, the amount of sample in relation to solvent volume was not large enough to be obstructive (230 ± 35 mg wet weight biomass was extracted in 10 ml solvent system). Debris from the biomass sample formed a thin, distinct layer between the two phases and the lower phase was easily accessible using a glass

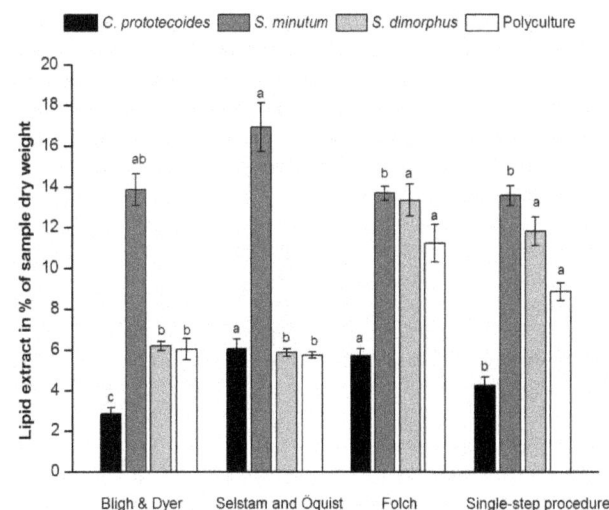

Figure 1. Gravimetric yields of total lipids extracted by the previous methods and the single-step procedure. Bars show mean yields from the different algae expressed as a percentage of the algal sample dry weight (mean ± SE, n=4; experiment 2). Different letters above bars of the same alga indicate a significant difference at α=0.05.

Pasteur pipette. If handled gently the small layer of debris did not readily mix with the chloroform and the lower phase could easily be recovered without prior filtration.

Multiple step extraction, i.e. to repeatedly rinse the sample with fresh solvent, is commonly performed to increase yields but single-step extraction has been shown to achieve equally high or only slightly smaller recoveries [26,27]. To save time and conserve resources, only single-step extractions were carried out in the present study.

3. Comparison of techniques for cell disruption

Manual homogenization of individual samples was the most time consuming step of the extraction procedure. While Bligh and Dyer [11] employed a blender, Folch et al. [10] used a Potter-Elvehjelm homogenizer. The latter is widely used for small scale homogenization of suspended cell cultures, but any means of homogenization should suffice. The aim of Folch et al. and Bligh and Dyer was to extract lipids from more or less solid animal tissues [10,11]. Plant tissues are generally more rigid and may require homogenization down to a particle size of 300 μm or smaller [13]. Unicellular microalgae are already small in particle size but neutral lipids, most notably triacylglycerols stored within the algae, could nonetheless be solubilized by disrupting the cells. However, while employing a chloroform-methanol solvent system, complete cell disruption is not needed as the lipids are extracted across the cell wall [28]. Nevertheless, a cell disruptive treatment step could still have an impact on lipid extractability and yield from microalgae [28].

In this study, different methods of cell disruption allowing simultaneous treatment of samples were assessed as alternatives to the Potter-Elvehjelm homogenizer. Microwave treatments have previously been employed to facilitate solvent extraction of lipids [29,30] and as a means to permeabilize the thick and rigid cell walls of green microalgae for staining purposes [31]. Lee et al. [30] investigated different methods of cell disruption for solvent extraction of lipids from green microalgae and acquired the highest yields when microwave treatment was employed. Though microwave cell disruption techniques are relatively novel, ultra-sonication is a proven and popular method for disruption of microalgal cells [32] and may be the preferred method of disruption for protein extraction [33]. However, none of the investigated cell disruption techniques produced substantially higher yields (Fig. 2) and the only differences of statistical significance were achieved when treating *S. dimorphus* with microwaves or the Potter-Elvehjelm homogenizer, which increased yields by approximately 24% compared to the untreated control and to sonication (Fig. 2). In addition, Ryckebosch et al. [27] reported that employing a similar total lipid extraction method could slightly increase yields from *S. obliquus* by cell disruption, but concluded that no cell disruption was generally necessary for lipid extraction from lyophilized microalgae. Indeed, none of the cell disruption methods used in this study provided a general gain in yields.

4. The single-step procedure

Yields of the single-step procedure closely resembled those of the previous Folch method (Fig. 1 and in experiment 1 from frozen algae biomass data not shown). To investigate if the modification to the previous protocol affected the relative recovery of individual lipid compounds, fatty acid profiles of extracts from *S. minutum* were determined and a semi-qualitative GC/MS analysis was employed in full scan mode to screen for undefined compounds. Between the single-step procedure and the previous method, there were no differences in presence of undefined compounds

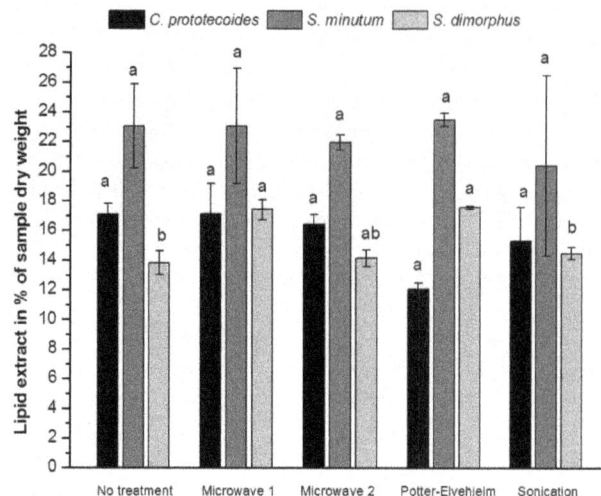

Figure 2. Comparison of cell disruption techniques to increase yields. Bars show gravimetric yields of total lipids extracted by the single-step procedure from different algal species. Yields are expressed as a percentage of the algal sample in dry weight (mean ± SE, n = 2; experiment 1). Different letters above bars of the same alga indicate a significant difference at α = 0.05.

determined by overlay analysis of chromatograms normalized against sample concentrations and matched according to retention time. Quantitative FA profiles, obtained from *S. minutum* using the single-step procedure, were compared by regression analysis to those obtained using the previous method by Folch et al. [10]. The profiles of the single-step procedure closely resembled those of the previous method, producing a regression coefficient (r^2) of 0.985. Qualitative FA profiles had a regression coefficient r^2 equal to 0.991 when comparing the most representative fatty acids extracted (Table 1 and experiment 1 data not shown). Also qualitative FA profiles between the single-step procedure and the methods by Bligh and Dyer [11] and Selstam and Öquist [22] were highly correlated (Table 1) with a regression coefficient r^2 equal to 0.99 and 0.992 respectively. Furthermore, the single-step procedure produced the same gravimetrical yields of total lipids from dried microalgae as from fresh or frozen microalgal paste if the dried biomass was homogenized before or during the extraction.

To validate the recovery of total lipids using the single-step procedure, a known amount of vegetable oil (olive oil; 1.2 – 1.8 mg) was added to the extracting solution.

As determined gravimetrically, the procedure achieved complete recovery of the vegetable oil, showing an average recovery of 91% ± 4.7 SE.

In an additional experiment performed on quadruplicate samples the single-step procedure had 3.5 times higher total lipids yield than a method based on hexane extraction [19] for the green alga *S. minutum* (data not shown).

In a recent study an extraction methodology similar to the single-step procedure but favouring a system of 1:1 (v/v) chloroform-methanol was proposed for extraction of lipids from microalgae [27]. This ratio was investigated and compared to the 2:1 (v/v) ratio with the same setup as in the single-step procedure. Gravimetrical yields appeared significantly higher but blank control extractions showed that the 1:1 system left residues after evaporating the solvents. The remnants could not be explained by any isolated individual constituent of the solvent system. Note that

Table 1. The fatty acid profile of *Selenastrum minutum* in extracts obtained by the different methods.

	Bligh and Dyer			Selstam and Öquist			Folch et al.			Single-step		
	%			%			%			%		
C16:0	23.86	±	0.43	24.26	±	0.66	25.05	±	0.43	23.92	±	0.19
C18:0	1.46	±	0.18	1.49	±	0.08	1.77	±	0.13	1.26	±	0.06
C18:1	20.43	±	0.24	21.49	±	0.20	20.68	±	0.17	19.62	±	0.15
C18:2	9.47	±	0.11	9.04	±	0.22	7.64	±	0.12	7.77	±	0.06
C18:3	38.46	±	0.56	37.53	±	0.70	37.17	±	0.39	40.06	±	0.25
C24:0	2.35	±	0.09	2.23	±	0.13	3.47	±	0.26	3.45	±	0.05

Presence of the most representative fatty acids in total lipid extracts of *Selenastrum minutum* obtained by the previous extraction methods and by the single-step procedure developed in this study. Individual FA presented in per cent of total FA (mean ± SE, n = 3; experiment 2).

the 2:1 system evaluated in this paper, employing the same constituents, did not leave significant residues.

The gravimetric yield of the single-step procedure was however dependent on the sample size relative to the solvent system volume as extraction from larger sizes of sample resulted in smaller relative yields. A sample size limit assessment of the 10 ml solvent system revealed that up to 300 mg of wet microalgal paste corresponding to ca 30 mg in dry weight could be extracted (data not shown). This limitation was less pronounced when extracting lipids from dried biomass which allowed a 3 times larger maximum sample size (i.e. 90 mg). Below these limits, sample size has a small relative effect on gravimetric yield but lipid profiles should remain intact. Nevertheless, for mutual comparisons, samples should be normalized with regard to size prior to extraction to avoid inaccurate results.

Only green freshwater microalgae were assessed in this study and the presented method should be used with caution in works on other microalgae.

Conclusions

The single-step procedure is suitable for total lipid extraction and may be applied fo screening of algae for qualitative-quantitative analyses of total fatty acids. The method presented in this work had at least five times higher sample throughput when compared to the previous methods.

Acknowledgments

We are grateful to Lucia Kovacova and Pedro Gómez Requeni, Dept of Food Science, Swedish University of Agricultural Sciences Uppsala and Markus Axelsson at ALcontrol Laboratories in Linköping for fatty acids analyses. The help from the staff at Umeå Energi and Umeva personal was very much appreciated. We thank Dr. John Ball Anita Norman and Andrew Allen for language editing.

Author Contributions

Conceived and designed the experiments: MA FG. Performed the experiments: MA FG. Analyzed the data: MA FG. Contributed reagents/materials/analysis tools: MA FG. Wrote the paper: MA FG.

References

1. Chisti Y (2007) Biodiesel from microalgae. Biotechnology Advances 25: 294–306.
2. Benemann JR (1992) Microalgae aquaculture feeds. Journal of Applied Phycology 4: 233–245.
3. Brown MR, Jeffrey SW, Volkman JK, Dunstan GA (1997) Nutritional properties of microalgae for mariculture. Aquaculture 151: 315–331.
4. Khozin-Goldberg I, Iskandarov U, Cohen Z (2011) LC-PUFA from photosynthetic microalgae: occurrence, biosynthesis, and prospects in biotechnology. Applied Microbiology and Biotechnology 91: 905–915.
5. Vílchez C, Forján E, Cuaresma M, Bédmar F, Garbayo I, et al. (2011) Marine carotenoids: Biological functions and commercial applications. Marine Drugs 9: 319–333.
6. Cardozo KHM, Guaratini T, Barros MP, Falcão VR, Tonon AP, et al. (2007) Metabolites from algae with economical impact. Comparative Biochemistry and Physiology 146: 60–78.
7. Milledge JJ (2011) Commercial application of microalgae other than as biofuels: a brief review. Reviews in Environmental Science and Biotechnology 10: 31–41.
8. Spolaore P, Joannis-Cassan C, Duran E, Isambert A (2006) Commercial applications of microalgae. Journal of Bioscience and Bioengineering 101: 87–96.
9. Singh J, Gu S (2010) Commercialization potential of microalgae for biofuels production. Renewable and Sustainable Energy Reviews 14: 2596–2610.
10. Folch J, Lees M, Stanley GHS (1957) A simple method for the isolation and purification of total lipides from animal tissues. The Journal of Biological Chemistry 226: 497–509.
11. Bligh EG, Dyer WJ (1959) A rapid method of total lipid extraction and purification. Canadian Journal of Biochemistry and Physiology 37(8): 911–917.
12. Christie WW (1993) Preparation of lipid extracts from tissues. Advances in Lipid Methodology 2: 195–213.
13. Phillips KM, Ruggio DM, Amanna KR (2008) Extended validation of a simplified extraction and gravimetric determination of total fat to selected foods. Journal of Food Lipids 15: 309–325.
14. Iverson SJ, Lang SLC, Cooper MH (2001) Comparison of the Bligh and Dyer and Folch methods for total lipid determination in a broad range of marine tissue. Lipids 36 (11): 1283–1287.
15. Fishwick MJ, Wright AJ (1977) Comparison of methods for the extraction of plant lipids. Phytochemistry 16: 1507–1510.
16. Hu Q, Sommerfeld M, Jarvis E, Ghirardi M, Posewitz M, et al. (2008) Microalgal triacylglycerols as feedstocks for biofuel production: perspectives and advances. The Plant Journal 54: 621–639.
17. Guckert JB, Cooksey KE, Jackson LL (1988) Lipid solvent systems are not equivalent for analysis of lipid classes in the microeukaryotic green alga, Chlorella. Journal of Microbiological Methods 8: 139–149.
18. Halim R, Gladma B, Danquah MK, Webley PA (2011) Oil extraction from microalgae for biodiesel production. Bioresource Technology 102: 178–185.
19. Hara A, Radin NS (1978) Lipids extraction of tissues with a low-toxicity solvent. Analytical Biochemistry 90: 420–426.
20. Sheng J, Vannela R, Rittmann BE (2011) Evaluation of methods to extract and quantify lipids from *Synechocystis* PCC 6803. Bioresource Technology 102: 1697–1703.
21. Cheng KC, Ren M, Ogden KL (2013) Statistical optimization of culture media for growth and lipid production of *Chlorella protothecoides* UTEX 250. Bioresource Technology 128: 44–48.
22. Selstam E, Öquist G (1985) Effects of frost hardening on the composition of galactolipids and phospholipids occurring during isolation of chloroplast thylakoids from needles of scots pine. Plant Science 42: 41–48.
23. Appelqvist LÅ (1968) Rapid methods of lipid extractions and fatty acid methyl ester preparation for seed and leaf tissue with special remarks on preventing the accumulation of lipids contaminants. Arkiv För Kemi, Royal Swedish Academy of Science (Kungliga Svenska Vetenskapsakademien) 28(36): 551–570.

24. Fredriksson Eriksson S, Pickova J (2007) Fatty acids and tocopherol levels in M. Longissimus dorsi of beef cattle in Sweden – A comparison between seasonal diets. Meat Science 76: 746–754.

25. Smedes F, Thomasen T K (1996) Evaluation of the Bligh and Dyer lipid determination method. Marine Pollution Bulletin 32: 681–688.

26. Smedes F, Askland TK (1999) Revisiting the development of the Bligh and Dyer total lipid determination method. Marine Pollution Bulletin 38 (3): 193–201.

27. Ryckebosch E, Muylaert K, Foubert I (2012) Optimization of an analytical procedure for extraction of lipids from microalgae. Journal of the American Oil Chemists' Society 89: 189–198.

28. Ranjan A, Patil C, Moholkar VS (2010) Mechanistic assessment of microalgal lipid extraction. Industrial and Engineering Chemistry Research 49: 2979–2985.

29. Carrapiso AL, García C (2000) Development in lipid analysis: some new extraction techniques and in situ transesterification. Lipids 35 (11): 1167–1177.

30. Lee JY, Yoo C, Jun SY, Ahn CY, Oh HM (2010) Comparison of several methods for effective lipid extraction from microalgae. Bioresource Technology 101: 75–77.

31. Chen W, Sommerfeld M, Hu Q (2011) Microwave-assisted Nile red method for in vivo quantification of neutral lipids in microalgae. Bioresource Technology 102: 135–141.

32. Simon RD (1974) The use of an ultrasonic bath to disrupt cells suspended in volumes of less than 100 μliters. Analytical Biochemistry 60: 51–58.

33. Meijer EA, Wijffels RH (1998) Development of a fast, reproducible and effective method for the extraction and quantification of proteins of micro-algae. Biotechnology Techniques 12 (5): 353–358.

7

Defining the Alloreactive T Cell Repertoire Using High-Throughput Sequencing of Mixed Lymphocyte Reaction Culture

Ryan O. Emerson[1⑨], James M. Mathew[2,3⑨], Iwona M. Konieczna[2], Harlan S. Robins[4], Joseph R. Leventhal[2]*

1 Adaptive Biotechnologies Corporation, Seattle, Washington, United States of America, 2 Department of Surgery, Comprehensive Transplant center, Northwestern University, Chicago, Illinois, United States of America, 3 Department of Microbiology-Immunology, Northwestern University, Chicago, Illinois, United States of America, 4 Public Health Sciences Division, Fred Hutchinson Cancer Research Center, Seattle, Washington, United States of America

Abstract

The cellular immune response is the most important mediator of allograft rejection and is a major barrier to transplant tolerance. Delineation of the depth and breadth of the alloreactive T cell repertoire and subsequent application of the technology to the clinic may improve patient outcomes. As a first step toward this, we have used MLR and high-throughput sequencing to characterize the alloreactive T cell repertoire in healthy adults at baseline and 3 months later. Our results demonstrate that thousands of T cell clones proliferate in MLR, and that the alloreactive repertoire is dominated by relatively high-abundance T cell clones. This clonal make up is consistently reproducible across replicates and across a span of three months. These results indicate that our technology is sensitive and that the alloreactive TCR repertoire is broad and stable over time. We anticipate that application of this approach to track donor-reactive clones may positively impact clinical management of transplant patients.

Editor: Jay Reddy, University of Nebraska-Lincoln, United States of America

Funding: The authors have no support or funding to report.

Competing Interests: ROE has full-time employment and equity ownership at Adaptive Biotechnologies Corporation. HSR has consultancy, patents & royalties, and equity ownership at Adaptive Biotechnologies Corporation. There are no patents, products in development or marketed products to declare.

* Email: jleventh@nmh.org

⑨ These authors contributed equally to this work.

Introduction

Cellular immune response is the most important mediator of transplant rejection and a major barrier to transplant tolerance [1–3]. It is largely mediated by memory T cell populations specific for allo-peptides presented either on allo-MHC (direct antigen presentation) or on self-MHC (indirect antigen presentation) [3–5]. Positive selection in the thymus requiring immature T cells to have some binding affinity for self-HLA means that a significant proportion of mature T cells also have off-target specificity for allo-HLA alleles. Negative selection removes T cells specific for self-peptides presented on self-HLA, buts leaves T cells specific for self-peptides presented on allo-HLA [6–12]. The production of the alloreactive T cell repertoire is further complicated by molecular mimicry. Thus, in one well-studied example a public T cell response specific to EBV in the context of HLA-B*08:01 has been shown to exhibit cross-reactivity with a self-peptide presented by HLA-B*44:02 [13–16]. These cross-reactive T cells have been observed in HLA-B*08:01/HLA-B*44:02 mismatched lung allografts, suggesting direct clinical relevance for this mode of T cell alloreactivity [17]. Even in individuals with no history of allo-HLA sensitization, viral exposure or vaccine administration can create HLA cross-reactive memory T cells [18–22].

Many studies have identified public and private alloreactive T cell clones that can be primed by a variety of immunogenic events. However, while public T cell clones may play an important role in specific exposures they represent a very small proportion of the entire T cell repertoire; investigating private T cell specificities allows for a much broader view of the alloreactive T cell repertoire but private T cell responses must be measured anew in each subject.

It is our hypothesis that the alloreactive T cell repertoire can be studied by performing mixed lymphocyte reaction cultures [23,24], followed by molecular analysis of clonotypes thus generated. The availability of high-throughput sequencing of rearranged T cell receptor genes, which act as unique molecular tags for each clonal population, now allows for unprecedented depth and accuracy in the characterization of T cell repertoires. Here, we employ this high-throughput TCR sequencing to test our hypothesis by thoroughly interrogating the alloreactive T cell repertoire between three pairs of healthy adult subjects as well as the persistence of alloreactive T cell clones across biological replicates and across time.

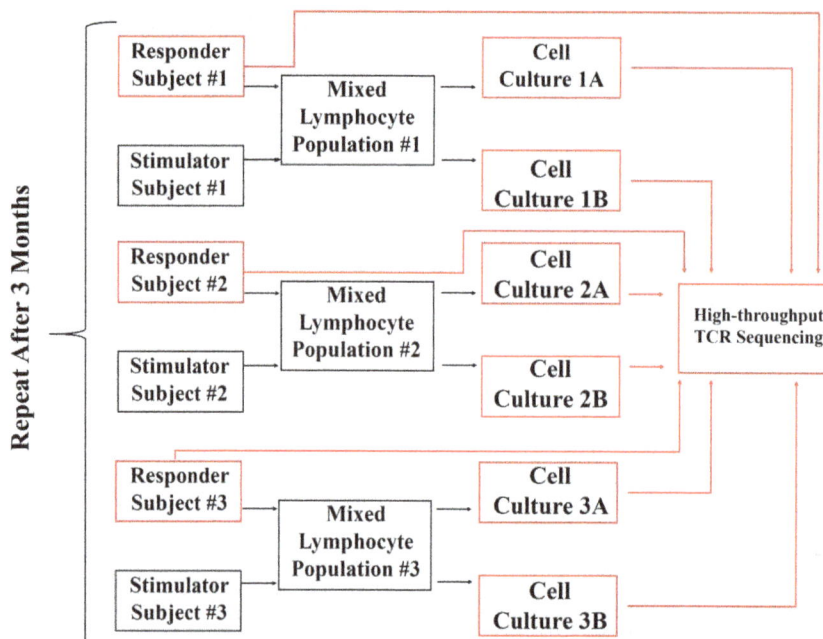

Figure 1. Experimental design. We assayed three pairs of healthy adult subjects using mixed lymphocyte reaction cultures. For each pair, lymphocytes from a responder subject were mixed with inactivated lymphocytes from a stimulator subject and cultured in duplicate. Uncultured freshly isolated PBMC from the responder as well as proliferating T cell populations from the duplicate cultures were subjected to high-throughput sequencing: we sequenced nine samples in total across the three pairs of subjects. Three months later, the experiments were repeated to generate nine more samples for high-throughput TCRβ sequencing.

Methods

Subjects

Human peripheral blood samples were obtained from laboratory volunteers under a protocol following written informed consent approved and supervised by a Northwestern University Institutional Review Board. These healthy volunteers were HLA-typed by the Northwestern HLA laboratory using molecular methods (reverse sequence specific oligonucleotide probe hybridization).

Mixed Lymphocyte Reaction (MLR) Culture and Alloreactive Responding Cell Isolation

Peripheral blood mononuclear cells (PBMC) were isolated using Ficoll-Hypaque. The responder cells were labeled with CFSE and the stimulator cells labeled with PKH26 as described previously [25,26]. The responders and stimulators were matched for 1 HLA-DR antigen to mimic the minimum requirement for some clinical transplants [27]. The PKH26 labeled stimulator cells were also irradiated at 3000 rads. The responder and stimulator cells were cultured in bulk in 15% normal AB serum containing RPMI 1640 culture medium (NAB-CM) at 1×10^6/ml each. After 7 days these were harvested and the proliferating responders were then sorted on FACSAria (BD, San Jose, CA) by gating on the CFSE dim or negative cells after gating out both CFSE high non-proliferating and the very few PKH26+ stimulator cells that still survived.

In parallel, flow cytometric analysis of the above MLR cultures was performed to determine which subsets of responder cells proliferated in response to allostimulation, using fluorochrome conjugated monoclonal antibodies. The data were acquired on an FC500 flow cytometer (Beckman-Coulter) and analyzed for cell subsets by gating on the CFSE dim or negative cells after gating out both CFSE high non-proliferating and the very few PKH26+

stimulator cells [25,26]. Additionally, standard 7-day ^{3}H-thymidine incorporation assays were also performed to monitor the strength of the MLR responses as described previously [25,26].

High-Throughput TCRβ Sequencing

Genomic DNA was extracted from cell samples using Qiagen DNeasy Blood extraction Kit (Qiagen, Gaithersburg, MD, USA). We sequenced the CDR3 region of rearranged TCRβ genes; the TCRβ CDR3 region was defined according to the IMGT collaboration [28]. TCRβ CDR3 regions were amplified and sequenced using previously-described protocols [29,30]. Briefly, a multiplexed PCR method was employed using a mixture of 60 forward primers specific to TCR Vβ gene segments and 13 reverse primers specific to TCR Jβ gene segments. Reads of 87 bp were obtained using the Illumina HiSeq System. Raw HiSeq sequence data were preprocessed to remove errors in the primary sequence of each read, and to compress the data. A nearest neighbor algorithm was used to collapse the data into unique sequences by merging closely related sequences, to remove both PCR and sequencing errors.

PCR Template Abundance Estimation

To estimate the average read coverage per input template in our PCR and sequencing approach, we employed a set of approximately 850 unique types of synthetic TCR analog, comprising each combination of Vβ and Jβ gene segments [31]. These molecules were included in each PCR reaction at very low concentration so that only some types of synthetic template were observed. Using the known concentration of the synthetic template pool, we simulated the relationship between the number of observed unique synthetic molecules and the total number of synthetic molecules added to the reaction (this is very nearly one-to-one at the low concentrations we employed). These molecules

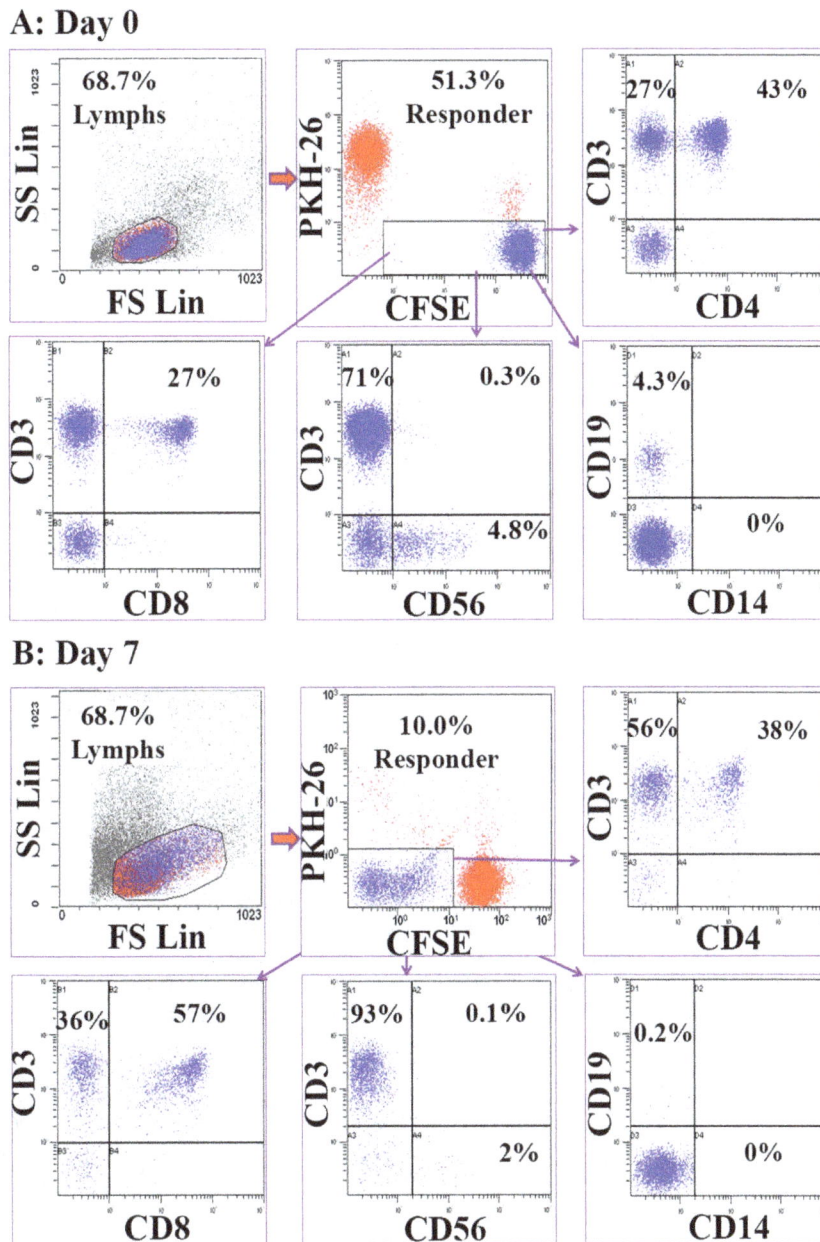

Figure 2. Alloreactive Cellular Subset Profile Generated in MLR. Bulk MLRs were prepared as described in Materials and Methods. The cellular makeup of responder cell populations were delineated at the onset and after 7 days in culture using fluorochrome coupled monoclonal antibodies. The cells were analyzed first by gating on lymphocytes and then after gating either on total CFSE positive responder cells (A: Day 0) or on CFSE diluted proliferating responder cells (B: Day 7).

then allowed us to calculate for each PCR reaction the mean number of sequencing reads obtained per molecule of PCR template, and thus to estimate the number of T cells in the input material bearing each unique TCR rearrangement.

Results and Discussion

Isolation of the Alloreactive T Cell Repertoire

In order to study the breadth, clonal structure and dynamics of the alloreactive T cell repertoire, we performed a one-way mixed lymphocyte culture using CFSE-labeled responder cells and PKH26-labeled stimulator cells on each of three pairs of healthy adult subjects [25,26], with cell culture performed in duplicate.

Three months after the first experiment, we repeated this cell culture protocol for the same three pairs of subjects. In total, we generated 18 samples of T cells, comprising six samples from each pair of subjects: uncultured total PBMC and purified proliferating T cells from duplicate MLR, at baseline and after three months (Figure 1 summarizes experimental design).

For each MLR reaction, after 7 days the proliferating responders were sorted by gating on the CFSE dim or negative cells after gating out both CFSE high non-proliferating and the very few PKH26+ stimulator cells that still survived (Figure 2A). The proliferating cells consisted of 40.3±4.7% CD3+CD4+ and 57.2±5.1% CD3+CD8+ T cells as well as minor subset of CD56+ NK cells (Figure 2B). Each population of uncultured PBMC or

Table 1. Summary of TCRβ sequencing results.

Sample	T cells assayed (estimated)[a]	Unique TCRβ sequences	Sequencing reads[b]
Fresh PBMC sample #1, 0 months	4,336,812	750,211	51,160,577
Fresh PBMC sample #2, 0 months	4,774,312	1,375,340	46,370,325
Fresh PBMC sample #3, 0 months	4,016,260	991,848	33,633,101
Fresh PBMC sample #1, 3 months	713,990	264,159	17,437,692
Fresh PBMC sample #2, 3 months	1,847,987	1,046,492	23,507,950
Fresh PBMC sample #3, 3 months	2,197,064	1,061,154	18,766,880
Proliferated MLR responder #1A, 0 months	1,885,973	33,677	23,366,016
Proliferated MLR responder #1B, 0 months	1,997,723	33,387	26,098,554
Proliferated MLR responder #2A, 0 months	1,575,201	79,174	24,704,053
Proliferated MLR responder #2B, 0 months	1,527,643	68,505	13,832,785
Proliferated MLR responder #3A, 0 months	3,372,150	58,382	37,022,643
Proliferated MLR responder #3B, 0 months	3,190,902	53,316	23,126,368
Proliferated MLR responder #1A, 3 months	640,366	57,778	12,741,642
Proliferated MLR responder #1B, 3 months	587,681	53,260	9,806,707
Proliferated MLR responder #2A, 3 months	1,022,417	68,565	10,736,335
Proliferated MLR responder #2B, 3 months	522,273	53,337	10,679,864
Proliferated MLR responder #3A, 3 months	685,126	64,615	9,788,942
Proliferated MLR responder #3B, 3 months	760,990	67,586	10,999,866
	35,654,870	**6,180,786**	**403,780,300**

[a]see Methods.
[b]the total number of 87-bp sequencing reads generated.

proliferating T cells was subjected to amplification and high-throughput sequencing of the CDR3 region of TCRβ, which somatically rearranges during T cell maturation and acts as a unique molecular tag for each clonal population of T cells. Sequencing results are presented in Table 1.

Size of the Alloreactive T Cell Repertoire

To determine the number of T cell clonal lineages involved in the alloreactive T cell response, we analyzed the number of unique CDR3 sequences observed in the proliferated T cell samples in comparison to uncultured bulk T cells from the same subjects. We defined alloreactive T cell clones as those observed in at least 10 cells in the proliferated sample and unobserved in the uncultured T cell sample, or T cells whose frequency in the proliferated sample was at least ten-fold higher than in the uncultured T cell sample. We defined two sets of alloreactive T cell clones: low-abundance alloreactive clones (below the threshold of detection in the subject's baseline T cell repertoire) and high-abundance alloreactive clones (present at measurable frequency in the subject's baseline T cell repertoire). On average, we observed

14,000 alloreactive T cell clones in each experiment; 84% of alloreactive T cell clones were low-abundance before proliferation, but in total low-abundance clones made up 40% and high-abundance clones made up 60% of the alloreactive T cell repertoire when weighting by post-proliferation clonal abundance (Table 2). While the number of proliferated low-abundance clones varied considerably, variation in the number of high-abundance (thus, presumably antigen-experienced) T cell clones between subjects was much smaller, at about 2,000 clones in each of the six experiments. These data indicate that thousands of different clonal populations of T cells comprise the alloreactive T cell repertoire.

Reproducibility of the Alloreactive T Cell Repertoire

To assay the consistency of the alloreactive T cell repertoire, we examined the persistence of each T cell clone. After defining high-abundance and low-abundance alloreactive T cells, we compared the set of alloreactive T cell clones generated in duplicate cell culture experiments (Figure 3). In each subject, essentially all clones that were highly expanded in proliferated cell culture assorted to the high-abundance subset (i.e., were present at

Table 2. Size of the alloreactive T cell repertoire.

	Mean (N = 6)	SD	% of proliferated T cells
Number of alloreactive clones	13750	6823	100%
Low-abundance pre-culture[a]	11610	6494	40.0%
High-abundance pre-culture[b]	2140	539	60.0%

[a]unobserved in pre-culture sample and ≥10 T cells after MLR.
[b]present in pre-culture sample and ≥10× enriched after MLR.

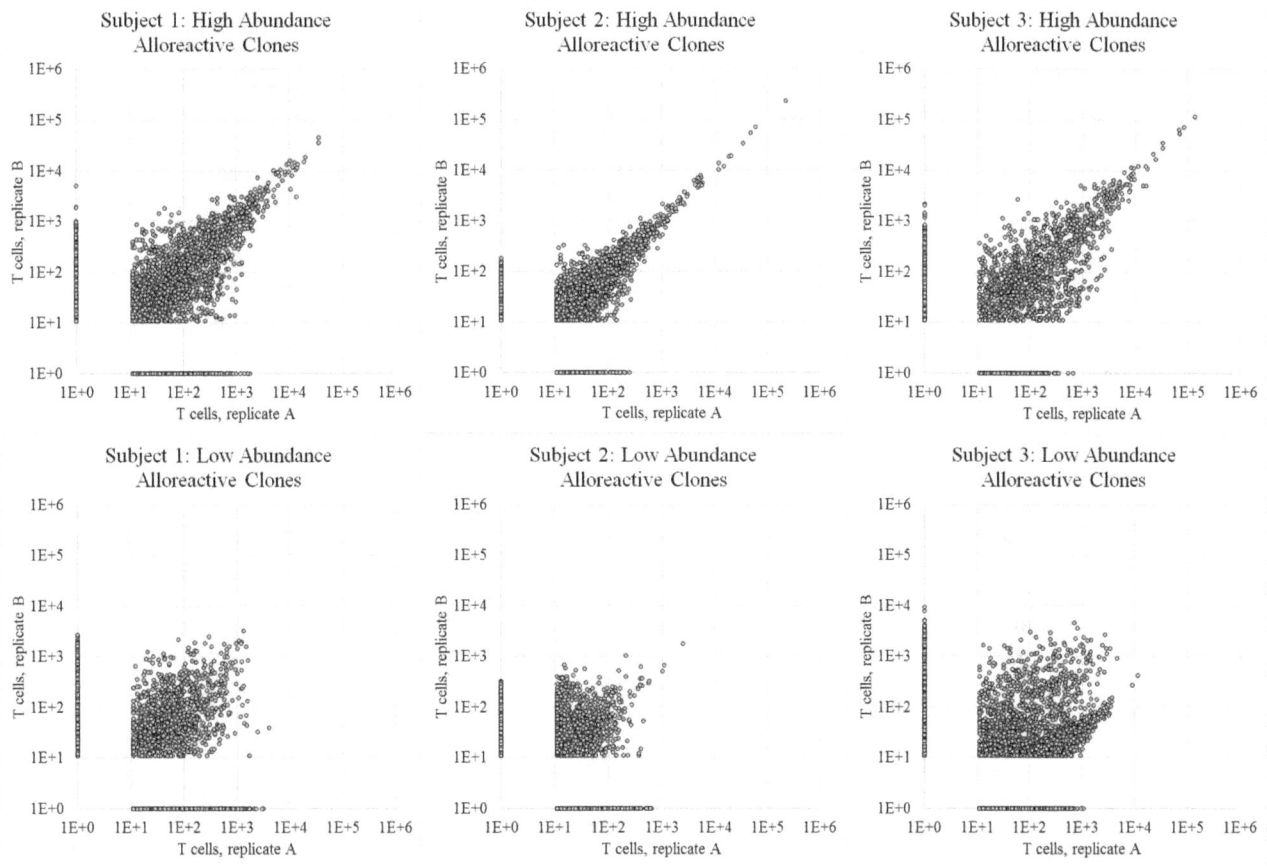

Figure 3. T cell clonal frequency among biological replicates of mixed lymphocyte culture. Above are six scatter plots showing the number of T cells bearing each unique CDR3 sequence in replicate mixed lymphocyte culture experiments performed on three pairs of healthy adult subjects. Each column corresponds to one pair of subjects; the top row of plots show T cell clones that were previously observed in a pre-MLR sample of peripheral T cells (high-abundance), and the bottom row of plots show T cell clones unobserved in a pre-MLR sample of peripheral T cells (low-abundance). Each point represents a unique T cell clone, and points are plotted at (# of observed T cells +1), so that clones unobserved in one sample are plotted on the axes.

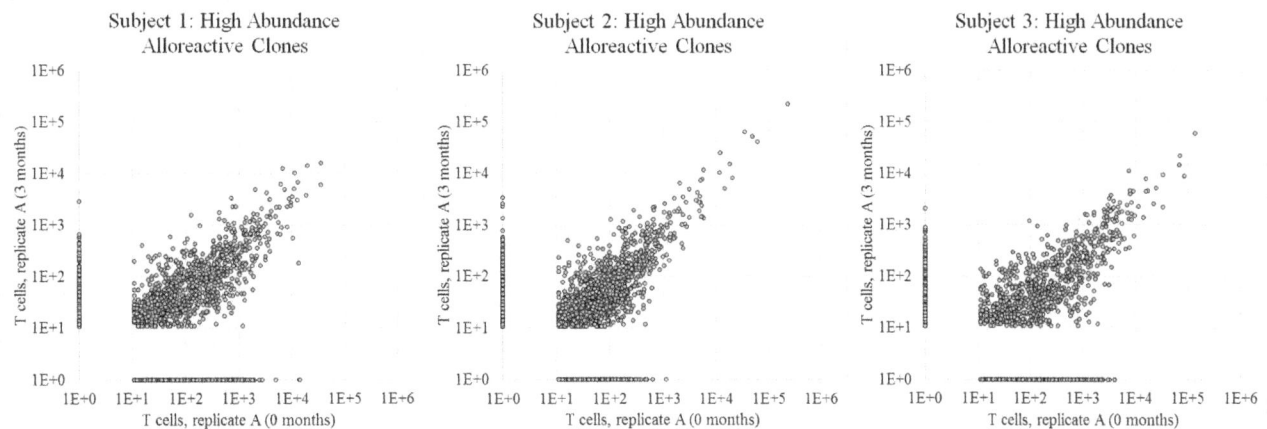

Figure 4. T cell clonal frequency among temporal replicates of mixed lymphocyte culture. Above are three scatter plots showing the number of T cells bearing each unique CDR3 sequence in replicate mixed lymphocyte culture experiments performed three months apart on each of three pairs of healthy adult subjects. Considering only T cell clones previously observed in a pre-MLR T cell sample from each time-point and enriched at least ten-fold after mixed lymphocyte culture, each point represents a unique T cell clone and points are plotted at (# of observed T cells +1) so that clones unobserved in one sample are plotted on the axes.

Table 3. TCR overlap between biological & temporal replicate mixed lymphocyte culture experiments.

	Biological replicates (N = 2)[a]	Temporal replicates (N = 4)[b]
High-abundance pre-culture[c]		
Subject 1	0.96	0.78
Subject 2	0.98	0.93
Subject 3	0.98	0.89
Average	**0.97**	**0.87**
Low-abundance pre-culture[d]		
Subject 1	0.54	0.15
Subject 2	0.43	0.06
Subject 3	0.67	0.08
Average	**0.55**	**0.10**

[a]MLR Cultured in duplicate.
[b]MLR performed at three months apart.
[c]Present in pre-culture sample and ≥10× enriched after MLR.
[d]Unobserved in pre-culture sample and ≥10 T cells after MLR.

appreciable frequency in the peripheral T cell repertoire to begin with). Reproducibility between duplicate cell culture experiments was high among this set of abundant and highly alloreactive T cell clones (average r^2 among three subjects = 0.96), indicating that when presented with identical stimuli these clonal populations of T cells responded in a very reproducible manner.

Since our replicate cell culture experiments did not address the stability of the alloreactive T cell repertoire over time, we repeated the T cell isolation and duplicate MLR experiments with the same three pairs of subjects three months after our initial experiment. Specifically, we hypothesized that high-abundance alloreactive clones, which we presume to represent memory T cells due to their frequency in the peripheral T cell repertoire, should be stable over time and thus should remain in the alloreactive T cell compartment. Figure 4 presents the high-abundance T cell repertoire after three months in each pair of subjects. Many T cell clones identified as part of the high-abundance alloreactive T cell repertoire at baseline were observed in the high-abundance alloreactive T cell repertoire three months later, at similar clonal frequencies (Figure 4; average $r^2 = 0.78$). To quantify similarity between sets of T cells, we calculated a TCR overlap metric (the proportion of T cells belonging to clones found in both samples) [29]. Table 3 presents the TCR overlap between duplicate cell culture experiments and between experiments spaced three months apart. While duplicate cell culture experiments generated more concordant sets of alloreactive T cell clones than experiments from different time-points, overlap between different time-points was nonetheless quite high (mean overlap = 0.97 for duplicate experiments vs. 0.87 across time-points). We hypothesize that the lower overlap over time might be due to the emergence of naïve T cell clones of exceptional size which would not be expected to persist in the periphery and/or the noise in the estimation of absolute cellular abundance could have caused a subset of low-abundance clones to be erroneously classified as high-abundance in our experiment [32–34].

The low-abundance alloreactive T cell clones, however, showed lower reproducibility between duplicate cell culture experiments (Table 3, bottom) and appeared to be considerably more transient. Comparisons between biological duplicates were much more concordant than comparisons between time-points (mean overlap = 0.55 for duplicate experiments vs. 0.10 across time-points). Several hypotheses may explain why T cell clones were not

reproducibly found in the low-abundant alloreactive T cell compartment; first, the lower overlap between biological replicates is mostly due to sample error (most unique T cell lineages are at very low abundance, and we cannot reliably find a T cell clone in two biological replicates unless it is present in at least several cells); second, the even lower reproducibility after three months can be attributed to a preponderance of newly emerged naïve T cell clones among this subset; lastly, these clones may represent memory T cell populations that did not persist at detectable levels in the periphery over the intervening time [32–34].

Taken together, our TCR repertoire analysis is highly sensitive and reproducible. Further, our results indicated that a majority of the alloreactivity observed between three pairs of healthy adults was attributable to a set of several thousand T cell clones, present at reasonably high frequency in the peripheral T cell repertoire, whose alloreactive potential remained stable over at least several months. While we cannot conclusively demonstrate using TCR sequencing that the proliferating clones identified are specifically alloreactive, our screening algorithm (requiring a T cell clone to represent a 10× higher proportion of the proliferated than the fresh sample) should ensure that only a minimal number of nonspecifically-proliferating clones are identified. Likewise, we do not anticipate that our method will reliably identify all T cell clones which will react to and infiltrate an allograft. However, we expect neither tracking of some irrelevant T cell clones nor failure to track some genuinely alloreactive clones should compromise clinical utility so long as a sufficient number of truly alloreactive clones are also identified and tracked, providing a means of quantitating the host cellular immune response to the allograft.

We anticipate that application of our approach to transplantation could have a positive impact in the clinical management of patients. This is to be achieved by performing donor-specific MLR at transplant to pre-define the donor-reactive T cell repertoire, and then tracking their presence, abundance and dynamics in recipient primary tissues (e.g. peripheral blood, allograft biopsies, urine) during the post-transplant period. Such an approach will make this technology utilizable in both living donor and deceased donor transplants. We speculate that foreknowledge of the alloreactive T cell repertoire could thus be combined with post-transplant immune profiling in the recipient peripheral blood for non-invasive monitoring of cellular allograft rejection. Conversely, absence of the donor reactive clones from the post-transplant repertoire would indicate immune tolerance.

Author Contributions

Conceived and designed the experiments: ROE JMM HSR JRL. Performed the experiments: ROE JMM IMK. Analyzed the data: ROE JMM. Wrote the paper: ROE JMM.

References

1. Adams AB, Williams MA, Jones TR, Shirasugi N, Durham MM, et al. (2003) Heterologous immunity provides a potent barrier to transplantation tolerance. J Clin Invest 111: 1887–1895.

2. Brook MO, Wood KJ, Jones ND (2006) The impact of memory T cells on rejection and the induction of tolerance. Transplantation 82: 1–9.

3. D'Orsogna LJ, Roelen DL, Doxiadis, II, Claas FH (2012) TCR cross-reactivity and allorecognition: new insights into the immunogenetics of allorecognition. Immunogenetics 64: 77–85.

4. Burrows SR, Khanna R, Silins SL, Moss DJ (1999) The influence of antiviral T-cell responses on the alloreactive repertoire. Immunol Today 20: 203–207.

5. Welsh RM, Selin LK (2002) No one is naive: the significance of heterologous T-cell immunity. Nat Rev Immunol 2: 417–426.

6. Borbulevych OY, Piepenbrink KH, Gloor BE, Scott DR, Sommese RF, et al. (2009) T cell receptor cross-reactivity directed by antigen-dependent tuning of peptide-MHC molecular flexibility. Immunity 31: 885–896.

7. Colf LA, Bankovich AJ, Hanick NA, Bowerman NA, Jones LL, et al. (2007) How a single T cell receptor recognizes both self and foreign MHC. Cell 129: 135–146.

8. Ely LK, Burrows SR, Purcell AW, Rossjohn J, McCluskey J (2008) T-cells behaving badly: structural insights into alloreactivity and autoimmunity. Curr Opin Immunol 20: 575–580.

9. Griesemer AD, Sorenson EC, Hardy MA (2010) The role of the thymus in tolerance. Transplantation 90: 465–474.

10. Macedo C, Orkis EA, Popescu I, Elinoff BD, Zeevi A, et al. (2009) Contribution of naive and memory T-cell populations to the human alloimmune response. Am J Transplant 9: 2057–2066.

11. Marrack P, Kappler J (1988) T cells can distinguish between allogeneic major histocompatibility complex products on different cell types. Nature 332: 840–843.

12. Schild H, Rotzschke O, Kalbacher H, Rammensee HG (1990) Limit of T cell tolerance to self proteins by peptide presentation. Science 247: 1587–1589.

13. Argaet VP, Schmidt CW, Burrows SR, Silins SL, Kurilla MG, et al. (1994) Dominant selection of an invariant T cell antigen receptor in response to persistent infection by Epstein-Barr virus. J Exp Med 180: 2335–2340.

14. Burrows SR, Khanna R, Burrows JM, Moss DJ (1994) An alloresponse in humans is dominated by cytotoxic T lymphocytes (CTL) cross-reactive with a single Epstein-Barr virus CTL epitope: implications for graft-versus-host disease. J Exp Med 179: 1155–1161.

15. Burrows SR, Silins SL, Moss DJ, Khanna R, Misko IS, et al. (1995) T cell receptor repertoire for a viral epitope in humans is diversified by tolerance to a background major histocompatibility complex antigen. J Exp Med 182: 1703–1715.

16. Gras S, Burrows SR, Kjer-Nielsen L, Clements CS, Liu YC, et al. (2009) The shaping of T cell receptor recognition by self-tolerance. Immunity 30: 193–203.

17. Mifsud NA, Nguyen TH, Tait BD, Kotsimbos TC (2010) Quantitative and functional diversity of cross-reactive EBV-specific CD8+ T cells in a longitudinal study cohort of lung transplant recipients. Transplantation 90: 1439–1449.

18. Amir AL, D'Orsogna LJ, Roelen DL, van Loenen MM, Hagedoorn RS, et al. (2010) Allo-HLA reactivity of virus-specific memory T cells is common. Blood 115: 3146–3157.

19. Danziger-Isakov L, Cherkassky L, Siegel H, McManamon M, Kramer K, et al. (2010) Effects of influenza immunization on humoral and cellular alloreactivity in humans. Transplantation 89: 838–844.

20. D'Orsogna LJ, Roelen DL, Doxiadis, II, Claas FH (2011) Screening of viral specific T-cell lines for HLA alloreactivity prior to adoptive immunotherapy may prevent GvHD. Transpl Immunol 24: 141.

21. D'Orsogna LJ, van Besouw NM, van der Meer-Prins EM, van der Pol P, Franke-van Dijk M, et al. (2011) Vaccine-induced allo-HLA-reactive memory T cells in a kidney transplantation candidate. Transplantation 91: 645–651.

22. Wang T, Chen L, Ahmed E, Ma L, Yin D, et al. (2008) Prevention of allograft tolerance by bacterial infection with Listeria monocytogenes. J Immunol 180: 5991–5999.

23. Bain B, Lowenstein L (1964) Genetic Studies On The Mixed Leukocyte Reaction. Science 145: 1315–1316.

24. Bain B, Vas MR, Lowenstein L (1964) The Development Of Large Immature Mononuclear Cells In Mixed Leukocyte Cultures. Blood 23: 108–116.

25. Levitsky J, Leventhal JR, Miller J, Huang X, Chen L, et al. (2012) Favorable effects of alemtuzumab on allospecific regulatory T-cell generation. Human Immunology 73: 141–149.

26. Levitsky J, Miller J, Huang X, Chandrasekaran D, Chen L, et al. (2013) Inhibitory Effects of Belatacept on Allospecific Regulatory T-Cell Generation in Humans. Transplantation 96: 689–696 610.1097/TP.1090b1013e31829f31607.

27. Mathew JM, Garcia-Morales RO, Carreno M, Jin Y, Fuller L, et al. (2003) Immune responses and their regulation by donor bone marrow cells in clinical organ transplantation. Transplant Immunology 11: 307–321.

28. Yousfi Monod M, Giudicelli V, Chaume D, Lefranc MP (2004) IMGT/JunctionAnalysis: the first tool for the analysis of the immunoglobulin and T cell receptor complex V-J and V-D-J JUNCTIONs. Bioinformatics 20 Suppl 1: i379–385.

29. Emerson RO, Sherwood AM, Rieder MJ, Guenthoer J, Williamson DW, et al. (2013) High-throughput sequencing of T-cell receptors reveals a homogeneous repertoire of tumour-infiltrating lymphocytes in ovarian cancer. J Pathol 231: 433–440.

30. Robins HS, Campregher PV, Srivastava SK, Wacher A, Turtle CJ, et al. (2009) Comprehensive assessment of T-cell receptor beta-chain diversity in alphabeta T cells. Blood 114: 4099–4107.

31. Carlson CS, Emerson RO, Sherwood AM, Desmarais C, Chung MW, et al. (2013) Using synthetic templates to design an unbiased multiplex PCR assay. Nat Commun 4: 2680.

32. Jenkins MK, Chu HH, McLachlan JB, Moon JJ (2010) On the composition of the preimmune repertoire of T cells specific for Peptide-major histocompatibility complex ligands. Annu Rev Immunol 28: 275–294.

33. Jenkins MK, Moon JJ (2012) The role of naive T cell precursor frequency and recruitment in dictating immune response magnitude. J Immunol 188: 4135–4140.

34. Surh CD, Sprent J (2008) Homeostasis of naive and memory T cells. Immunity 29: 848–862.

8

The Aggregation of Four Reconstructed Zygotes is the Limit to Improve the Developmental Competence of Cloned Equine Embryos

Andrés Gambini[1,2], **Adrian De Stefano**[1], **Romina Jimena Bevacqua**[1,2], **Florencia Karlanian**[1], **Daniel Felipe Salamone**[1,2*]

1 Laboratory of Animal Biotechnology, Faculty of Agriculture, University of Buenos Aires, Buenos Aires, Argentina, **2** National Institute of Scientific and Technological Research, Buenos Aires, Argentina

Abstract

Embryo aggregation has been demonstrated to improve cloning efficiency in mammals. However, since no more than three embryos have been used for aggregation, the effect of using a larger number of cloned zygotes is unknown. Therefore, the goal of the present study was to determine whether increased numbers of cloned aggregated zygotes results in improved *in vitro* and *in vivo* embryo development in the equine. Zona-free reconstructed embryos (ZFRE's) were cultured in the well of the well system in four different experimental groups: I. 1x, only one ZFRE per microwell; II. 3x, three per microwell; III. 4x, four per microwell; and IV. 5x, five ZFRE's per microwell. Embryo size was measured on day 7, after which blastocysts from each experimental group were either a) maintained in culture from day 8 until day 16 to follow their growth rates, b) fixed to measure DNA fragmentation using the TUNEL assay, or c) transferred to synchronized mares. A higher blastocyst rate was observed on day 7 in the 4x group than in the 5x group. Non-aggregated embryos were smaller on day 8 compared to those aggregated, but from then on the *in vitro* growth was not different among experimental groups. Apoptotic cells averaged 10% of total cells of day 8 blastocysts, independently of embryo aggregation. Only pregnancies resulting from the aggregation of up to four embryos per microwell went beyond the fifth month of gestation, and two of these pregnancies, derived from experimental groups 3x and 4x, resulted in live cloned foals. In summary, we showed that the *in vitro* and *in vivo* development of cloned zona-free embryos improved until the aggregation of four zygotes and declined when five reconstructed zygotes were aggregated.

Editor: Jason Glenn Knott, Michigan State University, United States of America

Funding: This work was partially funded by SIDUS Company. The funders had no role in study design, data collection and analysis, decision to publish, or preparation of the manuscript.

Competing Interests: The authors received funding from SIDUS Company.

* Email: mailto:salamone@agro.uba.ar

Introduction

To date, many equine clones have been reported; however, cloning efficiency remains low [1–9]. Research is hampered in this species by the limited number of slaughterhouses and low recovery rates of oocytes by transvaginal aspiration.

As a means to improve cloning efficiency, the strategy of embryo aggregation has been applied in several species. These include the mouse [10], [11], bovine [12–14], pig [15] and horse [8]. These studies have reported benefits of embryo aggregation for *in vitro* and/or *in vivo* embryo development. In addition, embryo aggregation has been successfully used for chimera production [16–23], and to improve the establishment of parthenogenetic stem cells and the expression of imprinted genes [24].

The aggregation of two or three embryos at the onset of cloned embryo development could compensate for epigenetic defects of individual cells with different reprogramming status. This appears to be one reason for the improved developmental competence of aggregated embryos [10,25]. Furthermore, despite the fact that aggregated embryos are larger on day 7, after day 8 they are no different from non-aggregated embryos [8]. To elucidate this intriguing fact, mechanisms such as apoptosis must be studied. Apoptosis is a cellular mechanism that controls cell numbers during embryonic development and levels of apoptotic cells can be used as an indicator of embryo quality [26–30]. To date, apoptosis levels have been measured in the horse by the TUNEL assay for both ICSI embryos and for those produced *in vivo* [31], [32].

The aim of this work was to determine whether the number of aggregated zygotes changes the aggregation strategy efficiency. Up to five equine clones were aggregated and the *in vitro* and *in vivo* embryo development were evaluated. In addition, the effect of aggregation on embryo quality was measured by evaluating blastocyst size, DNA fragmentation levels, *in vitro* embryo growth beyond day 8 and the establishment of pregnancies and cloned foal production.

Materials and Methods

Chemicals

Except otherwise indicated, all chemicals were obtained from Sigma Chemicals Company (St. Louis, MO, USA).

Animal Welfare

All the research protocols were in accordance with the recommendations of the guidelines stated in the Guide for the Care and Use of Agricultural Animals in Agricultural Research and Teaching. The study design was approved by the Ethics and Animal Welfare Committee for the Faculty of Agriculture, University of Buenos Aires under number CEyBAFAUBA2014/1. All efforts were made to minimize animal suffering. All animals were housed at "Don Antonio" equine center in Buenos Aires, Argentina. Trained people provided daily care and feeding, and horses had permanent *ad-libitum* access to water. Recipient palpations, ultrasounds, hormone treatments and embryo transfer procedures were always performed by trained veterinarians.

Cell culture

Fibroblasts were obtained by culture of skin biopsies from an Argentinean Polo Pony (donor cell A) and a Show-jumping horse (donor cell B). They were cultured in Dulbecco's modified Eagle's Medium (DMEM; 11885, Gibco, Grand Island, NY, USA) supplemented with 10% fetal bovine serum (FBS; 10499-044, Gibco), 1% antibiotic–antimycotic (ATB; 15240-096, Gibco), and 1 µl/ml insulin-transferrin-selenium (ITS; 51300-044, Gibco) in 6.5% CO_2 in humidified air at 39°C. After establishment of the primary culture, fibroblasts were expanded, frozen in DMEM with 20% FBS and 10% DMSO, and stored in liquid nitrogen. Donor cells were induced into quiescence by being grown to confluence. Cells were trypsinized before use and resuspended in TALP-H with 10% FBS.

Oocyte collection and *in vitro* maturation

Slaughterhouse ovaries were collected and transported to the laboratory within 4–7 h, at 26–28°C. Equine oocyte recovery was performed by a combination of scraping and washing of all visible follicles using a syringe filled with DMEM/Nutrient Mixture F-12 medium (DMEM/F12; D8062), supplemented with 20 IU mL-1 heparin (H3149). Oocytes were matured for 24–26 h in 100 µl microdrops of bicarbonate-buffered TCM-199 (31100-035; Gibco) supplemented with 10% FBS, 2.5 µL/mL ITS, 1 mM sodium pyruvate (P2256), 100 mM cysteamine (M9768), 0.1 mg/mL of follicle-stimulating hormone (NIH-FSH-P1, Folltropin; Bioniche, Belleville, ON, Canada) and 1% ATB, under mineral oil (M8410). Maturation conditions were 6.5% CO_2 in humidified air at 39°C.

Cumulus and zona pellucida removal

Cumulus cells were removed by a combined treatment of pipeting oocytes in 0.05% Trypsin-EDTA (25300, Gibco) and vortexing them for 2 minutes in hyaluronidase [H4272; 1 mg/mL in Hepes-buffered Tyrodes medium containing albumin, lactate and pyruvate (TALP-H)]. Oocytes were individually observed under stereoscopic microscopy to confirm the presence of the first polar body.

In order to prepare the metaphase II oocytes for enucleation, the zona pellucida was removed by incubating oocytes for 3–6 min in 1.5 mg/ml pronase (P8811) in TALP-H on a warm plate. Zona-free oocytes (ZF-oocytes) were rinsed in TALP-H and placed in a microdrop of Synthetic Oviductal Fluid (SOF), supplemented with 2.5% FBS and 1% ATB, until enucleation.

Oocyte enucleation

Aspiration of the metaphase plate was performed in a microdrop of TALP-H containing 0.5 µg/ml of cytochalasin B (C6762). A blunt pipette was used for the aspiration, and a closed holding pipette to support the oocyte during the procedure. In order to observe the metaphase plate under UV light, ZF-oocytes were incubated (5 min), prior to enucleation, in a microdrop of SOF containing 1 µg/mL Hoechst bisbenzimide 33342 (H33342). Zona-free enucleated oocytes (ZFE-oocytes) were kept in a SOF microdrop until nuclear transfer.

Nuclear transfer and cloned embryo reconstruction

Zona-free enucleated oocytes were individually washed for a few seconds in 50 µl drops of 1 mg/ml phytohemagglutinin (L8754) dissolved in TCM-Hepes, and then dropped over a donor cell resting on the bottom of a 100 µl TALP-H drop; consequently these two structures were attached. Formed cell couplets were washed in fusion medium [0.3 M mannitol (M9546), 0.1 mM $MgSO_4$ (M7506), 0.05 mM $CaCl_2$ (C7902), 1 mg/ml polyvinyl alcohol (P8136)], and then fused in a fusion chamber containing 2 ml of warm fusion medium. A double direct current pulse of 1.2 kV/cm V, each pulse for 30 µs, 0.1 s apart was utilized for fusion. Couplets were individually placed in a 10 µl drop of SOF medium supplemented with 2.5% FBS and incubated under mineral oil, at 39°C in 5% CO_2 in air. Twenty minutes after the first round of fusion, non-fused couplets were re-fused.

Chemical activation

Two hours after the first round of fusion, zona-free reconstructed embryos (ZFRE's) were subjected to chemical activation. Chemical activation was achieved by a 4 min treatment in TALP-H containing 8.7 mM ionomycin (I24222; Invitrogen, Carlsbad, CA, USA) followed by a 4 h individual culture in a 5 µl drop of SOF supplemented with 1 mM 6-dimethylaminopurine (D2629) and 5 mg/ml cycloheximide (C7698).

In vitro embryo culture until day 8 and embryo aggregation

In vitro culture of ZFRE's was carried out in microwells containing 50 µl microdrops of DMEM/F12 medium under mineral oil. These microwells were produced using a heated glass capillary lightly pressed to the bottom of a 35 x 10 mm Petri dish. Four different experimental groups were set up according to the number of ZFRE's placed per each microwell: I. Group **1x**: one ZFRE per microwell (non aggregated embryos), II. Group **3x**: three ZFRE's per microwell, III. Group **4x**: four ZFRE's per microwell, IV. Group **5x**: five ZFRE's per microwell. Culture conditions were 5% O_2, 5% CO_2 and 90% N_2 in a humidified atmosphere at 38.5°C. Half of the medium was renewed on Day 3, with DMEM/F-12 HAM medium containing 10% FBS, and 1% ATB. A similar ratio of ZFRE's/culture medium was maintained for all experimental groups. Cleavage was assessed 72 h after activation, and rates of blastocyst formation and their diameter were recorded at Day 7 and Day 8 when the embryos were either fixed for TUNEL assay, maintained in *in vitro* culture or transferred to synchronized mares.

In vitro embryo culture beyond day 8

Fourteen derived B donor cell blastocysts from all experimental groups were kept in *in vitro* culture from day 8 until day 16–17 unless they collapsed earlier. One blastocyst from the 4x experimental group was found collapsed on day 15 and was not included for apoptosis analysis. On day 12, blastocysts were placed in a fresh 100 µl microdrop of DMEM/F12 medium containing 15% FBS and 1% ATB. Blastocyst diameters were measured daily using a millimeter eyepiece. At day 16, embryos were fixed for TUNEL assay.

Embryo fixing and TUNEL assay

DNA fragmentation was evaluated using the DeadEnd Fluorometric TUNEL System (Promega G3250, Madison, WI, USA). Embryos were fixed in 4% paraformaldehyde in DPBS, washed in BSA (A7906) solution (1 mg BSA/ml DPBS), permeabilized with 0.5% Triton X-100 in DPBS for 15 min at room temperature, and rinsed again in BSA solution. After three washes, embryos were incubated in the dark for 2 h at $39°C$ in a buffer consisting of equilibration buffer and a nucleotide mix containing fluorescein-dUTP and terminal deoxynucleotidyl transferase. Negative controls lacked the terminal deoxynucleotidyl transferase. The nuclei were counterstained with 0.5% propidium iodide for 30 min at room temperature. Embryos were washed in BSA solution and mounted on a glass slide in 70% v/v glycerol under a coverslip. Embryos were analyzed on a Nikon Confocal C.1 scanning laser microscope. An excitation wavelength of 488 nm was selected for detection of fluorescein-12-dUTP and a 544 nm wavelength to excite propidium iodide. Images of serial optical sections were recorded every 1.5–2 µm vertical step along the Z-axis of each embryo. Three-dimensional images were constructed using EZ-C1 3.9 software (Nikon Corporation, Japan). Total cell numbers and DNA-fragmented nuclei were counted manually for day 8 embryos. Due to the large number of cells of day 16 embryos, cells of five different areas of each day 16 blastocyst were counted using the Image J software (1.47 version, Wayne Rasband National Institutes of Health, USA).

Embryo transfer and production of cloned foals

Blastocyst transfers to recipients were performed during the breeding season. Mares aged 3 to 10 years were examined 2–3 times/week by transrectal ultrasound (5MHz linear probe, Aloka 500) to determine the phase of their estrous cycle. Prostaglandin F2 Alfa (Ciclase, Sintex, Buenos Aires, Argentina) and human chorionic gonadotropin (Ovusyn, Sintex, Buenos Aires, Argentina) were used to synchronize the day of ovulation. Transcervical embryo transfer was performed 5 to 7 days after ovulation, with one or two day 7 blastocysts. Blastocysts were transported in a 0.5 cc straw containing DMEM/F12, and the shipping container was held at $36°C$ for the 3 h transportation interval. Pregnancies were diagnosed by transrectal ultrasound 15 days after ovulation. At day 300 of gestation pregnant mares were moved to an equine hospital (KAWELL, Equine Rehabilitation Center, Solís, Argentina) where they were monitored until parturition.

DNA comparison

The cloned foals were confirmed by an external laboratory (Laboratorio de Genética Aplicada de la Sociedad Rural Argentina, ISAG code 84535). Twenty eight loci were compared using hair samples from each foal and its respective donor animal.

Statistical analysis

Differences among treatments in each experiment were determined using GraphPad Prism software version 5. Blastocyst rates, embryo size and pregnancy rates were analyzed by Chi-square or Fisher's exact test. TUNEL-positive cells were evaluated with the Kruskal-Wallis non parametric test and Dunn's post test. The effect of treatment on in vitro embryo growth rates was assessed by one-way within subjects (repeated measures) analysis of variance. Multiple observations of embryo growth rates in the same experimental units (embryo/days) constituted the within subject factor. Post-hoc pairwise comparisons of mean growing rates/days were performed by the Tukey Honestly Significant Differences test.

Results

In vitro embryo development of aggregated cloned equine embryos until day 8

A total of 765 ZFRE's were produced and cultured in vitro in four different experimental groups. Cleavage and blastocyst rates on day 7 and day 8 per embryo (microwell) and per ZFRE were recorded for all experimental groups (**Table 1**). A significant improvement of blastocyst rates per embryo was observed on day 7 when numbers of aggregated zygotes were up to 4/well. Furthermore, aggregation did not involve the use of additional oocytes to obtain blastocysts, since no significant differences from the control group were observed on day 7. On day 7, there were fewer blastocysts in the 5x group per number of ZFRE's compared to the 4x experimental group but no significant differences were found from the control group. The aggregation of four zygotes resulted in the best rate of in vitro embryo development, whereas when five reconstructed zygotes were aggregated embryo development rates decreased. Additional data are available in Table S1 and Table S2 showing the in vitro cloned equine embryo developmental competence per somatic donor cell. Embryo aggregation showed a similar effect between donor cells. Diameters of day 7 blastocysts are shown in **Table 2**. Cloned blastocyst size was smaller when no embryo aggregation was used. During zona-free embryo development some blastomeres were observed to be sloughed off into the microwell. This situation was observed in all experimental groups.

In vitro development of cloned blastocysts beyond day 8

In vitro growth rates of 14 cloned blastocysts (donor cell B) were determined daily. Growth patterns were similar between aggregated and non-aggregated groups. The number of embryos analyzed per group were: 1x: (n = 3), 3x: (n = 4), 4x: (n = 4) and 5x (n = 3). Each embryo within each experimental group derived from different replicates. Mean embryo sizes per day ±SD were: Day 8, 130.12 µm ±32.88; Day 9, 219.12 µm ±76.32; Day 11, 489.44 µm ±265.32; Day 12, 713.72 µm ±391.91; Day 13, 954.55 µm ±327.34; Day 14, 1664.87 µm ±713.54; Day 15, 2212.20 µm ±783.32 and day 16, 2677.18 µm ±977.43. **Figure 1** shows an equine zona-free cloned blastocyst from the 3x experimental group during in vitro embryo culture.

DNA fragmentation levels in cloned equine aggregated embryos on day 8 and day 16 by TUNEL assay

DNA fragmentation levels (mean ± SEM) of day 8 and day 16 cloned embryos of all experimental groups are shown in **Table 3**. There were no differences between groups, on day 8, in the levels of fragmented DNA, with an average of 10% positive TUNEL cells seen in all embryos (**Figure 2**). TUNEL-positive cells in day 8 embryos were: Group 1x (n = 6): 9.26%, 10.26%, 12.41%, 13.04%, 17.39% and 25.58%; Group 3x (n = 3): 8.97%, 9.43% and 10.29%; Group 4x (n = 5): 3.62%, 9.04%, 10.61%, 11.63% and 11.84%; and Group 5x (n = 2): 8.86% and 11.16%. There were no differences between groups on day 16 in the levels of fragmented DNA, with an average of 2.5% positive TUNEL cells seen in all cloned embryos (**Figure 3**). TUNEL-positive cell percentage of total cells counted per day 16 embryos and experimental group were: Group 1x (n = 3): 4.23%, 3.64% and

Table 1. Effects of equine cloned embryo aggregation on *in vitro* development until day 8.

Experimental groups	No. of ZFRE's (%)	No. of embryos (well)	No. of cleaved (%)	Blastocyst production					
				Day 7			Day 8		
				No.	% per Embryo	% per ZFRE	No.	% per Embryo	% per ZFRE
1x	131	131	99 (75.57)	13	9.92[a]	9.92[ac]	20	15.27[a]	15.27
3x	228	76	193 (85.4)	26	34.21[b]	11.40[ac]	40	52.63[b]	17.54
4x	292	73	229 (78.16)	42	57.53[c]	14.38[bc]	52	71.23[b]	17.80
5x	115	23	90 (78.26)	7	30.43[b]	6.08[a]	15	65.22[b]	13.04
Total	**765**	**303**	**611 (79.87)**	**88**	**29.04**	**11.50**	**127**	**41.91**	**16.60**

Values with different superscripts in a column are significantly different (Chi-square Test P<0.05) (*a, b, c*). ZFRE's: Zona-free reconstructed embryos.

4.42%; Group 3x (n = 2): 3.35% and 0.33%; Group 4x (n = 3): 0.15%, 0.88% and 2.03%; and Group 5x (n = 3): 1.80%, 2.91% and 4.79%. Additional data are available in Figure S1.

In vivo development of aggregated cloned equine embryos

Embryo transfer, pregnancy and survival rates for all experimental groups are shown in **Table 4**. Early pregnancy rates were higher when aggregated embryos were transferred; however, no statistical differences were found in pregnancy rates between non-aggregated and aggregated groups. Only pregnancies resulting from the aggregation of up to four embryos per well survived beyond the fifth month of gestation. One of the pregnant mares from the 3x experimental group showed clinical signs of Equine Metabolic Syndrome, dying in the last month of gestation. The cloned fetus presented normal vital parameters until the death of the mare. The two cloned foals obtained in this study derived from experimental groups 3x and 4x (**Figure 4**). Their gestation times were normal.

Both foals needed neonatology assistance and they responded positively to treatment (oxygen, antibiotics and parental nutrition). The cloned foal derived from experimental group 3x presented a high degree of angular and flexural forelimb deformities.

Discussion

This study analyzed the effect of an increase in the numbers of aggregated zygotes on *in vitro* and *in vivo* cloned embryo development in the equine. Embryo aggregation has previously proved to enhance the efficiency of cloning. However, to date, studies on mammalian cloned embryo aggregation have focused on determining the effect of the aggregation of a maximum of three zygotes on *in vitro* and *in vivo* embryo production [8], [10–14], [25], [33–36].

In vitro embryo development of aggregated embryos until day 8

The aggregation of three and four embryos per microwell improved blastocyst rates on a per embryo basis. Unexpectedly, blastocyst development in the 5x experimental group at day 7 was not improved over that of the 4x experimental group, and blastocyst rates were similar to those obtained in the 3x experimental group. Furthermore, as we have previously demonstrated [8], embryo aggregation also enhanced blastocyst diameters on day 7. If the positive effects of embryo aggregation are due to an epigenetic compensation and/or an increase in embryo cell number [10], [13], [15], [25], then an increase in the number of aggregated zygotes should correlate to an improvement in *in vitro* embryo development. Some of the reported benefits of embryo aggregation are related to a normalization of gene expression and developmental potential [10], [11], [37], an increase in cell number [10], [12], [33] and a consequent improvement in blastocyst development rates [10], [13], [33]. Additionally, the lower results of the 5x experimental group indicate that these benefits are related to the number of aggregated zygotes. Possible mechanisms to explain this could be associated to an altered microenvironment inside the microwell, limited capability of the embryo to incorporate cells during embryo development or to alterations in cell cycle or cell death.

Each microwell provides a particular microenvironment reported to be beneficial for the embryo [38–40]. Embryo aggregation may induce modifications in this microenvironment leading to positive or negative effects depending on the number of reconstructed zygotes placed per microwell. In addition, the effect

Table 2. Effects of equine cloned embryo aggregation on *in vitro* embryo size at day 7.

Experimental groups	No. blastocyst	Blastocyst diameter				
		80–119 μm (%)	120–169 μm (%)	180–219 μm (%)	230–269 μm (%)	≥270 μm (%)
1x	13	8 (61.54)[a]	4 (30.77)[ac]	1 (7.69)[a]	0 (0)[a]	0 (0)[a]
3x	25	6 (24.00)[b]	4 (16.00)[a]	11 (44.00)[b]	3 (12.00)[a]	1 (4.00)[a]
4x	35	10 (28.57)[b]	7 (20.00)[a]	14 (40.00)[b]	3 (8.57)[a]	1(2.86)[a]
5x	7	0 (0)[b]	5 (71.43)[bc]	1 (14.29)[ab]	1 (14.29)[a]	0 (0)[a]
Total	80	26 (32.50)	20 (25.0)	25 (31.25)	7 (8.75)	2 (2.50)

Values with different superscripts in a column are significantly different (Fisher's exact test P<0.05) (*a, b*).

of embryo culture density has been reported to alter embryo development [40], [41]; furthermore in the bovine, the distance between individual embryos in culture has been shown to influence preimplantation development [42]. Therefore, placing an excessive number of embryos per microwell could negatively affect the microenvironment of the microwell and consequently the developmental capability of aggregated embryos. In addition, when aggregation is performed, the individual capability of each ZFRE to produce a blastocyst is lost. The maximum blastocyst rate per ZFRE when 5x aggregation is performed is 20%. As this maximum value was not reached, we do not consider this a reason

for the reduced developmental competence of this experimental group.

The high number of initial reconstructed zygotes of the 5x group could impact negatively on embryo development. It has been suggested that cell proliferation is a competitive process that allows recognition and elimination of defective cells during the early stages of development [43]. However, this situation could be altered with an excessive number of embryo cells. Furthermore, the developmental kinetics of *in vitro*-produced embryos is related to their developmental competence [44], and can be affected by the volume of the cytoplasm of the initial oocyte [45]. Even though

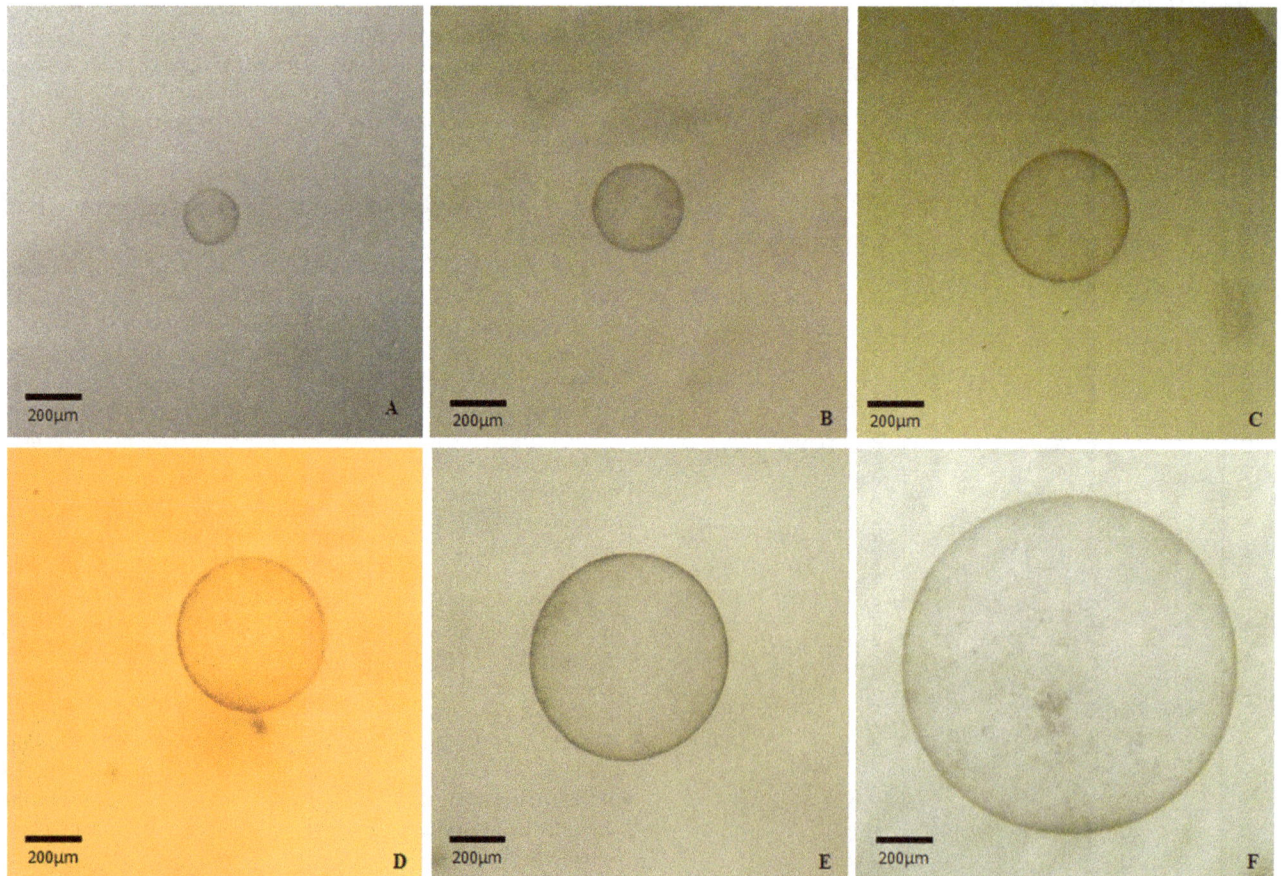

Figure 1. Photographs of an equine cloned aggregated blastocyst in *in vitro* embryo culture beyond day 8. An equine cloned zona free blastocyst placed in a 100 μl drop of DMEM/F12 medium derived from the experimental group 3x during *in vitro* embryo culture from day 8 until day 16. (**A**) Day 9, 195.21 μm. (**B**) Day 10, 314.28 μm. (**C**) Day 11 444.75 μm. (**D**) Day 12, 498.76 μm. (**E**) Day 13, 672.17 μm. (**F**) Day 14, 1127.98 μm.

Table 3. Evaluation of DNA fragmentation levels in equine aggregated and non-aggregated cloned blastocysts at day 8 and day 16.

Blastocysts	Experimental group	No.	TUNEL+ cells (Mean ±SEM)	Evaluated cells (Mean ±SEM)	TUNEL+/evaluated cells (Mean ±SEM)
Day 8	1x	6	26.83±3.99	217.3±51.92	14.64±2.45
	3x	3	21.80±5.70	228.4±52.88	9.37±0.25
	4x	5	39.20±8.38	427.8±65.28	9.34±1.51
	5x	2	25.00±1.00	252.0±19.00	10.01±1.15
Day 16	1x	3	38.67±15.30	963.7±383.6	4.09±0.23
	3x	2	58.00±54.00	2163±962.5	1.83±1.5
	4x	3	48.67±22.38	4618±1207	1.02±0.54
	5x	3	56.33±12.20	2038±788.0	3.16±0.87

No significant differences were detected within blastocyst day (Kruskal-Wallis test P<0.05).

embryo aggregation does not imply an increase in the cytoplasm-nucleus ratio, an increased embryo volume could also alter embryo developmental kinetics and the developmental competence of aggregated embryos. In addition, a study suggested that the length of the cell cycle can be regulated in the early post-implantation mouse chimeric embryo in order to compensate for increased pre-implantation cell numbers induced by aggregation [46].

In vitro embryo development beyond day 8

Day 8 cultured blastocysts expanded and their cell numbers increased; however, no statistical differences were observed among experimental groups. Previous studies in mice revealed that size regulation occurred in aggregated chimeric embryos during the early post-implantation stage [46]. Moreover, despite the abnormal proportions of ICM and trophectoderm in 4x aggregated chimeric mice blastocysts, the proportions of the tissues derived from them is already normal by day 5 [47]. Our observations together with those previously reported [8] support the notion that size regulation of aggregated cloned equine embryos occurs in the pre-implantation stages.

In vitro embryo development after day 8 allows the study of embryo developmental competence in a controlled environment avoiding the disadvantages associated with embryo transfer. Among domestic animals, in vitro development of embryos after reaching the blastocyst stage has been studied in the bovine [48–54] and porcine [51]. In the present study, the zona-free cloned blastocyst maintained its spherical shape (see **Figure 1**). It has been suggested that the embryo capsule is largely responsible for maintaining the spherical shape of the conceptus in the equine after day 6 [55]. Nevertheless, a capsule could not be clearly identified by microscopy in any experimental group. Thus, if present, it must be deficient. Alterations in early embryonic coats have been reported for equine ICSI [56] and cloned embryos [8] and also for rabbit [57]. Hence, improvements in in vitro culture conditions are necessary to allow for the normal formation of early embryonic coats in this species.

DNA fragmentation levels of aggregated cloned equine embryos

Regardless of the fact that aggregated embryos began their development with more cells, DNA fragmentation levels in cloned blastocysts on day 8 were the same among experimental groups,

with an average of 10% of cells being apoptotic. Thus, at this stage of development, the beneficial effects of embryo aggregation would not be related to an anti-apoptotic phenomenon in the horse. This contrasts with recent observations in pigs that indicate that embryo aggregation has an anti-apoptotic effect due to fewer numbers of apoptotic cells [34]. Species differences in embryo physiology such as in the number of embryonic cells could be related to the effects of embryo aggregation. On the other hand, size regulation of aggregated chimeric mouse embryos was reported to be not related to cell death since more than 90% of the cells were synthesizing DNA and were presumably viable [46].

Apoptosis may have detrimental effects if either the number of apoptotic cells or the proportion of these cells are elevated [58]. In the equine, in vivo embryos recovered on day 6 did not show apoptotic cells, while 4% of cells had fragmented DNA in in vitro embryos produced by ICSI [32]. Consequently, the higher proportion of apoptotic cells observed in our study could be a reason for the lower developmental competence generally reported for in vitro-produced equine embryos. A very interesting observation was that day 16 embryos had a similar proportion of apoptotic cells to day 8 embryos independently of aggregation. Therefore, in vitro embryo culture in DMEM/F12 medium allows embryo cell proliferation without inducing apoptosis.

In vivo embryo development of aggregated embryos

In this study, viable pregnancies resulted from the aggregation of three and four reconstructed zygotes. In our previous publication, embryo aggregation increased pregnancy rates for cloned horses, being higher for 2x and even higher for 3x aggregations [8]. Embryo aggregation of up to three zygotes also improved pregnancy rates and in vivo embryo development in the mouse [10] and bovine [60]. Nevertheless, the aggregation of more than four cloned embryos did not improve pregnancy rates in the present study. These observations agree with suggestions previously reported [61] that the yield of live-born pups from chimeric aggregation of five or more embryos would be low. On the other hand, the abnormalities detected in the 3x cloned foals of the present study have been reported in 50% of cloned foals [59], indicating that angular and flexural limb deformities are not induced by zona removal or embryo aggregation.

In conclusion, the data presented in this paper indicate that cloned embryo aggregation in the equine results in increased blastocyst rates at day 7 for the aggregation of up to 4 zygotes.

PI+ Cells (red)	TUNEL+ Cells (green)	Merge

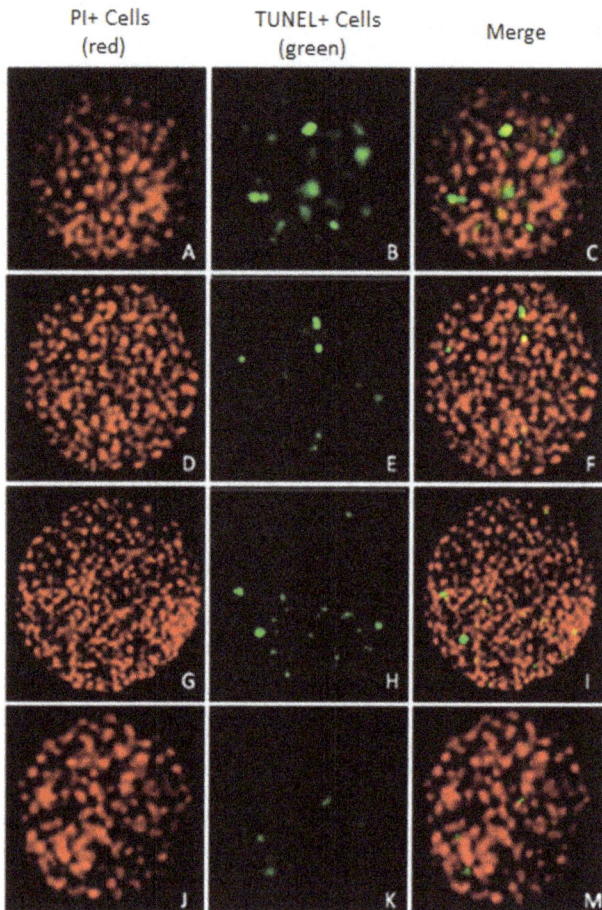

Figure 2. Photomicrographs of day 8 equine cloned embryo expression of TUNEL. (**A, B, C**) Day 8 Non-aggregated cloned equine embryo, 40x zoom. (**D, E, F**) Day 8 3x aggregated cloned embryo, 40x zoom. (**G, H, I**) Day 8 4x aggregated cloned embryo, 40x zoom. (**J, K, M**) Day 8 5x aggregated cloned embryo, 40x zoom.

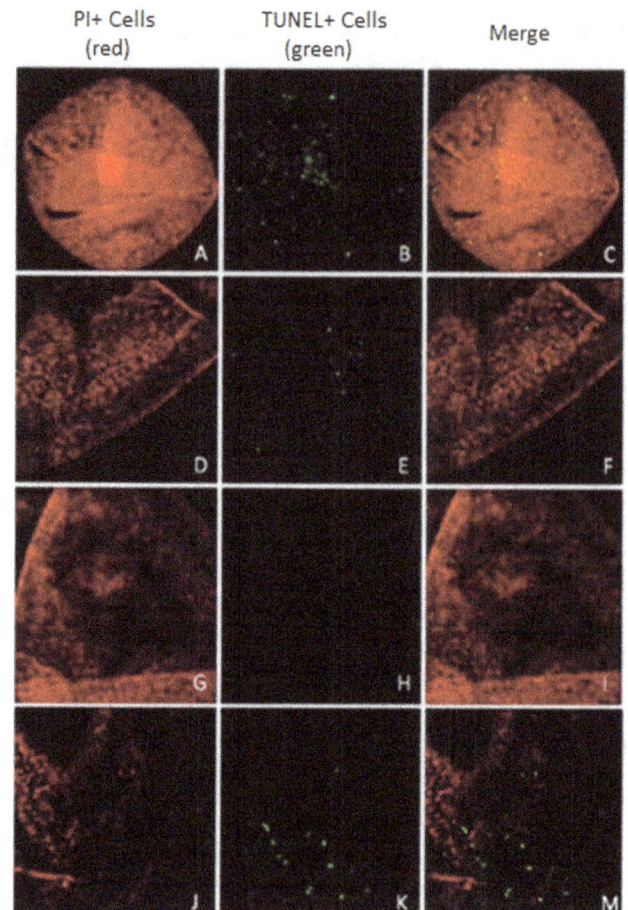

Figure 3. Photomicrographs of day 16 equine cloned embryo expression of TUNEL. (**A, B, C**) Day 16 Non-aggregated cloned equine embryo, 20x zoom. (**D, E, F**) Day 16 3x aggregated cloned embryo, 20x zoom. (**G, H, I**) Day 16 4x aggregated cloned embryo, 20x zoom. (**J, K, M**) Day 16 5x aggregated cloned embryo, 20x zoom.

Beyond four reconstructed zygotes, blastocyst rates do not continue to increase. Aggregated cloned embryos were initially larger, but *in vitro* embryo size compensated after day 8 as we have previously reported. A similar proportion of apoptotic cells was observed on day 8 in all experimental groups, and this phenomenon does not appear to be responsible for the observed compensation. Only aggregated embryos from groups 3x and 4x produced cloned offspring. This is the first report to show that the *in vitro* and *in vivo* development of cloned zona-free embryos can be improved when up to four zygotes are aggregated.

Table 4. Effects of equine cloned embryo aggregation on *in vivo* development.

Experimental Groups	No. of Recipient	Pregnant recipients (%)	Pregnancy Dynamic			No. offspring (%)
			1st month no. (%)	5th month no. (%)	8th month no. (%)	
1x	10	1 (10.00)	1 (100.00)	0 (0)	0 (0)	0 (0)
3x	17	3 (17.64)	2 (66.66)	2 (66.66)	2 (66.66)*	1 (33.33)
4x	11	2 (18.18)	1 (50.00)	1 (50.00)	1 (50.00)	1 (50.00)
5x	5	0 (0)	0 (0)	0 (0)	0 (0)	0 (0)
Total	44	6 (13.95)	4 (66.66)	3 (50.00)	3 (50.00)	2 (33.33)

No significant differences were found (Fisher's exact test P<0.05). *One pregnant mare died of Equine Metabolic Syndrome.

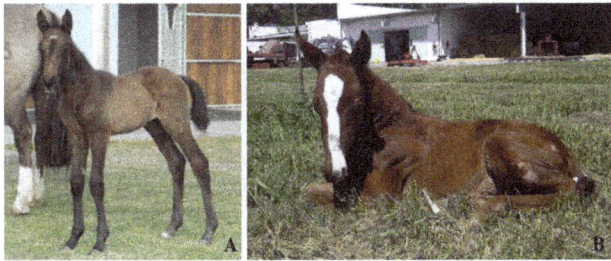

Figure 4. Photographs of equine cloned foals derived from aggregated embryos. (**A**) Equine cloned foal derived from 4x experimental group, born on the 18th of September, 2013. (**B**) Equine cloned foal derived from 3x experimental group, born on the 12th of January, 2013.

Supporting Information Legends

Figure S1   Scatter plot of TUNEL positive cells proportion in cloned equine blastocysts. (A) Day 8 blastocysts mean TUNEL-positive cells of groups 1x, 3x, 4x and 5x. (B) Day 16

References

1. Woods GL, White KL, Vanderwall DK, Li GP, Aston KI, et al. (2003) A mule cloned from fetal cells by nuclear transfer. Science 301: 1063.
2. Galli C, Lagutina I, Crotti G, Colleoni S, Turini P, et al. (2003) Pregnancy: a cloned horse born to its dam twin (letter). Nature 424: 635.
3. Lagutina I, Lazzari G, Duchi R, Colleoni S, Ponderato N, et al. (2005) Somatic cell nuclear transfer in horses: effect of oocyte morphology, embryo reconstruction method and donor cell type. Reproduction 130: 559–567.
4. Hinrichs K, Choi YH, Love CC, Chung YG, Varner DD (2006) Production of horse foals via direct injection of roscovitine-treated donor cells and activation by injection of sperm extract. Reproduction 131: 1063–1072.
5. Hinrichs K, Choi YH, Varner DD, Hartman DL (2007) Production of cloned horse foals using roscovitine-treated donor cells and activation with sperm extract and/or ionomycin. Reproduction 134: 319–325.
6. Lagutina I, Lazzari G, Duchi R, Turini P, Tessaro I, et al. (2007) Comparative aspects of somatic cell nuclear transfer with conventional and zona-free method in cattle, horse, pig and sheep. Theriogenology 67: 90–98.
7. Choi YH, Hartman DL, Fissore RA, Bedford-Guaus SJ, Hinrichs K (2009) Effect of sperm extract injection volume, injection of PLCzeta cRNA, and tissue cell line on efficiency of equine nuclear transfer. Cloning Stem Cells 11: 301–308.
8. Gambini A, Jarazo J, Olivera R, Salamone DF (2012) Equine cloning: *in vitro* and *in vivo* development of aggregated embryos. Biol Reprod.87: 15 1–9.
9. Choi YH, Norris JD, Velez IC, Jacobson CC, Hartman DL, et al. (2013) A viable foal obtained by equine somatic cell nuclear transfer using oocytes recovered from immature follicles of live mares. Theriogenology 79: 791–796.
10. Boiani M, Eckardt S, Leu NA, Scholer HR, McLaughlin KJ (2003) Pluripotency deficit in clones overcome by clone-clone aggregation: epigenetic complementation? EMBO J 22: 5304–5312.
11. Balbach ST, Esteves TC, Brink T, Gentile L, McLaughlin KJ, et al. (2010) Governing cell lineage formation in cloned mouse embryos. Dev Biol 343: 71–83.
12. Zhou W, Xiang T, Walker S, Abruzzese RV, Hwang E, et al. (2008) Aggregation of bovine cloned embryos at the four-cell stage stimulated gene expression and *in vitro* embryo development. Mol Reprod Dev 75: 1281–1289.
13. Ribeiro ES, Gerger RP, Ohlweiler LU, Ortigari I Jr, Mezzalira JC, et al. (2009) Developmental potential of bovine hand-made clone embryos reconstructed by aggregation or fusion with distinct cytoplasmic volumes. Cloning and Stem Cells 11: 377–386.
14. Akagi S, Yamaguchi D, Matsukawa K, Mizutani E, Hosoe M, et al. (2011) Developmental ability of somatic cell nuclear transferred embryos aggregated at the 8-cell stage or 16- to 32-cell stage in cattle. J Reprod Dev 57: 500–506.
15. Terashita Y, Sugimura S, Kudo Y, Amano R, Hiradate Y, et al. (2011) Improving the quality of miniature pig somatic cell nuclear transfer blastocysts: aggregation of SCNT embryos at the four-cell stage. Reprod Domest Anim 46: 189–196.
16. Hillman N, Sherman MI, Graham C (1972) The effect of spatial arrangement on cell determination during mouse development. J Embryol Exp Morphol 28: 263–278.
17. Boediono A, Suzuki T, Li LY, Godke RA (1999) Offspring born from chimeras reconstructed from parthenogenetic and in vitro fertilized bovine embryos. Mol Reprod Dev 53: 159–170.
18. Lee SG, Park CH, Choi DH, Kim HS, Ka HH, et al. (2007) *In vitro* development and cell allocation of porcine blastocysts derived by aggregation of *in vitro* fertilized embryos. Mol Reprod Dev 74: 1436–1445.
19. Yang F, Hao R, Kessler B, Brem G, Wolf E, et al. (2007) Rabbit somatic cell cloning: effects of donor cell type, histone acetylation status and chimeric embryo complementation. Reproduction 133: 219–230.
20. Ohtsuka M, Miura H, Gurumurthy CB, Kimura M, Inoko H, et al. (2012) Fluorescent transgenic mice suitable for multi-color aggregation chimera studies. Cell Tissue Res 350: 251–260.
21. He W, Kong Q, Shi Y, Xie B, Jiao M, et al. (2013) Generation and developmental characteristics of porcine tetraploid embryos and tetraploid/diploid chimeric embryos. Genomics Proteomics Bioinformatics 11: 327–333.
22. Hiriart MI, Bevacqua RJ, Canel NG, Fernández-Martin R, Salamone DF (2013) Production of chimeric embryos by aggregation of bovine egfp eight-cell stage blastomeres with two-cell fused and asynchronic embryos. Theriogenology 80: 357–364.
23. Nakano K, Watanabe M, Matsunari H, Matsuda T, Honda K, et al. (2013) Generating porcine chimeras using inner cell mass cells and parthenogenetic preimplantation embryos. PLoS One 8: e61900.
24. Shan ZY, Wu YS, Shen XH, Li X, Xue Y, et al. (2012) Aggregation of pre-implantation embryos improves establishment of parthenogenetic stem cells and expression of imprinted genes. Dev Growth Differ 54: 481–488.
25. Eckardt S, McLaughlin KJ (2004) Interpretation of reprogramming to predict the success of somatic cell cloning. Anim Reprod Sci 83: 97–108.
26. Hardy K (1997) Cell death in the mammalian blastocyst. Mol Hum Reprod 3: 919–925.
27. Hardy K, Stark J, Winston RM (2003) Maintenance of the inner cell mass in human blastocysts from fragmented embryos. Biol Reprod 68: 1165–1169.
28. Hao Y, Lai L, Mao J, Im GS, Bonk A, et al. (2003) Apoptosis and *in vitro* development of preimplantation porcine embryos derived *in vitro* or by nuclear transfer. Biol Reprod 69: 501–507.
29. Maddox-Hyttell P, Gjorret JO, Vajta G, Alexopoulos NI, Lewis I, et al. (2003) Morphological assessment of preimplantation embryo quality in cattle. Reproduction 61: 103–116.
30. Melka MG, Rings F, Hölker M, Tholen E, Havlicek V, et al. (2010) Expression of apoptosis regulatory genes and incidence of apoptosis in different morphological quality groups of *in vitro*-produced bovine pre-implantation embryos. Reprod Domest Anim 45: 915–921.
31. Moussa M, Tremoleda JL, Duchamp G, Bruyas JF, Colenbrander B, et al. (2004) Evaluation of viability and apoptosis in horse embryos stored under different conditions at 5 degrees C. Theriogenology 61: 921–932.
32. Pomar FJ, Teerds KJ, Kidson A, Colenbrander B, Tharasanit T, et al. (2005) Differences in the incidence of apoptosis between *in vivo* and *in vitro* produced blastocysts of farm animal species: a comparative study. Theriogenology 63: 2254–2268.
33. Tecirlioglu RT, Cooney MA, Lewis IM, Korfiatis NA, Hodgson R, et al. (2005) Comparison of two approaches to nuclear transfer in the bovine: hand-made cloning with modifications and the conventional nuclear transfer technique. Reprod Fertil Dev 17: 573–585.
34. Misica-Turner PM, Oback FC, Eichenlaub M, Wells DN, Oback B (2007) Aggregating embryonic but not somatic nuclear transfer embryos increases cloning efficiency in cattle. Biol Reprod 76: 268–278.

blastocysts mean TUNEL-positive cells of groups 1x, 3x, 4x and 5x.

Table S1   Effects of equine cloned embryo aggregation on *in vitro* development until day 8. Donor Cell A.

Table S2   Effects of equine cloned embryo aggregation on *in vitro* development until day 8. Donor Cell B.

Acknowledgments

The authors wish to thank Dr. Rafael Fernandez-Martin and DVM María Belén Rodriguez for reading and discussing the manuscript. Also thanks to Dr Elizabeth Crichton for English revision.

Author Contributions

Conceived and designed the experiments: AG DFS. Performed the experiments: AG AD FK RJB. Analyzed the data: AG DFS. Contributed reagents/materials/analysis tools: DFS. Wrote the paper: AG DFS.

35. Oback B (2008) Climbing mount efficiency-small steps, not giant leaps towards higher cloning success in farm animals. Reprod Domest Anim 43: 407–416.

36. Siriboon C, Tu CF, Kere M, Liu MS, Chang HJ, et al. (2014) Production of viable cloned miniature pigs by aggregation of handmade cloned embryos at the 4-cell stage. Reprod Fertil Dev 26: 395–406.

37. Kurosaka S, Eckardt S, Ealy AD, McLaughlin KJ (2007) Regulation of blastocyst stage gene expression and outgrowth interferon tau activity of somatic cell clone aggregates. Cloning Stem Cells 9: 630–641.

38. Vajta G, Peura TT, Holm P, Paldi A, Greve T, et al. (2000) Method for culture of zona-included or zona-free embryos: the Well of the Well (WOW) system. Mol Reprod Dev 55: 256–264.

39. Taka M, Iwayama H, Fukui Y (2005) Effect of the Well of the Well (WOW) system on in vitro culture for porcine embryos after intracytoplasmic sperm injection. J Reprod Dev 51: 533–537.

40. Hoelker M, Rings F, Lund Q, Ghanem N, Phatsara C, et al. (2009) Effect of the microenvironment and embryo density on developmental characteristics and gene expression profile of bovine preimplantative embryos cultured in vitro. Reproduction 137: 415–425.

41. Sananmuang T, Phutikanit N, Nguyen C, Manee-In S, Techakumphu M, et al. (2013) In vitro culture of feline embryos increases stress-induced heat shock protein 70 and apoptotic related genes. J Reprod Dev 59: 180–188.

42. Gopichandran N, Leese HJ (2006) The effect of paracrine/autocrine interactions on the in vitro culture of bovine preimplantation embryos. Reproduction 131: 269–277.

43. Sancho M, Di-Gregorio A, George N, Pozzi S, Sánchez JM, et al. (2013) Competitive interactions eliminate unfit embryonic stem cells at the onset of differentiation. Dev Cell 26: 19–30.

44. Balbach ST, Esteves TC, Houghton FD, Siatkowski M, Pfeiffer MJ, et al. (2012) Nuclear reprogramming: kinetics of cell cycle and metabolic progression as determinants of success. PLoS One 7: e35322.

45. Li J, Li R, Villemoes K, Liu Y, Purup S, et al. (2013) Developmental potential and kinetics of pig embryos with different cytoplasmic volume. Zygote 15: 1–11.

46. Lewis NE, Rossant J (1982) Mechanism of size regulation in mouse embryo aggregates. J Embryol Exp Morphol 72: 169–81.

47. Rands GF (1986) Size regulation in the mouse embryo. I. The development of quadruple aggregates. J Embryol Exp Morphol 94: 139–148.

48. Bertolini M, Beam SW, Shim H, Bertolini LR, Moyer AL, et al. (2002) Growth, development, and gene expression by in vivo and in vitro–produced Day 7 and 16 embryos. Mol Reprod Dev 63: 318–328.

49. Vajta G, Hyttel P, Trounson AO (2000) Post-hatching development of in vitro produced bovine embryos on agar and collagen gels. Anim Reprod Sci 60-61: 208.

50. Brandão DO, Maddox-Hyttel P, Løvendahl P, Rumpf R, Stringfellow D, et al. (2004) Post hatching development: a novel system for extended in vitro culture of bovine embryos. Biol Reprod 71: 2048–2055.

51. Vejlsted M, Du Y, Vajta G, Maddox-Hyttel P (2006) Post-hatching development of the porcine and bovine embryo–defining criteria for expected development in vivo and in vitro. Theriogenology 65: 153–165.

52. Alexopoulos NI, French AJ (2009) The prevalence of embryonic remnants following the recovery of post-hatching bovine embryos produced in vitro or by somatic cell nuclear transfer. Anim Reprod Sci 114: 43–53.

53. Machado GM, Ferreira AR, Guardieiro MM, Bastos MR, Carvalho JO, et al. (2013) Morphology, sex ratio and gene expression of day 14 in vivo and in vitro bovine embryos. Reprod Fertil Dev 25: 600–608.

54. Machado GM, Ferreira AR, Pivato I, Fidelis A, Spricigo JF, et al. (2013) Post-hatching development of in vitro bovine embryos from day 7 to 14 in vivo versus in vitro. Mol Reprod Dev 80: 936–947.

55. Allen WR, Stewart F (2001) Equine placentation. Reprod Fertil Dev 13: 623–634.

56. Tremoleda JL, Stout TAE, Lagutina I, Lazzari G, Bevers MM, et al. (2003) Effects of in vitro production on horse embryo morphology, cytoskeletal characteristics, and blastocyst capsule formation. Biol Reprod 69: 1895–1906.

57. Fischer B, Mootz U, Denker HW, Lambertz M, Beier HM (1991) The dynamic structure of rabbit blastocyst coverings. III. Transformation of coverings under non-physiological developmental conditions. Anat Embryol (Berl) 183: 17–27.

58. Levy RR, Cordonier H, Czyba JC, Goerin JF (2001) Apoptosis in preimplantation mammalian embryo and genetics. Int J Anat Embryol 106: 101–108.

59. Johnson AK, Clark-Price SC, Choi YH, Hartman DL, Hinrichs K (2010) Physical and clinicopathologic findings in foals derived by use of somatic cell nuclear transfer: 14 cases (2004-2008). J Am Vet Med Assoc 236: 983–990.

60. Pedersen HG, Schmidt M, Sangild PT, Strobech L, Vajta G, et al. (2005) Clinical experience with embryos produced by handmade cloning: work in progress. Mol Cell Endocrinol 234: 137–143.

61. Petters RM, Mettus RV (1984) Survival rate to term of chimeric morulae produced by aggregation of five to nine embryos in the mouse, Mus musculus. Theriogenology 22: 167–174.

9

Characterization of the CD14^{++}CD16^{+} Monocyte Population in Human Bone Marrow

Manuela Mandl[1], Susanne Schmitz[1¤], Christian Weber[1,2], Michael Hristov[1]*

1 Institute for Cardiovascular Prevention, Ludwig-Maximilians-University (LMU), Munich, Germany, **2** Munich Heart Alliance, Munich, Germany

Abstract

Numerous studies have divided blood monocytes according to their expression of the surface markers CD14 and CD16 into following subsets: classical CD14^{++}CD16^{-}, intermediate CD14^{++}CD16^{+} and nonclassical CD14^{+}CD16^{++} monocytes. These subsets differ in phenotype and function and are further correlated to cardiovascular disease, inflammation and cancer. However, the CD14/CD16 nature of resident monocytes in human bone marrow remains largely unknown. In the present study, we identified a major population of CD14^{++}CD16^{+} monocytes by using cryopreserved bone marrow mononuclear cells from healthy donors. These cells express essential monocyte-related antigens and chemokine receptors such as CD11a, CD18, CD44, HLA-DR, Ccr2, Ccr5, Cx3cr1, Cxcr2 and Cxcr4. Notably, the expression of Ccr2 was inducible during culture. Furthermore, sorted CD14^{++}CD16^{+} bone marrow cells show typical macrophage morphology, phagocytic activity, angiogenic features and generation of intracellular oxygen species. Side-by-side comparison of the chemokine receptor profile with unpaired blood samples also demonstrated that these rather premature medullar monocytes mainly match the phenotype of intermediate and partially of (non)classical monocytes. Together, human monocytes obviously acquire their definitive CD14/CD16 signature in the bloodstream and the medullar monocytes probably transform into CD14^{++}CD16^{-} and CD14^{+}CD16^{++} subsets which appear enriched in the periphery.

Editor: Nathalie Signoret, University of York, United Kingdom

Funding: This work was supported by the Deutsche Forschungsgemeinschaft, (FOR809, HR18/1-1, SFB 1123 A1) August-Lenz-Stiftung. The funder had no role in study design, data collection and analysis, decision to publish, or preparation of the manuscript.

Competing Interests: The authors have declared that no competing interests exist.

* Email: michael.hristov@med.uni-muenchen.de

¤ Current address: Center of Allergy & Environment (ZAUM), Technical University of Munich, Munich, Germany

Introduction

Current knowledge defines three major monocyte subsets in human peripheral blood based on the expression patterns of CD16 (FcγRIII) and the LPS-receptor CD14: classical CD14^{++}CD16^{-}, intermediate CD14^{++}CD16^{+} and nonclassical CD14^{+}CD16^{++} monocytes [1]. These subsets differ essentially in their chemokine receptor expression, phagocytic activity and tissue distribution during inflammation or steady-state conditions [2]. Hence, the monocyte subsets are differentially involved in the pathophysiology of inflammation, atherosclerosis and regeneration after injury [3–5]. In particular, the unique features of the intermediate in contrast to the nonclassical subset (both previously referred as CD16^{+} monocytes) has become recently more evident and especially in cardiovascular disease this subset was shown to predict independently cardiovascular events at follow-up [6–8]. Thus, human monocyte subsets may represent a novel prognostic marker or therapeutic targets in clinical medicine and detailed investigation of their characteristics in bone marrow (BM) as compared to peripheral blood appears important for better understanding of the subset-specific development, maturation and functional specialization. However, the experimental research on resident human monocytes in BM together with the plasticity of subset development and trafficking to the periphery is highly restricted and only few published data exist. Therefore the aim of our study is to provide novel evidence on medullar CD14/CD16 monocytes by using commercially available cryopreserved BM samples from healthy donors as an alternative to fresh BM and to reveal further the effects of freezing/thawing on monocyte number, function and chemokine receptor expression.

Materials and Methods

Cells

Cryopreserved bone marrow mononuclear cells (BMCs; 25×10^{6} or 100×10^{6}) from ten healthy single donors were obtained by Lonza (Cologne, Germany). The BMCs were thawed and maintained overnight prior all further experiments in *Iscove's Modified Dulbecco's Medium* (IMDM; PAA, Pasching, Austria) supplemented with 10% FCS and antibiotics in 24-well plates (1×10^{6} cells per well). After written informed consent was obtained, 20 ml of citrate anticoagulated peripheral blood were collected by aseptic venipuncture from ten healthy volunteers. The procedure was approved by the Ethics Committee of the Medical Faculty of the LMU Munich. PBMCs were separated by *Biocoll* (Biochrom AG, Berlin, Germany) density gradient centrifugation. Some fresh PBMCs were immediately analyzed by flow cytometry as described below while the remaining cells were frozen in RPMI 1640 medium (PAA) with 10% FCS and 5% dimethyl sulfoxide

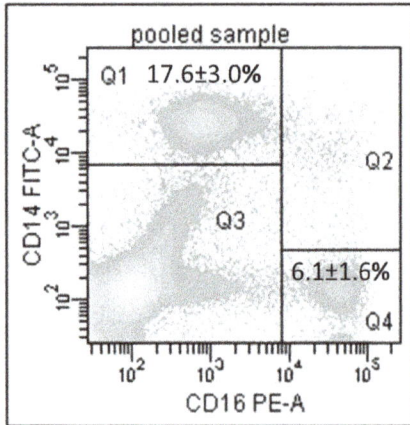

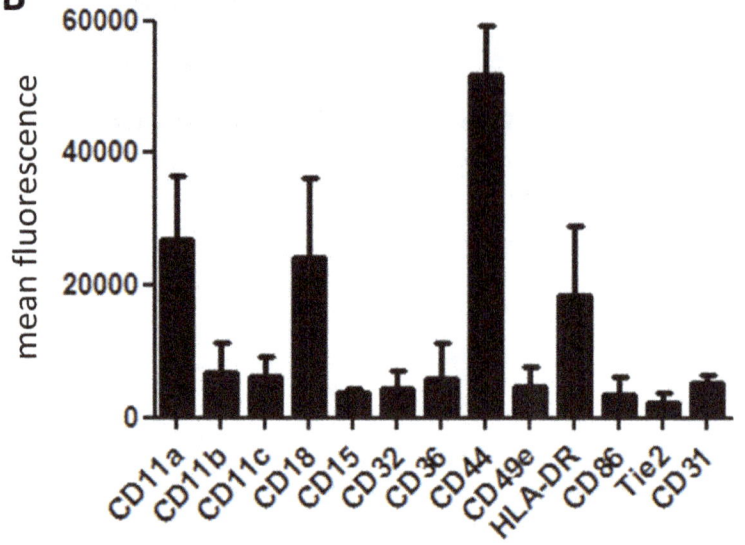

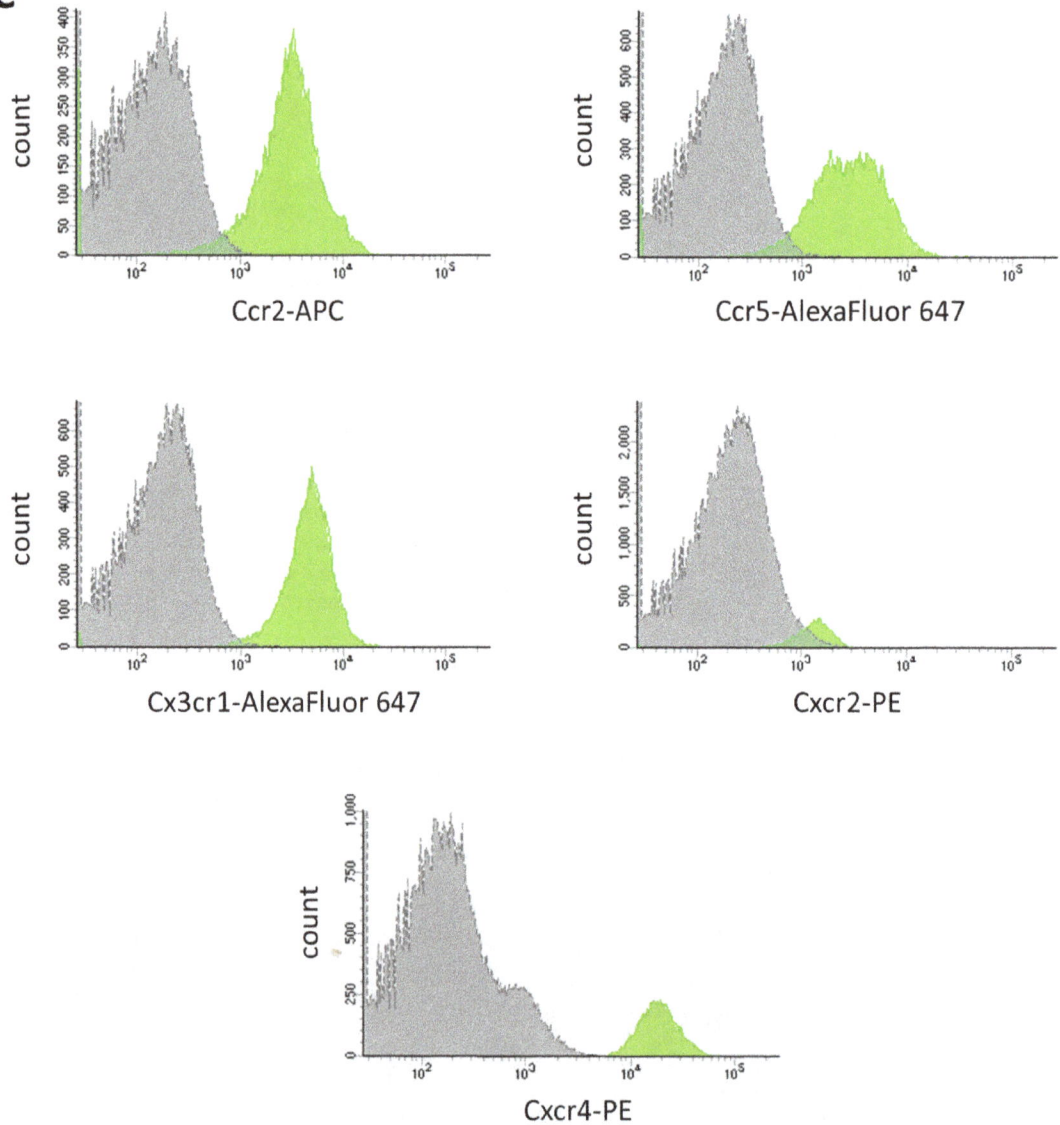

Figure 1. Characterization of CD14^{++}CD16$^+$ bone marrow monocytes by flow cytometry. (A) Representative contour plot of pooled BMC sample from three single donors showing a clear separation of two populations: CD14^{++}CD16$^+$ (Q1) and CD14$^-$CD16^{++} (Q4). Gating was performed first in FSC/SSC dot plot following gating on CD45$^+$ events in a SSC/CD45 dot plot. Finally, the expression of CD14 and CD16 was evaluated inside the CD45$^+$ population. (B,C) Expression of monocyte-related antigens and representative histogram overlays with isotype control (gray filled) of crucial chemokine receptors on CD14^{++}CD16$^+$ BMCs as analyzed by flow cytometry. Data are from 3 to 6 individual donors.

(DMSO; Sigma-Aldrich, Taufkirchen, Germany). The cells were cooled to $-80°C$ in a *CoolCell Freezing Container* (BioCision, Mill Valley, CA) before being transferred to liquid nitrogen for storage. The cryopreserved PBMCs were thawed/cultured after one week of storage in the same fashion as the BMCs.

Flow cytometry analysis of surface antigens

Cells were collected by thorough resuspension and the flow cytometry analysis was performed as 4-color experiment by using the *Lyoplate Screening Panel* (BD, Heidelberg, Germany) following staining with anti-human CD45-PerCP (clone 2D1; BD), CD14-FITC (clone HCD14) and CD16-PE (clone 3G8) mAbs (both Biolegend, Fell, Germany). The control tube included anti-human CD45/CD14/CD16 mAbs and the AlexaFluor 647-conjugated secondary Ab of the screening panel. Further anti-human mAbs used for multicolor flow cytometry were as follows: CD117-APC (clone 104D2), CD16-Pacific Blue (clone 3G8), Ccr2-APC mouse IgG2a (clone K036C2, 100 µg/ml), Ccr5-AlexaFluor 647 rat IgG2a (clone HEK/1/85a, 100 µg/ml), Cx3cr1-AlexaFluor 647 rat IgG2b (clone 2A9-1, 100 µg/ml), Cxcr2-PE mouse IgG1 (clone 5E8, 50 µg/ml), Cxcr4-PE mouse IgG2a (clone 12G5, 50 µg/ml; all Biolegend); CD15-eFluor450 (clone HI98) and CD56-APC (clone MEM188; both eBioscience, Vienna, Austria); Tie2-APC (clone 83715; R&D Systems, Wiesbaden, Germany). Matching fluorescent isotype control Abs (clones MOPC-21/173 and RTK2758/4530; Biolegend) were used at same concentration as the respective mAb. Sample tubes were acquired on a FACSCanto II with Diva v8.0 software (BD) and appropriate CS&T bead calibration setting. The fluorescence compensation was automatically calculated using *OneComp eBeads* (eBioscience). At least 10.000 events were recorded within the target population. The specific mean fluorescence intensity (MFI) was assessed by subtracting the background of the respective isotype control. To determine chemokine receptor expression over time some BMCs were cultured in RPMI 1640 medium with 10% FCS and antibiotics on 6-well plates ($4×10^6$ cells per well). The flow cytometry analysis for expression of CD45, CD14, CD16, Ccr2, Ccr5 and Cx3cr1 was performed after one and seven days of culture as described above.

Functional assays

Single-donor BMCs (10^7) or pooled BMCs ($75×10^6$) of three individual donors were sorted (FACSAria III; BD) after staining for CD14/CD16/CD45 and plated on 24-well plates or 4-well

chamber slides in RPMI 1640 medium with 10% FCS and antibiotics. Phagocytosis of FITC-conjugated latex beads (1:1000; Sigma-Aldrich) and hydroethidine (10 µg/ml; Invitrogen, Darmstadt, Germany) staining for spontaneous intracellular superoxide production in PBMC- or sorted BMC-derived macrophages were determined by flow cytometry after 24 hours in culture. Cytoskeletal filaments in cultured BMCs were visualized by fluorescence microscopy (Leica DM6000B, Wetzlar, Germany) after cell fixation/permeabilization (*Cytofix/Cytoperm*; BD) following incubation with anti-α-tubulin-FITC (Sigma-Aldrich) mAb and rhodamine-phalloidin (Biotium, Hayward, CA). *Vectashield* with DAPI (Vector Laboratories, Peterborough, UK) was used as mounting medium and to stain cell nuclei. In another experimental setup the sorted BMCs were cultured in angiogenesis µ-slide (IBIDI, Martinsried, Germany) on *Matrigel* (BD) supplemented with 50 ng/ml VEGF and 100 ng/ml Cxcl12 (both from Peprotech, Hamburg, Germany).

Statistics

Data were analyzed using *Prism 5* software (GraphPad Inc., La Jolla, CA) and presented as mean±SD. Data distribution was assessed by the *Shapiro-Wilk* normality test. Accordingly, we have used for comparison between two groups the unpaired *t test* or nonparametric *Mann-Whitney* test and for comparison between three groups one-way ANOVA with *Newman-Keuls* post-test. Differences with $p<0.05$ were considered statistically significant.

Results

Flow cytometry analysis of thawed human BMCs after staining for CD14, CD16 and CD45 showed clear separation of two scatter populations (Fig. 1A): a larger CD14^{++} population (Q1: 17.6±3.0% of CD45$^+$ cells) with low-to-intermediate expression of CD16 (CD16$^+$) and a smaller CD14$^-$CD16^{++} population (Q4: 6.1±1.6% of CD45$^+$ cells). The CD14$^-$CD16^{++} cells in Q4 were also CD15$^-$CD56^{++} thus probably referring to NK and NK-T cells (data not shown).

Further analysis of the CD14^{++}CD16$^+$ population revealed expression of CD11a, CD18, CD44 and HLA-DR together with other essential monocyte-related antigens (Fig. 1B;S1). There was also expression of the chemokine receptors Ccr2, Ccr5, Cx3cr1, Cxcr2 and Cxcr4 as shown in Fig. 1C. Moreover, the expression of Ccr2 on medullar macrophages was significantly up-regulated after 7 days in culture (Table 1).

Table 1. Chemokine receptor expression on cultured medullar macrophages from six single donors as analyzed by flow cytometry.

Receptor	1 day	7 days	p-value
Ccr2	97±29%	351±192%	0.04
Ccr5	152±31%	340±136%	0.06
Cx3cr1	113±41%	328±159%	0.06

The MFI values of CD14^{++}CD16$^+$ monocytes before plating served as control (= 100%).

A

B

C

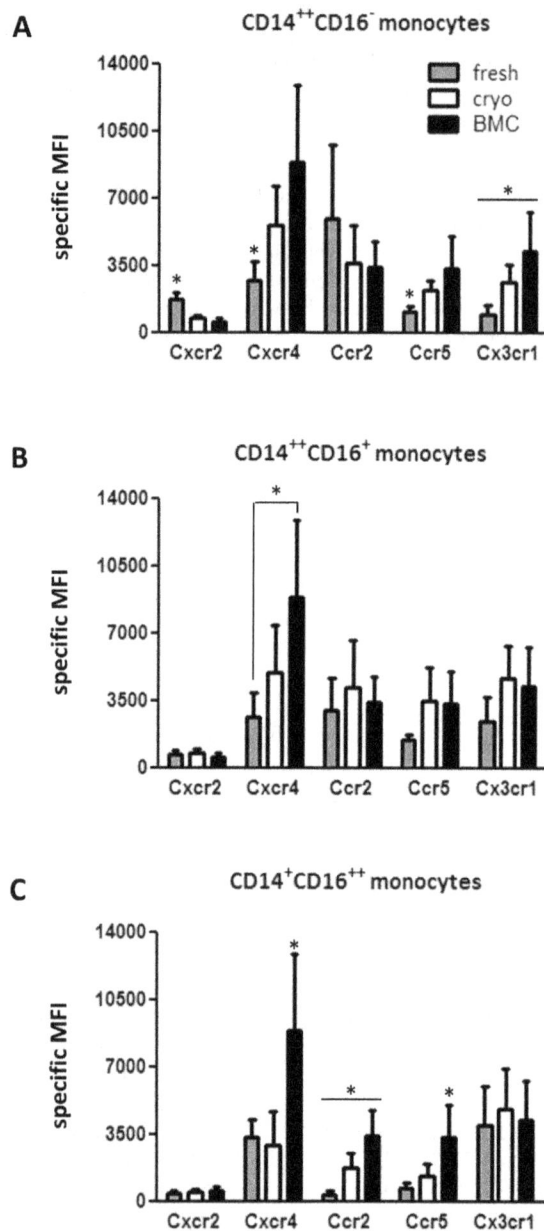

Figure 2. Side-by-side comparison of chemokine receptor expression on medullar with frozen or fresh blood monocytes. PBMCs from different donors were separated by density gradient centrifugation. Chemokine receptor expression was analyzed and compared on fresh monocyte subsets (gray bars) versus cryopreserved cells of the same donor (open bars) or unpaired medullar monocytes (black bars). *p<0.05, n=6–10.

Side-by-side examination of chemokine receptor expression on CD14++CD16+ BM monocytes versus unpaired cryopreserved blood samples showed very similar levels for Ccr2, Ccr5 and Cx3cr1 as compared to intermediate monocytes together with almost identical levels of Ccr2 on classical and Cx3cr1 on nonclassical monocytes (Fig. 2A–C; Table S1). Further comparison with unpaired fresh blood monocytes revealed identical expression of Cx3cr1 as to the nonclassical subset (Fig. 2C). Overall, the BM monocytes showed highest levels for Cxcr4 (Fig. 2A–C) and more than 20× higher expression of the hematopoietic precursor marker CD117 (data not shown) whereas fresh classical monocytes had highest levels of Cxcr2 and Ccr2 (Fig. 2A–C). Another analysis of paired fresh versus cryopreserved blood monocytes demonstrated significant decrease of Cxcr2 next to increased expression of Cxcr4, Ccr5 and Cx3cr1 on classical monocytes while thawed nonclassical monocytes had higher expression only of Ccr2 as compared to freshly processed cells (Fig. 2A,C). Finally, cryopreservation failed to affect the percentages of all blood monocyte subsets (Table 2).

Functional analysis demonstrated phagocytic activity by 58.8±19.9% of sorted CD14++CD16+ BMCs as well as spontaneous generation of reactive oxygen species (ROS) at 61.5±8.9% after 24 hours in culture (Fig. 3A,D). By comparison, macrophages originating from frozen PBMCs showed phagocytic capacity of 51.7±9.5% and spontaneous superoxide production at 41.9±7.5% (Fig. 3B,E) while freshly isolated macrophages revealed 70.7±8.1% of bead uptake and 18.0±2.3% of ROS production after 24 hours of culture (Fig. 3C,F). Microscopic immunofluorescence analysis of cultured CD14++CD16+ BMCs further confirmed typical *fried egg-like* macrophage morphology with α-tubulin/F-actin-rich protrusions (Fig. 3G,H). The CD14++CD16+ BMCs displayed also angiogenic properties as they express CD31, Tie2 and Cxcr4, and fuse to clusters with shape change in Matrigel (Fig. 1B,C;3I).

Discussion

To our best knowledge, this is the first study characterizing CD14/CD16 monocyte subsets in cryopreserved human BM of healthy donors as a possible alternative to freshly prepared cells. We found a major pool of CD14++CD16+ BMCs that seems to correspond to intermediate blood monocytes. These cells clearly feature monocytes/macrophages as illustrated by their phenotype and function. Another important analysis mapped the effect of freezing/thawing on chemokine receptor expression, subset proportion and function of bloodstream monocytes. A limitation of our study is the lack of direct validation and comparison with fresh marrow samples, ideally obtained from the same donor. However, in contrast to venous blood it is not conceivable to collect routinely fresh BM of healthy volunteers for experimental research. Consequently, only one prior work has evaluated the immunophenotype of monocyte subsets in BM aspirates originat-

Table 2. Percentages of blood monocyte subsets before and after freezing/thawing (n = 7–10).

Subset	fresh	thawed	p-value
CD14++CD16-	80.1±7.0%	82.8±3.0%	0.54
CD14++CD16+	3.7±2.0%	3.8±1.5%	0.96
CD14+CD16++	6.2±2.8%	6.3±1.7%	0.93

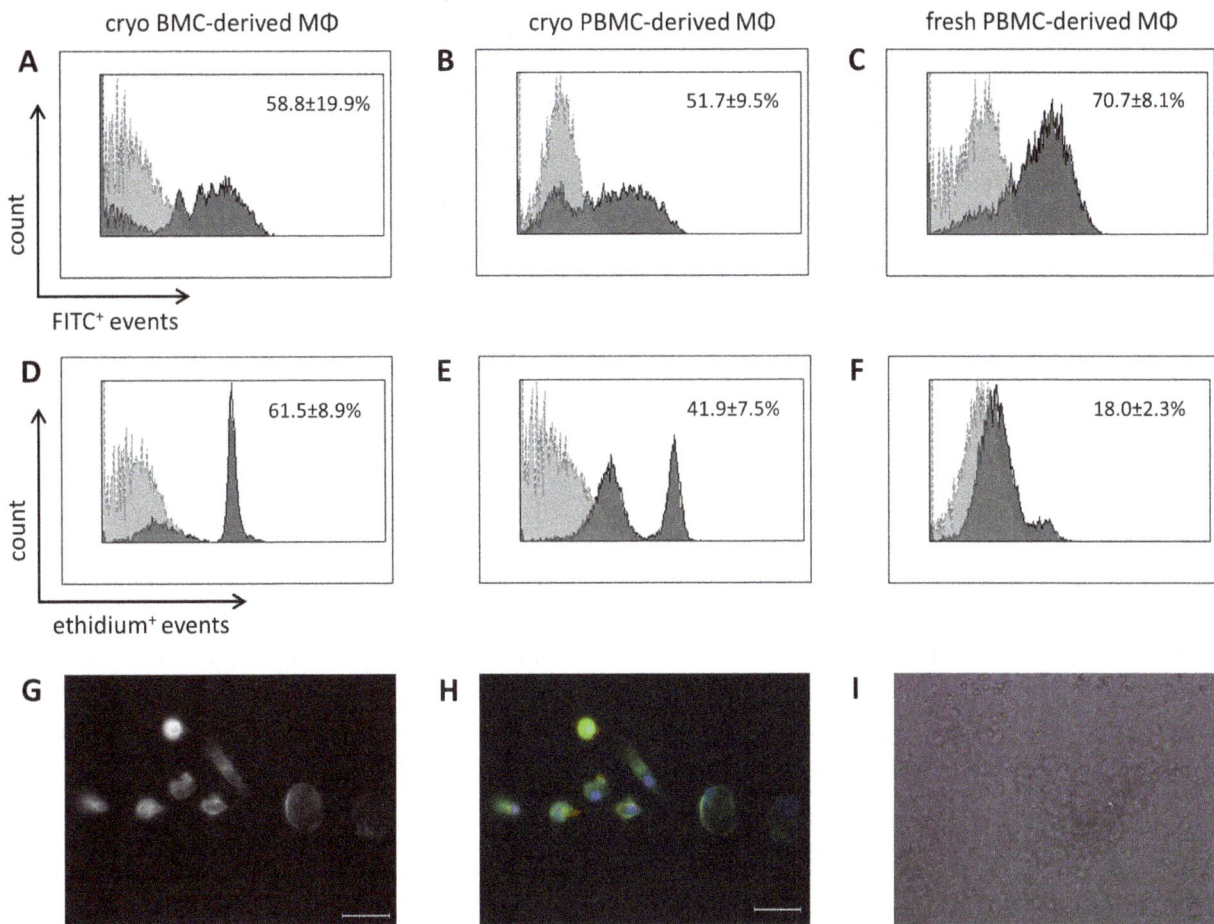

Figure 3. Functional analyses of CD14^{++}CD16^{+} BMC- and PBMC-derived macrophages. Representative flow cytometry histogram overlays for uptake of FITC-conjugated beads (A–C) and hydroethidine staining (D–F) in macrophages cultured from CD14^{++}CD16^{+} BMCs, fresh or cryopreserved PBMCs of three single donors. The gray filled histogram shows cells without beads or dye, respectively. (G,H) Phase contrast and immunofluorescent overlay of CD14^{++}CD16^{+} BMC-derived macrophages which were cultured in chamber slides and stained for α-tubulin-FITC/rhodamine-phalloidin after 7 days as described in methods. Cell nuclei were counterstained with DAPI. The scale bar indicates 40 μm. (I) Cluster formation and shape change of CD14^{++}CD16^{+} BM cells on Matrigel after 72 hours. Data in (G–I) are representative images of triplicates from pooled sample of three independent donors.

ing from healthy individuals with suspected peripheral lymphomas up to now [9].

Compared to humans, the recruitment of mouse monocyte subsets *in vivo* was investigated in detail by using several models of receptor knock-out, BM transplantation, cell depletion or adoptive transfer [10–12]. Up to now, there are two major murine monocyte subsets: the Gr1^{+}/Ly6C^{hi} cells are similar to human CD14^{++}CD16^{-} monocytes and Gr1^{-}/Ly6C^{low} cells are considered as counterparts of the CD14^{+}CD16^{++} monocytes [3,10,11]. During egress from BM, the CC-chemokine Ccl2 (MCP-1) plays a central role by either inducing Ccr2-dependent chemokinesis or chemotaxis of Ccr2^{+}Ly6C^{hi} mouse monocytes [11]. Our data demonstrated comparable expression levels of Ccr2 between medullar and thawed CD14^{++} blood monocytes. Remarkably, we found higher amount of Cxcr4 on BM monocytes in comparison with frozen or fresh bloodstream monocytes. As human CD14^{++}CD16^{+} BMCs express also the adhesion molecule CD44 they may be accessorily attracted by ligands for CD44 and Cxcr4 such as glycosaminoglycans and Cxcl12 to move towards the vascular niche of the BM and to enter the bloodstream. According to their origin BM monocytes still express the precursor marker CD117 which was absent on their terminally differentiated

counterparts in the bloodstream. Further analysis of chemokine receptors revealed very similar expression profile between medullar and frozen intermediate monocytes. Fresh classical monocytes showed highest levels of Ccr2 while the expression of Cx3cr1 was similar between medullar and CD16^{++} blood subsets (fresh or frozen). Of note, the expression of Ccr2 on BMC-derived macrophages was inducible in culture. Thus, human monocytes probably acquire their differential chemokine receptor signature (e.g. high levels of Ccr2 on the classical subset) during or after recruitment to the periphery with simultaneous decline of Cxcr4 and CD117 in terms of differentiation. It is well conceivable that such maturation may occur more rapid *in vivo* than *in vitro*, especially during inflammatory conditions or infection with accompanying monocytosis and preferential shift of classical monocytes.

Our results next reveal higher expression of some chemokine receptors on blood monocytes shortly after thawing while the ratio of receptor expression together with subset frequencies remains unaffected as compared to fresh monocytes. Accordingly, cryopreservation may affect at least transiently the corresponding chemokine receptors on BMCs as well. Furthermore, macrophages that originated from frozen blood or BM monocytes showed similar phagocytic capacity but had significantly increased ROS production as compared to freshly obtained blood macrophages. One possible explanation for this difference could be transient cell activation following cryopreservation. In line with our findings previous studies have already reported that cryopreservation can rather influence activation status than frequency or function of mononuclear cells. In particular, human T-cells, monocytes or circulating angiogenic cells can be thawed without considerable alteration of their phenotype and function [13–15]. Cryopreservation is also routinely used for storage of autologous $CD34^+$ cells suggesting that cryopreserved blood cells usually match fine fresh cells [16]. Hereby, even if the cryopreserved cells in our study were not fully equal in phenotype and ROS generation to fresh cells (at least shortly after thawing) they may represent another alternative in terms of availability, storage, transportation and functionality.

Surprisingly, we also found that nonclassical monocytes do not certainly persist as population with clear CD14/CD16 boundaries among thawed BMCs. As mentioned above it is rather unlike that cryopreservation *per se* could impact the abundance of nonclassical monocytes since their proportion remains well-preserved after freezing/thawing as confirmed also by others [17]. As further shown in a previous study, the nonclassical monocytes in fresh marrow samples were not primary identified as separate $CD14^+CD16^{++}$ scatter population but first on the basis of their lower Ccr2 expression among gated $CD16^+$ events [9]. Hence, the proper detection of nonclassical monocytes in BM appears debatable and may indeed result from "contamination" with variable amounts of peripheral blood during the BM aspiration.

In contrast to the $Ccr2^{++}$ classical and intermediate subsets in the periphery, the nonclassical monocytes are $Cx3cr1^{++}$ but dimly express Ccr2 and function mainly as weak phagocytes that patrol the microvasculature in LFA-1 dependent manner to sense pathogens or cell remnants at steady-state [18]. Moreover, nonclassical monocytes express the angiopoietin receptor Tie2 and are further considered as essential paracrine players in tumor angiogenesis [5,19]. Similar pro-angiogenic characteristics were

recently described for the intermediate subset as well [8]. Since the $CD14^{++}CD16^+$ BMCs express similar levels of Cx3cr1 and display pro-angiogenic properties, it seems that nonclassical monocytes could arise from this mixed population as well but specialize first in the periphery after coming across pathogens at homeostatic conditions, *e.g.* during patrolling the microvasculature of gut and lung.

In summary, a common $CD14^{++}CD16^+$ monocyte pool featuring mostly intermediate monocytes is primary enriched in human BM but depleted in peripheral blood where this subset exists at lowest percentage. The medullar monocytes seems to require Ccr2, Cxcr4 and possibly CD44 for mobilization and may give rise to more specialized peripheral $Ccr2^{++}$ classical monocytes with implication during inflammation next to $Cx3cr1^{++}$ nonclassical monocytes that are rather recruited at steady-state. The maturation of medullar monocytes obviously associates with decrease in the expression of Cxcr4 and c-kit. Thus, it is conceivable that human monocytes acquire their differential CD14/CD16 signature first in the bloodstream.

Supporting Information

Figure S1 Representative histogram overlays for surface markers of the BD Lyoplate screening panel.

Table S1 Specific MFI values of chemokine receptor expression on medullar (n = 6) and blood monocyte subsets (n = 10).

Author Contributions

Conceived and designed the experiments: MH. Performed the experiments: MM SS MH. Analyzed the data: MM SS MH. Contributed reagents/materials/analysis tools: MH CW. Wrote the paper: MH MM.

References

1. Ziegler-Heitbrock L, Ancuta P, Crowe S, Dalod M, Grau V, et al. (2010) Nomenclature of monocytes and dendritic cells in blood. Blood 116: e74–80.
2. Wong KL, Tai JJ, Wong WC, Han H, Sem X, et al. (2011) Gene expression profiling reveals the defining features of the classical, intermediate, and nonclassical human monocyte subsets. Blood 118: e16–31.
3. Hristov M, Weber C (2011) Differential role of monocyte subsets in atherosclerosis. Thromb Haemost 106: 757–62.
4. Rogacev KS, Ulrich C, Blömer L, Hornof F, Oster K, et al. (2010) Monocyte heterogeneity in obesity and subclinical atherosclerosis. Eur Heart J 31: 369–76.
5. De Palma M, Murdoch C, Venneri MA, Naldini L, Lewis CE (2007) Tie2-expressing monocytes: regulation of tumor angiogenesis and therapeutic implications. Trends Immunol 28: 519–24.
6. Rogacev KS, Cremers B, Zawada AM, Seiler S, Binder N, et al. (2012) CD14++ CD16+ monocytes independently predict cardiovascular events: a cohort study of 951 patients referred for elective coronary angiography. J Am Coll Cardiol 60: 1512–20.
7. Tapp LD, Shantsila E, Wrigley BJ, Pamukcu B, Lip GY (2012) The CD14++ CD16+ monocyte subset and monocyte-platelet interactions in patients with ST-elevation myocardial infarction. J Thromb Haemost 10: 1231–41.
8. Zawada AM, Rogacev KS, Rotter B, Winter P, Marell RR, et al. (2011) SuperSAGE evidence for CD14++CD16+ monocytes as a third monocyte subset. Blood 118: e50–61.
9. Shantsila E, Wrigley B, Tapp L, Apostolakis S, Montoro-Garcia S, et al. (2011) Immunophenotypic characterization of human monocyte subsets: possible implications for cardiovascular disease pathophysiology. J Thromb Haemost 9: 1056–66.
10. Weber C, Zernecke A, Libby P (2008) The multifaceted contributions of leukocyte subsets to atherosclerosis: lessons from mouse models. Nat Rev Immunol 8: 802–15.
11. Shi C, Pamer EG (2011) Monocyte recruitment during infection and inflammation. Nat Rev Immunol 11: 762–74.
12. Soehnlein O, Drechsler M, Döring Y, Lievens D, Hartwig H, et al. (2013) Distinct functions of chemokine receptor axes in the atherogenic mobilization and recruitment of classical monocytes. EMBO Mol Med 5: 471–81.
13. Van Hemelen D, Oude Elberink JN, Heimweg J, van Oosterhout AJ, Nawijn MC (2010) Cryopreservation does not alter the frequency of regulatory T cells in peripheral blood mononuclear cells. J Immunol Methods 353: 138–140.
14. Hiebl B, Fuhrmann R, Franke RP (2008) Characterization of cryopreserved CD14+-human monocytes after differentiation towards macrophages and stimulation with VEGF-A(165). Clin Hemorheol Microcirc 39: 221–8.
15. Sofrenovic T, McEwan K, Crowe S, Marier J, Davies R, et al. (2012) Circulating angiogenic cells can be derived from cryopreserved peripheral blood mononuclear cells. PLoS One 7: e48067.
16. Bakken AM (2006) Cryopreserving human peripheral blood progenitor cells. Curr Stem Cell Res Ther 1: 47–54.
17. Berg KE, Ljungcrantz I, Andersson L, Bryngelsson C, Hedblad B, et al. (2012) Elevated CD14++CD16− monocytes predict cardiovascular events. Circ Cardiovasc Genet 5: 122–31.
18. Cros J, Cagnard N, Woollard K, Patey N, Zhang SY, et al. (2010) Human CD14dim monocytes patrol and sense nucleic acids and viruses via TLR7 and TLR8 receptors. Immunity 33: 375–86.
19. Venneri MA, De Palma M, Ponzoni M, Pucci F, Scielzo C, et al. (2007) Identification of proangiogenic TIE2-expressing monocytes (TEMs) in human peripheral blood and cancer. Blood 109: 5276–85.

10

Survival of Skin Graft between Transgenic Cloned Dogs and Non-Transgenic Cloned Dogs

Geon A Kim[1], Hyun Ju Oh[1], Min Jung Kim[1], Young Kwang Jo[1], Jin Choi[1], Jung Eun Park[1], Eun Jung Park[1], Sang Hyun Lim[2], Byung Il Yoon[3], Sung Keun Kang[2], Goo Jang[1], Byeong Chun Lee[1]*

1 Department of Theriogenology & Biotechnology, College of Veterinary Medicine, Seoul National University, Seoul, Republic of Korea, 2 Central Research Institutes, K-stem cell, Seoul, Republic of Korea, 3 Laboratory of Histology and Molecular Pathogenesis, College of Veterinary Medicine, Kangwon National University, Chuncheon, Gangwon-do, Republic of Korea

Abstract

Whereas it has been assumed that genetically modified tissues or cells derived from somatic cell nuclear transfer (SCNT) should be accepted by a host of the same species, their immune compatibility has not been extensively explored. To identify acceptance of SCNT-derived cells or tissues, skin grafts were performed between cloned dogs that were identical except for their mitochondrial DNA (mtDNA) haplotypes and foreign gene. We showed here that differences in mtDNA haplotypes and genetic modification did not elicit immune responses in these dogs: 1) skin tissues from genetically-modified cloned dogs were successfully transplanted into genetically-modified cloned dogs with different mtDNA haplotype under three successive grafts over 63 days; and 2) non-transgenic cloned tissues were accepted into transgenic cloned syngeneic recipients with different mtDNA haplotypes and vice versa under two successive grafts over 63 days. In addition, expression of the inserted gene was maintained, being functional without eliciting graft rejection. In conclusion, these results show that transplanting genetically-modified tissues into normal, syngeneic or genetically-modified recipient dogs with different mtDNA haplotypes do not elicit skin graft rejection or affect expression of the inserted gene. Therefore, therapeutically valuable tissue derived from SCNT with genetic modification might be used safely in clinical applications for patients with diseased tissues.

Editor: Pascale Chavatte-Palmer, INRA, France

Funding: This study was supported by Rural Development Administration (#PJ008975022014), Korea Institute of Planning and Evaluation for Technology (#311062-04-3SB010), NATURE CELL (#2014-0082), Research Institute for Veterinary Science, Nestle Purina PetCare, Natural Balance Korea, and the BK21 plus program. The funders had no role in study design, data collection and analysis, decision to publish, or preparation of the manuscript.

Competing Interests: The authors received funding from NATURE CELL CO., LTD, Nestle Purina PetCare, and Natural Balance Korea. There are no further patents, products in development or marketed products to declare.

* Email: bclee@snu.ac.kr

Introduction

Somatic cell nuclear transfer (SCNT) produces genetically identical cloned animals [1]. Moreover, canine SCNT combined with transgenic technologies can make genetically identical cloned dogs with functional genetic modifications that could be used for gene therapy [2]. For example, transgenic cloned dogs could be used in replacement of diseased (malfunctioning/worn out) organs. However, tissues derived from transgenic cloned dogs, reprogrammed from somatic cells with enucleated oocytes, had not yet investigated whether they are immunologically identical tissues or cell sources of transplantation. Especially, effects of red fluorescent protein (RFP) expression using genetically identical animal models derived from SCNT have not been described and this is a critical subject since RFP has been used as a potential marker for clinical trials of gene therapy [3–5].

In addition, SCNT uses oocytes from animals unrelated to the prospective transplant recipient, oocyte-derived mitochondrial DNA (mtDNA) derived antigen could lead to rejection problems in kidney transplant [6] or not in skin transplant [7,8]. Although tissues derived from SCNT, using the recipient's somatic cells as nuclear donors, provide identical genetics, the absence of immune rejection has not yet been confirmed in cloned dogs.

To our knowledge, no previous report has mentioned *in vivo* skin immune responses against tissue expressing foreign gene or the capable effects of mitochondrial derived minor antigen in cloned animals. Here, we firstly evaluated the anti-foreign gene or minor antigen derived immune responses in cloned dogs with the following design: (1) for investigation of mtDNA derived antigen compatibility, skin graft was performed between transgenic cloned dogs with different mtDNA haplotypes; (2) furthermore, skin graft was also performed between transgenic cloned dogs and non-transgenic cloned dogs for examination of immunogenicity of foreign gene.

Materials and Methods

1. Animals

Two genetically identical cloned female beagles (C1, C2) were generated by SCNT using a beagle fetal fibroblast cell line (BF3) described in a previously study [9]. Transgenic cloned female beagles (R1, R2, R3 and R5) were also produced by SCNT using BF3 transfected with RFP [2].

Non-related controls (Co1, Co2) were healthy age-matched normal female beagles purchased from commercial kennels (Marshall Beijing Biotech Ltd., Beijing, China). All animals used in this study were cared for in accordance with recommendations described in "The Guide for the Care and Use of Laboratory Animals" published by the Institutional Animal Care and Use Committee (IACUC) of Seoul National University (approval number; SNU-110915-2). Dog housing facilities and the procedures performed met or exceeded the standards established by the Committee for Accreditation of Laboratory Animal Care. All surgery was performed under isoflurane anesthesia, and all efforts were made to minimize suffering.

2. DNA extractions and PCR reaction

Blood was collected from two control beagles and six female cloned beagles 4 years of age for DNA extractions, blood typing and blood cross-matching. Approximately 10 ml of blood were collected from the jugular vein into tubes containing EDTA as anticoagulant and used for peripheral blood mononuclear cell isolation and DNA extraction, and 3 ml of blood in plain tubes were collected to provide serum samples for antibody levels. Blood samples were kept at 38°C to maintain cell viability.

Freshly retrieved non-coagulated blood samples were mixed with RBC lysis buffer (Invitrogen, Carlsbad, CA, USA) at room temperature for 15 min. Genomic DNA was isolated according to the manufacturer's protocol. Extracted DNA samples were stored at −30°C. DLA class I (MHC class I) and II (MHC class II) typing analysis was performed by means of PCR and sequencing. The polymorphic exon 2 and exon 3 of the DLA-88 gene was amplified using PCR primers [10]. The polymorphic exon 2 of the DRB1, DQA and DQB genes was also amplified using PCR primers [11]. For PCR, Maxime PCR PreMix kit (iNtRON Biotechnology, Inc., Gyeongi, Korea) was used. In each PCR tube, 1 μl of genomic DNA, 1 μl (10 pM/μl) of forward primer, 1 μl (10 pM/μl) of reverse primer and 17 μl of sterilized distilled water were added according to the manufacturer's instructions. These components were then mixed and centrifuged briefly. PCR was done using a PCR machine (Biometra, Goettingen, Germany). PCR amplification was carried out for 1 cycle with denaturing at 94°C for 5 min, and subsequently for 30 cycles with denaturing at 94°C for 40 sec, annealing at 63°C (DLA-DRB1), 55°C (DLA-DQA1) and 66°C (DLA-DQB1) for 40 sec, extension at 72°C for 40 sec, and a final extension at 72°C for 5 min. Amplified PCR product was run on the gel by gel electrophoresis (Mupid-exu, Submarine electrophoresis system, Advance, Japan) at 100 V for 20 min. A 2% agarose gel was prepared using agarose (Invitrogen) and 1X TAE buffer. The stain (RedSafe, iNtRON Biotechnology Inc.) was used at a concentration of 2.5 μl per 50 ml of gel. After running gels, images were made under ultraviolet light. PCR product was sequenced directly using the Big Dye Terminator kit (Applied Biosystems, Foster City, CA, USA). Sequencing was performed on an automated DNA sequencer model 377 or capillary model 3110 (Applied Biosystems).

3. Sequencing of Mitochondrial DNA haplotype

For mitochondrial DNA analysis, the oligonucleotide primers were synthesized over the hypervariable regions (forward, 5'-CCTAAGACTTCAAGGAAGAAGC-3'; reverse, 5'-TTGACT-GAATAGCACCTTGA-3') of the complete nucleotide sequence of canine mtDNA (GenBank accession no. U96639). Isolated genomic DNA sample were dissolved in 50 ul TE buffer and used for PCR amplifications. It were performed in a 50 μl volume containing 5 μl of 10× reaction buffer containing 1.5 mM MgCl2, 0.2 mM dNTPs, 0.2 μM each primer, 1.5 U Taq DNA polymerase (Intron, Kyunggi, Korea). Starting denaturing for 1 cycle at 95°C for 3 minutes, subsequently denaturation at 94°C for 30 seconds, annealing at 57°C for 30 seconds, extension at 72°C for

Table 1. Mitochondrial DNA sequences of non-transgenic cloned dog (C2) and four transgenic cloned dogs (R1, R2, R3 and R5).

Sample	Nucleotide positions																					
	15435	15483	15508	15526	15595	15611	15612	15620	15627	15632	15639	15643	15650	15652	15781	15800	15814	15815	15912	15955	16025	16083
Reference[1]	G	C	C	C	C	C	T	T	A	T	T	A	T	G	C	T	C	T	C	C	T	A
C2	G	C	C	T	T	T	T	A	A	G	G	G	T	A	C	T	C	T	C	T	T	A
R1	G	C	C	C	C	T	C	T	A	T	T	A	T	G	C	T	T	T	C	C	T	A
R2	G	C	C	C	C	T	T	A	A	A	A	A	T	G	C	T	T	T	C	C	C	A
R3	G	C	C	C	C	T	T	A	G	A	A	A	T	G	C	T	T	T	C	C	T	A
R5	G	T	C	T	T	T	T	A	G	A	A	A	T	G	C	T	T	T	C	C	T	A

GenBank accession number :U96639 (Kim et al., 1998).

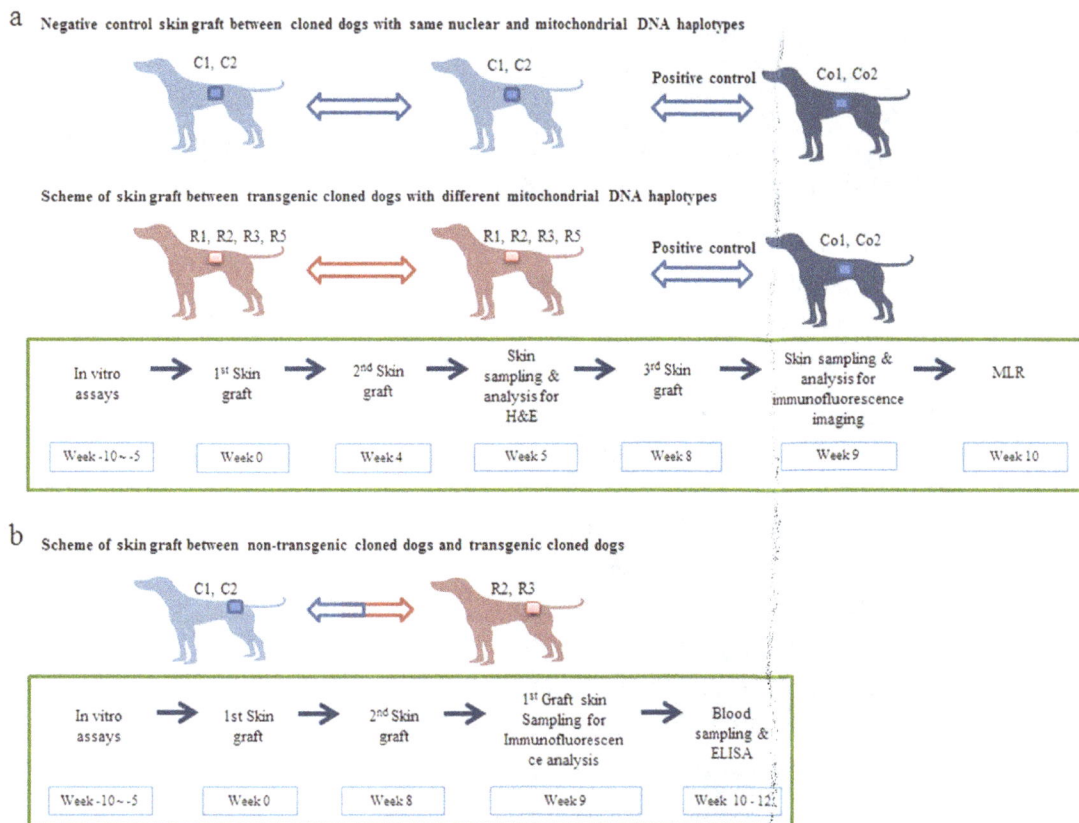

Figure 1. Experimental design and image analysis result between cloned dogs. (a) Experimental design and timeline of skin graft between cloned dogs with different mitochondrial haplotypes. As negative control, auto grafts as well as cloned dogs with same mtDNA haplotype (C1, C2) were used. Before skin graft, all *in vitro* assays were performed. For H&E staining, immunofluorescence imaging, 1st skin graft fragments were analyzed. (b) Experimental design and timeline between transgenic cloned dogs and non-transgenic cloned dogs. Before skin graft, all *in vitro* assays were performed. All dogs were tested twice for each skin graft, then skin samplings were performed. For immunofluorescence imaging, 1st skin graft fragments were analyzed. RFP expression were monitored until 63 days after skin graft.

30 seconds of 35 cycles, and a final extension at 72°C for 3 minutes were carried out. After purification of PCR products using a Gel Extraction Kit (Qiagen, Hilden, Germany), they were sequenced with an ABI3100 instrument (Applied Biosystems). Their identities with mtDNA were confirmed by BLAST search (http://blast.ncbi. nlm.nih.gov/).

4. Blood crossmatching and blood typing

Blood collection was performed from the jugular vein of all cloned dogs (R1, R2, R3, R5, C1 and C2) into an evacuated tube containing EDTA as anticoagulant. Collected samples were submitted to a commercial laboratory kit (Antech Diagnostics, Phoenix, AZ, USA). Blood type was confirmed using the tube agglutination method with antiserum; consisting of 6 types of monoclonal antibodies for canine blood typing [12].

The blood crossmatching test was done on EDTA-treated blood using the tube agglutination method. Isolated RBCs of all dogs were washed 3 times with 0.9% saline, and a 4% RBC suspension was made from the washed cells. RBC suspensions from cloned beagles (C1) were combined with equal volumes of another cloned beagle's serum (C2) and the reverse reaction was also performed. All mixtures were incubated at 37°C for 20 min, centrifuged and then assessed for hemolysis or agglutination. Agglutination was evaluated by comparing the color of supernatant in the test tube with those of the control sample. Each sample was shaken until all red blood cells in the "button" at the bottom of the tube had

become suspended. Again, the degree of RBC clumping of the test sample was compared with that of the auto-mixture of RBC and plasma. When the plasma was clear, no clumping of RBCs was detected at 400× magnification, these results were considered as negative. A positive result showed agglutination resembling stacked coins. Images were obtained using a microscope, the ProgRes Capture camera system, and the ProgRes Capture 2.6 software (JENOPTIK, Jena, Germany).

5. Peripheral blood mononuclear cell isolation and mixed lymphocyte reactions

Blood was collected from two control dogs and six female cloned dogs before and 10 weeks after skin graft. EDTA-treated whole blood was transferred to 50 ml conical centrifuge tubes. An equal volume of phosphate buffered solution (PBS, Gibco, Carlsbad, CA, USA) was mixed with the sample prior to the isolation process. Peripheral blood mononuclear cells (PBMC) were isolated from EDTA-treated blood using lymphocyte separation medium on a Ficoll-paque gradient (Ficoll-Paque Plus, GE Healthcare, Pittsburgh, PA, USA). Mixed lymphocyte reactions were modified from the previous reports [13–15]. Washed cells were diluted in culture medium (RPMI1640, Gibco) supplemented with 10% FBS to 2×10^6 cells/ml. To stimulate proliferation of lymphocytes, PBMCs were preincubated with 2 ug/ml of phytohemagglutin for 24 h before mix reaction. Then 50 ul of this cell suspension was added into each well of a 96-well

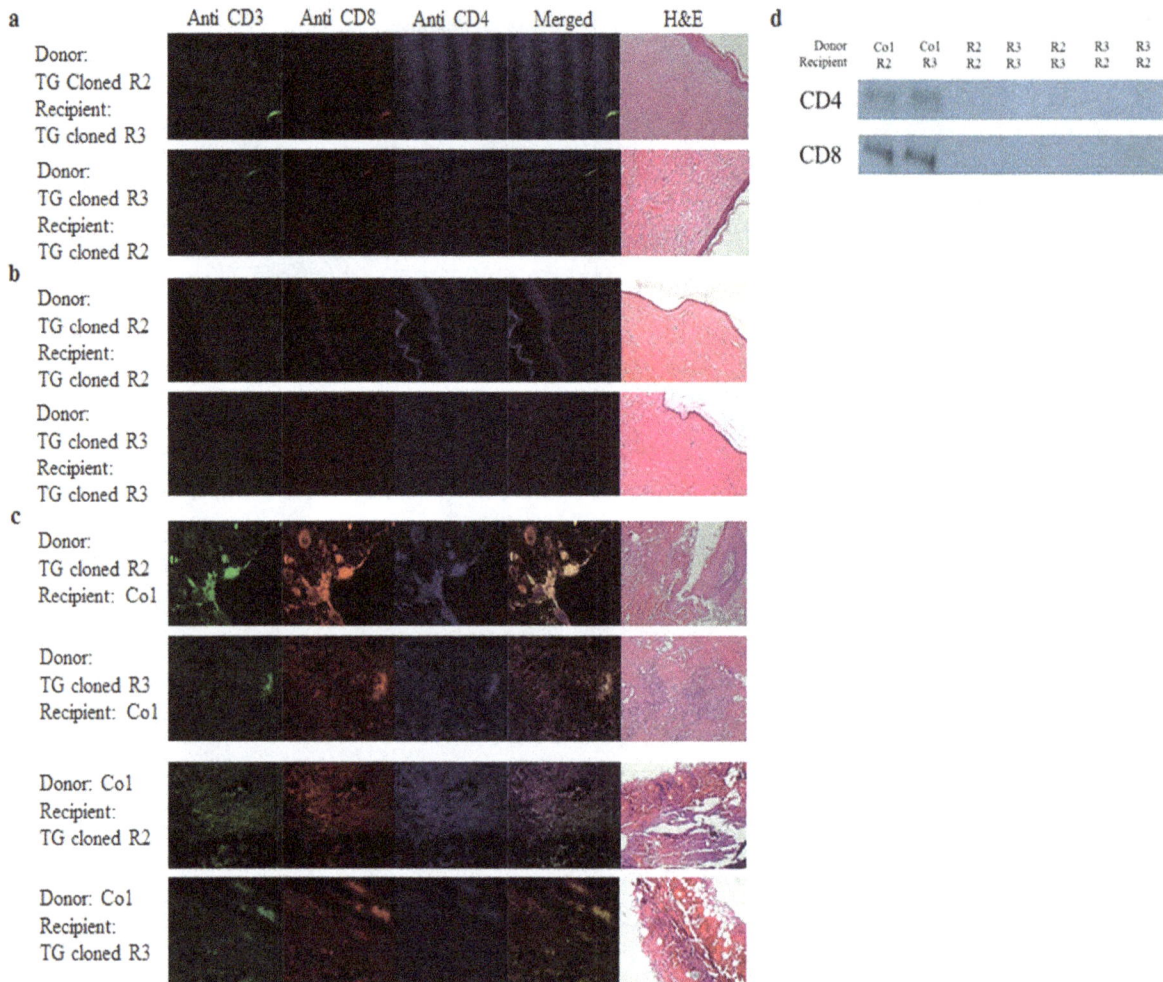

Figure 2. Absence of *in vivo* immunogenicity in skin grafts between cloned dogs with different mitochondrial DNA sequences. No evidence with infiltration of T cell was detected in the skin segments transplanted into the recipient dogs with different mitochondrial haplotypes (a). Sections from skin segments of autografts were used as negative controls (b). Sections from skin segments of cloned dogs transplanted into control dogs were used as positive controls (c). Western blot analysis confirms high protein levels of CD4 and CD8 in the positive controls, whereas CD4 and CD8 expression intensities were significantly lower in allograft of cloned dogs with different mitochondrial haplotypes (d). Upper lane indicates the donor dog and lower lane means recipient dog.

microplate except for the wells required for the blank and cultured at 37.5°C in a water-saturated atmosphere containing 5% CO_2. Each cell combination was tested in quadruplicate in a flat-bottomed micro plate containing 0.1 ml of culture medium per well. The mixture was cultured for 5 days and then pyrimidine analogue, bromodeoxyuridine labeling reagent (Cell proliferation ELISA, Roche Applied Science, Indianapolis, IN, USA) was added and re-incubated for 24 h. After removing the labeling medium, results are expressed as absorbance units at 450 nm wavelength read by a micro plate reader, Sunrise (Tecan Sunrise, Hayward, CA, USA). Time-course kinetics was studied by harvesting on day 7 of culture.

6. DNA walking

For confirmation of the transgene (RFP) location, PCR was performed with a DNA Walking SpeedUP Kit (Seegene Inc., Seoul, Korea) and products were gel purified (QIAquick PCR purification kit; QIAGEN, Valencia, CA, USA), and DNA strands were directly sequenced (Macrogen, Seoul, Korea; http://www.

macrogen.com) using a custom-synthesized primer (5′-TCACA-GAAGTATGCCAAGCGA-3′). The sequences, except for known sequences, including primers of each product were aligned by sequence homology analysis using the Basic Local Alignment Search Tool (BLAST) at the National Center for Biotechnology Information (NCBI) GenBank (http://blast.ncbi.nlm.nih.gov/).

7. Skin graft

For skin graft procedures, experimental dogs were anesthetized with ketamine hydrochloride (6 mg/kg) after pretreatment with xylazine (0.05 mg/kg), and were maintained with 2% isoflurane in oxygen. A flank skin segment 1.5 cm×1.5 cm was excised from each donor dog. Simultaneously, the same sized skin piece was excised from recipient dogs, and the excised skin was grafted by suturing into the graft bed of the same region of an anesthetized recipient dog. Bandages were changed every day after surgery and the grafts were observed weekly.

For examination of effects mtDNA haplotypes differences among cloned dogs, skin grafts of three times were performed

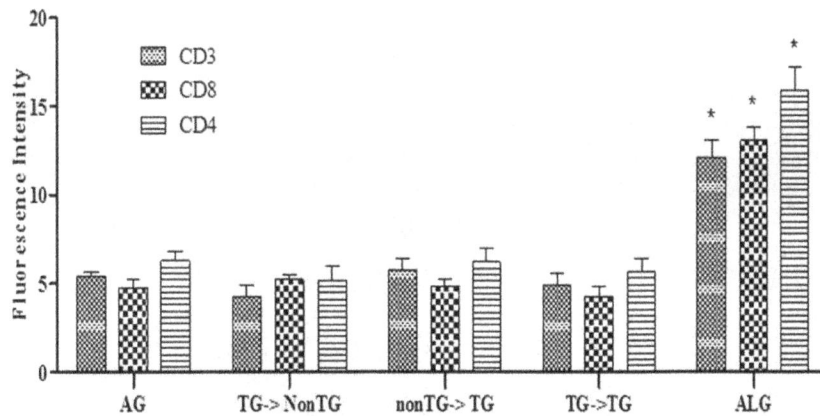

Figure 3. Expression levels of CD3, CD4 and CD8 of skin grafts between cloned dogs using fluorescence image analysis. Immunological response level of CD3, CD4 and CD8 were similar in AG (autograft), TG-> NonTG (donor: transgenic dog, recipient: non-transgenic cloned dogs), NonTG->TG (donor: non-transgenic dog, recipient: transgenic cloned dogs) and TG cloned dogs (donor: transgenic dog, recipient: transgenic cloned dogs). However, Both of ALG (allograft) between TG dogs and non-related control dogs and allograft between non-TG dogs and non-related control dogs shows significantly higher intensity of immunological response (p<0.05). Results are presented as mean ± SEM. Replication number is at least 8 times.

every 4 weeks between non-transgenic cloned dogs with same mtDNA haplotype and between transgenic cloned dogs with disparate mtDNA haplotypes. Accepted tissues were maintained until 9 weeks after skin graft. Biopsies of skin were performed after 63 days after first skin graft. A flank skin segment of 1st graft with size of 0.5 cm × 1.5 cm including donor and recipient tissue were excised for H&E staining at 5 weeks of skin graft and remnant tissue were excised for immunofluorescence imaging and western blot at later.

8. Histological and immunofluorescence analysis

Immuno-staining of canine skin immune cells was carried out on formaldehyde-fixed sections using a rabbit monoclonal antibody to CD3 (1:100, ab94756, Abcam, Cambridge, MA, USA), visualized with an anti-rabbit polyclonal DyLight 488 (1:200, ab96895, Abcam) antibody. In these sections, CD4 and CD8 cells were counterstained with a CD4 (1:100, LS c122857, Lifespan Bioscience Inc., Seattle, WA, USA) and CD8 (1:200, ab22505, Abcam) specific antibody detected with a DyLight 405 (1:200, 3069-1, Abcam) and DyLight 649 (1:200, ab98389, Abcam) coupled secondary antibody. Skin sections were also processed for assessing expression of RFP using rabbit polyclonal RFP antibody (1:200, ab62341, Abcam) and visualized with an anti-rabbit polyclonal DyLight 488 (1:200, ab96895, Abcam) antibody. Sections were counterstained with 4', 6'-diamidino-2-phenylindole (DAPI).

Histology was done by fixing skin fragment in 4% neutral formalin and embedding in paraffin; sections were stained with standard hematoxylin and eosin (H&E) procedures. Fluorescent and bright field images were obtained with a Leica DMI 6000B microscope using a DFC350 camera and LAS software (Leica Microsystems Pty Ltd., North Ryde, Australia) and analyzed by a computer-assisted image analysis system (Metamorph version 6.3r2; Molecular Devices Corporation, PA, USA). To maintain a constant threshold for each image and to compensate for subtle variability of the immune-fluorescent imaging, we only counted cells that were at least 70% lighter than the average level of each positive control image after background subtraction. All image analytical procedures described above were performed blind without knowledge of the experimental scheme.

9. Western blot

Skin fragments of graft was excised and homogenized in PRO-PREP protein extraction solution (iNtRON Biotechnology, Inc.) using a tissue homogenizer. After measuring protein concentration using Nanodrop 2000 (Thermo fischer scientific, Seoul, Korea), equal amounts of proteins were loaded on 10% SDS-PAGE. Proteins were electrophoresed and blotted onto polyvinylidene fluoride membranes. The membranes were blocked with 5% skim milk in TBS with 0.1% Tween-20 and incubated with primary antibodies for 2 hours at room temperature. Monoclonal CD4 and CD8 antibodies were used as markers for immune rejection. Subsequently, membranes were incubated with goat anti-mouse IgG, anti-rat IgG (Pierce, Rockford, IL, USA) with horse radish peroxidase conjugation for 1 h at room temperature. Then, WEST-one[TM] Western blot detection system (iNtRON Biotechnology, Inc.) was added and visualized after exposing the membrane to X-ray film.

10. Statistical Analysis

The data of mixed lymphocyte reaction, image analysis of immunocytochemistry and western blot were analyzed using one-way ANOVA and a protected least significant different (LSD) test using general linear models to determine differences among experimental groups. Data were analyzed using GraphPad Prism software (GraphPad Software Inc., San Diego, CA, USA). Absorbance mean values were considered significantly different when the P-value was less than 0.05. The observations of mixed lymphocyte reaction among experimental groups were replicated at least 8 times.

Results and Discussions

It has been reported that immune rejection can occur when tissues of genetically identical SCNT cloned animals were transplanted to each other, due to the tissues having different maternally-derived antigens [6,16,17]. Antigens derived from mtDNA in accelerated skin rejection in syngeneic rodent recipients [18,19]. It has also been generally assumed that genetically-engineered tissues with insertion of a foreign gene could invoke immune-rejection by the recipient even in inbred mice [20]. Using embryonic stem cells derived from SCNT, the complete rescues of

DAPI	RFP	Merged

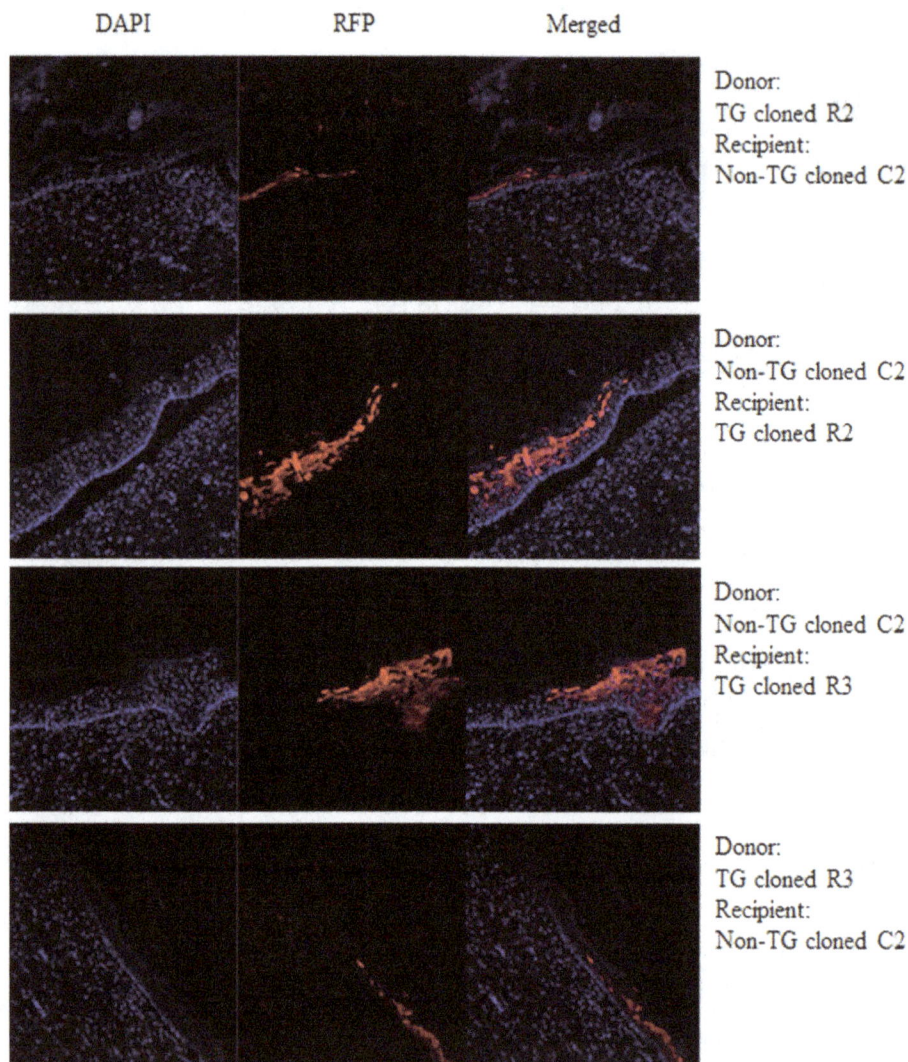

Donor:
TG cloned R2
Recipient:
Non-TG cloned C2

Donor:
Non-TG cloned C2
Recipient:
TG cloned R2

Donor:
Non-TG cloned C2
Recipient:
TG cloned R3

Donor:
TG cloned R3
Recipient:
Non-TG cloned C2

Figure 4. Maintenance of foreign gene expression between transgenic cloned beagle with foreign genes and non-transgenic cloned dogs. No expression of foreign gene in non-transgenic dog (C2) recipient was maintained in skin graft of transgenic cloned dog (R2). The limit between donor and recipient were not changed until 63 days after skin graft.

genetic defect with genetically-engineered cell therapy were not observed [21]. Engraftment of hematopoietic precursor cells differentiated from SCNT or induced pluripotent stem cells (iPSCs) was only successful in the absence of natural killer cells and immunogenicity of iPSCs was reported [21–23].

In the present study, cloned dogs produced by SCNT had different mtDNA haplotypes (Table 1), because canine SCNT used oocytes obtained from several oocyte donor dogs and the oocyte mtDNA was still present after the SCNT procedure. To examine the immunogenicity of skin tissue derived from syngeneic grafts exhibiting different mtDNA haplotypes, we initially performed *in vitro* molecular typing of dog leukocyte antigen (DLA), mixed lymphocyte reaction (MLR) and blood cross-matching using cells derived from cloned dogs with different mtDNA haplotypes (Fig. S1). Despite the different mtDNA haplotypes, they had no effects on *in vitro* immunological compatibility.

To gain insights into the therapeutic applicability of canine skin tissues with different mtDNA haplotypes, skin grafting between

cloned dogs was performed to determine immunological compatibility *in vivo* (Fig. 1). Whereas allogeneic Co 1 (non-related control dogs) skin fragments were rapidly rejected in R2 (transgenic cloned dog) and R3 (transgenic cloned dog) recipients with massive infiltration of CD4+, CD8+ T cells, infiltration, edema and perivascular inflammation 7 days after 2nd skin graft, skin tissues of R2 and R3 were accepted in R3 and R2 recipients as well as autografts, without any evidence of immune rejection (Fig. 2). Likewise, skin segments from cloned dogs with different mtDNA sequences did not induce immune rejection in the recipient cloned dogs (Fig. S2). In MLR of 10 weeks after 3rd skin graft, we couldn't detect any sign of mtDNA derived minor antigen immunogenicity with no significant differences compared to those of MLR before skin graft (data not shown).

In mice, mtDNA encoded proteins could elicit rejection by innate immunity in a setting where the genomic DNA matched [24,25]. Furthermore, kidneys transplanted between cloned pigs differing in some mtDNA genes rejected those grafts [6]. Therefore, different antigenicity of grafts from different tissues could be also considered. In our experiment, despite a high level of diversity of mtDNA haplotypes heteroplasmy among domestic dogs [26], skin grafts were successfully accepted in at least 20 donor-recipient combinations. In cattle and pigs, it was shown that SCNT-derived tissues were not rejected by the immune system of the nucleus donor after SCNT in skin graft [7,27,28]. Our findings suggest that differences of canine mtDNA haplotypes could not elicit skin graft rejection among cloned dogs, as previously observed in cattle and pigs.

We also showed genetic identity between tissues of non-transgenic cloned dogs (C1, C2) derived from beagle fibroblasts (BF3) [9] and tissues of transgenic cloned dogs (R1, R2, R3 and R5) derived from BF3 transfected with RFP (Table S1.) [2]. Immunological compatibility between these dogs was completely established through *in vitro* tests such as DLA typing and MLR (Fig. S1). Skin tissues of non-transgenic cloned dogs were transplanted into transgenic cloned dogs and *vice-versa*. Skin tissues derived from cloned dogs were transplanted with no immune rejection, as determined by T cell infiltration of peri-graft skin sections after 7 days of 2nd skin graft. Despite insertion of the foreign gene RFP in transgenic cloned dogs, skin tissue from RFP transgenic cloned dogs was completely accepted in non-transgenic cloned dog recipients (Fig. 3, Fig. S3). These finding indicate that foreign gene insertion in cloned dogs did not induce a T cell-dependent skin graft rejection response in syngeneic recipients. It has been suggested that the nuclear reprogramming process in SCNT could result in surface expression of proteins and molecules unknown to the immune system of the graft recipients. In this regard, in inbred mice, enhanced GFP (eGFP) skin transplantation causes an acute reaction [29]. It was proved that eGFP also induce immune responses that interfere with its applicability in gene insertion of mouse [30]. However, our results suggest that inserted foreign gene, RFP has no immunological effects on the antigens of transgenic cloned dogs against to the non-transgenic cloned dogs. It also suggested that non-transgenic cloned dogs produced by SCNT using transfected cells have no immune regulatory effect on the host immune system and that the canine SCNT process did not result in surface expression of immunogenic molecules. Nonetheless, the possibility of immune rejection of other foreign genes, for example, pathogenically relevant transgene in clinical science remains to be confirmed.

Finally we examined whether functional expression of RFP was maintained in skin tissue grafts. During the course of this experiment, the expression level of RFP positive skin tissues were maintained for at least 63 days after surgery and RFP positive cells were detected in the epidermis, hair follicles and sebaceous glands (Fig. 4 and Table S1). It has been suggested that the co-expression of selection markers can limit or abrogate the persistence of expression of therapeutic genes [31,32]. The potential success of gene therapy or production of transgenic cloned dogs may depend on long-term transgene expression to cure or slow down the progression of disease. In addition, there were no host immune responses to the skin grafts among transgenic dogs and non-transgenic cloned dogs, and it appears that the level and duration of RFP transgene expression was not affected. This also indicates possible successful of therapeutic transplantation of tissues or cells derived from transgenic cloned dogs. In addition, the insertion site of the RFP gene into genomic DNA is not the same in all experimental dogs (Table S2). If the RFP gene insertion site can affect the immune response, it should affect the results of syngeneic skin grafting. However, no immune rejection was apparent in skin grafts with different transgene insertion sites. Our findings indicate that SCNT-derived somatic cells with or without foreign genes can be accepted in syngeneic recipients.

Our study established that tissues derived from canine SCNT can be accepted in syngeneic recipients despite different mtDNA haplotypes. We also provide evidence that skin segments containing a foreign gene are sufficiently acceptable to syngeneic recipients with or without the foreign gene. Taken together, these data indicate that SCNT using transgenic technology can support immunological compatibility between genetically engineered tissues and patients and thereby help to accelerate clinical therapeutic research and its applications.

Supporting Information

Figure S1 Immunological feature of transgenic dogs and non-transgenic dogs. (a) Molecular typing of dog leukocyte antigen, DLA-88 (MHC class I), DRB, DQA1, DQB1(MHC class II) polymorphic region in all cloned dogs (C1, C2, R1, R2, R3, and R5). (b) *In vitro* immunogenicity test using mixed lymphocyte reaction between all experimental dogs before skin graft. (c) Blood typing. (d) Analysis of blood crossmatching in all cloned dogs and control dogs.

Figure S2 Fluorescence image analysis of skin grafts between cloned dogs with different mtDNA haplotypes

Figure S3 Absence of *in vivo* immune rejection between non-transgenic dogs and transgenic dogs. (a) Positive control of skin graft, as donor skin segments were derived from non-related control dogs (Co1, Co2), they were completely rejected in the graft bed in transgenic cloned dogs (R2, R3). (b) However, skin grafts between a transgenic cloned dog, R2 and a non-transgenic cloned dog, C1 showed no apparent immune rejection. Similarly, as shown in (c) R2 - C2, (d) R3-C1, (e) R3-C2, there was no immune rejection in these grafts as well. (f) Western blot analysis of the skin graft between cloned dogs confirmed the expression of CD4 and CD8 protein only in the graft between cloned dogs and non-related control dogs.

Figure S4 Foreign gene expression between skin graft of two transgenic dogs (R2, R3). Red fluorescent protein expression in skin graft was maintained after 63 days skin graft in syngenic graft beds.

Table S1 Genetic background for microsatellite analysis of two non-transgenic cloned dogs and four transgenic cloned dogs.

Table S2 Insertion site of foreign gene, RFP in transgenic cloned dogs.

Acknowledgments

We thank Won Woo Lee for critical reading of the manuscript. We would also like to thank Dr, Barry D. Bavister for his valuable editing of the manuscript.

Author Contributions

Conceived and designed the experiments: GAK HJO MJK SKK GJ BCL. Performed the experiments: GAK YKJ JC JEP EJP SHL. Analyzed the data: GAK HJO BIY BCL. Contributed reagents/materials/analysis tools: JEP BIY BCL. Wrote the paper: GAK HJO SKK BCL.

References

1. Lee BC, Kim MK, Jang G, Oh HJ, Yuda F, et al. (2005) Dogs cloned from adult somatic cells. Nature 436: 641.
2. Hong SG, Kim MK, Jang G, Oh HJ, Park JE, et al. (2009) Generation of red fluorescent protein transgenic dogs. Genesis 47: 314–322.
3. Chang RS, Suh MS, Kim S, Shim G, Lee S, et al. (2011) Cationic drug-derived nanoparticles for multifunctional delivery of anticancer siRNA. Biomaterials 32: 9785–9795.
4. Lee CY, Li JF, Liou JS, Charng YC, Huang YW, et al. (2011) A gene delivery system for human cells mediated by both a cell-penetrating peptide and a piggyBac transposase. Biomaterials 32: 6264–6276.
5. Kinoshita Y, Kamitani H, Mamun MH, Wasita B, Kazuki Y, et al. (2010) A gene delivery system with a human artificial chromosome vector based on migration of mesenchymal stem cells towards human glioblastoma HTB14 cells. Neurol Res 32: 429–437.
6. Kwak HH, Park KM, Teotia PK, Lee GS, Lee ES, et al. (2013) Acute rejection after swine leukocyte antigen-matched kidney allo-transplantation in cloned miniature pigs with different mitochondrial DNA-encoded minor histocompatibility antigen. Transplant Proc 45: 1754–1760.
7. Martin MJ, Yin D, Adams C, Houtz J, Shen J, et al. (2003) Skin graft survival in genetically identical cloned pigs. Cloning Stem Cells 5: 117–121.
8. Theoret CL, Dore M, Mulon PY, Desrochers A, Viramontes F, et al. (2006) Short- and long-term skin graft survival in cattle clones with different mitochondrial haplotypes. Theriogenology 65: 1465–1479.
9. Hong SG, Jang G, Kim MK, Oh HJ, Park JE, et al. (2009) Dogs cloned from fetal fibroblasts by nuclear transfer. Anim Reprod Sci 115: 334–339.
10. Burnett RC, DeRose SA, Wagner JL, Storb R (1997) Molecular analysis of six dog leukocyte antigen class I sequences including three complete genes, two truncated genes and one full-length processed gene. Tissue Antigens 49: 484–495.
11. Kennedy LJ (2007) 14th International HLA and Immunogenetics Workshop: report on joint study on canine DLA diversity. Tissue Antigens 69 Suppl 1: 269–271.
12. Ogawa H, Galili U (2006) Profiling terminal N-acetyllactosamines of glycans on mammalian cells by an immuno-enzymatic assay. Glycoconj J 23: 663–674.
13. Gluckman JC (1980) [Modification of mixed lymphocyte reactivity between DLA-identical dog sibs, after in vivo sensitization]. C R Seances Acad Sci D 290: 105–108.
14. Kolb HJ, Rieder I, Grosse-Wilde H, Scholz S, Kolb H, et al. (1975) Canine marrow grafts in donor-recipient combinations with one-way nonstimulation in mixed lymphocyte culture. Transplant Proc 7: 461–464.
15. Widmer MB, Bach FH (1972) Allogeneic and xenogeneic response in mixed leukocyte cultures. J Exp Med 135: 1204–1208.
16. Do M, Jang WG, Hwang JH, Jang H, Kim EJ, et al. (2012) Inheritance of mitochondrial DNA in serially recloned pigs by somatic cell nuclear transfer (SCNT). Biochem Biophys Res Commun 424: 765–770.
17. Hiendleder S (2007) Mitochondrial DNA inheritance after SCNT. Adv Exp Med Biol 591: 103–116.
18. Chan T, Fischer Lindahl K (1985) Skin graft rejection caused by the maternally transmitted antigen Mta. Transplantation 39: 477–480.
19. Lindahl KF, Burki K (1982) Mta, a maternally inherited cell surface antigen of the mouse, is transmitted in the egg. Proc Natl Acad Sci U S A 79: 5362–5366.
20. Andersson G, Illigens BM, Johnson KW, Calderhead D, LeGuern C, et al. (2003) Nonmyeloablative conditioning is sufficient to allow engraftment of EGFP-expressing bone marrow and subsequent acceptance of EGFP-transgenic skin grafts in mice. Blood 101: 4305–4312.
21. Rideout WM 3rd, Hochedlinger K, Kyba M, Daley GQ, Jaenisch R (2002) Correction of a genetic defect by nuclear transplantation and combined cell and gene therapy. Cell 109: 17–27.
22. Hanna J, Wernig M, Markoulaki S, Sun CW, Meissner A, et al. (2007) Treatment of sickle cell anemia mouse model with iPS cells generated from autologous skin. Science 318: 1920–1923.
23. Zhao T, Zhang ZN, Rong Z, Xu Y (2011) Immunogenicity of induced pluripotent stem cells. Nature 474: 212–215.
24. Ishikawa K, Toyama-Sorimachi N, Nakada K, Morimoto M, Imanishi H, et al. (2010) The innate immune system in host mice targets cells with allogenic mitochondrial DNA. J Exp Med 207: 2297–2305.
25. Loveland B, Wang CR, Yonekawa H, Hermel E, Lindahl KF (1990) Maternally transmitted histocompatibility antigen of mice: a hydrophobic peptide of a mitochondrially encoded protein. Cell 60: 971–980.
26. Webb KM, Allard MW (2009) Mitochondrial genome DNA analysis of the domestic dog: identifying informative SNPs outside of the control region. J Forensic Sci 54: 275–288.
27. Lanza RP, Chung HY, Yoo JJ, Wettstein PJ, Blackwell C, et al. (2002) Generation of histocompatible tissues using nuclear transplantation. Nat Biotechnol 20: 689–696.
28. Oiso N, Fukai K, Kawada A, Suzuki T (2013) Piebaldism. J Dermatol 40: 330–335.
29. Lu F, Gao JH, Mizuro H, Ogawa R, Hyakusoku H (2007) [Experimental study of adipose tissue differentiation using adipose-derived stem cells harvested from GFP transgenic mice]. Zhonghua Zheng Xing Wai Ke Za Zhi 23: 412–416.
30. Stripecke R, Carmen Villacres M, Skelton D, Satake N, Halene S, et al. (1999) Immune response to green fluorescent protein: implications for gene therapy. Gene Ther 6: 1305–1312.
31. Riddell SR, Elliott M, Lewinsohn DA, Gilbert MJ, Wilson L, et al. (1996) T-cell mediated rejection of gene-modified HIV-specific cytotoxic T lymphocytes in HIV-infected patients. Nat Med 2: 216–223.
32. Bonini C, Ferrari G, Verzeletti S, Servida P, Zappone E, et al. (1997) HSV-TK gene transfer into donor lymphocytes for control of allogeneic graft-versus-leukemia. Science 276: 1719–1724.

11

Equations for Lipid Normalization of Carbon Stable Isotope Ratios in Aquatic Bird Eggs

Kyle H. Elliott[1]*, Mikaela Davis[2], John E. Elliott[3]

1 Department of Biological Sciences, University of Manitoba, Winnipeg, Canada, **2** Department of Biological Sciences, Simon Fraser University, Burnaby, Canada, **3** Science & Technology Branch, Environment Canada, Delta, Canada

Abstract

Stable isotope ratios are biogeochemical tracers that can be used to determine the source of nutrients and contaminants in avian eggs. However, the interpretation of stable carbon ratios in lipid-rich eggs is complicated because ^{13}C is depleted in lipids. Variation in ^{13}C abundance can therefore be obscured by variation in percent lipids. Past attempts to establish an algebraic equation to correct carbon isotope ratios for lipid content in eggs have been unsuccessful, possibly because they relied partly on data from coastal or migratory species that may obtain egg lipids from different habitats than egg protein. We measured carbon, nitrogen and sulphur stable isotope ratios in 175 eggs from eight species of aquatic birds. Carbon, nitrogen and sulphur isotopes were enriched in lipid-extracted egg samples compared with non extracted egg samples. A logarithmic equation using the C:N ratio and carbon isotope ratio from the non extracted egg tissue calculated 90% of the lipid-extracted carbon isotope ratios within ±0.5‰. Calculating separate equations for eggs laid by species in different habitats (pelagic, offshore and terrestrial-influenced) improved the fit. A logarithmic equation, rather than a linear equation as often used for muscle, was necessary to accurately correct for lipid content because the relatively high lipid content of eggs compared with muscle meant that a linear relationship did not accurately approximate the relationship between percent lipids and the C:N ratio. Because lipid extraction alters sulphur and nitrogen isotope ratios (and cannot be corrected algebraically), we suggest that isotopic measurement on bulk tissue followed by algebraic lipid normalization of carbon stable isotope ratio is often a good solution for homogenated eggs, at least when it is not possible to complete separate chemical analyses for each isotope.

Editor: André Chiaradia, Phillip Island Nature Parks, Australia

Funding: KHE received financial support via an NSERC Vanier scholarship, ACUNS Garfield Weston Foundation Northern Studies Award and Jennifer Robinson Memorial Award (Arctic Institute of North America). M. Davis and J. Elliott were funded by Environment Canada's Chemical Management Plan. The funders had no role in study design, data collection and analysis, decision to publish, or preparation of the manuscript.

Competing Interests: The authors have declared that no competing interests exist.

* E-mail: urialomvia@gmail.com

Introduction

Stable isotope analysis is a useful technique for tracing the origin of nutrients in tissues with applications in environmental chemistry, paleoecology, migration biology and diet reconstruction [1–10]. Stable isotope analysis applied to egg tissue is particularly useful for understanding where resources are derived for reproduction (capital vs. income breeding) and to account for variation in toxic contamination within the egg due to diet [8–14].

Agencies in a number of countries systematically collect and archive bird eggs for toxicological and chemical analyses because eggs obtained early in the season can be re-laid, collection of 10–20 eggs has little impact on bird populations numbering in the thousands or millions, and the lipid-rich matrix accumulates many of the lipophilic toxins of interest [9,15–21]. Archived egg specimen banks allow retrospective analysis of toxic contaminants. Stable isotope analysis helps tease apart whether changes in contamination on archived tissue occur due to changes in diet or changes in toxin abundance [17,22–24].

The most common stable isotope ratios used by ecologists are those involving carbon (^{13}C:^{12}C, measured relative to the PeeDee Belemnite standard and denoted $\delta^{13}C$) and nitrogen (^{15}N:^{14}N, measured relative to pure air and denoted $\delta^{15}N$) ratios. Carbon isotopes can be used to identify habitat, as $\delta^{13}C$ varies systematically with degree of aquatic and anthropogenic input. Nitrogen isotopes are primarily used to determine trophic level, as $\delta^{15}N$ increases predictably with trophic level. Increasingly, sulfur isotope ratios (^{34}S:^{32}S, measured relative to the Vienna Cañon Diablo Troilites standard and denoted $\delta^{34}S$) are also used to distinguish nutrients originating from marine environments. Stable isotope ratios change in a systematic fashion as nutrients are assimilated from an animal's food (i.e. the discrimination factor). An accurate knowledge of the discrimination factor is necessary to quantitatively predict nutrient origin [5,25].

Discrimination factors differ among tissue types, with lipids being more depleted in ^{13}C than protein because lipid biosynthesis preferentially incorporates ^{12}C compared with protein [26–31]. In particular, isotopic fractionation during the oxidation of pyruvate to acetyl coenzyme A, the main precursor to fatty acids, preferentially incorporates ^{12}C [32]. Thus, despite originating from the same resource, the stable isotope ratio of consumer tissue will be more depleted in ^{13}C if the lipid content of that tissue is high.

There are two methods for correcting for lipid content [31,33–36]. First, lipids can be extracted chemically prior to measurement, although lipid extraction can alter $\delta^{15}N$ by washing out nitrogenous compounds. Measurement of $\delta^{13}C$ in lipid-extracted tissue and $\delta^{15}N$ in non-extracted tissue overcomes that issue, but doubles cost and work load. Furthermore, the lipid extraction process can be time-consuming and can involve the use of hazardous chemicals, such as chloroform. Second, provided an index of lipid content is available, isotope ratios can be corrected for lipid content analytically without extraction. Often, the ratio of total weight of carbon to total weight of nitrogen within the sample (C:N ratio) is used as an index of lipid content because nitrogen abundance is high in proteins and low in lipids, and C:N ratios are readily calculated from data obtained during stable isotope analysis [31,33–36].

Lipid normalization models are equations that use the C:N ratio or percent lipids to calculate the value of $\delta^{13}C$ that would have been present in a tissue following lipid extraction from the value measured in non-extracted tissue. Such models are accurate in muscle tissue in a wide variety of animals [31,35,37,38]. Furthermore, equations developed from the Post et al. dataset [31] accurately predicted 85% of the data points in a different dataset [35], suggesting that equations are robust across datasets.

Using lipid normalization models in eggs, however, has been more problematic (Table 1). For instance, $\delta^{13}C$ in eggs collected from three species of wild birds varied widely between lipid-extracted and non-extracted tissue, and was not predicted by C:N ratio [38,39]. Similarly, the mean-square error for eggs was four times greater than for muscle (once three outliers and seven points with incomplete lipid extraction were excluded for muscle; no such points were apparently excluded for eggs, [38]). Waterfowl, in particular, show a different or inconsistent trend between lipid-extracted and non-extracted tissue [35,39]. One potential reason for variation among studies is that coastal birds (e.g. eagles, waterfowl) may incorporate variable amounts of freshwater-derived rather than marine-derived lipids into their tissues, that variation may be independent of protein source, and freshwater-derived lipids are depleted in $\delta^{13}C$ compared to marine-derived lipids [38,39].

Both linear and non-linear equations have been used to describe the relationship between the C:N ratio and the effect of lipid extraction on tissue $\delta^{13}C$ values. For tissues that have a relatively low percent lipids, such as muscle, a linear equation appears to hold (most studies show no or only slight non-linearity [31,35,38]; but see [40] that found support for a non-linear equation). However, a nonlinear relationship, whereby the effect of lipid extraction is smaller at higher percent lipid content, appears to work better for eggs, which usually have higher percent lipid content than muscles [35]. The mathematical necessity of a decelerating, non-linear equation for tissues with high percent lipids is evident in the fact that a mixture of pure fatty acids (100% lipids) would have a nitrogen content of zero and an infinite C:N ratio. A linear model would then imply that the $\delta^{13}C$ value would be depleted infinitely, when in fact $\delta^{13}C$ would merely be depleted by 5–6‰.

Because of the paucity of data on the effect of lipid-extraction on the eggs of non-waterfowl [35,38], we examined the effect of lipid extraction on the eggs of six species of marine seabirds and an aquatic raptor. Those seabirds were chosen in the 1980s as part of a toxic contaminant monitoring program in Pacific Canada. The species were chosen because they feed in nearshore, continental shelf and offshore environments, and, therefore, can be used to monitor contaminants in those habitats [9,41]. In the context of the current study, the variation in habitat (freshwater vs. nearshore vs. offshore) allows us to examine the effect of habitat on the predictability of lipid normalization equations. By including multiple individuals per species, from seven different species, we can examine both intra-specific and inter-specific variability in the relationship between lipid-extracted and non-lipid-extracted $\delta^{13}C$ relative to the ratio of bulk C:N. In addition, we examined the effect of a second lipid extraction process on $\delta^{13}C$ of an eighth seabird species. The goal of the manuscript is to develop an equation for normalizing for the effect of lipids on whole egg homogenate, to examine whether that relationship is linear or

Table 1. Difference between lipid-extracted and non-extracted samples for bird egg tissue for carbon ($\Delta\delta^{13}C$), nitrogen ($\Delta\delta^{15}N$) and sulphur ($\Delta\delta^{34}S$).

Species	N	Solvent	$\Delta\delta^{13}C$	$\Delta\delta^{15}N$	$\Delta\delta^{34}S$	Relationship	Reference
Bald eagle	109	Diethyl ether	2.0±1.2‰	0.0±0.3‰		None	Ricca et al. 2007[1]
King eider	18	Chloroform-methanol	4.1±1.2‰	1.2±0.3‰	2.3±1.1‰	None	Oppel et al. 2010[2]
Spectacled eider	15	Chloroform-methanol	2.8±0.4‰	1.0±0.7‰		None	Oppel et al. 2010[2]
Snow goose	11	Chloroform-methanol	1.9‰	0.5‰		None	Ehrich et al. 2011
32 arctic species	1–4	Chloroform-methanol	3.3‰	0.6‰		Nonlinear	Ehrich et al. 2011
Glaucous-winged gull	19	Chloroform-methanol	4.5±0.7‰	0.5±0.5‰		Linear	Our study[2]
Ancient murrelet	6	Petroleum ether	2.0±0.3‰	0.7±0.4‰		Linear	Our study
Double-crested cormorant	10	Petroleum ether	1.3±0.3‰	0.4±0.2‰		Linear	Our study
Great blue heron	2	Petroleum ether	1.0±0.3‰	0.5±0.2‰		None	Our study
Leach's storm-petrel	68	Petroleum ether	2.3±0.4‰	0.9±0.5‰	−0.1±0.9‰	Linear	Our study
Osprey	12	Petroleum ether	1.0±0.6‰	0.8±0.2‰		None	Our study
Pelagic cormorant	26	Petroleum ether	1.8±0.5‰	0.9±0.4‰		Nonlinear	Our study
Rhinoceros auklet	51	Petroleum ether	2.2±0.2‰	1.3±0.3‰	1.6±1.7‰	Nonlinear	Our study
All species (except gulls)	175	Petroleum ether	2.0±0.6‰	0.9±0.5‰	0.5±1.5‰	Nonlinear	Our study

Uncertainty represents SD, where given in published studies. "Relationship" shows whether the relationship was reported to be non-significant ("None"), significant and linear ("Linear") or significant and non-linear ("Non-linear").
[1]Compared percent lipids rather than C:N ratio.
[2]Analyzed whole yolk; all other studies examined egg homogenate.

nonlinear, and to determine whether the relationship is species- or habitat-specific.

Materials and Methods

We randomly selected archived egg contents from the Environment Canada's Specimen Bank [42]. We did not control for laying order (storm-petrels and auklets only lay a single egg), or days since lay, which can affect stable isotope ratios in eggs [43,44]. As we collected eggs randomly during early incubation, we do not believe those issues created systematic bias. Eggs were collected from a nearshore marine seabird (double-crested cormorant *Phalacrocorax auritus*), three seabirds that forage primarily on the continental shelf during the breeding season (ancient murrelet *Synthliboramphus antiquus*, rhinoceros auklet *Cerorhinca monocerata*, pelagic cormorant *Phalacrocorax pelagicus*), one offshore-foraging marine seabird (Leach's storm-petrel *Oceanodrama leucorhoa*), a freshwater aquatic raptor (osprey *Pandion haliaetus*) and an aquatic bird that also uses the terrestrial environment (great blue heron *Ardea herodias*). For the purposes of analysis, we classified bird species into three habitat types: offshore (Leach's storm-petrel), continental shelf (rhinoceros auklet and pelagic cormorant) and those in habitats influenced by the terrestrial environment, such as lakes or estuaries (ancient murrelet, osprey, double-crested cormorant and great blue heron). Egg contents were homogenized and frozen until analysis.

After thawing, samples were freeze-dried. A small sample (1 mg) was removed, encapsulated in tin, and sent to the Stable Isotope Facility at University of California, Davis (http://stableisotopefacility.ucdavis.edu). On a separate sample, we extracted the lipids using a Soxhlet apparatus with petroleum ether as the solvent [45]. Specifically, a thimble filled with dried sample was placed in a Soxhlet extractor and washed with petroleum ether at $94°C$ for 8 hours. The solvent was then distilled off and the residue dried for 60 minutes in a drying oven. We chose petroleum ether rather than chloroform:methanol as our solvent because several Canadian government institutions are considering the restrictions on the use of chloroform, and we wished our methods to be accessible to future researchers. After extraction, a small sample (1 mg) was removed, encapsulated, and sent to the same facility. Likewise, small samples (3 mg) of lipid-extracted and non-extracted tissue were sent to the Environmental Isotope Laboratory at University of Waterloo (http://www.uweilab.ca) for sulphur analysis.

Samples were analyzed for $^{13}C/^{12}C$ and $^{15}N/^{14}N$ isotopes using a PDZ Europa ANCA-GSL elemental analyzer (EA) interfaced to a PDZ Europa 20-20 isotope ratio mass spectrometer (IRMS; Sercon Ltd., Cheshire, UK). Samples were combusted at $1000°C$ in a reactor packed with chromium oxide and silvered copper oxide. Following combustion, oxides were removed in a reduction reactor (reduced copper at $650°C$). The helium carrier then flowed through a water trap (magnesium perchlorate). N_2 and CO_2 were separated on a Carbosieve GC column ($65°C$, 65 mL/min) before entering the IRMS. Samples were analyzed for $^{34}S/^{32}S$ using a Europa Roboprep-20/20 EA-IRMS. During analysis, samples were interspersed with several replicates of at least two different laboratory standards. The final delta values were expressed relative to international standards Vienna PeeDee Belemnite, Vienna Cañon Diablo Troilite and air for carbon, sulphur and nitrogen, respectively.

To examine the effect of a different lipid extraction process on variation in $\delta^{13}C$, we also collected egg yolk from glaucous-winged gulls (*Larus glaucescens*), a nearshore marine seabird that also uses the terrestrial environment. We used chloroform:methanol extraction to remove lipids from egg yolk. Lipid-extracted and non-extracted egg yolks were encapsulated and sent to the UC Davis Stable Isotope Laboratory for carbon and nitrogen analyses. All data for both projects are presented in Table S1.

We used R 2.14.2 for all statistical analyses. We used linear regression and non-linear regression based on previous published logarithmic models [35] to examine the relationship between $\Delta\delta^{13}C$ ($\delta^{13}C_{\text{lipid extracted}}$ - $\delta^{13}C_{\text{non extracted}}$), $\Delta\delta^{15}N$ ($\delta^{15}N_{\text{lipid extracted}}$ - $\delta^{15}N_{\text{non extracted}}$), $\Delta\delta^{34}S$ ($\delta^{34}S_{\text{lipid extracted}}$ − $\delta^{34}S_{\text{non extracted}}$) and the C:N ratio. Linear models are appropriate for tissues with low percent lipids because as long as percent lipids is low a linear model approximates the nonlinear function; for tissues with high percent lipids (e.g., eggs), the C:N ratio does not change linearly with percent lipids (because the denominator, nitrogen, approaches zero) necessitating a non-linear function [35,40]. Because models with an increasing number of parameters will necessarily increase fit while including spurious relationships (i.e., a model with 175 parameters would explain 100% of the variation in our 175-sample data set), we used Akaike's information criterion to select the most parsimonious relationship. We considered both linear and non-linear functions, and functions that included species and habitat (offshore, shelf and terrestrial-influenced), and their interactions, as co-variates. We then examined whether each of the 175 values were predicted within 0.5‰ by each model, with each model recalculated to exclude that point. Because many authors will only examine these relationships within a single species, we also calculated species-specific regressions to determine the likelihood of finding a relationship given only a single species with a more limited range of lipid content. We also correlated $\delta^{13}C$lipid-extracted and $\delta^{13}C$non-extracted. We removed eight values, all from the same Soxhlet run, because the C:N ratio after lipid extraction was >5.0. All raw data are available in Table S1.

Results

Lipid-extracted egg samples were more enriched in carbon, sulphur and nitrogen than non-extracted egg samples (Table 1). Across the entire dataset, there was strong support for a logarithmic relationship between $\Delta\delta^{13}C$ and the C:N ratio (Table 2, Fig. 1). Relationships between $\Delta\delta^{15}N$ and the C:N ratio, and $\Delta\delta^{34}S$ and the C:N ratio, were weaker, but still supported (Table 2). For the entire dataset of 175 eggs, 22 (12%) of the lipid-extracted egg values differed by more than 0.5‰ in $\delta^{13}C$ when calculated using the linear model, 19 (10%) differed by more than 0.5‰ for the logarithmic model and 12 (7%) differed by more than 0.5‰ from the complete model with habitat and species interactions included. Nonetheless, the best-supported model included separate terms in the intercept (but not the slope) for each habitat (Table 2). In contrast, 39 (22%) of the lipid-extracted egg values differed by more than 0.5‰ compared with the linear "waterfowl" equation from Ehrich et al. (2011; their Table 2, "Linear") and 173 (99%) of the values compared with the linear "non-waterfowl" equation. Lipid-extracted and algebraically-corrected $\delta^{13}C$ values were highly correlated ($R^2 = 0.990$; compared with $R^2 = 0.969$ for the non-algebraically corrected values). Average repeatability (standard deviation in ‰, N = 19) for duplicates (separately homogenized) was 0.26 for $\delta^{34}S$, 0.12 for $\delta^{13}C$ and 0.31 for $\delta^{15}N$.

There was a statistically-significant relationship between $\Delta\delta^{13}C$ and the C:N ratio for all of the species with sample sizes greater than two (ancient murrelet: $t_5 = 4.34$, P = 0.007; double-crested cormorant: $t_8 = 3.73$, P = 0.006; Leach's storm-petrel: $t_{66} = 6.43$, P<0.0001; pelagic cormorant: $t_{24} = 7.32$, P<0.0001; rhinoceros

Table 2. Ranking of models used to describe the difference between lipid-extracted and non-extracted bird egg tissue.

Model	ΔAIC	Equation
$\Delta\delta^{13}C$		
Species	102.38	
Linear	54.48	
Non-linear	38.08	$-4.46\pm0.35+7.32\pm0.40$ * Log (C:N Ratio)
Linear+Species	17.16	
Linear+Habitat	11.10	
Non-linear+Species+Non-linear*Species	10.21	
Non-linear+Species	6.99	
Non-linear+Habitat+Non-linear*Habitat	3.97	
Non-linear+Habitat	0.00	$-3.65\pm0.34+6.03\pm0.40$ * Log (C:N Ratio)+0.32 ± 0.07 (If Offshore)+0.50 ± 0.07 (If Pelagic)
$\Delta\delta^{15}N$		
Linear	25.97	
Non-linear	24.84	$-1.47\pm0.49+2.72\pm0.55$ * Log (C:N Ratio)
Habitat	22.01	
Species	8.84	
Non-linear+Habitat	8.43	
Non-linear+Species+Nonlinear*Species	7.03	
Linear+Species	1.53	
Non-linear+Species	0.00	$-1.66\pm0.74+2.50\pm0.76$ * Log (C:N Ratio)-0.02 ± 0.24 (If double-crested cormorant)+0.27 ± 0.37 (If great blue heron)+0.26 ± 0.18 (If Leach's storm-petrel)+0.56 ± 0.24 (If osprey)+0.43 ± 0.20 (If pelagic cormorant)+0.62 ± 0.18 (If rhinoceros auklet)
$\Delta\delta^{34}S$		
Non-linear	8.57	
Linear[5]	8.50	$4.69\pm3.51+-0.50\pm0.43$ (C:N Ratio)
Non-linear+Species+Species*Non-linear	1.61	
Linear+Species+Species*Linear	1.59	
Species	0.58	
Non-linear+Species	0.02	
Linear+Species	0.00	$0.93\pm3.23+-0.12\pm0.38$ * (C:N Ratio)+1.68 ± 0.50 (If rhinoceros auklet)

Equations are shown for the most parsimonious complete models and most-parsimonious species- and habitat-independent models. Habitat classifications were "terrestrially-influenced" (default), "offshore" (continental shelf) or "pelagic" (beyond the shelf).

auklet: $t_{49}=6.56$, $P<0.0001$; gull yolk: $t_{17}=4.44$, $P=0.0004$), except ospreys ($t_{10}=-0.13$, $P=0.90$).

Discussion

Lipid extraction enriched egg $\delta^{13}C$ values by ~2‰, $\delta^{15}N$ values by ~1‰ and $\delta^{34}S$ by ~0.5‰. The degree of enrichment in egg $\delta^{13}C$ values ($\Delta^{13}C$) was correlated with the percent lipids, as inferred by the C:N ratio. Indeed, across the entire dataset, and within all marine species, the C:N ratio strongly predicted $\Delta^{13}C$. As more than 90% of the samples were estimated within ±0.5‰ by the best-fit algebraic equation (logarithm model), our algebraic equation is a robust method of calculating lipid-extracted $\delta^{13}C$ values given $\delta^{13}C$ values measured on non-extracted egg tissue from marine birds.

Our models for egg tissue were as accurate as those for muscle tissue [31,35]. In contrast, most past attempts at providing a method for calculating lipid-extracted values for egg tissues have been unsuccessful (Table 1). Those methods focused on migratory or coastal species, such as bald eagles, eiders, snow geese and other waterfowl (Table 1). For migratory birds—at least those that are capital breeders—egg lipids can be derived from energy reserves

obtained on non-breeding grounds with different $\delta^{13}C$ signatures; egg protein, in contrast, may be derived from the breeding grounds [35,39,46]. Differences in $\delta^{13}C$ may be particularly large for coastal birds switching between freshwater and marine prey bases [17], or for other species showing dietary switches during egg-laying [47,48], even if they are non-migratory. For instance, the ospreys whose eggs were included in this study were tracked via satellites to wintering grounds in both marine and freshwater environments [49]. If a portion of their egg lipid $\delta^{13}C$ values were derived from wintering grounds and their protein $\delta^{13}C$ values were mainly derived from breeding grounds, that would explain why the lipid-extracted and non-extracted samples were consistently different [35,39]. We therefore concur with Oppel et al. [39] that algebraic correction for the effect of lipids on $\delta^{13}C$ in eggs is accurate for birds relying on resources acquired concurrently with reproduction for egg synthesis, such as income breeders, but not for coastal or migratory birds that may bring lipid reserves from a habitat with a different $\delta^{13}C$ value. Specifically, algebraic corrections are applicable where lipids and proteins would be expected to be acquired in isotopically similar landscapes, with capital and income breeding being extremes of a continuum with

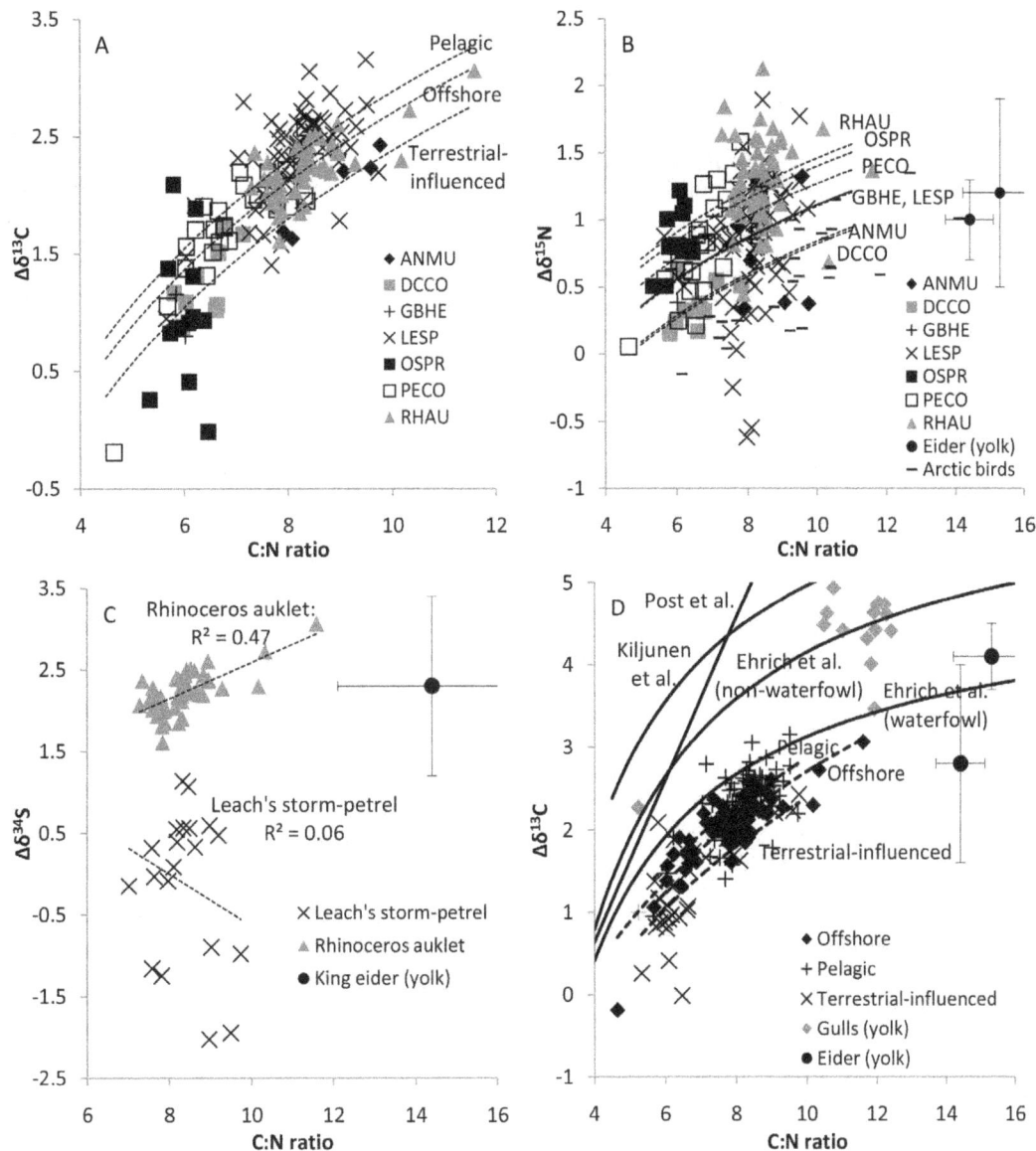

Figure 1. Difference between lipid-extracted and non-extracted stable isotope ratios for bird egg tissue. Specifically (A) carbon ($\Delta\delta^{13}C$), (B) nitrogen ($\Delta\delta^{15}N$) and (C) sulphur ($\Delta\delta^{34}S$) increases with ratio of carbon to nitrogen by weight (C:N ratio) across seven aquatic bird species: ancient murrelet (ANMU), double-crested cormorant (DCCO), great blue heron (GBHE), Leach's storm-petrel (LESP), osprey (OSPR), pelagic cormorant (PECO) and rhinoceros auklet (RHAU). Also shown are results from studies listed in Table 1 (eider average with SD bars shown, Arctic birds) and best-fit habitat- and species-dependent regression models listed in Table 2. (D) $\Delta\delta^{13}C$ for groups within our study compared with arithmetic lipid-correction models proposed by Post et al. [29], Ehrich et al. [33] (filled lines) and within our own study (dashed lines).

actual reproductive strategies ranging from high to low reliance on stored nutrients [50,51]. It is also important to emphasize that our results were primarily for egg homogenate, which has a higher proportion of protein (mostly albumin) than yolk. Arithmetic correction for yolk may be more problematic because of the lower proportion of protein in yolk and may account for problems encountered by previous authors working with yolk [39].

Stage of embryonic development also may have impacted the relationship between C:N ratio and $\Delta\delta^{13}C$. As the embryo develops, there may be differential uptake of varying nutrients and, therefore, variation in $^{13}CO_2$ expiration. For instance, chicken yolk $\delta^{15}N$ was depleted by 1.0‰ after 15 days of development and chicken albumin $\delta^{13}C$ was depleted by 0.2‰ after 3 days of development [43]. In principle, if ^{13}C was excreted differentially to ^{12}C in yolk relative to plasma, that could increase

the variance in $\Delta\delta^{13}C$ relative to lipid content. As we did not know the age of the embryo, this issue could play a role in our dataset, although collectors attempted to obtain eggs shortly after laying.

At the early stages of development (prior to significant growth of the embryo), the egg interior primarily consists of the cysteine-rich protein albumin (egg white) and lipid-rich yolk. Many authors have separated the two components and analyzed stable isotopes separately on each component [39,43]. As we were interested in establishing lipid normalization equations for retrospective analyses of specimen banks for toxicological assays, where specimens have usually been homogenized, we did not examine each component separately, except for the gull eggs. Interestingly, our lipid-extracted C:N ratio, 4.33 (SD = 0.50), is almost identical to that of egg albumin (4.28 based on the albumin amino-acid sequence), suggesting following lipid extraction on egg homogenate we were left almost entirely with albumin or with a matrix that was very similar in makeup. In particular, simple statements that pure protein should have a C:N ratio of ~3.0 [30] seem erroneous as the C:N ratio will depend on the particular amino acid composition of the protein.

The influence of habitat and breeding strategy on $\Delta\delta^{13}C$

Pelagic foragers (Leach's storm-petrels) had higher $\Delta\delta^{13}C$ than nearshore foragers (rhinoceros auklets and pelagic cormorants), which had higher $\Delta\delta^{13}C$ than birds foraging in terrestrial-influenced habitats (ancient murrelets, double-crested cormorants, great blue herons and ospreys). Ancient murrelets may forage farther offshore than pelagic cormorants, but for the current classification scheme they appeared to group with those species feeding in terrestrial-influenced habitats. Perhaps there is a larger difference in the carbon isoscape in winter (where lipids were derived) relative to breeding (where non-lipids were derived) habitat for birds feeding offshore than for those feeding in the terrestrial environment. Alternatively, within the capital-income breeding strategy continuum, Leach's storm-petrels may rely more on endogenous stores for egg production and the terrestrial-influenced species may rely more on exogenous stores for egg production [48,50–52]. Ehrich et al. [35] found higher $\Delta\delta^{13}C$ for terrestrial birds (along with one seabird and several shorebirds) than waterfowl. Either variation in habitat or breeding strategy (endogenous or capital vs. exogenous or income) may explain why $\Delta\delta^{13}C$ differed between terrestrial birds and waterfowl in that study.

The effect of solvent

The solvent used during lipid extraction is known to effect measurement of $\delta^{13}C$ and $\delta^{15}N$, at least in fish and invertebrates [45,52]. The ideal solvent would extract all lipids, and only lipids, but in practice both lipids and non-lipids vary in polarity; non-polar lipids, such as triglycerides, are more soluble in non-polar solvents (e.g. petroleum ether, polarity index ~0.1) and polar lipids, such as phospholipids, are more soluble in polar solvents (e.g. 2:1 chloroform:methanol, polarity index ~4.4). Thus, no solvent will extract all lipids and all solvents will extract some non-lipids, such as hydrophobic amino acid or carbohydrate derivatives. In one study, $\delta^{13}C$ in lipid-extracted tissue varied among three different solvents, with chloroform-methanol (Bligh-Dyer method, [54]) causing systematic errors on one type of invertebrate tissue [45]. In contrast, another study on fish tissue found that chloroform-methanol extraction was somewhat more effective than other techniques [53].

We believe a solvent effect is present for bird eggs. The arithmetic correction for $\Delta\delta^{13}C$ proposed by Ehrich et al. [35] is quite effective for non-waterfowl (gulls in our study) and waterfowl [39] in other studies that used chloroform:methanol as the solvent (Fig. 1d). The Ehrich et al. [36] equation calculates higher $\Delta\delta^{13}C$ values than our equations, which would be consistent with chloroform-methanol being a stronger solvent (extracting a greater overall portion of lipids, but also extracting some non-lipids). Furthermore, pure gull albumin had a C:N ratio of 3.72±0.06 and lipid-extracted whole egg had a C:N ratio of 3.72±0.04 in the gull dataset using the chloroform-methanol extraction process, whereas lipid-extracted whole egg had a C:N ratio of 4.33±0.50 in the seabird dataset using the petroleum ether extraction process. Thus, we believe that chloroform-methanol extractions remove more, or at least different, lipids (and other molecules) than petroleum ether, and therefore cause a larger $\Delta\delta^{13}C$.

Why use lipid normalization equations?

Our lipid normalization equations will be of use in accounting for the effect of lipids in toxicological studies of purely marine birds, although we urge the establishment of separate equations for

each species. There are many reasons why lipid extraction is sub-optimal: chloroform is not allowed at many laboratories for health reasons, the Soxhlet apparatus used in the current studies cost over $10 000, the apparatus created a bottleneck of only eight samples processed each day, $\delta^{13}C$ (and $\delta^{15}N/\delta^{34}S$) will vary among extraction techniques and, finally, to avoid overestimation of $\delta^{15}N$ or $\delta^{34}S$ (which cannot be corrected algebraically) it would be necessary to complete two different analyses, doubling expenses. Lastly, lipid extraction removes at least some amino acids, which can insert some (unwanted) variability into diet estimation. For studies of nutrient allocation or prey use, very accurate and unbiased measurements of $\delta^{13}C$ may be necessary, but for toxicological studies that primarily use stable isotope ratios as correlative, predictive variables, the algebraic approach may be sufficient; across all species, lipid-extracted and algebraically-corrected values were highly correlated ($R^2 = 0.99$) so that in correlative studies essentially no information would be lost by using the algebraic approach. Furthermore, where the algebraic equations do not provide an accurate index of lipid extracted $\delta^{13}C$ because lipids and proteins are derived from different environments, the algebraic equations will give the value more associated with lipophilic contaminants; the stable isotope ratios measured on lipid-extracted tissue (albumin) will not reflect the diet associated with the prey from which the lipophilic contaminants are derived. We, therefore, urge the use of algebraic corrections, rather than lipid extraction, in toxicological studies.

What equation to use?

The most-widely used equation is the Post et al. [31] equation, which is accurate for muscle from many different organisms [31,35]. Muscle is relatively low in lipids, so the non-linearity apparent in our dataset is not a significant issue (i.e. the "Post" equation is similar to the "Ehrich" equation at low C:N ratio, Fig. 1d). Because egg tissue can have much higher levels of lipids, a non-linear equation is necessary (Fig. 1a,d, Table 2). To estimate lipid-extracted egg tissue values (using a petroleum ether extraction process), we therefore suggest using the equation

$$\delta^{13}C_{\text{lipid-extracted}} = \delta^{13}C_{\text{non-extracted}} + 1.47$$
$$- 2.72 * Log_{10}(C:N \text{ Ratio})$$

which can be increased or decreased depending on the habitat of the study species (following Table 1). The C:N Ratio is the ratio of the weight of carbon in the sample to the weight of nitrogen in the sample, often reported in µg. The equation is not species-specific, as we did not find that species was an important term in our models.

Supporting Information

Table S1 Stable isotope data analyzed in the manuscript.

Acknowledgments

We thank J. Roth for access to the Soxhlet apparatus. M. Guigueno and D., J. and L. Jamieson completed the lipid extraction and sample encapsulation. R. McDonald and three anonymous reviewers provided useful comments on a previous version of this manuscript.

Author Contributions

Conceived and designed the experiments: KHE JEE MD. Performed the experiments: KHE MD. Analyzed the data: KHE. Contributed reagents/materials/analysis tools: JEE. Wrote the paper: KHE.

References

1. Roth JD (2003) Variability in marine resources affects arctic fox population dynamics. J Anim Ecol 72: 668–676.
2. West JB, Bowen GJ, Cerling TE, Ehleringer JR (2006) Stable isotopes as one of nature's ecological recorders. Trends Ecol Evol 21: 408–414.
3. Chiaradia A, Forero MG, Hobson K, Swearer S, Hume F, et al. (2012) Diet segregation between two colonies of little penguins *Eudyptula minor* in southeast Australia. Austr Ecol 37: 610–619.
4. Choy ES, Gauthier M, Mallory ML, Smol JP, Lean D, et al. (2010) An isotopic investigation of mercury accumulation in terrestrial food webs adjacent to an Arctic seabird colony. Sci Tot Environ 408: 1858–1867.
5. Jardine TD, Kidd KA, Fisk AT (2006) Applications, considerations, and sources of uncertainty when using stable isotope analysis in ecotoxicology. Environ Sci Toxicol 40: 7501–7511.
6. Elliott KH, Crump D, Gaston AJ (2010) Sex-specific behavior by a monomorphic seabird represents risk partitioning. Behav Ecol 21: 1024–1032.
7. Wolf N, Carleton SA, Martinez del Rio CM (2009) Ten years of experimental animal isotopic ecology. Funct Ecol 23: 17–26.
8. Inger R, Bearhop S (2008) Applications of stable isotope analyses to avian ecology. Ibis 150: 447–461.
9. Elliott JE, Elliott KH (2013) Tracking marine pollution. Science 340: 556–558.
10. Elliott KH, Cesh LS, Dooley JA, Letcher RJ, Elliott JE (2009) PCBs and DDE, but not PBDEs, increase with trophic level and marine input in nestling bald eagles. Sci Tot Environ 407: 3867–3875.
11. Guigueno MF, Elliott KH, Levac J, Wayland M, Elliott JE (2012) Differential exposure of alpine ospreys to mercury: melting glaciers, hydrology or deposition patterns? Environ Int 40: 24–32.
12. Elliott JE, Levac J, Guigueno M, Shaw P, Wayland ME, et al. (2012) Factors influencing legacy pollutant accumulation in alpine osprey: biology, topography or melting glaciers?. Environ Sci Tech 46: 9681–9689.
13. Day RD, Roseneau DG, Vander Pol SS, Hobson KA, Donard OF, et al. (2012) Regional, temporal, and species patterns of mercury in Alaskan seabird eggs: Mercury sources and cycling or food web effects? Environ Poll 166: 226–232.
14. Ito M, Kazama K, Niizuma Y, Minami H, Tanaka Y, et al. (2012) Prey resources used for producing egg yolks in four species of seabirds: insights from stable-isotope ratios. Ornithol Sci 11: 113–119.
15. Braune BM, Hobson KA, Malone BJ (2005) Regional differences in collagen stable isotope and trace element profiles in populations of long-tailed duck breeding in the Canadian Arctic. Sci Tot Environ 346: 156–168.
16. Braune BM, Trudeau S, Jeffrey DA, Mallory ML (2011) Biomarker responses associated with halogenated organic contaminants in northern fulmars (*Fulmarus glacialis*) breeding in the Canadian Arctic. Environ Poll 159: 2891–2898.
17. Hebert CE, Weseloh DVC, Idrissi A, Arts MT, O'Gorman R, et al. (2008) Restoring piscivorous fish populations in the Laurentian Great Lakes causes seabird dietary change. Ecology 89: 891–897.
18. Holmström KE, Järnberg U, Bignert A (2005) Temporal trends of PFOS and PFOA in guillemot eggs from the Baltic Sea, 1968–2003. Environ Sci Tech 39: 80–84.
19. Mallory ML, Braune BM (2012) Tracking contaminants in seabirds of Arctic Canada: Temporal and spatial insights. Mar Poll Bull 64: 1475–1484.
20. Vander Pol SS, Becker PR, Kucklick JR, Pugh RS, Roseneau DG, et al. (2004) Persistent organic pollutants in Alaskan murre (*Uria* spp.) eggs: Geographical, species, and temporal comparisons. Environ Sci Tech 38: 1305–1312.
21. Crosse JD, Shore RF, Jones KC, Pereira MG (2012) Long-term trends of PBDEs in gannet (*Morus bassanus*) eggs from two UK colonies. Environ Poll 161: 93–100.
22. Elliott JE (2005) Chlorinated hydrocarbon contaminants and stable isotope ratios in pelagic seabirds from the North Pacific Ocean. Arch Environ Contamin Toxicol 49: 89–96.
23. Elliott JE (2005) Trace metals, stable isotope ratios, and trophic relations in seabirds from the North Pacific Ocean. Environ Toxicol Chem 24: 3099–3105.
24. Vo ATE, Bank MS, Shine JP, Edwards SV (2011) Temporal increase in organic mercury in an endangered pelagic seabird assessed by century-old museum specimens. Proc Natl Acad Sci U S A 108: 7466–7471.
25. Bond AL, Diamond AW (2011) Recent Bayesian stable-isotope mixing models are highly sensitive to variation in discrimination factors. Ecol Appl 21: 1017–1023.
26. Sotiropoulos MA, Tonn WM, Wassenaar LI (2004) Effects of lipid extraction on stable carbon and nitrogen isotope analyses of fish tissues: potential consequences for food web studies. Ecology of Freshwater Fish 13: 155–160.
27. Sweeting CJ, Polunin NVC, Jennings S (2004) Tissue and fixative dependent shifts of delta C-13 and delta N-15 in preserved ecological material. Rapid Comm Mass Spectr 18: 2587–2592.
28. Sweeting CJ, Polunin NVC, Jennings S (2006) Effects of chemical lipid extraction and arithmetic lipid correction on stable isotope ratios of fish tissues. Rapid Comm in Mass Spectr 20: 595–601.
29. Bodin N, Le Loc'h F, Hily C (2007) Effect of lipid removal on carbon and nitrogen stable isotope ratios in crustacean tissues. J Exp Mar Biol Ecol 341: 168–175.
30. Hussey NE, Olin JA, Kinney MJ, McMeans BC, Fisk AT (2012) Lipid extraction effects on stable isotope values ($\delta^{13}C$ and $\delta^{15}N$) of elasmobranch muscle tissue. J Exp Mar Biol Ecol 434: 7–15.
31. Post DM, Layman CA, Arrington DA, Takimoto G, Quattrochi J, et al. (2007) Getting to the fat of the matter: models, methods and assumptions for dealing with lipids in stable isotope analyses. Oecologia 152: 179–189.
32. De Niro MJ, Epstein S (1977) Mechanism of carbon isotope fractionation associated with lipid synthesis. Science 197: 261–263
33. Kiljunen M, Grey J, Sinisalo T, Harrod C, Immonen H, et al. (2006) A revised model for lipid-normalizing delta C-13 values from aquatic organisms, with implications for isotope mixing models. J Appl Ecol 43: 1213–1222.
34. Mintenbeck K, Brey T, Jacob U, Knust R, Struck U (2008) How to account for the lipid effect on carbon stable-isotope ratio (delta C-13): sample treatment effects and model bias. J Fish Biol 72: 815–830.
35. Ehrich D, Tarroux A, Stien J, Lecomte N, Killengreen S, et al. (2010) Stable isotope analysis: modeling lipid normalization for muscle and eggs from arctic mammals and birds. Meth Ecol Evol 2: 66–76.
36. Tarroux A, Ehrich D, Lecomte N, Jardine TD, Bêty J, et al. (2010) Sensitivity of stable isotope mixing models to variation in isotopic ratios: evaluating consequences of lipid extraction. Meth Ecol Evol 1: 231–241.
37. Lesage V, Morin Y, Rioux È, Pomerleau C, Ferguson SH, et al. (2010) Stable isotopes and trace elements as indicators of diet and habitat use in cetaceans: predicting errors related to preservation, lipid extraction, and lipid normalization. Mar Ecol Prog Ser 419: 249–265.
38. Ricca MA, Miles AK, Anthony RG, Deng X, Hung SSO (2007) Effect of lipid extraction on analyses of stable carbon and stable nitrogen isotopes in coastal organisms of the Aleutian archipelago. Can J Zool 85: 40–48.
39. Oppel S, Federer RN, O'Brien DM, Powell AN, Hollmen TE (2010) Effects of lipid extraction on stable isotope ratios in avian egg yolk: is arithmetic correction a reliable alternative? Auk 127: 1–7.
40. Logan JM, Jardine TD, Miller TJ, Bunn SE, Cunjak RA, et al. (2008) Lipid corrections in carbon and nitrogen stable isotope analyses: comparison of chemical extraction and modelling methods. J Anim Ecol 77: 838–846.
41. Elliott JE, Noble DG, Norstrom RJ, Whitehead PE, Simon M, et al. (1992) Patterns and trends of organic contaminants in Canadian seabirds, 1968–1990. In Persistent Pollutants in the Marine Environment, C.H. Walker and D.R. Livingston (Eds.) Pergamon Press. Oxford, pp. 181–194.
42. Elliott JE (1984) Collecting and archiving wildlife specimens in Canada. In: R.A. Lewis, N. Stein, and C.W. Lewis (eds.) Environmental Specimen Banking and Monitoring as Related to Banking, Martinus Nijhoff, The Hague, pp. 45–64.
43. Sharp CM, Abraham KF, Burness G (2009) Embryo development influences the isotopic signatures of egg components in incubated eggs. Condor 111: 361–365.
44. Hobson KA, Thompson JE, Evans MR, Boyd S (2005) Tracing nutrient allocation to reproduction in Barrow's goldeneye. J Wildl Manage 69:1221–1228.
45. Schlechtriem C, Focken U, Becker K (2003) Effect of different lipid extraction methods on $\delta^{13}C$ of lipid and lipid-free fractions of fish and different fish feeds. Isotopes in Environ Health Stud 39: 135–140.
46. Yohannes E, Valcu M, Lee RW, Kempenaers B (2010) Resource use for reproduction depends on spring arrival time and wintering area in an arctic breeding shorebird. J Avian Biol 41: 580–590.
47. Jacobs SR, Elliott KH, Gaston AJ, Weber JM (2009) Fatty acid signatures of female Brünnich's guillemot *Uria lomvia* suggests reliance on local prey for replacement egg production. J Avian Biol 40: 327–336.
48. Morrissey CA, Elliott JE, Ormerod SJ. 2010. Diet shifts during egg laying: implications for measuring contaminants in bird eggs. Environ Poll 158: 447–454.
49. Elliott JE, Morrissey CA, Henny CJ, Inzunza ER, Shaw P (2007) Satellite telemetry and prey sampling reveal contaminant sources to Pacific Northwest ospreys. Ecol Appl 17: 1223–1233.
50. Thomas VG (1989) Body condition, ovarian hierarchies, and their relation to egg formation in anseriform and galliform species. Proc Int Ornithol Congr XIX: 353–363.
51. Jönsson KI (1997) Capital and income breeding as alternative tactics of resource use in reproduction. Oikos : 57–66.
52. Hobson KA, Jehl JR (2010) Arctic waders and the capital-income continuum: further tests using isotopic contrasts of egg components. J Avian Biol 41: 465–572.
53. Logan JM, Lutcavage ME (2008) A comparison of carbon and nitrogen stable isotope ratios of fish tissues following lipid extractions with non-polar and traditional chloroform/methanol solvent systems. Rapid Comm Mass Spectr 22: 1081–1086.
54. Bligh EG, Dyer WJ (1959) A rapid method of total lipid extraction and purification. Can J Biochem Physiol 37: 911–917.

12

A Recommended Numbering Scheme for Influenza A HA Subtypes

David F. Burke*, Derek J. Smith

Department of Zoology, University of Cambridge, Cambridge, United Kingdom

Abstract

Comparisons of residues between sub-types of influenza virus is increasingly used to assess the zoonotic potential of a circulating strain and for comparative studies across subtypes. An analysis of N-terminal cleavage sites for thirteen subtypes of influenza A hemagglutinin (HA) sequences, has previously been described by Nobusawa and colleagues. We have expanded this analysis for the eighteen known subtypes of influenza. Due to differences in the length of HA, we have included strains from multiple clades of H1 and H5, as well as strains of H5 and H7 subtypes with both high and low pathogenicity. Analysis of known structures of influenza A HA enables us to define amino acids which are structurally and functionally equivalent across all HA subtypes using a numbering system based on the mature HA sequence. We provide a list of equivalences for amino acids which are known to affect the phenotype of the virus.

Editor: Paul Digard, University of Edinburgh, United Kingdom

Funding: Funding provided by (DJS) Bill & Melinda Gates Foundation Global Health (http://www.gatesfoundation.org/) Grant # OPPGH5383, (DJS) European Union FP7 program ANTIGONE (http://cordis.europa.eu/programme/rcn/852_en.html) (278976) and (DJS) National Institute of Allergy and Infectious disease (http://www.niaid.nih.gov) Contract HHSN266200700010C. The funders had no role in the study design, data collection, analysis, decision to publish or preparation of the manuscript.

Competing Interests: The authors have declared that no competing interests exist.

* Email: dfb21@cam.ac.uk

Introduction

Increasingly, amino acid changes in HA, resulting from either natural evolution or experimental design, are compared to amino acids within another subtype. A common example are those mutations that have been shown to confer binding to human glycans. In strains from the H3 subtype, these are Gln226Leu and Gly228Ser whereas in strains from the H5 subtype these mutations are positions 222 and 224. Although simple 'rules-of-thumb' can be derived, such as the subtracting four from the H3 numbering to get the position in H5 viruses, this is not always straightforward, as typified by the recent focus on H7 viruses. The HA of H7 strains contain many amino acid insertions and deletions (indels) relative to viruses from the other subtypes. For amino acids close to the receptor binding site, such as the aforementioned mutations, the H7 numbering differs from H3 numbering by nine residues (Gln217 and Gly219). However, two other mutations of concern, His103Tyr and Thr315Ile, which were recently shown to facilitate the aerosol transmission of avian A/H5N1 viruses between mammals [1–2], lie in the N and C termini of HA1, respectively. Due to the indels in these regions, the equivalent amino acids in H7 strains differ by three (Gln100) and six (Thr309) amino acids, respectively. As shown for H7, the conversion of residue numbering between subtypes varies depending on the region of HA being compared. Yet another complication arises due to genetic changes within a subtype which, although uncommon, do occur. Over one-fifth of the avian H5N1 strains in the Middle East sequenced to date have a deletion between amino acids positions 128 and 130 (mature HA H5N1 numbering). This deletion was also found in human seasonal H1 strains after 1995 but was not present in early H1 strains or any of the H1pdm strains currently circulating [3]. Similarly, a clade of H7 strains circulating in North America and Canada since 1996 has been shown to have eight amino acids deleted, located surprisingly close to the receptor binding site [4]. Conversion rules thus also depend upon the lineage of the subtypes that are being compared.

Nobusawa and colleagues previously predicted the N-terminal sequence for thirteen subtypes of HA based on the likely signal peptide cleavage site of the N-terminal signal peptide [5], thus providing a numbering scheme based on the mature sequence of HA. Although widely cited, not all publications use this numbering. For example, only two (3M6S and 3ZTN) out of the thirteen currently available crystal structures of HA of the vaccine strain of H1pdm (A/California/04/2009) start with the mature HA sequence (Asp-Thr-Leu-Cys-Ile). Alternative structures include six (3AL4, 4JTV and 4JU0) or ten (3LZG, 3UBE, 3UBN, 3UBQ and 4F3Z) additional N-terminal amino acids. This variation in N-terminal numbering, in addition to subtype specific differences caused by indels, can increase confusion in interpreting amino acid equivalences. To avoid inaccuracies, it is important to have a scheme to define and compare numbering between subtypes.

Here we report an updated prediction of the proteolytic cleavage sites for all subtypes. We analyse known structures of HA to enable us to define amino acids which are structurally and functionally equivalent across the eighteen currently known subtypes of influenza A. Combining both of these results, we are able to compile a list of equivalences for amino acids which are

Table 1. Predicted signal peptide cleavage sites for all HA subtypes.

Subtype	Representative strain	Signal Peptide	N-terminal sequence of mature protein
H1	A/United Kingdom/1/1933	MKARLLVLLCALAATDA	DTICIGYHANNS
H2	A/Singapore/1/1957	MAIIYLILLFTAVRG	DQICIGYHANNS
H3	A/Aichi/2/1968	MKTIIALSYIFCLPLG	QDLPGNDNSTATLCLGHHAVPN
H4	A/swine/Ontario/01911-2/1999	MLSIAILFLLIAEGSS	QNYTGNPVICLGHHAVSN
H5	A/Vietnam/1203/2004	MEKIVLLFAIVSLVKS	DQICIGYHANNS
H6	A/chicken/Taiwan/0705/1999	MIAIIVIATLAAAGKS	DKICIGYHANNS
H7	A/Netherlands/219/2003	MNTQILVFALVASIPTNA	DKICLGHHAVSN
H8	A/turkey/Ontario/6118/1968	MEKFIAIAMLLASTNA	YDRICIGYQSNNS
H9	A/swine/Hong Kong/9/1998	MEAASLITILLVVTASNA	DKICIGYQSTNS
H10	A/mallard/bavaria/3/2006	MYKIVVIIALLGAVKG	LDKICLGHHAVAN
H11	A/duck/England/1/1956	MEKTLLFAAIFLCVKA	DEICIGYLSNNS
H12	A/duck/Alberta/60/1976	MEKFIILSTVLAASFA	YDKICIGYQTNNS
H13	A/gull/Maryland/704/1977	MALNVIATLTLISVCVHA	DRICVGYLSTNS
H14	A/mallard/Astrakhan/263/1982	MIALILVALALSHTAYS	QITNGTTGNPIICLGHHAVEN
H15	A/duck/Australia/341/1983	MNTQIIVILVLGLSMVRS	DKICLGHHAVAN
H16	A/black-headed-gull/Turkmenistan/13/1976	MMIKVLYFLIIVLGRYSKA	DKICIGYLSNNS
H17	A/little-yellow-shouldered bat/Guatemala/060/2010	MELIILLILLNPYTFVLG	DRICIGYQANQN
H18	A/flat-faced bat/Peru/033/2010	MITILILVLPIVVG	DQICIGYHSNNS

The N-terminal signal peptide cleavage site of HA was predicted using the *signalP* [7] for all HA subtypes. Most subtypes are cleaved close to a highly conserved aspartic acid. Three subtypes lacking this aspartic acid are cleaved at a glutamine resulting in a longer HA sequence.

known to affect the phenotype of the virus for all known HA subtypes.

Materials and Methods

Representative sequences of HA for each subtype were downloaded from the Influenza Research Database (IRD). Potential N-terminal cleavage sites were predicted using the *signalP* [6–7] web-server. The amino acid sequence N-terminal to the predicted cleavage site was removed from each sequence. If a crystal structure was available, these were aligned based on their structural similarity using Pymol [8]. We then aligned the remaining sequences to the sequences of the other subtypes using FUGUE [9]. In general, amino acids in protein secondary structures (α-helices, β-strands) which are inaccessible to solvent or involved in interactions with other amino acids, are more conserved than those in loop regions or those exposed to solvent. Thus, amino acid insertions or deletions are more likely to occur solvent exposed regions or in regions without well-defined secondary structures. FUGUE uses knowledge of these differences in evolutionary constraints, in addition to sequence conservation, to aid its sequence alignment. This structure-based sequence alignment was subsequently manually adjusted based on inspection of the structures to accurately reflect structural similarity of loop regions.

Results

We have re-analysed the predicted N-terminal signal peptide cleavage sites of subtypes H1 to H13 and have extended this analysis to include subtypes H14 to H18. Table 1 shows the signal

peptide and N-terminal amino acid sequence of the mature protein based on the cleavage sites predicted using *signalP* [7–8], for each of the HA subtypes. More than half of all subtypes are predicted to be cleaved at an aspartic acid which is three amino acids N-terminal to a completely conserved cysteine. In agreement with Nobusawa, three subtypes are predicted to be cleaved at the amino acid preceding this aspartic acid at either a leucine (H10) or a tyrosine (H8 and H12). Three subtypes, H3, H5 and H14, lack the aspartic acid and are predicted to be cleaved at a glutamine, resulting in a longer mature N-terminal region. The signal peptide contains a stretch of about 10 hydrophobic amino acids that have a tendency to form a single alpha-helix, albeit with little sequence conservation between subtypes. In total, between 16 and 19 amino acids are removed from the N-terminal sequence to facilitate the movement of the virus through the ER membrane.

To define amino acids which are structurally equivalent across subtypes, we compared the available protein structures of all subtypes of HA to produce a sequence alignment based on the structural similarity of HA. For those subtypes without an HA structure (H4, H6, H8, H10–H18), we aligned their sequences to those of the other subtypes using an algorithm which considers structural features in addition to sequence conservation (see Material & Methods) [9]. The structure-based sequence alignment of HA1 is shown in figure 1. The subtypes have been ordered according to their phylogenetic grouping [10] and coloured according to sequence conservation [11]. We have highlighted those regions of HA which show significant differences in structure between strains of different subtypes. These are typically loops between secondary structures and are regions which contain insertions and deletions. Amino acids in these regions should only

```
H1post1995  ----------DTICIGYHANNSTDTVDTVLEKNVTVTHSVNLLEDSHNGKLCLLKGIAPLQLGNCSVAGWILGNPECELLISKESWSYIVETPNPENGTC  90
H1N1pdm     ----------DTLCIGYHANNSTDTVDTVLEKNVTVTHSVNLLEDKHNGKLCKLNGVAPLHLGKCNIAGWILGNPECESLLSTASSWSYIVETPSSDNGTC  90
H2          ----------DQICIGYHANNSTEKVDTILERNVTVTHAKDILEKTHNGKLCKLNGIPPLELGDCSIAGWLLGNPECDRLLSVPEWSYIMEKENPRDGLC  90
H5          ----------DQICIGYHANNSTEQVDTIMEKNVTVTHAQDILEKAHNGKLCSLNGVKPLILRDCSVAGWLLGNPMCDEFLNVPEWSYIVEKPNPINGLC  90
H5c221      ----------DQICIGYHANNSTEQVDTIMEKNVTVTHAQDILEKTHNGKLCNLDGVKPLILRDCSVAGWLLGNPMCDEFLNVPEWSYIVEKINPANDLC  90
H6          ----------DKICIGYHANNSTTQVDTILEKNVTVTHSVELLENQKEERFCKIMNKSPLDLRECTIEGWILGNPKCDLLLGDQSWSYIVERPTAQNGIC  90
H8          ----------YDRICIGYQSNNSTVNTLIEQNVPVTQTMELVETEKHPAYCNTDLGAPLELRDCKIEAVIYGNPKCDIHLKDQGWSYIVERPSAPEGMC  91
H9          ----------DKICIGYQSTNSTETVDTLTETNVPVTHAKELLHTEHNGMLCATNLGHPLILDTCTIEGLIYGNPSCDLLLGGREWSYIVERPSAVNGMC  90
H11         ----------DEICIGYLSNNSTDKVDTIIENNVTVSSVELVETEHTGSFCSINGKQPISLGDCSFAGWILGNPMCDELIGKTSWSYIVEKPNPTNGIC  90
H12         ----------YDKICIGYQTNNSTETVNTLSEQNVPVTQVEELVHGGIDPILCGTELGSPLVLDDCSLEGLILGNPKCDLYLNGREWSYIVERPKEMEGVC  91
H13         ----------DRICVGYLSTNSSERVDTLLENGVPVTSSIDLIENTHTGTYCSLNGVPVHLGDCSFEGWIVGNPSCATNINIREWSYLIEDPAAPHGLC  90
H16         ----------DKICIGYLSNNSTDVDTLTENGVPVTSSVDLVETNHTGTYCSLNGISPIHLGDCSFEGWIVGNPSCATNINIREWSYLIEDPNAPNKLC  90
H17         ----------DRICIGYQANQNNQTVNTLLEQNVPVTGAQEILETNHNGKLCSLNGVPPLDLQSCTLAGWLLGNPNCDNLLEAEEWSYIKINENAPDDLC  90
H18         ----------DQICIGYHSNNSTQTVNTLLESNVPVTSSHSILEKEHNGLLCKLKGKAPLDLIDCSLPAWLMGNPKCDELLTASEWAYIKEDPEPENGIC  90
H3          QDLPGNDNSTATLCLGHHAVPNGTLVKTITDDQIEVTNATELVQSSSTGKICNN-PHRILDGIDCTLIDALLGDPHCDVFQNE-TWDLFVERSK-AFSNC  97
H4          ---QNYTGNPVICLGHHAVSNGTMVKTLTDDQIEVVTAQELVESQHLPELCPS-PLRLVDGQTCDIVNGALGSPGCDHLNGA-EWDVFIERPT-AVDTC  93
H7          ----------DKICLGHHAVANGTKVNTLTERGVEVVNATETVERTNVPRICSK-GKRTVDLGQCGLLGTIIGPPQCDQFLEF-SADLIIERRE-GSDVC  87
H10         ----------LDKICLGHHAVANGTIVKTLTNEQEEVTNATETVESTSLNRLCMK-GRNHKDLGNCHPIGMLIGTPACDLHLTG-TWDTLIEREN-AIAYC  88
H14         -QITNGTTGNPIICLGHHAVENGTSVKTLTDNHVEVVSAKELVETNHTDELCPS-PLKLVDGQDCDLINGALGSPGCDRLQDT-TWDVFIERPT-AVDTC  96
H15         ----------DKICLGHHAVTNGTKVNTLTEKGVEVVNATETVEITGINKVCTK-GKKAVDLGSCGILGTIIGPPQCDSHLKF-KADLIIERRN-SSDIC  87

            1.......10........20........30........40........50........60........70........80........90.......100


H1post1995  YPGYFADYEELREQLSSVSSFERFEIFPKESSWPNHTVT-GVSASCSH-NGKSSFYRNLLWLTGKN--GLYPNLSKSYVNNKEKEVLVLWGVHHPPNIGN  186
H1N1pdm     YPGDFIDYEELREQLSSVSSFEKFEIFPKTSSWPNHDSNKGVTAACPH-AGAKSFYKNLIWLVKKG--NSYPKLSKSYINDKGKEVLVLWGIHHPSTSAD  187
H2          YPGSFNDYEELKHLLSSVKHFEKVKILPKD-RWTQHTTT-GGSRACAV-SGNPSFFRNMVWLTKKE--SNYPVAKGSYNNTSGEQMLIIWGVHHPNDETE  185
H5          YPGDFNDYEELKHLLSSTNHFEKIQIIPRS-SWSNHEASSGVSSACPY-QGRSSFFRNVVWLIKKN--NAYPTIKRSYNNTNQEDLLVLWGIHHPNDAAE  186
H5c221      YPGNFNDYEELKHLLSRINHFEKIQIIPKN-SWSDHEAS-GVSSACPY-QGRSSFFRNVVWLTKKD--NAYPTIKRSYNNTNQEDLLVLWGIHHPNDAAE  185
H6          YPGALNEVEELKALIGSGERVERFEMFPKS-TWAGVDTSSGVTNACPSYTIGSSFYRNLVWLIKTNS-AAYPVIKGTYNNTGNQPILYFWGVHHPPNTGV  188
H8          YPGSVENLEELRFVFSSAASYKRIRLFDYS-RWN-VTRS-GTSKACNASTGGQSFYRSINWLTKKKP-DTYDFNEGAYVNNEDGDIIFLWGIHHPPDTKE  187
H9          YPGNVENLEELRSLFSSASSYQRIQIFPDT-IWN-VSYS-GTSKACS-----DSFYRSMRWLTQKN--NAYPIQDAQYTNNRGKSILFMWGIHHPPTDTV  180
H11         YPGTLESEEELRLKFSGVLEFNKFEVFTSN-GWGAVNSGVVTAACKF-GGSNSFFRNMVWLIHQS--GTYPVIKRTFNNTKGRDVLIVWGIHHPATLTE  186
H12         YPGSIENQEELRSLFSSIKKYERVKMFDFT-KWN-VTYT-GTSKACNTSNQGSFYRSMRWLTLKS--GQFPVQTDEYKNTRDSDIVFTWAIHHPPTSDE  186
H13         YPGELNNNGELRHLFSGIRSFSRTELIPPT-SWG-EVLD-GTTSACRDNTGTNSFYRNLVWLTEKN--NRYPVSKTYNNTTGRDVLVLWGIHHPPVSVDE  185
H16         YPGELDNNGELRHLFSGVNSFSRTELINPS-KWG-NVLD-GVTASCLD-RGASSFYRNLVWLVKQKI-GEYPVVKGEYNNTTGRDVLVLWGIHHPDTETT  185
H17         FPGNFENLQDLLLEMSGVQNFTKVKLFNPQ-SMTGVTTN-NVDQTCPF-EGKPSFYRNLNWIQGNSG---LPFNIEIKNPTSNFLLLLWGIHNTKDAAQ  183
H18         FPGDFDSLEDLILLVSNTDHFRKEIIDMT-RFSDVTTN-NVDSACPYDTNGASFYRNLNWQQNKG---KQLIFHYQNSENNPLILIVGVHQTSNAAE  184
H3          YPYDVPDYASLRSLVASSGTLEFITEGF---TWTGVTQN-GGSNACKR-GPGSSFFSRLNWLTKSG--STYPVLNVTMPNNDNFDKLYIWGIHHPSTNQE  190
H4          YPFDVPDYQSLRSILANNGKFEFIAEEF---QWNTVKQN-GKSGACAR-ANVNDFFNRLNWLTKSDG-NAYPLQNLTKVNNGDYARLYIWGIHHPSTDTE  187
H7          YPGKFVNEEALRQILRESGGIDKETMGF---TYSGIRTN-GATSACRR-SGSSFYAEMKWLLSNTDNAAFPQMTKSYKNTRKDPALIWGIHHSGSTTE  181
H10         YPGATVNEEALRQKIMESGGISKISTGF---TYGSSINSAGTTKACMR-NGGNSFYAELKWLVSKSKGQNFPQTTNTYRNTDTAEHIIMWGIHHPSSTQE  184
H14         YPFDVPDYQSLRSILASSGSLEFIAEQF---TWNGVKVD-GSSSACLR-GGRNSFFSRLNWLWLKATN-GNYGPINVTKENTGSYVRLYLWGVHHPSSTNE  190
H15         YPGKFTNEEALRQIIRESGGIDKEPMGF---RYSGIKTD-GATSACKR-TVSSFSEMKWLLSSKANQVFPQLNQTYRNNKEFALIVWGVHHSSSLDE  181

            .......110.......120.......130.......140.......150.......160.......170.......180.......190.......200


H1post1995  QRALYHTENAYVSVVSSHYSRRFTPEIAKRPKVRDQEGRINYYWTLLEPGDTIIFEANGNLIAPWYAFALSR---------GFGSGIITSNAPMDECDAK  277
H1N1pdm     QQSLYQNADTYVFVGSSRYSKKFKPEIAIRPKVRDQEGRMNYYWTLLEPGDKITFEATGNLVVPRYAFAMER--NAGSGIIISDTPVHDCNTT  278
H2          QRTLYQNVGTYVSVGTSTLNKRSPVEIATRPKVNGLGSRMEFSWTLLDMWDTINFESTGNLIAPEYGFKISK---------RGSSGIMKTEGTLENCETK  276
H5          QTKLYQNPTTYVSVGTSTLNQRSVPEIAIRPKVNGQGSRMEFFWTILKPNDAINFESNGNFIAPEYAYKIVK---------KGDSAIMKSGLEYGNCNTK  277
H5c221      QTRLYQNPTTYISVGTSTLNQRSVPEIATRSKVNGQGSRMEFFWTILKSNDAINFESNGNFIAPENAYKIVK---------KGDSTIMKSELEYGNCNTK  276
H6          QDTLYGSGERYVRMGTDSMNFAKSPEIAERPVVNGQRGRIDYYWSVLKPGETLNVESNGNLIAPWYAYKFVS---------TNKKGAVFKSNLPIENCDAT  280
H8          QTNLYKNANTLSSVTTNTINRSFQPNIGPRPLVRGQQGRMDYYWGILKRGETLKIRTNGNLIAPEFGYLLKG---------ESYGRIIQNEDIPIGNCNTK  279
H9          QTNLYTRTDTTTSVTTEDINRTFKPVIGPRPLVNGLHGRIDYYWSVLKPGQTLRVRSNGNLIAPWYGHILSG---------ESHGRILKTDLNSGNCVVQ  271
H11         HQDLYKKDSSYVAVGSETYNRRFTPEINTRPRVNGQQAGRMTFYWKIVKPGESITFESNGAFLAPRYAFEIVS---------VGNGKLFRSELNIESCSTK  277
H12         QVKLYKNPDTLSSVTTEINRSFKPNIGPRPLVRGQQGRMDYYWAVLKPGQTVKIQTNGNLIAPEYGHLITG---------KSHGRILKNNLPMGQCVTE  277
H13         TKTLYVNSDPYTLVSTKSWSEKYKLETGVRPGYNGQRSWMKIYWSLIHPGEMITFESNGGFLAPRYGYIIEE---------YGKGRIFQSRIRMSRCNTK  276
H16         ATNLYVNKNPYTLVSTKEWSKRYELEIGTRIGD-GQRSWMKLYWHLMHPGERIMFESNGGLIAPRYGYIIEK---------YGTGRIFQSGVRMAKCNTK  275
H17         QRNLYGNDYSYTIFNFGEKSEEFRPDIGQRDEIKAHQDRIDYYWGSLPAQSTLRIESTGNLIAPEYGFYYKR---------KEGKGGLMKSKLPISDCSTK  275
H18         QNTYYGSQTGSTTITIGEETNTYPLVISESSILNGHSDRINYFWGVVNPNQNFSIVSTGNFIWPEYGYFFQK---------TTNISGIIKSSEKISDCDTI  276
H3          QTSLYVQASGRVTVSTRRSQQTIIPNIGSRPWVRGLSSRISIYWTIVKPGDVVVINSGNLIAPRGYFKMRT---------GKSSIMRSDAPIDTCISE  280
H4          QTNLYKNNPGRVTVSTQTSQTSVVPNIGSRPWVRGLSSRISFYWTIVEPGDLVFNTIGNLIAPRGHYKLNS---------QKKSTILNTAVPIGSCVSK  278
H7          QTKLYGSGNKLITVGSSNYQQSFVPSPGARPQVNGQSGRIDFHWLMLNPNDTVTFSFNGAFIAPDRASFLR---------GKSMGIQSSVQVDANCEGD  271
H10         KNDLYGTQSLSISVGSSTYQNNFVPVVGARPQVNGQSGRIDFHWTLVQPGDNITFSHNGGLIAPSRVSKLI---------GRGLGIQSDAPIDNNCESK  274
H14         QTDLYKVATGRVTVSTRSDQISIVPNIGSRPRVRNQSGRISIYWTLVNPGDSIIFNSINGNLIAPRGHYKISK---------STKSTVLKSDKRIGSCTSP  281
H15         QNKLYGAGNKLITVGSSKYQQSFSPSPGDRPKVNGQAGRIDFHWMLLDPGDTVTFTFNGAFIAPDRATFLRSNAPSGVEYNGKSLGIQSDAQIDESCEGE  281

            .......210.......220.......230.......240.......250.......260.......270.......280.......290.......300


H1post1995  CQTPQGAINSSLPFQNVHPVTIGECPKYVRSAKLRMVTGLRNIPSIQS----R  326
H1N1pdm     CQTPKGAINTSLPFQNIHPITIGKCPKYVKSTKLRLATGLRNIPSIQS----R  327
H2          CQTPLGAINTTLPFHNVHPLTIGECPKYVKSEKLVLATGLRNVPQIES----R  325
H5          CQTPMGAINSSMPFHNIHPLTIGECPKYVKSDRLVLATGLRNVPQRET----R  326
H5c221      CQTPIGAINSSMPFHNIHPLTIGECPKYVKSNRLVLATGLRNSPQRERRRKKR  329
H6          CQTIAGVLRTNKTFQNVSPLWIGECPKYVKSESLRLATGLRNIPQIKT----R  329
H8          CQTYAGAINSSKPFQNASRHYMGECPKYVKKASLRLAVGLRNTPSVEP----R  328
H9          CQTERGGLNTTLPFHNVSKYAFGNCPKYVGVKSLKLAVGLRNVPARSS----R  320
H11         CQTEIGGINTNKSFHNVHRNTIGDCPKYVNVKSLKLATGPRNVPAIAS----R  326
H12         CQLNEGVMNTSKPFQNTSKHYIGKCPKYIPSGSLKLAIGLRNVPQVQD----R  326
H13         CPTSVGGINTNRTFQNIDKNALGDCPKYIKSGQLKLATGLRNVPAISN----R  325
H16         CQTSLGGINTNKTFQNIKNALDCPKYIKSGQLKLATGLRNVPIPIGE---R  325
H17         CQTPLGALNSTLPFQNVHQQTIGNCPKYVKATSLMLATGLRNNPQMEG----R  324
H18         CQTKIGAINSTLPFQNIHQNAIGDCPKYVKAQELVLATGLRNNPIKET----R  325
H3          CITPNGSIPNDKFQNVNKITYGACPKYVKQNTLKLATGMRNVPEKQT----R  329
H4          CHTDKGSISTTKPFQNISRISIGDCPKYVKQGSLKLATGMRNIPEKAT----R  327
H7          CYHSGGTIISNLPFQNINSRAVGKCPKYVKQESLMLATGMKNVPEIPKG----R  321
H10         CFWRGGSINTNLPFQNVSPRTVGQCPKYVNKKSLMLATGMRNVPEIMQG----R  324
H14         CLTDKGSIQSDKPFQNVSRIAIGNCPKYVKQGSLMLATGMRNIPGKQA----R  330
H15         CFYSGGTINSPLPFQNIDSWAVGRCPKYVKQSSLPLALGMKNVPEKIHT----R  331

            .......310.......320.......330.......340.......350...
```

Figure 1. Sequence alignment of HA for known sub-types. Alignment of mature HA sequence for all known HA sub-types. Additional strains have been included for sub-types which show variation in the length of HA. Sequences are ordered according to their phylogenetic classification as group 1 (magenta bar) or group 2 (orange bar) HA. The protein secondary structure elements, α-helices and β-strands, are highlighted with red bars and cyan arrows, respectively. A blue box highlights regions which have high structural variation across all subtypes. Amino acids within these regions should not be defined as equivalent between all sub-types. Each amino acid is coloured according to clustalx2 rules [11]. Briefly, glycine and proline are coloured orange and yellow, respectively. Conserved positively charged residues and negatively charged residues are coloured red and magenta, respectively. Conserved cysteines are coloured pink while conserved serine or threonine residues are in green. The remaining amino acids, if conserved are coloured blue. The sequences representative of each subtype are as follows: H1(A/United Kingdom/1/1933); H1pdm(A/California/04/2009); H2(A/Singapore/1/1957); H3(A/Aichi/2/1968); H4(A/swine/Ontario/01911/2/1999); H5(A/Vietnam/1203/2004); H5c221(A/chicken/Egypt/0915-NLQP/2009); H6(A/chicken/Taiwan/0705/1999); H7(A/Netherlands/219/2003); H8(A/turkey/Ontario/6118/1968); H9(A/swine/HongKong/9/1998); H10(A/mallard/bavaria/3/2006); H11(A/duck/England/1/1956); H12(A/duck/Alberta/60/1976); H13(A/gull/Maryland/704/1977); H14(A/mallard/Astrakhan/263/1982); H15(A/duck/Australia/341/1983); H16(A/black-headed-gull/Turkmenistan/13/1976); H17(A/little-yellow-shouldered-bat/Guatemala/060/2010); H18(A/flat-facedbat/Peru/033/2010).

be considered to be equivalent when comparing closely related subtypes.

As previously described, some subtypes show clade specific differences in the length of the amino acid sequence of HA. We have therefore distinguished in our analysis H1 strains post-1995 and strains from clade 2.2.1 of H5. Additionally, the insertion of many positively charged amino acids in the C-terminal of HA1 in some strains of H5 and H7 subtypes is well known to increase the pathology of viral infection in poultry, leading to high rates of fatality [12]. A consequence is that the numbering of positions C-terminal to the cleavage site (position 326 for low pathogenic strains of H5) will differ. For H5 and H7 subtypes, we therefore also include both low-pathogenic (H5N1:A/mallard/Italy/3401/2005; H7:A/Turkey/Italy/220158/2002) and high-pathogenic (H5N1:A/Vietnam/1203/2004; H7N7:A/Netherlands/219/2003) strains. The sequence alignment including all subtypes spanning both HA1 and HA2 is available as File S1.

From these alignments, we can now derive residue numbering in each subtype, of every position of HA, relative to its mature sequence. This list of equivalences for all residue positions and across all subtypes are available as File S2 and at http://www.antigenic-cartography.org/surveillance/evergreen/HAnumbering. Positions which are most often compared across subtypes are those which have been shown to be associated with changes in phenotype. In 2012, the WHO Collaborating Center for Influenza Reference and Research at the Centers for Disease Control and Prevention in Atlanta compiled an inventory of amino acid mutations found in H5N1 viruses http://www.cdc.gov/flu/avianflu/h5n1/inventory.htm). The equivalent residue numbering for these mutations in HA are listed in Table 2 for those subtypes which circulate in humans (H1, H3) or from which zoonoses frequently occur (H5, H7, H9).

Discussion

The length of the HA segment of influenza A shows substantial variation both between and within HA subtypes. This is caused by both changes in the length of the N-terminal signal peptide cleavage site and subtype specific amino acid insertions and deletions within the HA. These differences often makes it difficult to compare amino acid changes within HA of one subtype to those seen in another subtype.

We have re-assessed the predicted N-terminal signal peptide cleavage sites of all known subtypes (H1 to H18), confirming the previous definitions of the thirteen subtypes of HA previously reported by Nobusawa [5]. Using a structure-based approach we have analysed the structural and functional conservation of each position of HA across all subtypes. We have identified regions of HA which are structurally conserved across subtypes, including both low and highly pathogenic strains of H5 and H7 subtypes, and strains of H1 and H5 which show clade specific differences in

the length of HA. From this data we have defined equivalent residue numbering for each subtype.

It is often stated that amino acid positions are 'equivalent' but rarely is this term defined explicitly. In structural biology, when comparing structures of proteins with evolutionary divergent sequences, such as HA from different subtypes, segments of the structure can be described as being either structurally conserved regions (SCRs) or structurally variable regions (SVRs). SCRs have similar structural features, such as the shape of the peptide backbone and the orientation of the sidechain atoms, and these regions usually have high sequence conservation. Like many proteins, the conserved regions within HA are those which are critical for its function, such as the receptor-binding site, or those that are required for the correct folding or stability of the protein structure. Amino acids within these regions can be described as equivalent in the sense that they will adopt nearly identical conformations and form similar interactions with other amino acids or bio-molecules. It is equally important to appreciate the limitations of a sequence alignment. Most alignment algorithms are parameterised to favour as few insertions and deletions as possible and do not always reflect local structural similarity. It is possible to have regions of sequences aligned which show little structural similarity and thus should not be described as SCRs. However, it needs to be noted that the SCR designation is not an absolute. Whilst many SCRs can be conserved across highly divergent sequences (between influenza A and influenza B viruses, for example), it is possible to define SCRs which are only conserved between closely related sequences, such as only between group 1 sub-types of HA.

In contrast, SVRs are regions which have very little structural or functional similarity between two related proteins. These regions are usually in the solvent exposed turns of the protein structure. These are also the regions where insertions and deletions of amino acids frequently occur, since they can be accommodated without major disruption of the fold or function of the protein. Amino acids in these regions should not be described as equivalent and comparisons between sub-types has little biological relevance.

Many studies attempt to compare, and sometimes replicate, mutations seen in one subtype, such as H5, to those in another subtype. Careful consideration of the level of structural and functional conservation of that region (its equivalence), however, is crucial. This is especially important when inferring analogous mutations from subtypes belonging to a different phylogenetic group. We feel that the use of this set of residue numbering and analysis of structural conservation will facilitate cross-subtype comparisons and reduce confusion in reporting amino acid numbering.

Table 2. Equivalent amino acid numbering for subtypes currently circulating in humans or have pandemic potential.

Mutation	H1pdm	H3	H5	H7	H9	Phenotype	Reference
Tyr → His	7	17	7	7	7	Increase in fusion pH	[13]
His → Gln	8	18	8	8	8	Decrease in fusion pH; increased stability	[13]
Asn → Any	11	21	11	11	11	Loss of N-glycosylation; increased virulence	[14]
Glu→Lys	75	83	75	73	75	Increased virus binding to α2-6 glycans	[15]
His → Tyr	103	110	103	100	103	Increased stability	[2]
Ser→ Asn	122	126	121	116	121	Increased virus binding to α2-6 glycans	[16]
Ser→Pro	124	128	123	118	123	Increased virus binding to α2-6 glycans	[15]
Ala → Δ	130	Δ	129	Δ	Δ	Increased virus binding to α2-6 glycans	[17–18]
Ser → Ala	134	137	133	127	131	Increased virus binding to α2-6 glycans	[19]
Ala → Val	135	138	134	128	132	Increased virus binding to α2-6 glycans	[20]
Gly→Arg	140	143	139	132	Δ	Increased virus binding to α2-6 glycans	[15]
Ile→ Thr	152	155	151	144	145	Increased infectivity in SIAT Cells	[17–18]
Asn→ Asp	155	158	154	147	148	Loss of N-glycosylation; increased binding and transmission	[2]
Thr→ Ala	157	160	156	151	150	Loss of N-glycosylation; increased binding and transmission	[1]
Asn→Lys	183	186	182	177	176	Increased virus binding to α2-6 glycans	[15,21]
Asp→Gly	184	187	183	178	177	Increased virus binding to α2-6 glycans	[22]
Glu→Gly	187	190	186	181	180	Increased virus binding to α2-6 glycans	[22]
Thr→Ile	189	192	188	183	182	Increased virus binding to α2-6 glycans	[19]
Lys→Arg	190	193	189	184	183	Increased virus binding to α2-6 glycans	[16]
Gln→Arg/His	193	196	192	187	186	Increased virus binding to α2-6 glycans	[15,18,22]
Asn→Lys	194	197	193	188	187	Increased virus binding to α2-6 glycans	[15]
Val → Ile	211	214	210	205	204	Increased virus binding to α2-6 glycans	[18]
Gln→Leu	223	226	222	217	216	Increased virus binding to α2-6 glycans	[21]
Ser→Asn	224	227	223	218	217	Increased virus binding to α2-6 glycans	[21–23]
Gly→Ser	225	228	224	219	218	Increased virus binding to α2-6 glycans	[14–15,24]
Pro→Ser	236	239	235	230	229	Increased virus binding to α2-6 glycans	[18]
Glu→Lys	252	255	251	246	245	Increased virus binding to α2-6 glycans	[22]
Thr→Ile	316	318	315	309	309	Increase in fusion pH	[1]
Insertion of Arg or Lys	327	329	326	321	320	Poly-basic cleavage; increased pathogenicity	[25]
Lys → Ile	385	387	384	379	378	Increase in fusion pH; increased stability	[13,26]
Asn→ Lys	441	443	440	435	434	Increase in fusion pH; decreased stability	[13]
Asn → Asp	444	446	443	438	437	Increase in fusion pH	[27]
Arg → Lys	494	496	493	488	487	Increased virus binding to α2-6 glycans	[15]

Residue numbering is based on the mature sequence of HA1 across all subtypes for a set mutations shown to cause phenotypic differences. Positions where there is a deletion relative to other subtypes are represented by a "Δ".

Supporting Information Legends

File S1 Structure based sequence alignment for HA. The sequence alignment including all subtypes spanning both HA1 and HA2. This alignment includes a strain of seasonal H1N1 strain post-1995 (A/NewCaledonia/20/1999/H1N1) and strains of H5 (A/mallard/Italy/3401/2005/H5N1) and H7 (A/Turkey/Italy/220158/2002/H7N3) with low pathogenicity.

File S2 Equivalent amino acid numbering for all known HA subtypes. Residue numbering is based on the mature sequence of HA across all subtypes. The amino acid at each position for the representative strain of that subtype is also given.

Positions where there is a deletion relative to other subtypes are represented by a "Δ".

Acknowledgments

D.F.B acknowledges the use of the CamGrid distributed computing resource.

Author Contributions

Conceived and designed the experiments: DFB. Performed the experiments: DFB. Analyzed the data: DFB DJS. Contributed reagents/materials/analysis tools: DFB DJS. Wrote the paper: DFB DJS.

References

1. Imai M, Watanabe T, Hatta M, Das SC, Ozawa M, et al. (2012) Experimental adaptation of an influenza H5 HA confers respiratory droplet transmission to a reassortant H5 HA/H1N1 virus in ferrets. Nature 486: 420–428.
2. Herfst S, Schrauwen EJA, Linster M, Chutinimitkul S, de Wit E, et al. (2012) Airborne transmission of influenza A/H5N1 virus between ferrets. Science 336: 1534–1541.
3. McDonald NJ, Smith CB, Cox NJ (2007) Antigenic drift in the evolution of H1N1 influenza A viruses resulting from deletion of a single amino acid in the haemagglutinin gene. J Gen Virol 88: 3209–3213.
4. Suarez DL, Garcia M, Latimer J, Senne D, Perdue M (1999) Phylogenetic analysis of H7 avian influenza viruses isolated from the live bird markets of the Northeast United States. J Virol 73: 3567–3573.
5. Nobusawa E, Aoyama T, Kato H, Suzuki Y, Tateno Y, et al. (1991) Comparison of complete amino acid sequences and receptor-binding properties among 13 serotypes of hemagglutinins of influenza A viruses. Virology 182: 475–485
6. Von Heijne G, Gavel Y (1988) Topogenic signals in integral membrane proteins. Eur J Biochem 174: 671–678.
7. Petersen TN, Brunak S, von Heijne G, Nielsen H (2011) SignalP 4.0: discriminating signal peptides from transmembrane regions. Nat Methods 8: 785–786.
8. DeLano WL (2002) The PyMOL Molecular Graphics System. Schrödinger LLC
9. Shi J, Blundell TL, Mizuguchi K (2001) FUGUE: sequence-structure homology recognition using environment-specific substitution tables and structure-dependent gap penalties. J Mol Biol 310: 243–257
10. Medina RA, Garcia-Sastre A (2011) Influenza A viruses: new research developments. Nat Rev Microbiol 9: 590–603
11. Larkin M, Blackshields G, Brown N, Chenna R, McGettigan P, et al. (2007) ClustalW and ClustalX version 2. Bioinformatics 23: 2947–2948
12. Steinhauer DA (1999) Role of hemagglutinin cleavage for the pathogenicity of influenza virus. Virology 258: 1–20.
13. Reed ML, Yen H-L, DuBois RM, Bridges OA, Salomon R, et al. (2009) Amino acid residues in the fusion peptide pocket regulate the pH of activation of the H5N1 influenza virus hemagglutinin protein. J Virol 83: 3568–3580.
14. Deshpande KL, Fried VA, Ando M, Webster RG (1987) Glycosylation affects cleavage of an H5N2 influenza virus hemagglutinin and regulates virulence. Proc Natl Acad Sci U S A 84: 36–40.
15. Yamada S, Suzuki Y, Suzuki T, Le MQ, Nidom CA, et al. (2006) Haemagglutinin mutations responsible for the binding of H5N1 influenza A viruses to human-type receptors. Nature 444: 378–382.
16. Wang W, Lu B, Zhou H, Suguitan AL, Cheng X, et al. (2010) Glycosylation at 158N of the hemagglutinin protein and receptor binding specificity

synergistically affect the antigenicity and immunogenicity of a live attenuated H5N1 A/Vietnam/1203/2004 vaccine virus in ferrets. J Virol 84: 6570–6577.
17. Auewarakul P, Suptawiwat O, Kongchanagul A, Sangma C, Suzuki Y, et al. (2007) An avian influenza H5N1 virus that binds to a human-type receptor. J Virol 81: 9950–9955.
18. Watanabe Y, Ibrahim MS, Ellakany HF, Kawashita N, Mizuike R, et al. (2011) Acquisition of human-type receptor binding specificity by new H5N1 influenza virus sublineages during their emergence in birds in Egypt. PLoS Pathog 7: e1002068
19. Yang Z-Y, Wei C-J, Kong W-P, Wu L, Xu L, et al. (2007) Immunization by avian H5 influenza hemagglutinin mutants with altered receptor binding specificity. Science 317: 825–828.
20. Naughtin M, Dyason JC, Mardy S, Sorn S, Von Itzstein M, et al. (2011) Neuraminidase inhibitor sensitivity and receptor-binding specificity of Cambodian clade 1 highly pathogenic H5N1 influenza virus. Antimicrob Agents Chemother 55: 2004–2010.
21. Chutinimitkul S, Herfst S, Steel J, Lowen AC, Ye J, et al. (2010) Virulence-associated substitution D222G in the hemagglutinin of 2009 pandemic influenza A(H1N1) virus affects receptor binding. J Virol 84: 11802–11813
22. Chen LM, Blixt O, Stevens J, Lipatov AS, Davis CT, et al. (2012) In vitro evolution of H5N1 avian influenza virus toward human-type receptor specificity. Virology 422: 105–113.
23. Gambaryan A, Tuzikov A, Pazynina G, Bovin N, Balish A, et al. (2006) Evolution of the receptor binding phenotype of influenza A (H5) viruses. Virology 344: 432–438.
24. Stevens J, Blixt O, Tumpey TM, Taubenberger JK, Paulson JC, et al. (2006) Structure and receptor specificity of the hemagglutinin from an H5N1 influenza virus. Science 312: 404–410.
25. Bosch FX, Garten W, Klenk HD, Rott R (1981) Proteolytic cleavage of influenza virus hemagglutinins: primary structure of the connecting peptide between HA1 and HA2 determines proteolytic cleavability and pathogenicity of Avian influenza viruses. Virology 113: 725–735.
26. Zaraket H, Bridges O a, Duan S, Baranovich T, Yoon S-W, et al. (2013) Increased acid stability of the hemagglutinin protein enhances H5N1 influenza virus growth in the upper respiratory tract but is insufficient for transmission in ferrets. J Virol 87: 9911–9922.
27. Murakami S, Horimoto T, Ito M, Takano R, Katsura H, et al. (2012) Enhanced Growth of Influenza Vaccine Seed Viruses in Vero Cells Mediated by Broadening the Optimal pH Range for Virus Membrane Fusion. J Virol 86: 1405–1410.

13

Evolution of an Expanded Mannose Receptor Gene Family

Karen Staines, Lawrence G. Hunt, John R. Young, Colin Butter*

The Pirbright Institute, Compton, United Kingdom

Abstract

Sequences of peptides from a protein specifically immunoprecipitated by an antibody, KUL01, that recognises chicken macrophages, identified a homologue of the mammalian mannose receptor, MRC1, which we called MRC1L-B. Inspection of the genomic environment of the chicken gene revealed an array of five paralogous genes, *MRC1L-A* to *MRC1L-E*, located between conserved flanking genes found either side of the single *MRC1* gene in mammals. Transcripts of all five genes were detected in RNA from a macrophage cell line and other RNAs, whose sequences allowed the precise definition of spliced exons, confirming or correcting existing bioinformatic annotation. The confirmed gene structures were used to locate orthologues of all five genes in the genomes of two other avian species and of the painted turtle, all with intact coding sequences. The lizard genome had only three genes, one orthologue of *MRC1L-A* and two orthologues of *the MRC1-B* antigen gene resulting from a recent duplication. The Xenopus genome, like that of most mammals, had only a single *MRC1*-like gene at the corresponding locus. *MRC1L-A* and *MRC1L-B* genes had similar cytoplasmic regions that may be indicative of similar subcellular migration and functions. Cytoplasmic regions of the other three genes were very divergent, possibly indicating the evolution of a new functional repertoire for this family of molecules, which might include novel interactions with pathogens.

Editor: Michelle L. Baker, CSIRO, Australia

Funding: This work was funded by the BBSRC (Biotechnology and Biological Sciences Research Council) under grant number BBS/E/I/00001423 as part of the Institute Strategic Programme Grant to the Avian Viral Diseases Programme at the Pirbright Institute. The funders had no role in study design, data collection and analysis, decision to publish, or preparation of the manuscript.

Competing Interests: The authors have declared that no competing interests exist.

* Email: colin.butter@avianimmunology.org

Introduction

Recent evolution of the repertoire of molecules involved in the function of the immune system has resulted in substantial divergence in the composition and functions of the gene families to which these molecules belong. Even among mammals, different families of molecules may carry out equivalent functions in different species [1]. While the functions of many molecules in immunity are well conserved between mammalian and avian species, in other cases there is extensive divergence in molecular repertoires, with cytokines and chemokines providing examples [2]. These differences often involve gene duplication followed by functional diversification [3]. Thus evolution has led to variety in molecular details in spite of more conserved underlying mechanisms in solutions to the problems of infection. Variation in molecular repertoires may underlie some of the differences between species in host-pathogen interactions. An understanding of these differences will be essential to optimise approaches to immune protection.

The mannose receptor C-type 1 gene (*MRC1, CD206*) is the eponymous member of the mannose receptor family. Their gene products are type I transmembrane glycoproteins containing arrays of C-type lectin domains (CTLDs). The family also includes DEC205 (CD205), MRC2 (Endo180, CD280) and Phospholipase A_2 receptor (PLA$_2$R), each having important functions in immunity [4]. These receptors all have an N-terminal cysteine-rich domain (CysR) followed by a single fibronectin type II domain

(FNII), then either 8 (MRC1, MRC2 and PLA$_2$R) or 10 (DEC205) CTLDs separated by linker regions. They have a transmembrane domain and a short cytoplasmic tail containing motifs that signal endocytosis. In mammals, *DEC205* and *PLA$_2$R* genes are arranged in tandem on one chromosome, while the others are unlinked. In the three genes encoding 8 CTLDs, the 30 exon gene structure and the splicing phases of all introns are completely conserved. The CTLDs fall into two groups, one having an extra pair of cysteine residues at the N-terminal end of the domain (domains 2, 3, 4, 6, 8) [5]. While individual CTLDs generally have low affinities for carbohydrate ligands, the molecules can exhibit high affinities for complex carbohydrate by cooperative binding [6]. Only the fourth CTLD of human MRC1 retains strong enough binding to have lectin activity on its own [7].

The mannose receptor is a recycling endocytosis receptor, rapidly internalised via clathrin-coated vesicles and delivered to early endosomes, with the majority of the receptors in the intracellular location in the steady state [8]. Endocytosis of bound molecules underlies the primary function of the mannose receptor in the recognition of pathogen associated molecular patterns and their consequent uptake for engulfment and for antigen presentation [9]. A soluble form of the mammalian mannose receptor, produced by proteolytic cleavage [10], may also function in the delivery of antigens to lymphoid follicles [11]. Clearance by binding to the mannose receptor may also be involved in the regulation of levels of some hormones [12]. In chickens, the orthologue of mammalian PLA$_2$R acts as an Fc receptor (FcRY), the functional equivalent of mammalian FcRn, extending the range of its endocytic targets to immune complexes [13].

Binding of the mannose receptor by a virus may elicit immunomodulatory responses [14]. It may also facilitate viral entry in a cell either indirectly, as with HIV [15], or directly, as with Dengue [16]. In the mouse, binding of influenza virus by the mannose receptor, in addition to its more widespread binding to sialic acid, is important for virus entry into macrophages [17]. The virus replicates inside infected macropahges, but they do not release infective virus. Instead, the infection enhances the presentation of influenza virus antigens and stimulates the generation of pro-inflammatory cytokines. Thus the participation of the mannose receptor in allowing infection of macrophages contributes to innate and eventually to adaptive protection [18]. Reciprocally driven evolution of the virus and the mannose receptor in different species may thus be a significant contributor to differences in host-pathogen interactions.

Employing mass spectrometry of immunoprecipitated antigen, we identified a molecule recognised by a macrophage marker antibody, KUL01 [19], as a chicken homologue of MRC1. Inspection of neighbouring avian genome sequence revealed that the locus contained five tandemly repeated genes encoding similar molecules that are likely to have arisen through duplication, of a single ancestral MRC1 gene, in the avian lineage. Very different cytoplasmic sequences and differences in relative transcript levels in tissues indicate diversification of function among the duplicated genes.

Results

The KUL01 antibody recognises a homologue of the macrophage mannose receptor MRC1

KUL01 antibody bound to agarose beads was used to adsorb proteins from a lysate of the transformed chicken macrophage cell line HD11, which were analysed by SDS PAGE after elution at low pH. Specific bands, obtained from beads coated with KUL01 but not from those coated with control antibody, were excised, digested with trypsin and analysed my mass spectroscopy. The major specifically recognised molecule was a (doublet) band with an apparent molecular weight of 180 kDa (figure 1). By Mascot search of the NCBI non-redundant chicken proteins in the IPI database, a sufficient number of peptides from the tryptic digest of this band were identified as being derivable from the sequence IPI00814304 to unequivocally identify it as the source of antigen specifically adsorbed by KUL01 (figure S1 and table S1). It was annotated as being a chicken homologue of MRC1.

Genomic context of chicken MRC1 orthologues

The genomic context of the gene for the KUL01 antigen was inspected to see whether additional evidence from conserved gene order would support its identification as the orthologue of MRC1. Inspection of the region between orthologues of the highly conserved genes, SLC39A12 and STAM, that flank MRC1 in mammals, revealed multiple segments with similarity to the MRC1 gene. Existing annotation and EST data, together with manual examination, allowed the definition of five potential MRC1L genes. For convenience, these were labelled MRC1L-A to E in sequence in the direction of their transcription (which is inverse to the genome map). Annotations of this gene array from different sources varied widely, in detailed exon composition, splicing sites and numbers of genes. To evaluate the predicted gene models, a series of PCR primers were designed for amplification of segments of the predicted transcripts from RNA. PCR products were amplified from RNA from the HD11

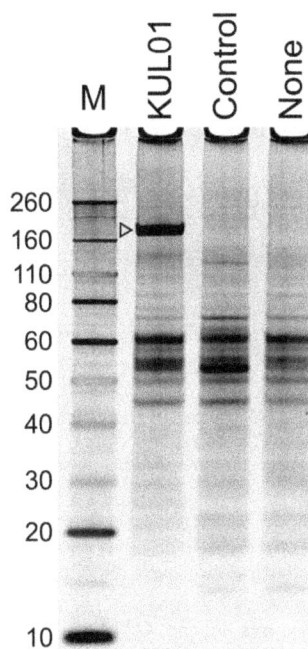

Figure 1. KUL01 specifically precipitates a molecule with apparent molecular weight 180 kDa. Track M contains molecular weight standards. The other tracks contain materials absorbed from a precleared lysate of the HD11 macrophage cell line, by agarose beads to which were attached either KUL01, an isotype matched control antibody, or no antibody, and eluted at low pH. The open arrowhead points to the band(s) specifically absorbed by the KUL01 antibody, which were analysed by mass spectroscopy.

transformed macrophage cell line, and from a cDNA library from RPRL Line 0 chicken spleen.

All predicted exons were amplified from spliced transcripts from the Line 0 chicken cDNA. All the transcript sequences confirmed in this way contained intact reading frames for MRC1-like proteins. These were submitted to the ENA database and received accession numbers HF569039, HF566127, HF569040, HF569041, HF569042, in order MRC1L-(A to E) and are provided in figure S2, together with their genomic locations. The MRC1L-B and –C genes are now correctly annotated in the ENSEMBL database (ENSGALT00000043091, ENSGALT00000014059). Annotation of the other genes is currently inaccurate, with errors as described in file S1 and are liable to change in subsequent database versions. Differences from the corresponding red jungle fowl genomic sequences are enumerated in table S2. The exon structures of the genes and their coding content are compared in figure 2. All encoded eight CTLDs. All except D also contained the exons encoding CysR and FNII receptor domains. That exception apart, the 30-exon structures are identical to that of the mammalian MRC1 genes, with all splice phases conserved and very similar exon lengths for all except the terminal exons.

One alternative splice acceptor site, for exon 8 of the MRC1L-E gene, resulting in the insertion of six amino acids, was found in a minority of the sequenced clones from Line 0 cDNA. While that was the only variant transcript in the Line 0 cDNAs, in the HD11 RNA, more frequent alternatively spliced transcripts were detected for MRC1L-E, most of which resulted in interruption of the open reading frame, so that no intact open reading frame for MRC1L-E was found in the HD11 cDNA. Thus it is possible the alternative splicing seen in HD11 was an artefact of the transformation of

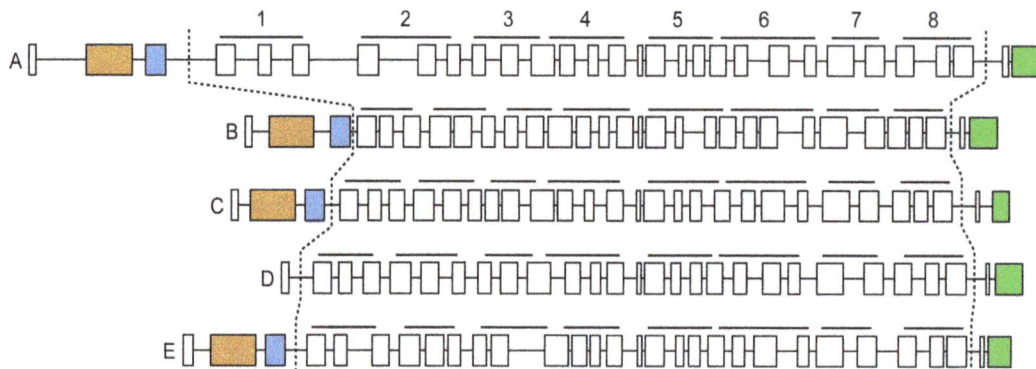

Figure 2. Structure of paralogous *MRC1* genes in the chicken genome. Exons are shown to scale as rectangles. Introns are drawn to 1/10 of the exon scale, except for the shortest which are expanded for visibility. Orange and blue exons are the CysR and FNII domains in all genes except D. The terminal green exon contains transmembrane and cytoplasmic regions. The central array of exons encodes the eight CTLDs indicated by the black bars above each gene.

these cells. The alternative spliced transcripts are illustrated in figure S3.

The locations of conserved features in the CTLDs of the chicken MRC1L genes are shown in figure 3. The tryptophan/hydrophobic/glycine/hydrophobic (WIGL) motif characteristic of the family [20] is present in all these domains of all genes, with minor variations. Four cysteine residues are also conserved in all these domains, an outer pair forming the disulphide bond spanning most of the domain, and an inner pair forming the disulphide bond stabilising the β3–β4 hairpin [21].

MRC1L genes in other species

The UCSC genome browser BLAT search [22,23] with individual and concatenated chicken genes was used to locate orthologous genes in genomes of other birds (turkey and zebrafinch), painted turtle, lizard and Xenopus (sequences in figure S4 and locations in figure S2). In all cases, the alignments with the highest scoring similarities were found between a pair of highly conserved orthologues of the same flanking genes, *SLC39A12* and *STAM*. Exons missing from these BLAT alignments were easily identified by manual inspection. The arrangements of these genes are compared with the orthologous region of the mouse genome in figure 4. The genes in the three birds are very similar, in structure and size of all five genes. The two gaps between the coding sequences of genes C, D and E are small compared with those between the upstream genes. The turtle appears to have a very similar set of five genes, although they occupy a segment of genome twice the length of that in the birds. The lizard genome contains only three genes, while the Xenopus genome, like the mammalian, contains only one *MRC1L* gene. The mouse genome, like that of other mammals, contains an additional gene, *TMEM236*, between the shared flanking genes.

Phylogenetic trees were constructed using a variety of sequence subsets and methods. The great majority of the results had similar topology to the tree depicted in figure 5. Several of the genes identified in other species were missing all or parts of exons in gaps in the genome assemblies. Some signal peptide exons were uncertain, and gene D lacked CysR and FNII domain exons. To avoid bias by these omissions, the tree shown was constructed using just those parts of the CTLDs that were available from all of the genes involved. All species had a single gene that was placed in the same clade as the mammalian *MRC1* gene in 100% of bootstrapped trees. For the avian and turtle genes, the same pattern of species was found for each gene, implying that these

arose by duplication before the divergence of these species. In contrast, the lizard lacked orthologues of genes C, D and E, but appeared to have two relatively similar genes of the gene represented in chickens by the KUL01 antigen. Thus the simplest consistent history of this gene family would be an original duplication of the ancestral MRC1 gene, giving rise to the *MRC1L-B* gene, followed in the shared avian and turtle ancestor by further duplications producing genes *C, D* and *E*, and in the lizard lineage by a second duplication of the *MRC1L-B* gene. Trees constructed using all the separated CTLDs generally gave the same pattern of species within a clade representing each domain, providing no evidence for domain reassortment. The majority produced the same topology as the tree shown, although bootstrap values were lower. Where the topologies differed, the bootstrap values were insufficient to support any contrary implications. A minority of alternative tree construction methods failed to place the lizard genes 2 and 3 in the MRC1L-B clade.

The cytoplasmic regions of the *MRC1L* gene products are compared in figure 6. The pattern of similarities between sequences are consistent with the evolutionary history that was implied by phylogenetic analysis. This part of the protein is highly conserved between the single mammalian *MRC1* gene and the other genes assigned to the same clade by analysis of the CTLDs. In these molecules, it contains potential motifs involved in targeting to the endocytic pathway, φxNxxY [24,25] and (DE)xxxLZ [25,26]. These motifs are shared by the genes that fall into the MRC1L-B clade that includes the KUL01 antigen, except for the replacement of tyrosine by histidine in the second of the two lizard genes in this clade. The group of genes including mammalian MRC1 also has a di-aromatic motif (YF) that may be involved in endosome sorting [27]. Although the latter is absent from the MRC1L-B orthologues, there are several other residues conserved between these two groups of proteins. In contrast, the cytoplasmic regions of the three downstream genes, found only in the bird and turtle genomes, are highly divergent between paralogues, although well conserved among orthologues. The product of MRC1L-C has only very short cytoplasmic sequences beyond the positively charged region expected to lie immediately inside the plasma membrane. Products of genes D and E have cytoplasmic sequences quite different from each other as well as from those of the MRC1L-A and MRC1L-B molecules. None of the downstream genes contain the endocytosis motifs conserved in the two upstream genes, although the MRC1L-D genes do have a

```
            β              α                   α                    β2    L1   L2    L3                  L4    β3       β4      β5
            Ω        Φ θ  C      θ θ O E Ωθ               ΦθGθ          Φ Ω  G  Ω Ω  W        P          EOCθ Ω    G WND C    Ω C
M1  LTGILYQINSKSAL---TWHQARASCKQQNADLLSVTEIHEQMYLTGLTSSL------SSGLWIGLNSLSVRSGWQWAGGSPFRYLNWLPGSP--------------SSEPGKS-CVSL--NPGKNAKWENLECVQKLGYICK
A1  LTNVQYQINSESAL---TWHQARKSCQQQKAELLSITELHEQTYLAGLTGKL------SSALWIGLNSLNFDSGWQWVCGAPFRYLNVVPGHP--------------SPEPGKI-CAAL--NPGKVAKWENWECNQKLGYICK
B1  LTETHYQINSNSLL---TWHQAKRSCQQQNAELLSVTNPHEEMFLLGLTSDLG-F---DAKLWTGLVRR-LDSSWEWTEGSPLRYLNWAPGNP--------------SVELLKM-CGTF--Q-GRNGKWENVACNQKLGYICK
C1  LRNAHYQINSESAL---TWHQARKSCQQQNAELLSITDIHEQTYLKELTEST------DSALWIGLNRLDLKSGWEWIGGTPFQYLNWAPGSP--------------SPESGKL-CVVL--NPETKAKWQNWECDQKLGYICK
D1  STGVLYQINSESAL---TWHQARKSCKQQNAELLSITEIHEQEYVGELIKKF------SFALWIGLNTLNFNSGWQWAGGSPFRYLNWAPGSP--------------FPAPGKI-CGTM--NPRQNAKWENQACNQRFGYICK
E1  LTGTFYQINFQSAL---TWHQARHSCKQQNAELLSVTEIHEQMYLRDLIDSN------RSPLWIGLNSLNLHSGWQWSGGTPFRYFNWAPGSP--------------SPEPDKL-CAVL--NPRTDAKWENRPCEQKVGYICK

M2  YAGHCYRIHREEKK---IQKYALQACRKEGGDLASIHSIEEFDFIFSQLGYEP-----NDELWIGLNDIKIQMYFEWSDGTPVTFTKWLPGEP--------------SHENNRQEDCVVM----KGKDGYWADRACEQPLGYICK
A2  YAGHCYIIHRDPK----IWKDALTSCRKEDGDLASIHNVEEYSFVISQLGYQP-----DDELWIGLNDLKVQMYFEWSDGTPVTYAKWLRGEP--------------THANNRQEDCVVM----KGKDGFWADHSCEKKIGYICK
B2  YAGHCYRIYRTPK----IWKQAQSSCRKEDGDLTSIHNVEEYSFIVSQLGYKP-----DDELWIGLNDFRFQMYFEWSDGTPVTYTKWQQRQP--------------THTPN-KADCIVM----NGEDGFWADSTCERKLGYICK
C2  YVDHCYKIFRETK----GWQEALTSCQNAGSHLASIQNFEEHSFIVSGLGYKP-----TDKLWIGLNDHKFQMFFEWSDGTPVTYTKWHLGEP--------------SSTNNRPEDCVMI----KGQDGYFADSNCEKKAGYVCK
D2  YASHCYSIQRESK----AWKDALTSCKRQGGDLASVHSITEYSFLVSQLGYMP-----TEELWLGLNDLKTHFYFEWSDGTPVTFTTWQRRHP--------------TYRNG-LEDCVVM----KGQDGYWATDVCDKQFGYICK
E2  YAGHCYVIHREPR----AWKDALMSCNESNGNLASIHNSEEHAFILSQLGYKA-----TDDLWIGMNDFSTQMYFEWSDETPVTYTKWLPGEP--------------THAVSGQEDCVVM----AGEDGYWADSDCDRKLGYICR

M3  HGFYCYLIGSTLS----TFTDANHTCTNEKAYLTTVEDRYEQAFLTSLVGLRP-----EKYFWTGLSDVQNKGTFRWTVDEQVQFTHWNADMP--------------GRKAGCV--AMKTGVACGLWDVLSCEEKAKFVCK
A3  HGFYCYFIGSTFV----TFSQANQTCERHQAYLATVQDRYEQAYLTSLVGLKT-----ERYFWIGLSDVEEKGTFRWANGEYVLFTHWNSEMP--------------GRKPGCV--AMRTGTAGGLWDVIKCEEKAKFLCK
B3  HGFYCYSIGQLPA----TFSEAKLICEENKAHLATVRDRYEQAYLTSIIGFKP-----VKYFWIGLSDMEEQGTFRWAGGDPVIFTHWNMGMP--------------GREPGCV--AMRTGTSACLWDILNCEEKNLFLCK
C3  YGTYCYFIGHVPA----TFSEANNTCKGEKGYLATVESRYEQAYLTSLVGLRP-----ERYFWIGLSDMEEQGTFRWSSGEDVSFTHWGAAIP--------------GSKPGCV--AMRTGTAAGLWDVLDCESKQKYICK
D3  YGFHCYLVGSALA----TFSDANKTCEQSKAYLATVETRNEQAFLISLTGLRS-----GKYFWLGLSDTEKRGMFKWTSGETPSFTHWNSAMP--------------GKEQGCV--AMGTGVSAGLWDVISCQETANFLCK
E3  HGSYCYLVGRAPV----TFSEAVKTCERIGGYLTTIEDRYEQAYLTSFVGLSS-----EKCFWIGLSNTEEQEIFKWETGEGVFYTNWNSAMP--------------GKEVGCV--ALRTGSAAGLWDVQNCELKAKFLCK

M4  KTSMCFKLYAKGKHEKKTWFESRDFCKAIGGELASIKSKDEQQVIWRLITSSGSY----HELFWLGLTYGSPSEGFTWSDGSPVSYENWAYGEP--------------NNYQNVEYCGELKGDP---GMSWNDINCEHLNNWICQ
A4  RISFCFKPFSKGE-QKKTWLESQEFCRTIGGDLASISGKDEQYVIWRSIANNGFY----HQHFWMGLYYLNPDDGFAWSDGSPVRYENWGFGEP--------------NNYQGIELCAEISGDS---SMLWNDRHCDYLYGWICQ
B4  QSSFCFKIFQRGREKMQTWIGARDFCRAIGGDLACIHSEEEQKLISS--LNKDYR----HVSYWMGLNALGSDGGFTWCDGSPVNFQKWANGEP--------------NNYDGNEKCCGVFYGYN---DMKWNDMFCEHMQDYVCQ
C4  TNKSCYKYFCRSDIKKKSWIEARDFCRQIGGDLATINNEEEKKMISR--GNSHCR----FERVWLGLFSLNPDEGFAWSDGSPVRYTGWS-DHP--------------RSSGGHMFCVFCEEQHDWVCQ
D4  HADSCFKFFVRDKNLKKNWFEAEEFCREIGGNLVTINSKEDQVLWQLALEKGLQ-----TQGFWMGLFLLNPDEGFTWIDGSPVIYENWDEDEP--------------NNDKGIEHCVMFNRSP---QMRWNDLYCEYLLNWICE
E4  STNSCFRTFVREKNHKKTWFEARDFCREIGGDLAAINSEEEQRVIEDLITKKLPS----SQLFWIGLQRLDPDGGLSWSDRSPVSYM---KTTP--------------FYDDPLENCGAISKEH---SISWINMHCEYSLDWICE

M5  YKDYQYYFSKEK----ETMDNARAFCKKNFGDLATIKSESEKKFLWKYINKNG-----GQSPYFIGMLIS-MDKKFIWMDGSKVDFVAWATGEP--------------NFANDDENCVTMYTNS----GFWNDINCGYPNNFICQ
A5  NEDRHYYFSTES----VPMEKGREFCKKNFGDLVVIDSETERKFLWRYILKNG-----KEDAYFICLQLS-VDQRTSWMDCTPVNYLAWAPHEP--------------NFANNDENCVVMYKNL----GFWNDINCGYPNPYICE
B5  YNHKEYYFSKEE----MPMEKAREYCKKNFGDLAIIENESERTFLWKY----TFYKD-RGNNFFIGLTVS-LDKTFRWIDGSTVNYVAWAPNEP--------------NFANNDENCVVMYTQT----GTWNDLNCGSVELFICE
C5  YKDKLYYISKEQ----VSMEEAQEFCRMNSADLAVISSNSERRFIQRALIKNDKYRT-ESEQYFIGLKIS-LDKTFSWIDGTPVTYVAWAPNEP--------------NFANNEEHCVVMFSKQ----GLWNDVNCGTTNRFVCE
D5  YEDKQYYFSRER----VPMEEARICQRNFADLVVIEDESERQFIWKYINRKRSGVFFQEESYFIGLFVS-SDQKLSWLGKTPVNYVAWAPEEP--------------NYSHNDENCVVMKEDF----GFWNDINCGLKNTFICE
E5  KGDKQYYFFSTES----TSMEKARTFCKNHRGDLAIIGDNNQRIFLWKYILKNG-----KLHSYLIGLILN-ADRQFRWVDGSTLHYAPWAQGEP--------------NFASAQEHCVVLDKKY----GLWNDVSCGHSHGFICE

M6  YKNKCFKIFGFANEEKKSWQDARQACKGLKGNLVSIENAQEQAFVTYHMRDST------FN-AWTGLNDINAEHMFLWTAGQGVHYTNWGKGYP---------GGRRSSLSYEDADCVVVIGGNSREAGTWMDDTCDSKQGYICQ
A6  FQNKCYKIFGSTEDERVTWHAARTACMNLGGNLATIPNEQVQAFLTFHMKDFL------TD-TWIGLNDINHELNFLWTDGTGVYFTNWAKGFP---------SGHLGSYSYNG-QADCVVMRNNPVKEAGKWADESCDNNRGYICQ
B6  FDNKCFKAFGLNENYTLTWHAARNNCITSGGNLATISKKENQAFLMSLLKNTA------TD-AWIGLNDINHEHTYLWTDGSPVVYTNWAK------------GSRSYYS-KDDCVYMKKNPIEQAGKWKDGDCKASKSYICQ
C6  FNNKCFKIFASNTTRKLAWHDAREVCIDLGGNLASVANEHAQAFVYYHLKDAT------TN-VWIGLNDINRESTFLWADGSTVSYTNWVEGAPETKQSFFDYYEYELLEDNITVETDCVFMTKSD----GKWRDDSCDNERGYICQ
D6  FQNKCYKIVGSREEERLTWYSARSACIEQGGNLASIHNAQVQAFLTFHLKDVT------DE-TWIGLNDK---HSYIWTDGSPYDYACWARGFP---------LGKYNRVGWKTDCIAMMIRSVNEAGKWENTDCHHNKSYICQ
E6  FKNQCYKFFGSQFQY---WYTANRDCISLGGHLATIQNEQVQAFLTYHLKDVL------YN-PWIGLNDIISELNFVWADGNTVSYTNWAPDSPKLYEPILYDSLHPEDGHNRMQYDCVSLK-TDYTDIGKWSDESCSKSSGYICQ

M7  YGKSSYSLMKLK----LPWHEAETYCKDHTSLLASILDPYSNAFAW-MKMHPF-----NVPIWIALNSNLTNNEYTWTDRWRVRYTNWGADEP--------------KLKSACV--YMD--VDGYWRTSYCNESFYFLCK
A7  YGNSSYLFIRTK----MNWEDARENCKRDQFDLSSILDPYSHSFLW-LKILKY-----GVPIWIGLNSNVTNGRYEWIDNWRMKYTKWAEGEP--------------KQKIGCV--YLD--ISGAWKTGSCNESYFSVCK
B7  YDDDRYAVINYK----MNWEEAQKNCKDQHADLASILDPYVEAYLW-LQTLKH-----GEPVWIGLNNTTHGLLWNGRRRSRYHNWASGEP--------------NKNAACA--YLD--LDGFWKTTSCNETFLSLCK
C7  YGDSSYLIVSSK----MQWEEARKNCQEQRAELASILDAYIHSFLW-IQMQKY-----GKPVWIGLKSNITRSYYKWTDNWKTRFTKWAAEEP--------------KKKNACV--YLD--IDGTWKTAPCKEMYFSVCK
D7  YGNSSYLIIPSK----MSWEERAKACREKSSELASISDYYSNIFLL-LQAAQY-----GEPLWIGINSNLSYGYYRWSDKRKIDFSNWHYEEP--------------KEKIACV--FLE--LSGEWKTAPCNEKHFSVCK
E7  SDGISYSVIHSK----MNWEEAQQSCNSNASELASILDPYSQSLLF-LIAQEY-----GQPMWIGLNSNLSYMTEGKYRWIDRWRLVYSKWSSGEP--------------KQTLACV--YLD--TDGTWKTASCKEKLFSICK

M8  FYGHCYYFESSFT---RSWGQASLECLRMGASLVSIETAAESSFLSYRVEPLKSK----TN-FWIGMFRN-VEGKWLWLNDNPVSFVNWKTGDP--------------SGERNDCVVLASSS----GLWNNIHCSSYKGFICK
A8  FRGHCYYVESSST---RNWAQASLECLRLGASLVSVEDSAEASFLTYIIEPLEGK----TSTFWTGMYRN-VDGEWLWLDNTAVNFVNWNTGEP--------------SPQQNEHCVEMYANS----GYWNNIYCTSYKGYVCK
B8  FRGHCYYVHTTSE---ASWPAASMMCIQMGASLVSIEDPAEMNFLLLYLSPFASD----NRKFWIGLFKN-IEGEWMWSDRSVVEFVNWEKGEP--------------TVMYDKHCVHMDVSS----GAWRNYYCSVDRNFICK
C8  YHGHCYYIEASAA---TSWAQASLKCTHLGATLVSVENVDESDFLIHTTQLLGNK----VGGFWIGLYRN-VDDQWLWLDNAVMDFVNWEEKE-------------SDEKHHCVEMTAPS----GYWDNTDCSSEKGFICK
D8  FRSHCYYFNPS-E---MSWVQSVTQCIQSGGMLTSVVDLAESNFLEEHADLYTSK----TSGFWIGLYRN-INGQLLWQDNSVLDFVNWGEAEP--------------LEEQHENEYCVQLSASS----GSWNSIPCSSRKGFICK
E8  FHGHCYHFEAVRK---KRWSQAHEECARLAADLLSVCDYTEANFVAETIKILHGK----SPNFWIGLKRD-DREQWVWTDKSELDFVNWQIGEP--------------ANRMHKDCGEVCALT----GFWNTNVCSFRKGYICK
```

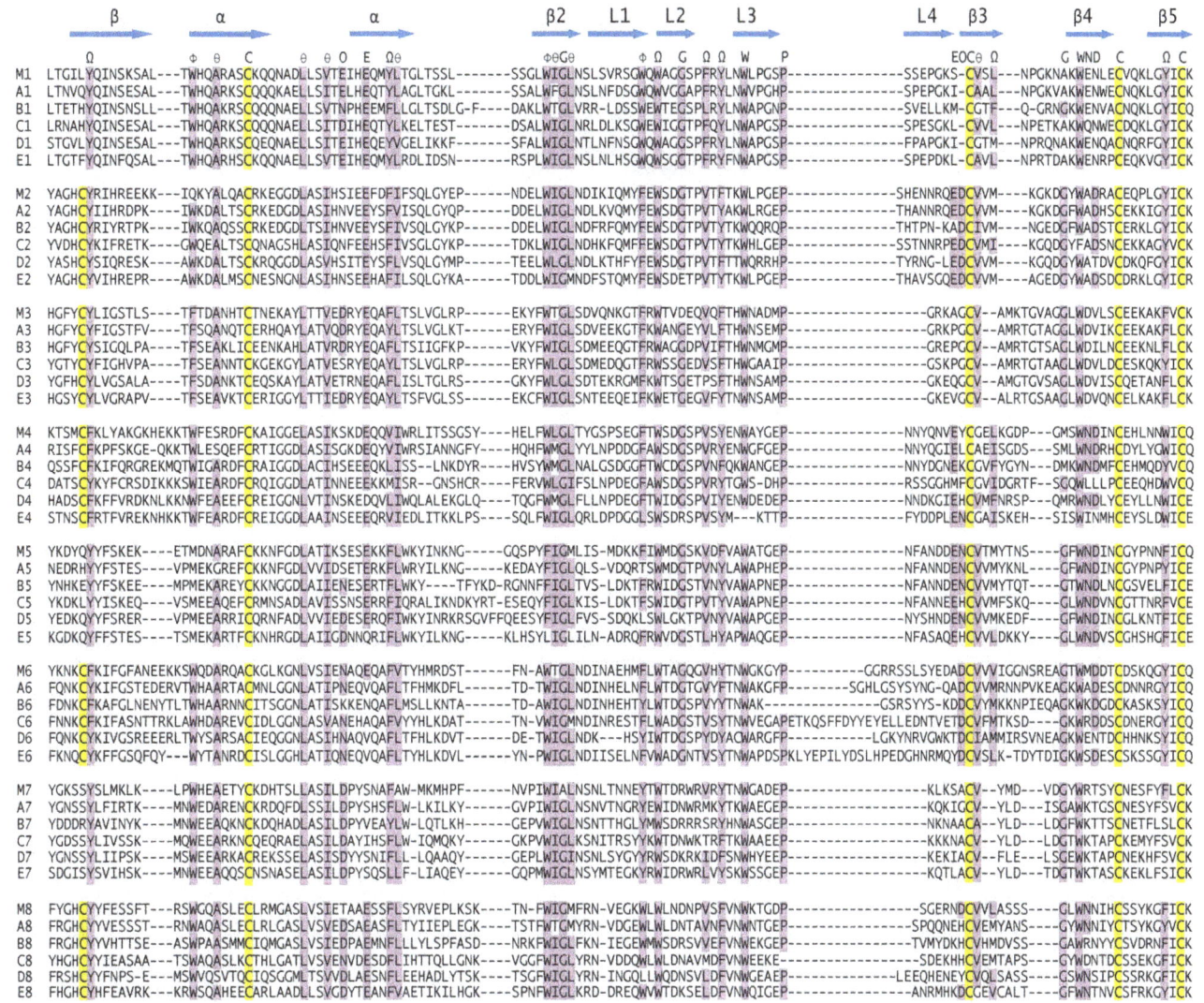

Figure 3. C type lectin domains of the avian MRC1 orthologue gene products. Sequences are labelled on the left, M being the mouse MRC1 sequence while the chicken genes are labelled A to E in genome order in the direction of their transcription, with sequential numbers to indicate the domains in order. Dashes indicate missing residues in the alignment. The short linker peptides between domains are omitted from this figure. Residues reported [20,51] to be conserved throughout the mannose receptor family are indicated above the sequences using the symbols Ω, aromatic or aliphatic; φ, aromatic; θ, aliphatic; C, E, G, P, W, N, D the standard amino acid codes; O, carbonyl oxygen containing (DNEQ). The corresponding residues in the sequences are shaded, yellow for cysteine and purple for the others. Additional cysteine residues in domains 2, 3, 4, 6 and 8 are also shaded. Likely locations of secondary structural features in the mouse sequence [52] are indicated by blue arrows above the sequence; β, beta strand; α alpha helix; L loop.

potential alternative endocytic pathway targeting motif YxxZ (FxxZ in the turtle) [28].

Transcription in tissues

Amplification of the spliced cDNAs for all five genes from the HD11 cell line suggested that all five genes might be transcribed in macrophages. PCR products were also obtained for all the genes from a spleen cDNA library. To obtain a more general picture of the pattern of transcript levels from these genes, quantitative PCR assays were developed for each and applied to compare levels of mRNA for each gene in different normal tissues. The mRNA levels, relative to 28S rRNA, found in various tissues from six Line 0 birds are shown in figure 7.

Exceptionally, the level of *MRC1L-A* transcript was highest in the liver whilst the level in the skin was highest for the other four genes, though only marginally so for *MRC1L-B*. Genes *C*, *D* and *E* had remarkably similar patterns of transcript levels in tissues, possibly indicating coordinated regulation. There was some variation between genes in the levels in different parts of the gut, although duodenum always had lower levels than distal regions of the digestive tract. Within the lymphoid tissues the highest level was always seen in the spleen. Relative transcript levels of all the genes were lowest in either kidney or liver. These assays are not calibrated to compare transcript levels between different genes.

The same assay was used to compare levels of expression, relative to 28S RNA, in several transformed cell lines (figure S5). Expression was clearly highest in the two macrophage cell lines, HD11 and MQ. Much lower levels of MRC1L-A were detected in some of the T cell derived cell lines.

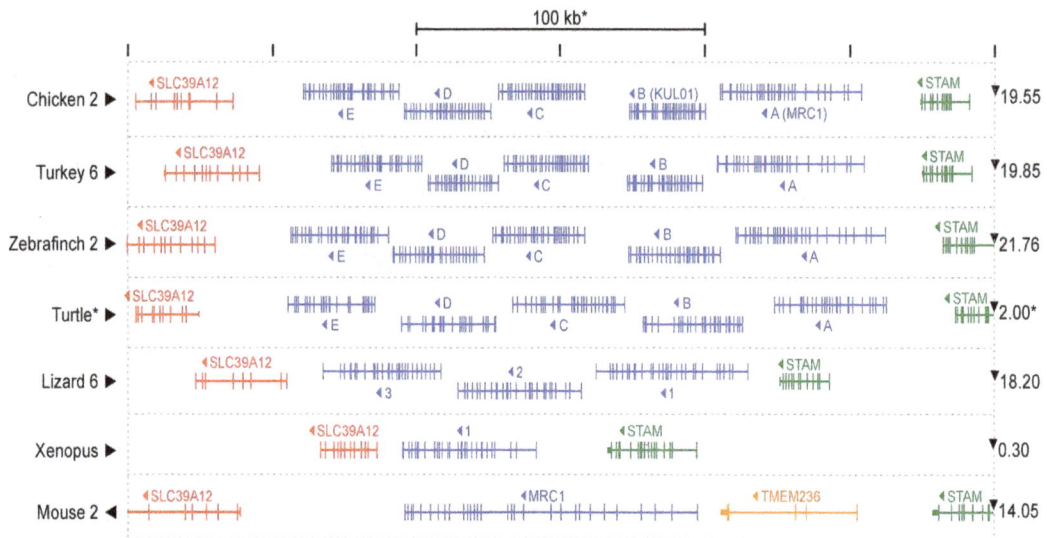

Figure 4. Arrangement the MRC1 orthologue locus in different species. Species are labelled at the left, with a numeral indicating the chromosome where that is known. Black arrowheads indicate the relative orientations of the reference genome maps. The conserved flanking genes SLC39A12 and STAM are indicated in red and green respectively. An additional gene TMEM236, found only in mammalian genomes, is coloured yellow. Predicted MRC1 paralogues are shown in blue. Vertical lines represent the exons of each gene. All the genomes are represented at the same scale, so that the region between vertical dotted lines is 300 kilobase pairs, except in the case of the Painted Turtle, where it represents 600 kilobase pairs. The location in megabase pairs of the right hand end of the map in the chromosome, or other map segment, is indicated at the right. The coding sequences of all genes shown run from right to left in this map, as indicated by arrowheads.

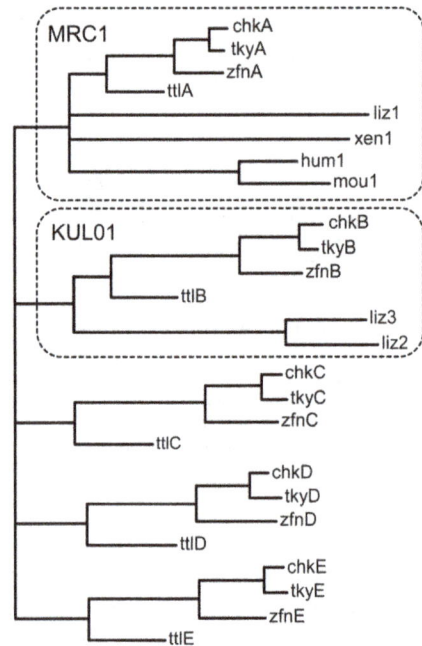

Figure 5. Evolutionary relationships of avian MRC1L genes. A maximum likelihood phylogenetic tree was constructed from predicted exons encoding all the CTLDs, using the Tamura-Nei model in the MEGA software, with 100 bootstrap datasets. All nodes with bootstrap values less than 100 were coalesced into multifurcations. Leaves are labelled with a three letter species code (chk, chicken (*Gallus gallus*); tky, turkey (*Maleagris gallopavo*); zfn, zebrafinch (*Taeniopygia guttata*); ttl, painted turtle (*Chrysemys picta bellii*); liz, lizard (*Anolis carolinensis*); xen, *Xenopus tropicalis*; hum, human (*Homo sapiens*); mou, mouse (*Mus musculus*); followed by either a letter or a number indicating the order of the genes in the direction of transcription. Clades representing orthologues of the MRC1 (human) and KUL01 (chicken) genes are surrounded by dotted lines.

```
hum1  KKRRVHLPQEG-AFENTLYFNSQSSPGTSDMKDLVGNIEQNEHSVI
mou1  KKRHALHIPQEA-TFENTLYFNSNLSPGTSDTKDLMGNIEQNEHAII
xen1  KRQKNKPPEDN-SFDNNLYFDGDRVPATHDTNILVENIEQNEHAIS
liz1  KKRNQHLVTEE-NFENSLYFNSNSAPGTSDTKDLVLNMEQNEHGAI
chkA  KRRKNRLATND-SFENNLYFNSD---GTGDTKDLVTNIERNEHATL
tkyA  KRRKNRLSTND-SFENNLYFNSD---GTGDTKDLVTNIERNEHATL
zfnA  KRRKNCLPTND-SFENNLYFNGDAVPGISDTKDLVTNIEQNEHATL
ttlA  RKRKNHMATND-SFENNLYFNTDAVPGTSDTKDLVDNIEQNEHAIL

liz2  RRRRGQPQTLG-GFDNSLYNKDRVVIPQKDPESLANNNEERLTSFRGNPS
liz3  KKQRKQQQNAA-GFNNSLHE-DNVIILQNDKELLVNNKAGD
chkB  KKRREQTTVTA-SFGNAIYCGTPDP-GTHESKCLVTNIEENEQAML
tkyB  KKRHEQTTVTA-SFGNAIYCGSPDP-GTHESKCLVTNIEENE
zfnB  KRRRDRQMFTA-GFDNAIYR-SDL--GTHESNCLVTNIEENE
ttlB  KKKRHNQLPTDVSFDNTLYCNRDAVPVASDSKYLVANIEQNEQAML

chkC  KRKRQNQLLRD
tkyC  KRKRQNQLSRD
zfnC  KRKRQNQLSTDSGNETLLR
ttlC  KKKRQNQL

chkD  KIKIQSEAERAVGQQSMLLEYSSALDRENDENDPASSKGGSERSDV
tkyD  KMKIQNEADQAAGQHSMLLEYSSAPARENNENNPASS
zfnD  KIKTGSGTGREVRRSSSQLEYSRALTAGDNGSGATNNKEKNEQSVV
ttlD  RKKRQNKLQTDASFNNVLLEHRDTVTGESDTKDSVDKKDQNEHTVI

chkE  RNKGQ-NQISISARMSGSEAALDIQEDDAHTNK
tkyE  RNKGQ-NQVSISARMSGSEAAVDIQGNDAHANK
zfnE  RRRKRQ-NLPHISTRMSGSEATVDIQVKDTHSDM
ttlE  KRKNQDNFLPIVTRMSDTKESVDNQEQDEHAVA
```

Figure 6. Alignments of cytoplasmic regions of MRC-like genes from various species. Gene names are as described in the legend to figure 5. Shaded residues show the locations of peptide motifs that may be involved in targeting to the endocytic pathway; green for the φxNxxY motif, red and blue for the (DE)xxxLZ motif, and purple for YxxZ (φ indicating a bulky hydrophobic residue and Z indicating a hydophobic residue). Light green shading indicates an overlapping potential di-aromatic endosome sorting motif in the MRC1 and MRC1L-A sequences.

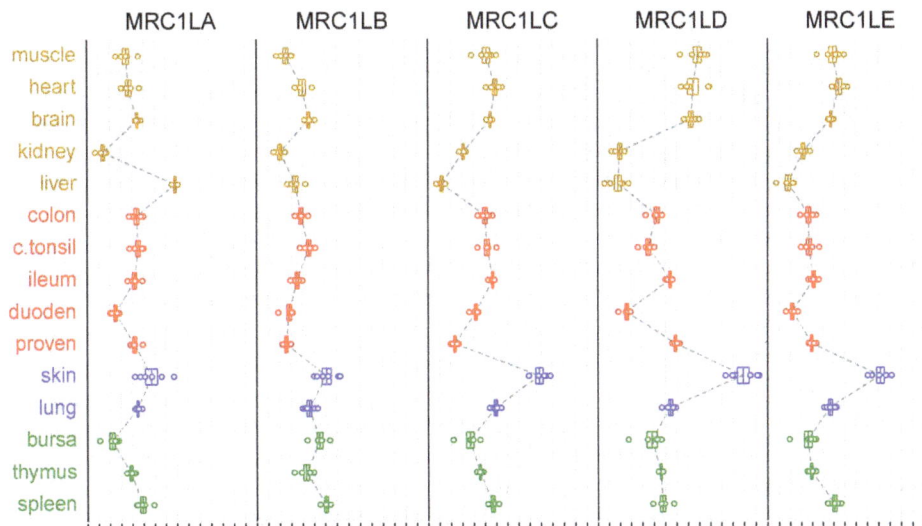

Figure 7. Relative levels of each MRC1 orthologue mRNA in different tissues, measured by quantitative PCR. Tissues, as labelled at the left, are grouped according to preponderance of immune function. For each gene, relative levels of mRNAs are plotted horizontally using a logarithmic scale with arbitrary origins. Circles are individual measurements from each of six birds. Boxes are centered on the means, and their ends indicate the standard errors of those means. All measurements were normalised relative to a constant level of 28S rRNA in each sample and adjusted to the log2 scale using the measured PCR efficiency of standard dilution series, before calculation of means and standard errors. Grey vertical lines and small scale bars at the bottom indicate two-fold differences in relative mRNA measurements.

Discussion

In the human genome, the region of chromosome 10 between the flanking markers *SLC39A12* and *STAM* is annotated as containing a tandemly repeated region, each repeat containing the genes *TMEM236* and *MRC1*. The repeated genes are part of a duplicated segment of about 200 kB with greater than 99% identity, separated by a large gap. There are only two BAC end pairs spanning the gap. In contrast all other mammals that we examined, including other primates, have only one copy of the *TMEM236* and *MRC1* genes between the same flanking marker orthologues, without the gap. While a very recent duplication in humans cannot be ruled out, it seems much more likely that this is a mis-assembled region in the human genome and thus that all mammals carry only a single *MRC1* gene. In species from other classes of terrestrial vertebrate, examination of the region of the genomes between the most highly similar homologues of the flanking markers revealed that some of these contained multiple, tandemly arranged diverged paralogues of *MRC1*. *Xenopus tropicalis* genomes contained only a single gene, the lizard *Anolis carolinensis* had three, while three birds and the painted turtle had five. This indicated duplication of the ancestral *MRC1* gene in the avian lineage and its precursors. The most likely sequence of events would have been an initial duplication producing the ancestors of chicken *MRC1L-A* and *MRC1L-B* genes, followed by a much more recent duplication of the latter in the lizard, and by further early duplications in the common ancestor of birds and turtles. In this context, it is of note that the phylogenetic position of the turtle has been the subject of much debate over a number of decades. Whilst a recent report based on an analysis of microRNAs suggested that turtles form a clade with lizards [29], subsequent reports place them in the archosaur lineage with birds and the crocodylia [30,31]. The more recent proposal is compatible with the simplest possible history of the MRC1 genes described in the present report.

Chicken orthologues of the adjacent *DEC205* [32] and *PLA₂R* genes, and of the *MRC2* gene, are found elsewhere in the genome. The additional genes in the *MRC1* locus are therefore not relocated orthologues of these genes.

All the identified genes in all the species examined had intact reading frames coding for proteins with the CTLD structure normally found in members of the mannose receptor family. All were found as spliced mRNAs in the chicken. Thus it is unlikely that any of the duplicated genes is a pseudogene, although differently spliced variants of the genes *D* and *E* transcripts were found in HD11 cDNA that had interrupted reading frames. The physical distances between the genes *C*, *D* and *E* were small, and the pattern of variation of their transcript levels in tissues was very similar. It may be that the transcription of these three genes is co-ordinately regulated by a shared set of upstream *cis*-acting elements. Indeed, the PCR amplifications used to confirm splice junctions would not have detected splicing between exons in different genes, so that the existence of splice variants that combine segments of the three genes, in a manner similar to the TWEPRIL transcripts from the TWEAK-APRIL genes in mouse [33], is not excluded.

The HD11 cell line contained mRNA for all five *MRC1L* genes, but peptides from protein immunoadsorbed by KUL01 included only those from MRC1L-B. This would be consistent with the KUL01 epitope being exclusive to MRC1L-B. However, the similarities between the MRC1L paralogues, while low, are sufficient that we could not exclude the possibility of recognition of the product of one or more of the other genes in the context where KUL01 is applied as a macrophage marker. To test this possibility we conducted two further experiments. As shown in figure S6, treatment of HD11 cells with transfection reagents including a small interfering RNA (siRNA) with 25/25 nucleotide identity to *MRC1L-B* cDNA sequence, caused 90% reduction in the median level of binding of fluorescently labelled KUL01 antibody, compared with the identical levels observed after the same treatment with either a control siRNA without or with no siRNA. The maximum similarity of the effective siRNA with the other *MRC1L* cDNA sequences, in either orientation, were 15/18 (A), 15/23 (C), 16/20 (D) and 17/22 (E). These are similar to the maximum similarity of the control siRNA with MRC1L-B cDNA (16/25), and would not generally be expected to be sufficient for cross-interference. However, since off-target interference effects have been reported with lower similarities, this observation does not completely exclude the possibility of cross reaction with the product of another MRC1L gene. In a second experiment, figure S7, we observed that that the KUL01 antibody only identified MRC1-B when expression plasmids coding for potential extracellular regions of all five MRCIL genes, as fusion proteins, were transfected into COS-7 cells. This provides compelling evidence that the KUL01 anybody binds the product of the MRC1L-B gene and not the remaining paralogues. Whilst the qRT-PCR analysis of MRC1L-B transcripts is consistent with the observed staining patterns reported with KUL01 across a number of immune-related tissues [19] it is not possible from the present data to infer the cellular distribution of the expression of the remaining MRC1L molecules, although, except for MRC1L-A in the liver, the similarity of the transcript profiles would be consistent with their predominant expression in the same cells as MRC1L-B.

In mammals, MRC1 is a multi-functional molecule. Being a pathogen-associated pattern recognition receptor, its involvements in uptake of antigen for presentation are important functions in innate and adaptive immune responses [9,34], but it also has roles in the clearance of hormones [12] and the regulation of circulating cytokine levels [35–37]. Cellular expression of the molecule is not restricted to macrophage alone but is also present on immature dendritic cells, reflecting its role in antigen capture [38].

The information presented here does not tell us whether a shared ancestor of birds and mammals had multiple MRC1L genes, with subsequent gene loss in the mammalian lineage, or whether it had a single gene that was subsequently duplicated only in the avian lineage. The former possibility would allow the hypothesis that the modern functions of mammalian MRC1 might have been distributed between the original paralogous genes. The latter model would have allowed the evolution of novel functional roles for the newly duplicated genes. The similarities between the cytoplasmic domains of MRC1L-A and MRC1L-B, especially with regard to trafficking signals, suggest biological functions similar to the mammalian MRC1, with the possibility of functional redundancy between these molecules. The very different cytoplasmic sequences of the other genes might reflect substantial functional divergence of these from the mammalian MRC1 genes.

The immune functions of MRC1 in the macrophage have given it an important role in determining the effectiveness of the response to influenza virus infection, at least in the lungs of mice. This presents a single interaction that is likely to be an effective target for evolution of viral virulence. If the additional genes in birds have similar functions in avian macrophages, then there is scope for redundant interactions with the virus that might be harder to evade. Expression of all these genes in macrophages is suggestive of conservation of these interactions. It will therefore be important to investigate whether these molecules have suitable carbohydrate binding activities, whether they are involved in endocytosis and phagocytosis, and whether modulation of their expression affects the susceptibility and response to influenza infection of avian macrophages. We have observed abortive replication of influenza in an avian macrophage cell line (KS and CB, Unpublished observations), which would allow a similar protective role for the MRC1L genes to that of MRC1 in the mouse, in generating effective responses. The involvement of multiple molecules, increasing redundancy in virus receptors, could increase the robustness of this immune mechanism in birds.

The known interaction of the mannose receptor with influenza virus in mice allows the hypothesis that a similar situation occurs in birds, facilitating infection of macrophages but leading to a protective innate immune response [18]. There are other enveloped avian viruses, including Marek's Disease Virus, Infectious Bronchitis Virus and Newcastle Disease Virus, that might be supposed to induce IFN-α by interaction with the mannose receptor [14].

Examples in which the mannose receptor acts as an innate pattern recognition molecule include the internalization of the yeast cell-wall particle zymosan [39], the phagocytosis of Pneumocystis by human alveolar macrophages [40] and Mycobacterium tuberculosis by the monocytic human cell line THP-1 [41]. The mannose receptor also appears to play a role in modulating the adaptive immune response through a role in myeloid plasticity [42]. However, the full repertoire of host-pathogen interactions allowed by the mannose receptor, and particularly the relevance of an expanded Mannose Receptor gene family, remains to be elucidated.

Materials and Methods

Ethics statement

All animal procedures were performed in accordance with the UK Animals (Scientific Procedures) Act 1986 [43]. This study was approved by the Pirbright Institute Ethical Review Panel and the UK Home Office under project licence 30/2683.

Experimental animals

RPRL (Regional Poultry Research Laboratory, East Lansing, MI.) Line O birds were obtained from the Compton specific pathogen free breeding facility, from parents negative for antibodies to specified pathogens, and were kept in controlled-environment isolation rooms with food and water provided ad libitum. For RNA preparations, tissue sections (approximately 500 mg), from birds between 4 and 5 weeks old, were collected into RNA Later stabilization fluid (Ambion, UK).

Antibodies, cells and cell lines

KUL01 is a monoclonal IgG1 antibody that recognises an antigen present on the surface of at least a subset of macrophages in chickens [19]. Purified antibody was purchased from Southern Biotech (Alabama).

The retrovirus-transformed macrophage-like cell line HD11 [44] was cultured in RPMI 1640 medium (Invitrogen), 10% FCS. Lines used in figure S5 are described in the figure legend.

Immunoprecipitation

Five to seven million HD11 cells pelleted at 200×g for 5 min were washed 3 times in PBS and resuspended in 500 μl of ice cold lysis buffer consisting of 20 mM TrisHCl, 100 mM NaCl, 0.5% v/v NP40 pH 7.6 to which 10 μl/ml HALT protease inhibitor cocktail and 10 μl/ml EDTA (Pierce Thermo product 87786) had been added. After vigorous mixing and incubation on ice for 30 minutes, Cell debris was then removed by centrifugation at 17,000×g for 15 minutes at 4°C and the lysate was stored at −80C.

Immunoprecipitations were carried out using a Thermo Scientific Pierce Immunoprecipitation kit (product number 1859011), following the manufacturer's instructions. Three hundred μg of antibody (Southern Biotech) was coupled to 100 μl AminoLink Plus Coupling Resin. Lysates were pre-cleared by two overnight incubations, mixing end over end at 4°C with agarose resin (Thermo Scientific) previously washed in lysis buffer, and then incubated for 2.5 hours with the immobilized antibody. After washing three times with 400 μl lysis buffer, bound proteins were eluted using five 100 µl aliquots of 0.1 M glycine. HCl, pH 2.8 including 0.5% (v/v) NP-40. Proteins were recovered from the pooled eluates by addition of trichloroacetic acid to 10% (w/v) and incubated on ice over night before pelleting in a microfuge at 17,000×g for 20 min at 4°C. Pellets were washed with 1 ml of ice cold 90% (v/v) acetone in water, then dried in a speed vac before resuspension and heating to ≥80°C in PAGE sample buffer including DTT for 10 min. PAGE was performed using 4–12% polyacrylamide Tris-Tricine gels in MES buffer and proteins visualised by rinsing the gels in water then incubating for 1–2 hours in Imperial stain (Thermo Scientific) followed by de-staining in water.

Peptide analysis

Bands of interest were excised from PAGE gels, cut into 1 mm cubes and individually placed in a covex 96 well microtitre plate. Reduction with DTT, alkylation with iodoacetamide and digestion using trypsin (Promega V511A) were all performed using a Hewlett Packard MassPREP robot. Digested extracts were transferred into low volume glass sample vials (Chromocol), dried in a speedvac then resuspended in 10 µl of 3% acetonitrile with 0.1% TFA. Liquid chromatography was carried out using a Waters NanoAquity UPLC system which supplied solvents A (0.1% formic acid in water) and B (0.1% formic acid in acetonitrile) to a 1.7 µm, 75 µm ×250 mm, BEH 130 C18 column (Waters) (HPLC solvents were all LC-MS grade from Fisher Scientific). Sample was concentrated onto a 180 µm ×20 mm, 5 µm Symmetry C18 trap (Waters) for 3 minutes at 15 µl/min, and separated at 250 nl/min using a gradient which ramped initially from 3–10% B over 1 minute then to 50% B over 41 minutes and to 85% B in 3 minutes followed by a wash step at this concentration for 2 minutes before re-equilibration at 3% B. Ionised peptides were analysed by a quadrupole time of flight (Q-ToF) Premier mass spectrometer (Waters) in data-dependent acquisition mode where a MS survey scan was used to automatically select multicharged peptides for further MS/MS fragmentation. From each survey scan up to four peptides were selected for fragmentation. MS/MS collision energy was dependent on precursor ions mass and charge state. A reference spectrum was collected every 10 seconds from Glu-fibrinopeptide B(785.8426 m/z), introduced via a reference sprayer. Raw MS/MS specta were processed using ProtenLynx Global Server (Waters) and were searched against the NCBInr database using the Mascot search algorithm.

RNA and quantitative PCR

RNA was extracted from 100 mg samples, of fifteen tissues from six birds of the same inbred line, using the Trizol Plus RNA Purification kit (Life Technologies), according to the manufacturer's instructions. Homogenisation was performed using a Mixer Mill MM300 (Retsch) and 3 mm stainless steel cone balls (Retsch). An on-column DNase digestion step was included (Purelink DNAse, Life Technologies). The majority of samples were diluted to have A_{260} approximately 1.0. Some samples with low RNA yields were used at up to ten-fold lower A_{260}.

Primers and probes for real-time quantitative PCR assay of 28S rRNA [45] and of the five predicted chicken macrophage mannose receptor mRNAs are detailed in table S3. The *MRC1L* cDNA primers and probes were designed so that the primers were entirely in different exons and the probe was approximately centred on an intron-exon boundary. *MRC1L* gene primers and probes were designed using Genscript primer design software (https://www.genscript.com/ssl-bin/app/primer) and Primer Express (Applied Biosystems, Foster City, California, USA). These

primers gave no detectable signal after 40 cycles with 2.5 ng chicken genomic DNA in the standard assay conditions.

Probes incorporated 5-carboxyfluorescein (FAM) at the 5′ end and N,N,N,N′ tetramethyl-6-carboxyrhodamine (TAMRA) at the 3′ end. Assays were carried out using the Superscript III platinum one-step qRT-PCR kit (Invitrogen). Amplification and detection of specific products were carried out with the 7500 Fast Real Time System (Taqman; Applied Biosystems) with the following cycle profile: 50°C for 5 min, 95°C for 2 min and then 40 cycles of 95°C for 3 sec and 60°C for 30sec.

To measure the PCR efficiencies, six 10-fold dilutions, of the highest expressing tissue for each MRC1-L assay, and of HD11 RNA for the 28S assay, were used in triplicate measurements. All *MRC1L* gene mRNA measurements were normalised to the levels of 28S ribosomal RNA in the samples using the equation $Xt = Ct - s(Ct′ - Q)/s′$ where Ct is the gene-specific threshold cycle, Ct′ is the threshold cycle for the 28S ribosomal RNA assay (on a constant dilution of the sample), s and s′ are slopes of linear regressions of threshold cycles (C_T) against $\log_{10}$(RNA) for target gene and 28S assays respectively. All sample Ct values were within the range of the standard plots. Details of the normalisation calculations and of statistical analyses confirming differential expression between tissues are provided in document S1 and document S2.

Sequencing and bioinformatics

Primers listed in table S4 were designed to amplify overlapping segments of the five predicted transcripts. Preparative PCR amplifications were carried out using methods described elsewhere [32], using templates of a line 0 chicken spleen cDNA library [46], freshly prepared total RNA from line 0 chicken spleen and total RNA from the cell line HD11. Amplified products excised from agarose gels were cloned into the pGEM-T-Easy vector (Promega). DNA prepared using the Qiagen QIAprep spin miniprep kit were used for sequencing. Sequencing reactions were performed by GATC Biotech. Sequence data were analysed using STADEN [47]. Multiple clones of PCR products were sequenced from each amplification to obtain the consensus sequence and to identify clones free from PCR errors.

Extensive use was made of the ClustalW [48]. The UCSC genome browser (http://genome.ucsc.edu; [49]) during the manual refinement of gene structures and in the preparation of figure 4. Assembly versions used were chicken, WUGSC 2.1/galGal3; turkey, TGC Turkey_2.01/melGal1; zebra finch, WUGSC 3.2.4/taeGut1; lizard, Broad AnoCar2.0/anoCar2; painted turtle, v3.0.1/chrPic1; Xenopus tropicalis, JGI 4.2/xenTro3; mouse, GRCm38/mm10; human, GRCh/hq19. Phylogenetic analyses were carried out using the MEGA package [50].

Supporting Information

Figure S1 Peptide sequences from MRC1L-B found in trypic digest of KUL01-adsorbed material.

Figure S2 Chicken MRC1L cDNA sequences and genomic locations of orthologues.

Figure S3 Alternative splicing in MRC1L-E cDNA.

Figure S4 Predicted sequences of MRC1L orthologues in other species.

Figure S5   Relative mRNA levels of MRC1L genes in chicken cell lines.

Figure S6   Suppression of KUL01 antigen expression by MRC1L-B specic siRNA.

Figure S7   KUL01 specifically recognises the MRC1L-B gene product in transfected COS cells.

Table S1   List of peptides from tryptyic digest of KUL01-adsorbed material.

Table S2   Differences between Line 0 cDNA sequence and genomic jungle fowl sequence.

Table S3   Primers and probes for TaqMan quantitative PCR.

Table S4   Primers used in amplifying chicken MRC1L cDNAs.

Document S1   Statistical analysis of qPCR data for MRC1L genes in different tissues.

Document S2   Statistical test for differences in MRC1L transcripts between tissues.

File S1   Chicken MRC1L genes: Links to ENSEMBL identifiers & Errors in release 75.

Acknowledgments

The authors would like to thank Dr John Hammond, for advice pertaining to phylogenetic analysis, and the dedicated staff of the Institute's animal services.

Author Contributions

Conceived and designed the experiments: KS LGH JRY CB. Performed the experiments: KS LGH JRY CB. Analyzed the data: KS LGH JRY CB. Contributed reagents/materials/analysis tools: JRY. Wrote the paper: KS LGH JRY CB.

References

1. Barten R, Torkar M, Haude A, Trowsdale J, Wilson MJ (2001) Divergent and convergent evolution of NK-cell receptors. Trends in Immunology 22: 52–57.
2. Kaiser P, Poh TY, Rothwell L, Avery S, Balu S, et al. (2005) A genomic analysis of chicken cytokines and chemokines. Journal of Interferon and Cytokine Research 25: 467–484.
3. Hughes AL (1994) The Evolution of Functionally Novel Proteins after Gene Duplication. Proceedings of the Royal Society B-Biological Sciences 256: 119–124.
4. Weis WI, Taylor ME, Drickamer K (1998) The C-type lectin superfamily in the immune system. Immunological Reviews 163: 19–34.
5. Taylor ME, Conary JT, Lennartz MR, Stahl PD, Drickamer K (1990) Primary Structure of the Mannose Receptor Contains Multiple Motifs Resembling Carbohydrate-Recognition Domains. Journal of Biological Chemistry 265: 12156–12162.
6. Taylor ME, Drickamer K (1993) Structural Requirements for High-Affinity Binding of Complex Ligands by the Macrophage Mannose Receptor. Journal of Biological Chemistry 268: 399–404.
7. Taylor ME, Bezouska K, Drickamer K (1992) Contribution to Ligand-Binding by Multiple Carbohydrate-Recognition Domains in the Macrophage Mannose Receptor. Journal of Biological Chemistry 267: 1719–1726.
8. Tietze C, Schlesinger P, Stahl P (1982) Mannose-Specific Endocytosis Receptor of Alveolar Macrophages - Demonstration of 2 Functionally Distinct Intracellular Pools of Receptor and Their Roles in Receptor Recycling. Journal of Cell Biology 92: 417–424.
9. Stahl PD, Ezekowitz RAB (1998) The mannose receptor is a pattern recognition receptor involved in host defense. Current Opinion in Immunology 10: 50–55.
10. Martinez-Pomares L, Mahoney JA, Kaposzta R, Linehan SA, Stahl PD, et al. (1998) A functional soluble form of the murine mannose receptor is produced by macrophages in vitro and is present in mouse serum. Journal of Biological Chemistry 273: 23376–23380.
11. Martinez-Pomares L, Gordon S (1999) Potential role of the mannose receptor in antigen transport. Immunology Letters 65: 9–13.
12. Fiete DJ, Beranek MC, Baenziger JU (1998) A cysteine-rich domain of the "mannose" receptor mediates GalNAc-4-SO4 binding. Proceedings of the National Academy of Sciences of the United States of America 95: 2089–2093.
13. Tesar DB, Cheung EJ, Bjorkman PJ (2008) The chicken yolk sac IgY receptor, a mammalian mannose receptor family member, transcytoses IgY across polarized epithelial cells. Molecular Biology of the Cell 19: 1587–1593.
14. Milone MC, Fitzgerald-Bocarsly P (1998) The mannose receptor mediates induction of IFN-alpha in peripheral blood dendritic cells by enveloped RNA and DNA viruses. Journal of Immunology 161: 2391–2399.
15. Nguyen DG, Hildreth JEK (2003) Involvement of macrophage mannose receptor in the binding and transmission of HIV by macrophages. European Journal of Immunology 33: 483–493.
16. Miller JL, Dewet BJM, Martinez-Pomares L, Radcliffe CM, Dwek RA, et al. (2008) The mannose receptor mediates dengue virus infection of macrophages. Plos Pathogens 4.
17. Reading PC, Miller JL, Anders EM (2000) Involvement of the mannose receptor in infection of macrophages by influenza virus. J Virol 74: 5190–5197.
18. Upham JP, Pickett D, Irimura T, Anders EM, Reading PC (2010) Macrophage Receptors for Influenza A Virus: Role of the Macrophage Galactose-Type

Lectin and Mannose Receptor in Viral Entry. Journal of Virology 84: 3730–3737.
19. Mast J, Goddeeris BM, Peeters K, Vandesande F, Berghman LR (1998) Characterisation of chicken monocytes, macrophages and interdigitating cells by the monoclonal antibody KUL01. Veterinary Immunology and Immunopathology 61: 343–357.
20. Weis WI, Kahn R, Fourme R, Drickamer K, Hendrickson WA (1991) Structure of the Calcium-Dependent Lectin Domain from a Rat Mannose-Binding Protein Determined by Mad Phasing. Science 254: 1608–1615.
21. Zelensky AN, Gready JE (2003) Comparative analysis of structural properties of the C-type-lectin-like domain (CTLD). Proteins-Structure Function and Genetics 52: 466–477.
22. Kent WJ (2002) BLAT - The BLAST-like alignment tool. Genome Research 12: 656–664.
23. Kent WJ, Sugnet CW, Furey TS, Roskin KM, Pringle TH, et al. (2002) The human genome browser at UCSC. Genome Research 12: 996–1006.
24. East L, Isacke CM (2002) The mannose receptor family. Biochimica Et Biophysica Acta-General Subjects 1572: 364–386.
25. Mellman I (1996) Endocytosis and molecular sorting. Annual Review of Cell and Developmental Biology 12: 575–625.
26. Pond L, Kuhn LA, Teyton L, Schutze MP, Tainer JA, et al. (1995) A Role for Acidic Residues in Di-Leucine Motif-Based Targeting to the Endocytic Pathway. Journal of Biological Chemistry 270: 19989–19997.
27. Schweizer A, Stahl PD, Rohrer J (2000) A di-aromatic motif in the cytosolic tail of the mannose receptor mediates endosomal sorting. Journal of Biological Chemistry 275: 29694–29700.
28. Sandoval IV, Bakke O (1994) Targeting of membrane proteins to endosomes and lysosomes. Trends Cell Biol 4: 292–297.
29. Hedges SB (2012) Amniote phylogeny and the position of turtles. Bmc Biology 10.
30. Chiari Y, Cahais V, Galtier N, Delsuc F (2012) Phylogenomic analyses support the position of turtles as the sister group of birds and crocodiles (Archosauria). Bmc Biology 10.
31. Crawford NG, Faircloth BC, McCormack JE, Brumfield RT, Winker K, et al. (2012) More than 1000 ultraconserved elements provide evidence that turtles are the sister group of archosaurs. Biology Letters 8: 783–786.
32. Staines K, Young JR, Butter C (2013) Expression of Chicken DEC205 Reflects the Unique Structure and Function of the Avian Immune System. PLoS One 8: e51799.
33. Pradet-Balade B, Medema JP, Lopez-Fraga M, Lozano JC, Kolfschoten GM, et al. (2002) An endogenous hybrid mRNA encodes TWE-PRIL, a functional cell surface TWEAK-APRIL fusion protein. Embo Journal 21: 5711–5720.
34. Prigozy TI, Sieling PA, Clemens D, Stewart PL, Behar SM, et al. (1997) The mannose receptor delivers lipoglycan antigens to endosomes for presentation to T cells by CD1b molecules. Immunity 6: 187–197.
35. Shibata Y, Metzger WJ, Myrvik QN (1997) Chitin particle-induced cell-mediated immunity is inhibited by soluble mannan - Mannose receptor-mediated phagocytosis initiates IL-12 production. Journal of Immunology 159: 2462–2467.
36. Vautier S, MacCallum DM, Brown GD (2012) C-type lectin receptors and cytokines in fungal immunity. Cytokine 58: 89–99.

37. Yamamoto Y, Klein TW, Friedman H (1997) Involvement of mannose receptor in cytokine interleukin-1 beta (IL-1 beta), IL-6, and granulocyte-macrophage colony-stimulating factor responses, but not in chemokine macrophage inflammatory protein 1 beta (MIP-1 beta), MIP-2, and KC responses, caused by attachment of Candida albicans to macrophages. Infection and Immunity 65: 1077–1082.

38. Sallusto F, Cella M, Danieli C, Lanzavecchia A (1995) Dendritic cells use macropinocytosis and the mannose receptor to concentrate macromolecules in the major histocompatibility complex class II compartment: downregulation by cytokines and bacterial products. J Exp Med 182: 389–400.

39. Underhill DM, Ozinsky A, Hajjar AM, Stevens A, Wilson CB, et al. (1999) The Toll-like receptor 2 is recruited to macrophage phagosomes and discriminates between pathogens. Nature 401: 811–815.

40. Zhang JM, Zhu JP, Bu X, Cushion M, Kinane TB, et al. (2005) Cdc42 and RhoB activation are required for mannose receptor-mediated phagocytosis by human alveolar macrophages. Molecular Biology of the Cell 16: 824–834.

41. Diaz-Silvestre H, Espinosa-Cueto P, Sanchez-Gonzalez A, Esparza-Ceron MA, Pereira-Suarez AL, et al. (2005) The 19-kDa antigen of Mycobacterium tuberculosis is a major adhesin that binds the mannose receptor of THP-1 monocytic cells and promotes phagocytosis of mycobacteria. Microbial Pathogenesis 39: 97–107.

42. Mishra PK, Morris EG, Garcia JA, Cardona AE, Teale JM (2013) Increased Accumulation of Regulatory Granulocytic Myeloid Cells in Mannose Receptor C Type 1-Deficient Mice Correlates with Protection in a Mouse Model of Neurocysticercosis. Infection and Immunity 81: 1052–1063.

43. UK Home Office (1986) Guidance on the Operation of the Animals (Scientific Procedures) Act.

44. Beug H vKA, Doderlein G, Conscience J-F, Graf T (1979) Chicken hematopoietic cells transformed by seven strains of defective avian leukemia viruses display three distinct phenotypes of differentiation. Cell 18: 375–390.

45. Moody A, Sellers S, Bumstead N (2000) Measuring infectious bursal disease virus RNA in blood by multiplex real-time quantitative RT-PCR. J Virol Methods 85: 55–64.

46. Tregaskes CA, Bumstead N, Davison TF, Young JR (1996) Chicken B-cell marker chB6 (Bu-1) is a highly glycosylated protein of unique structure. Immunogenetics 44: 212–217.

47. Staden R (1996) The Staden sequence analysis package. Molecular Biotechnology 5: 233–241.

48. Larkin MA, Blackshields G, Brown NP, Chenna R, McGettigan PA, et al. (2007) Clustal W and clustal X version 2.0. Bioinformatics 23: 2947–2948.

49. Meyer LR ZA, Hinrichs AS, Karolchik D, Kuhn RM, Wong M, et al. (2012) The UCSC Genome Browser database: extensions and updates 2013. Nucleic Acids Res Nov 15.

50. Tamura K, Peterson D, Peterson N, Stecher G, Nei M, et al. (2011) MEGA5: Molecular Evolutionary Genetics Analysis Using Maximum Likelihood, Evolutionary Distance, and Maximum Parsimony Methods. Molecular Biology and Evolution 28: 2731–2739.

51. Kim SJ, Ruiz N, Bezouska K, Drickamer K (1992) Organization of the Gene Encoding the Human Macrophage Mannose Receptor (Mrc1). Genomics 14: 721–727.

52. Harris N, Super M, Rits M, Chang G, Ezekowitz RAB (1992) Characterization of the Murine Macrophage Mannose Receptor - Demonstration That the down-Regulation of Receptor Expression Mediated by Interferon-Gamma Occurs at the Level of Transcription. Blood 80: 2363–2373.

14

Intraspecific Variation in Physiological Condition of Reef-Building Corals Associated with Differential Levels of Chronic Disturbance

Chiara Pisapia[1,2]*, Kristen Anderson[1], Morgan S. Pratchett[1]

1 ARC Centre of Excellence for Coral Reef Studies, James Cook University, Townsville, Australia, 2 AIMS@JCU Australian Institute of Marine Science, School of Marine Biology, James Cook University, Townsville, Australia

Abstract

Even in the absence of major disturbances (e.g., cyclones, bleaching), corals are subject to high levels of partial or whole-colony mortality, often caused by chronic and small-scale disturbances. Depending on levels of background mortality, these chronic disturbances may undermine individual fitness and have significant consequences on the ability of colonies to withstand subsequent acute disturbances or environmental change. This study quantified intraspecific variations in physiological condition (measured based on total lipid content and zooxanthellae density) through time in adult colonies of two common and widespread coral species (*Acropora spathulata* and *Pocillopora damicornis*), subject to different levels of biological and physical disturbances along the most disturbed reef habitat, the crest. Marked intraspecific variation in the physiological condition of *A. spathulata* was clearly linked to differences in local disturbance regimes and habitat. Specifically, zooxanthellae density decreased ($r^2 = 26$, df $= 5,42$, p$<$0.02, B $= -121255$, p $= 0.03$) and total lipid content increased ($r^2 = 14$, df $= 5,42$, p $= 0.01$, B $= 0.9$, p $= 0.01$) with increasing distance from exposed crests. Moreover, zooxanthellae density was strongly and negatively correlated with the individual level of partial mortality ($r^2 = 26$, df $= 5,42$, p$<$0.02, B $= -7386077$, p $= 0.01$). Conversely, *P. damicornis* exhibited very limited intraspecific variation in physiological condition, despite marked differences in levels of partial mortality. This is the first study to relate intraspecific variation in the condition of corals to localized differences in chronic disturbance regimes. The next step is to ascertain whether these differences have further ramifications for susceptibility to periodic acute disturbances, such as climate-induced coral bleaching.

Editor: Linsheng Song, Institute of Oceanology, Chinese Academy of Sciences, China

Funding: This project was supported by the ARC Centre of Excellence for Coral Reef Studies, Townsville, and AIMS@JCU, Townsville. The funders had no role in study design, data collection and analysis, decision to publish, or preparation of the manuscript.

Competing Interests: The authors have declared that no competing interests exist.

* E-mail: chiara.pisapia@my.jcu.edu.au

Introduction

Coral reefs are very dynamic ecosystems, impacted by a variety of natural and anthropogenic processes, which may vary in scale, frequency, and intensity [1]. Even in the absence of major disturbances (e.g., cyclones, bleaching or outbreaks of crown-of-thorns starfish), corals are still subject to a range of chronic, often small-scale disturbances that cause relatively high rates of background mortality (annual background mortality rates can generally vary from 1 to 30%: [2–5]). These background mortality agents (such as predation, competition and disease) are a normal part of the natural dynamics and turnover in coral populations and communities [6–8]. However, increases in prevalence and impact of chronic disturbances undermine the resilience of coral colonies and populations [4,5,8,9,10], which are subject to ever-increasing threats from climate change and other more direct anthropogenic disturbances [11,12].

Background mortality agents can trigger complex responses in corals that may affect colony physiological condition, alter demographic performance, especially growth [13–15] and reproduction [15,16,17] and they can therefore have significant consequences on the ability of colonies to withstand and survive periodic acute disturbances and environmental changes [18].

Intraspecific competition, for example, can substantially reduce fitness and growth rates of colonies engaged in competitive interactions [14]. Tanner 1997 documented a reduction in growth rates from 120 to 35% in *Acropora hyacinthus* when engaged in competitive interactions, and a decrease in growth from 45 to −16% in *Pocillopora damicornis*. Similarly, chronic predation can inflict a significant energetic cost to prey corals and may accelerate rates of coral decline following a disturbance [19]. Coral grazing fishes are a potentially important source of background coral mortality [19], even when they do not leave any visible signs of damage on coral colonies [20]. Rates of tissue removal from individual coral colonies can be considerable (16.75 ± 0.30 bites per 20 min, [19]) and this chronic removal of live tissue can have potentially important consequences for colony fitness. Similarly, sedimentation can affect coral physiological condition by exerting significant energetic costs due to the removal of particles from colonies and limit energy availability due to reduced light and photosynthetic activity [15,21]. *Siderastrea siderea* reduced linear extension rates from 3.5 mm to 3 mm three years following an oil spill, which caused increased sedimentation levels [21].

The physiological condition of a colony is largely determined by the energy available and by the partitioning of energy reserves among maintenance, growth, and reproduction [22]. Energy within a colony is a limited resource and it is distributed among costly life history processes. If a coral invests heavily in repairing tissues damaged by chronic predation or sedimentation, or is investing heavily in interspecific competition, then this will reduce resources available for growth and reproduction. Evidences of energy trade-offs have been widely documented in corals, with injury often causing a decline in growth [23,24] or fecundity [25]. Moreover, diversion of essential energy reserves may undermine the capacity of corals to withstand periodic acute disturbances, such as anomalous temperatures that cause widespread bleaching [22]. When injured, corals often divert energy towards regeneration of lost tissue, and species with high regenerative capacity (such as *Acropora* spp) being able to fully heal the injury in less than 80 days [23]. However, environmental stresses, large lesions and competition may impair regeneration and hence compromise survival [23,26]. The bare skeleton resulting from tissue loss can be colonized by algae, pathogens or bioeroders, which may undermine the integrity of the colony [27,28]. These organisms may later compete with the coral for food and space, or cause structural damage to the coral skeleton [27,28].

The capacity of corals to withstand ongoing disturbances is strongly size-dependent, with small colonies being more vulnerable to whole-colony mortality than larger ones [6]. Corals as modular organisms are made up of repeated units (polyps), each of which can function and survive as physiologically independent entities. However, partial mortality and the consequent decline in the total number of polyps that make up a colony can greatly reduce individual fitness and resilience [6,17,29]. A reduction in size results in fewer polyps available to support colony vital processes and will generally reduce survivorship [6,30], growth [23,24], reproduction and regeneration [15]. Large colonies have greater regenerative abilities [6,8], growth [31], are more fecund [17] and have lower rates of total mortality compared to smaller colonies [6,17]. Likelihood of survival in larger colonies is greater than smaller ones because there is a higher probability that part of the colony may remain unaffected [32]. Particularly, following a disturbance, big colonies can make a disproportionate contribution to population as they produce more eggs per unit area [17].

Intra-specific variation of corals in responses to stresses is largely due to genotypic and phenotypic variation among both corals and their zooxanthellae [33–35], however the disturbance history and current physiological condition of individual colonies may also play a critical role. The exhaustion of energy available to maintain vital processes represents a physiologically critical threshold for survival [36]. During a bleaching event for instance, a key determinant for survival and recovery of a coral is its amount of lipid reserves [22], as they can. When bleaching occurs the energy acquisition by the zooxanthellae stops, hence the coral must use its energy reserves accumulated in the form of lipids in order to survive [38–41]. So colonies in good physiological conditions, with a great magnitude of lipid reserves, are more likely to survive and recover from a bleaching event, than colonies with lower level of lipid reserves [22,42]. Also colonies which survived a previous disturbance and are potentially in good physiological condition, can substantially contribute to community recover through their growth and through their reproductive output [17,43,44].

The purpose of this study is to quantify intra-specific variation in physiological condition (specifically, total lipid content and zooxanthellae density) through time in adult colonies exposed to several biological and environmental factors. Variation in colony condition among individuals may account for differences in susceptibility to disturbances. Many studies have documented significant variation in the capacity of corals to withstand and

recover from major disturbances [4,8,45,46], but the underlying basis of this variation is still poorly understood. Most of these studies have focused on among-species variability for stress resistance. Hoegh-Guldberg [47] suggested that in the aftermath of climate change some coral species are more likely to adapt and survive better than others. But still little is known on intraspecific variability to environmental changes.

Methods

Ethics Statement

The activities for this study were conducted under permission from the Great Barrier.

Reef Marine Park Authority (Permit Number G12/35017.1). Visual censuses of fishes and benthic communities were conducted during this study; one coral branch per colony was collected in May and one in October.

Chronic Disturbances

This study was conducted at Lizard Island (14°40′S, 145°27′E) in the northern Great Barrier Reef, Australia. 24 colonies, ranging in size from 9 cm up to 35 cm diameter, of *Pocillopora damicornis* and *Acropora spathulata* were individually tagged and sampled in May and October 2012 to test for intraspecific differences in physiological condition. At the same time, detailed observations were undertaken to quantify intra-specific differences in background disturbance regimes (NB. There were no major bleaching events or other acute disturbances during the conduct of this study). Colonies at the same depth were selected from the reef crest in two different sites, one sheltered and one in the windward side of the island. For each coral colony we measured the distance from the reef crest (presumed to reflect colony physical position in respect of local hydrodynamic regime), proportional tissue loss attributable to coral competition and/or coral disease, and also rates of predation by corallivorous fishes. Variation in the level of predation among individual coral colonies was documented using GoPro cameras, deployed to record the total number of bites taken by all corallivorous fishes within replicate one-hour periods. The fish species and size were also recorded. Partial mortality was measured by quantifying the exact proportion of dead versus living tissue within the overall physical extent of each coral colony, using the software Image J. Growth rates were also calculated comparing colony surface area from pictures in May and in October for each individual colony.

Colony Physiological Condition

Colony condition was assessed based on total lipid content and zooxanthellae density. The size of lipid reserves is a good measure of colony condition because it represents an alternative source of fixed carbon, which can be allocated to vital processes such as growth or reproduction. Lipid reserves can also allow the host to meet its daily metabolic energy needs in absence of endosymbionts, such as during a bleaching event [42]. Similarly, the symbiotic relationship between the coral colonies and the symbionts makes zooxanthellae density a good proxy of coral condition [48]. Zooxanthellae density has been shown to decrease in response to chronic stresses such as exposure to both low and high temperature [49], sedimentation [50], disease [51], and water quality [52,53], and has been widely measured to assess coral condition in response to stimulants, as well as natural variation in environmental factors [52,53,54,55]. To measure both total lipid content and zooxanthellae density, one branch was collected from each of the tagged colony in May and October. To minimize within-branch variability in lipids, only central inner branches were collected [56].

Branches were fixed in 10% formalin seawater and decalcified in 5% Formic acid for 1 day followed by 10% formic acid for 5 days and then stored in 70% ethanol. To extract total lipids, coral branches were dried in the oven at 55°C for 24 h, weighed and placed in a solution of chloroform: methanol (2:1, v:v) to dissolve the lipids [57]. The tissues were redried at 55°C overnight and reweighed. The difference in weight was due to lipids loss, with total lipid content then expressed as percentage of dry weight. Total lipid content was analysed instead of lipid classes because of the total lipids, triacylglycerol and wax esters are the main storage lipids in corals, and can account for 40–73% of total lipids [58–60].

Zooxanthellae density (per unit surface area (cells/cm^2)) was quantified for each coral based on samples (5 mm ×5 mm) from the collected branch (4 replicates per branch). Each sample was homogenized and the ground solution was examined on a glass slide under a microscope and counts were normalized to coral surface area, following McCowan et al. [61].

Data Analyses

To test whether there were significant differences in partial mortality, in total lipid content, in zooxanthellae density, in competition and in the number of fish bites, between May and October, a series of paired t-tests were carried out for each variable. Proportional mortality of individual coral colonies was Arcsin transformed prior to analyses. A One-Way ANOVA was carried out to further investigate whether colonies exposed to predation had lower lipid content than colonies that did not receive any bite. To test whether physiological condition of coral colonies relates to biological and physical disturbance regimes, we used a stepwise Multiple Regression model, testing the extent to which i) partial mortality, 2) mean number of fish bites, 3) extent of coral competition, 4) colony size, 5) distance from crest, and 6) Site, explained intraspecific variation in either total lipid content or zooxanthelae density for each coral species. Separate analyses were carried out for total lipid content and zooxanthellae density. Bivariate correlations were also used to test for any relationship between zooxanthellae density and total lipid content in each coral species.

Results

Chronic Disturbances

Competitive interactions and partial mortality were constant between May and October in both coral species (Fig. 1). Only the number of fish bites differed significantly with time, being 23 times higher in October than in May for *P. damicornis* (paired t-test, p< 0.05). Some colonies received few bites in May (from 0 to 4 bites per hour) while they were exposed to high predation pressure in October (163 and 392 bites per hour). In *A. spathulata*, overall predation pressure was two times higher in October, but given marked intra-specific variation this was not statistically significant (Fig. 1). Bite rate varied among colonies in *P. damicornis*, ranging from 0 to >100 bites per hour among colonies. In both coral species, the colonies that received most bites in May were not the same ones that received most bites in October, while some colonies did not receive any bite in either May or October.

In *P. damicornis* the majority of the colonies were smaller than 1000 cm^2, with colony surface area ranging from 161 cm^2 to 679 cm^2, while in *A. spathulata* colony surface area ranged from 160

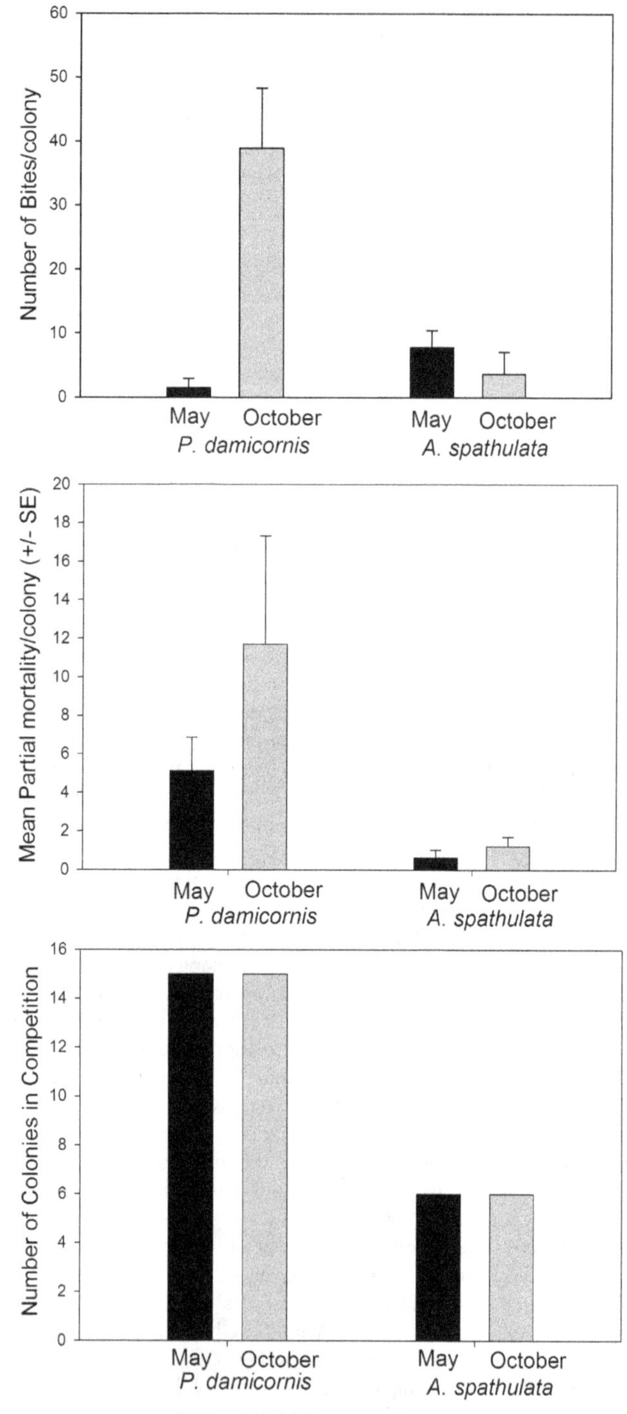

Figure 1. Chronic disturbance regimes in May and October in the two reef-building corals *P. damicornis* and *A. spathulata*. A) Predation – mean no. of bites taken per colony in replicate three-minute observations, where Go Pro cameras were used to record the total number of bites taken by all corallivorous fishes (mostly, butterflyfishes), B) Partial mortality –proportional of dead versus living tissue within the overall physical extent of each coral colony, C) – number of colonies engaged in competitive interactions.

cm^2 up to 1.830 cm^2. Colony growth (expressed as changes in colony surface area) from May to October in *A. spathulata* was 118.3 cm^2, while *P. damicornis* showed a negative growth rate (−10.3 cm^2) due to partial mortality.

Intraspecific Variation in Colony Condition

All the sampled corals survived the entire study period. Colony condition was found to vary between May and October in both coral species (Fig. 2). Specifically, a significant decline in total lipid content was observed in October compared to May (Table 1, 2; Fig. 2). In *A. spathulata* energy reserves in October were almost half compared to May (declined from 13.7 ($\pm$7.5) % to 7.8 ($\pm$1.8) %), while in *P. damicornis* the decline was two-fold during the same time (Fig. 2). Zooxanthellae density on the other hand, remained constant and did not change significantly between sampling periods in either coral species (Fig. 2). For *P. damicornis*, intraspecific variation in total lipid content was strongly correlated with zooxanthellae density (r = 99, df = 5,42, p<0.001), but no such relationship was found for *A. spathulata*.

P damicornis showed a high variation within colonies in partial mortality (between 0 and 20%), number of fish bites (between 0 and 392 bites per hour) and total lipid content (between 1.5 and 80% dw). By comparison, intraspecific variation in partial mortality and disturbance rates for *A. spathulata* were much smaller (Fig. 1).

In *P damicornis*, partial mortality, number of fish bites, competition, distance from crest and size were poor predictors of both lipid content (Multiple Regression total lipid content $r^2 = 11$, df = 5,42, p = 0.27; Table 1); and zooxanthellae density ($r^2 = 25$, df = 5, 42, p = 0.3; Table 1): the regressions explained only a very small proportion of the total variation (<12%). Conversely in *A.*

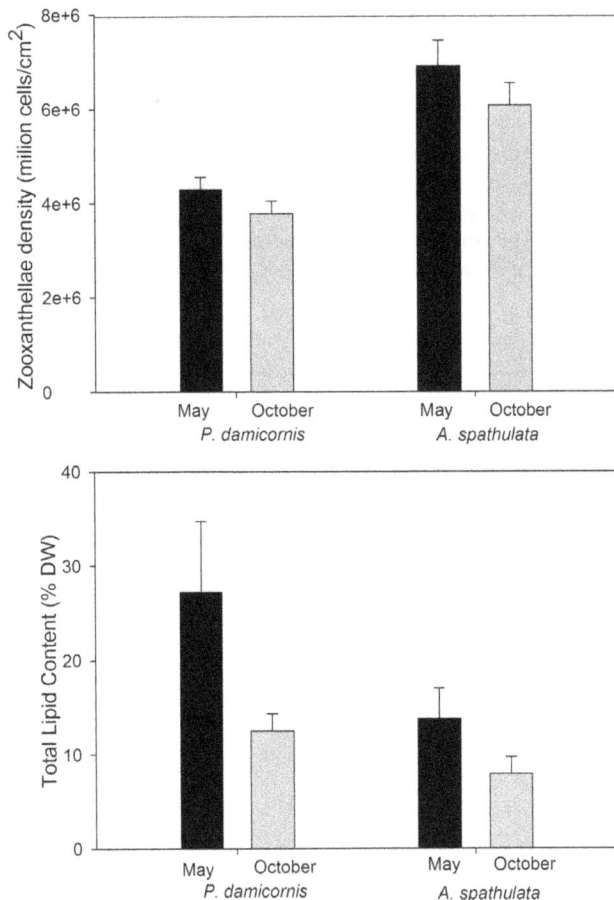

spathulata, partial mortality and distance from crest were found to have a significant effect on both total lipid content and zooxanthellae density (Multiple Regression total lipid content *A. spathulata* $r^2 = 14$, df = 5,42, p = 0.2; zooxanthellae density $r^2 = 26$, df = 5,42, p = 0.02; Table 2). In particular, total lipid content increased with distance from crest, while zooxanthellae density declined with increasing partial mortality and distance from crest (Table 2).

Discussion

This is the first study that attempts to relate intraspecific variation in physiological condition of scleractinian corals to small-scale differences in chronic disturbances, such as fish predation. It is well known that coral colonies living in close proximity may exhibit vastly different demographic rates [6,8,45,62], possibly reflective of differences in their disturbance history and subsequent energy allocation [27,63]. The difficulty in making this link is that very subtle differences in disturbance regimes, operating at any time in the lifetime of each coral, may lead to marked differences in contemporary condition and fitness of individual coral colonies. We acknowledge that the current study provides very limited insights on lifetime differences among closely positioned colonies, mainly due to the limited observational periods, and the range of factors that may be impacting on individual coral colonies. However, it is interesting that we saw no significant temporal shifts in rates of partial mortality, competition and predation between the two observational periods. The high degree of constancy in background mortality may be evidence that there is a high stability in terms of routine mortality.

Under low levels of background mortality, demographic models of scleractinian corals predict constant growth and fecundity of individual colonies, enabling rapid recovery following major acute disturbance [4,36,64]. However, in the present study, even within relatively constant rates of biological and physical disturbances, the incidence of injuries still varied among colonies. For instance, some coral colonies did not receive any fish bites in either May or October. Similarly, some colonies of *P. damicornis* that were not injured in May showed partial mortality in October, while some colonies never showed partial mortality. In the long term, these differences among colonies may likely be responsible for important inter-colony differences in condition and fitness. Importantly, variation in the disturbance history of individual colonies may have important ramifications for their long-term fate, especially during major disturbances (e.g., climate-induced coral bleaching).

Not unexpectedly, this study revealed marked intraspecific variation in the physiological condition of both *A. spathulata* and *P. damicornis*. However, these differences were only partially explained by inter-colony differences in rates of partial mortality, competition, predation, colony size or the position of the colony relative to the reef crest. Comparing to other studies, which documented a lipid level of 35% in tissue of *P. damicornis* [56,65], this study found a lower lipid content (27% dw). Conversely, zooxanthellae density was found to be higher (3.0 cells/cm^2 in May) than what reported in the literature (3.0 cells/cm^2, [66]). Also differences in colony condition, specifically in total lipid content, were greater among adjacent colonies of *P. damicornis* than when compared to colonies of *A. spathulata*, revealing intraspecific differences in physiological condition and in susceptibility to chronic disturbances. These differences suggest that coral physiological condition can be more variable than predicted with the outcome depending, in part, on flow, partial mortality, and position of the colony.

Predation rates on coral colonies were higher in October than in May in both coral species, especially in *P. damicornis*. Similarly,

Figure 2. Physiological condition, specifically A) total lipid content and B) zooxanthellae density, in May and October in the two reef-building corals *P. damicornis* and *A. spathulata*.

Table 1. Multiple Regression for zooxanthellae density and total lipid content in P. damicornis.

Zooxanthellae	B	StdErr of B	t(42)	p
Intercept				
Partial Mortality	−14	834221	−1.7	0.08
Competition	−44	312617	−1.4	0.1
Number of bites	1632	2745	0.5	0.5
Size	174	1044	0.16	0.8
Site	2338	374570	0.06	0.9
Distance from crest	68	61603	1.1	0.2
Lipid content	**B**	**StdErr of B**	**t(42)**	**p**
Intercept				
Partial Mortality	−23	13.7	−1.7	0.09
Competition	−3.6	5.1	−0.7	0.4
Number of bites	0.008	0.04	0.1	0.8
Size	0.003	0.01	−1.7	0.8
Site	3.96	6.14	0.64	0.5
Distance from crest	−1.6	1.01	−1.5	0.1

coral grazing parrotfishes have been shown to exhibit higher feeding rates in October compared to April on the GBR [67]. For parrotfishes, temporal differences in feeding rates have been previously attributed to differences the nutritional quality of colonies associated with gametogenesis [68]. For butterflyfishes, which tend to take very shallow bites [69], it is unlikely that gametogenesis of the corals would influence feeding behavior, but changes in the nutritional content may still occur within and among coral colonies. For instance mucus production can drive feeding preferences in butterflyfishes [70,71]. In October, colonies may have released more mucous as a stress response to environmental changes [72] and this discharge may have increased their desirability as food source. The observed differences in bite rates could also be due to seasonal differences in the metabolic rate of food demands of the fishes themselves.

Chronic disturbances were found to affect physiological condition only in. A. spathulata, which exhibited strong intraspecific variation that was explained to a large extent by inter-colony differences in biological disturbances and physical position, however these differences were not observed in P. damicornis. Even though both study species (A. spathulata and P. damicornis) are shallow, fast-growing, branching corals, they have slightly different

Table 2. Multiple Regression for zooxanthellae density and total lipid content in A. spathulata.

Zooxanthellae	B	StdErr of B	t(18)	p
Intercept				
Partial Mortality	−7386077	2909017	−2.5	0.01
Competition	−543263	551519	−0.98	0.3
Number of bites	22837	19454	1.17	0.24
Size	−487	643	−0.75628	0.4
Site	−510264	500279	−1.019	0.3
Distance from crest	−121255	55915	−2.16854	0.03
Lipid content	**B**	**StdErr of B**	**t(42)**	**p**
Intercept				
Partial Mortality	−2.1	18.5127	−0.11	0.9
Competition	0.7522	3.5098	0.21	0.8
Number of bites	−0.04	0.1238	0.33	0.7
Size	0	0.004	−0.05	0.9
Site	−1.87	3.2105	−0.58	0.5
Distance from crest	0.9	0.3558	2.57	0.01

life-histories strategies, which can explain observed differences. *P. damicornis* is a brooding, opportunistic coral which colonizes very disturbed habitats and it is one of the most resilient corals [46]. These characters may explain why *P. damicornis* was more resilient to chronic background disturbances than *A. spathulata*, which instead seems to dominate communities in relatively stable environments [46]. *A. spathulata* showed higher lipid reserves with increasing distance from the crest and lower symbiont density with increasing partial mortality and distance from crest. The observed increase in total lipid content with distance from crest may be due to the higher energetic cost of this reef habitat.

The reef crest is a shallow wave-exposed habitat, where water flow strongly influences organisms mechanically and physiologically with important consequences on community structure [73]. To avoid hydrodynamic dislodgment, colonies on the crest may need to invest more resources in growth to reach the dislodgment threshold [73], but since energy is limited within a colony, if more resources are allocated to increase colony size, less energy will be available to store. The findings from this study suggest that colonies in intermediate position between reef crest and reef flat have a better performance than conspecific on the crest. However, even though the reef crest is an energetic costly habitat, the high flow can positively affect colonies as they can benefit from it for feeding and escretion [74]. Together with light, flow is a critical abiotic factor affecting colony condition [74]. Colonies exposed to high flow generally have higher skeletal density, higher protein concentration, zooxanthellae density, chlorophyll content, and higher number and size of oocytes compared to colonies exposed to lower flow conditions [74]. Flow enhances zooxanthellae density and photosynthesis due to the enhanced nutrient supply [75], and can explain the decreasing zooxanthellae density with distance from crest found in this study.

The symbiotic relationship between the zooxanthellae and the host may be affected by a variety of internal and external factors and processes, the composition of which still has not been fully investigated [54,76]. Findings from this study suggest that increasing partial mortality and distance from crest may lead to a decline in density of *Symbiodinium*. Not many studies have shown differences in zooxanthellae density among reef habitats regardless of depth. For instance Strickland [76] did not find any difference in zooxanthellae density with increasing distance from the reef crest or location along the reef. Conversely, zooxanthellae within the same reef habitat have been shown to vary with environmental fluctuations and season cycles [54,77,78].

Despite consistency in levels of routine or background mortality, the lipid content within coral tissues consistently declined across all coral colonies between May and October in both *P. damicornis* and *A. spathulata*. The decline in total lipid content observed in October in both coral species may partly be explained by sustained and ongoing rates of background mortality, though the declines in may also reflect limited productivity during winter months, due to both reduced temperature and reduced day length [79]. Zooxanthellae supply corals with an excess of lipids and a limitation in their activity can results in a decline in lipid reserves [56,80]. Stimson [56] documented a decrease in total lipid content following about one month in *P. damicornis* due to light limitation. Corals tend to consume their lipid reserves when maintenance costs of a colony exceed carbon acquisition [22], during environmental unfavourable conditions such as limited light [6,58,81], during reproductive events [82–84] or whenever an increase in energy demand occurs such as the development of a tumor in coral tissue [85].

Large colonies generally have greater regenerative abilities [6,8], greater growth [31], are more fecund [17] and have lower rates of total mortality compared to smaller colonies [6,15].

Consequently we were expecting larger colonies to be more resilient to chronic background disturbances than smaller ones. Conversely, in the present study chronic disturbances had a similar effect on physiological condition of colonies regardless of the size, suggesting that larger colonies are not necessary more resilient than smaller colonies. Similar incidence of chronic disturbances on coral colonies regardless of the size also suggests a lack of size-specific susceptibility to agents of coral mortality [86]. Other studies documented a lack of differences in resilience between small and large colonies [87,88]. For instance *S. siderea* exposed to partial mortality continued to dedicate resources to reproduction even after the colony had shrunk below their size of maturation while larger colonies reduced their fecundity [88]. Often recent injuries play a bigger role than size in predicting colony fate [89]. Large colonies with higher partial mortality may die before small colonies with no injuries [89].

Extensive research effort has focused on understanding the ability of reef corals to withstand and absorb disturbances, thereby contributing to the persistence and resilience of coral colonies, populations and species [1,3,12,42,49,90,91]. Quantifying the effects of essentially routine and ongoing disturbances on colony condition and assess intraspecific differences in colony condition added to this understanding and it is critical because background mortality influences recovery capacity, time and vulnerability to future disturbances.

This study documented significant effects of partial mortality and distance from crest on zooxanthellae density in *A. spathulata* with important ecological consequences for recovery capacity in the aftermath of climate change. A reduction in performances arising from these sub-lethal stressors, is likely to reduce colony resilience and hence increase chances of whole-colony mortality so that colonies suffering from partial mortality may not survive a subsequent acute disturbance. The approach used here, investigating drivers of colony-condition and their energetic consequences for colony resilience, provides a strong framework for predicting resistance, recovery capacity and resilience of reef-building corals. If colonies in poor physiological conditions (e.g. less resilient) are more susceptible to bleaching, disease and other stressors, colonies capable of maintaining a higher physiological condition may have a distinct ecological advantage [22,92]. Consequently, colonies of *A. spathulata*, with high partial mortality rates and located on the reef crest, may have a lower potential to withstand and recover from environmental changes compared to conspecific with lower rates of partial mortality and located in intermediate habitats. The observed differences in physiological conditions could have a strong bearing on the selectivity of major disturbances and the capacity of corals to withstand major disturbances, and thereby adapt to changing conditions.

This study is the first to document significant intra-specific variation in background mortality and colony condition, the next step is to investigate whether this variation impacts individual vulnerability of corals. If so, this will provide strong incentive to reduce background levels of stresses (e.g. control all the factors that routinely injure colonies such as predation or anchoring) as a sure way to increase resilience of corals subject to inevitable increases in acute disturbances in association with global climate change.

Acknowledgments

We thank M. Chua for assistance in the field, M. Hoogenboom for assistance in the lab, and the staff at Lizard Island Research Stations for field and logistical support.

Author Contributions

Conceived and designed the experiments: CP KA MSP. Performed the experiments: CP KA MSP. Analyzed the data: CP KA MSP. Contributed reagents/materials/analysis tools: CP KA MSP. Wrote the paper: CP KA MSP.

References

1. Karlson RH, Hurd LE (1993) Disturbance, coral reef communities, and changing ecological paradigms. Coral Reefs 12: 117–125.
2. Stimson J (1985) The effect of shading by the table coral *Acropora hyacinthus* on understory corals. Ecology 66: 40–53.
3. Connell JH (1997) Disturbance and recovery of coral assemblages. Coral Reefs 16 Suppl: S101–S113.
4. Wakeford M, Done TJ, Johnson CR (2008) Decadal trends in a coral community and evidence of changed disturbance regime. Coral Reefs 27: 1–13.
5. Pratchett MS, Pisapia C, Sheppard C (2013) Background mortality rates for recovering populations of *Acropora cytherea* in the Chagos Archipelago, central Indian Ocean. Mar Environ Res 86: 29–34.
6. Hughes TP, Jackson JBC (1985) Population dynamics and life histories of foliaceous corals. Ecol Monogr 55: 141–166.
7. Knowlton N, Lang JC, Keller BD (1990) Case study of natural population collapse: post-hurricane predation on Jamaican staghorn corals. Washington: Smithsonian Institution Press.
8. Bythell JC, Gladfelter EH, Bythell M (1993) Chronic and catastrophic natural mortality of three common Caribbean reef corals. Coral Reefs 12: 143–152.
9. Bak RPM, Luckhurst BE (1980) Constancy and change in coral reef habitats along depth gradients at Curaçao. Oecologia 47: 145–155.
10. Harriot VJ (1985) Mortality rates of scleractinian corals before and during a mass bleaching event. Mar Ecol Prog Ser 21: 81–88.
11. Hughes TP, Baird AH, Bellwood DR, Card M, Connolly SR, et al. (2003) Climate change, human impacts, and the resilience of coral reefs. Science 301: 929–93.
12. Déath G, Fabricius KE, Sweatman H, Puotinen M (2012) The 27-year decline of coral cover on the Great Barrier Reef and its causes. Proc Natl Acad Sci 109: 17995–17999.
13. Cox EF (1986) The effects of a selective corallivore on growth rates and competition for space between two species of Hawaiian corals. J Exp Mar Biol Ecol 101: 161–174.
14. Tanner JE (1997) Interspecific competition reduces fitness in scleractinian corals. J Exp Mar Biol Ecol 214: 19–34.
15. Henry LA, Hart M (2005) Regeneration from injury and resource allocation in sponges and corals– a review. Int Rev Hydrobiol 90: 125–158.
16. Strauss SY, Agrawal AA (1999) The ecology and evolution of plant tolerance to herbivory. Trends Ecol Evol 14: 179–185.
17. Hall VR, Hughes TP (1996) Reproductive strategies of modular organisms: comparative studies of reef-building corals. Ecology 77: 950–963.
18. Rotjan RD, Dimond JL, Thornhill DJ, Leichter JJ, Helmuth B, et al. (2006) Chronic parrotfish grazing impedes coral recovery after bleaching. Coral Reefs 25: 361–368.
19. Cole AJ, Lawton RJ, Pratchett MS, Wilson SK (2011) Chronic coral consumption by butterflyfishes. Coral Reefs 30: 85–93.
20. Hourigan TF, Tricas TC, Reese ES (1988) Coral reef fishes as indicators of environmental stress in coral reefs. In: Soule DF, Kleppel GS, editors. Marine Organisms as Indicators. New York: Springer Verlag. 107–135.
21. Guzman HM, Burns KA, Jackson JBC (1994) Injury, regeneration and growth of Caribbean reef corals after a major oil spill in Panama. Mar Ecol Prog Ser 105: 231–241.
22. Anthony KRN, Hoogenboom MO, Maynard JA, Grottoli AG, Middlebrook R (2009) Energetics approach to predicting mortality risk from environmental stress: a case study of coral bleaching. Funct Ecol 23: 539–550.
23. Bak RPM (1983) Neoplasia, regeneration and growth in the reef-building coral *Acropora Palmata*. Mar Biol 77: 221–227.
24. Meesters EH, Noordeloos M, Bak RPM (1994) Damage and regeneration: links to growth in the reef-building coral *Montastrea annularis*. Mar Ecol Prog Ser 112: 119–128.
25. Kojis BL, Quinn NJ (1985) Puberty in *Goniastrea favulus* age or size limited? Proc 5th Int Coral Reef Congr 4: 289–293.
26. Meesters EH, Bos A, Gast GJ (1992) Effects of sedimentation and lesion position on coral tissue regeneration. Proc 7th Int Coral Reef Symp 2: 681–688.
27. Bak RPM, Brouns JJWM, Heys FML (1977) Regeneration and aspects of spatial competition in the scleractinian corals *Agaricia agaricites* and *Montastrea annularis*. Proc 3rd Int Coral Reef Symp 143–148.
28. Titlyanov EA, Titlyanova TV, Yakovleva IM, Nakano Y, Bhagooli R (2005) Regeneration of artificial injuries on scleractinian corals and coral/algal competition for newly formed substrate. J Exp Mar Biol Ecol 323: 27–42.
29. Bruckner AW, Hill RL (2009) Ten years of change to coral communities off Mona and Desecheo Islands, Puerto Rico, from disease and bleaching. Dis Aquat Organ 87: 19–31.
30. Babcock RC (1991) Comparative demography of three species of scleractinian corals using age and size-dependent classifications. Ecol Monogr 61: 225–244.
31. Hughes TP, Jackson JBC (1980) Do corals lie about their age? Some demographic consequences of partial mortality, fission, and fusion. Science 209: 713–715.
32. Jackson JBC (1979) Morphological strategies of sessile animals. In: Larwood G, Roser BR, editors. Biology and systematics of colonial organisms. London: Academic Press. 499–555.
33. Black NA, Voellmy R, Szmant AM (1995) Heat shock protein induction in *Montastrea faveolata* and *Aiptasia pallida* exposed to elevated temperatures. Biol Bull 188: 234–240.
34. Baker AC, Rowan R (1997) Diversity of symbiotic dinoflagellates (zooxanthellae) in scleractinian corals of the Caribbean and eastern Pacific. Proc 8th Int Coral Reef Symp 2: 1301–1306.
35. D'Croz L, Maté JL (2004) Experimental responses to elevated water temperature in genotypes of the reef coral *Pocillopora damicornis* from upwelling and non-upwelling environments in Panama. Coral Reefs 23: 473–483.
36. Gurney WSC, Middleton DAJ, Nisbet RM, McCauley E, Murdoch WM, et al. (1996) Individual energetics and the equilibrium demography of structured populations. Theor Popul Biol 49: 344–368.
37. Spencer-Davies P (1991) Effect of daylight variations on the energy budgets of shallow-water corals. Mar Biol 108: 137–144.
38. Szmant AM, Gassman NJ (1990) The effects of prolonged "bleaching" on the tissue biomass and reproduction of the reef coral *Montastrea annularis*. Coral Reefs 8: 217–224.
39. Fitt WK, McFarland FK, Warner ME, Chilcoat GC (2000) Seasonal patterns of tissue biomass and densities of symbiotic dinoflagellates in reef corals and relation to coral bleaching. Limnol Oceanogr 45: 677–685.
40. Grottoli AG, Rodrigues LJ, Juarez C (2004) Lipids and stable carbon isotopes in two species of Hawaiian corals, *Porites compressa* and *Montipora verrucosa*, following a bleaching event. Mar Biol 145: 621–631.
41. Rodrigues LJ, Grottoli AG (2007) Energy reserves and metabolism as indicators of coral recovery from bleaching. Limnol Oceanogr 52: 1874–1882.
42. Grottoli AG, Rodrigues LJ, Palardy JE (2006) Heterotrophic plasticity and resilience in bleached corals. Nature 440: 1186–1189.
43. Halford A, Cheal AJ, Ryan D, Williams D (2004) Resilience to large-scale disturbance in coral and fish assemblages on the Great Barrier Reef. Ecology 85: 1892–1905.
44. Connell JH, Hughes TP, Wallace CC, Tanner JE, Harms KE, et al. (2004) A long-term study of competition and diversity of corals. Ecol Monogr 74: 179–210.
45. Baird AH, Marshall PA (2002) Mortality, growth and reproduction in scleractinian corals following bleaching on the Great Barrier Reef. Mar Ecol Prog Ser 237: 133–141.
46. Darling ES, Alvarez-Philip L, Oliver TA, McClanahan TR, Côté IM (2012) Evaluating life-history strategies of reef corals from species traits. Ecol Lett 15: 1378–1386.
47. Hoegh-Guldberg O (1999) Climate change, coral bleaching and the future of the world's coral reefs. Mar Freshwater Res 50: 839–66.
48. Sheppard CRC, Davy SK, Pilling GM (2009) The biology of coral reefs. Oxford New York: Oxford University Press.
49. Baker AC, Glynn PW, Riegl B (2008) Climate change and coral reef bleaching: an ecological assessment of long-term impacts, recovery trends and future outlook. Estuar Coast Shelf S 80: 435–471.
50. Peters EC, Pilson MEQ (1985) A comparative study of the effects of sedimentation on symbiotic and asymbiotic colonies of the coral *Astrangia danae* Milne-Edwards and Haime 1849. J Exp Mar Biol Ecol 92: 215–230.
51. Cervino J, Goreau TJ, Nagelkerken I, Smith GW, Hayes R (2001) Yellow band and dark spot syndromes in Caribbean corals: distribution, rate of spread, cytology, and effects on abundance and division rate of zooxanthellae. Hydrobiologia 460: 53–63.
52. Cooper TF, Gilmour JP, Fabricius KE (2009) Bioindicators of changes in water quality on coral reefs: review and recommendations for monitoring programmes. Coral Reefs 28: 589–606.
53. Cooper TF, Ulstrup KE (2009) Mesoscale variation in the photophysiology of the reef building coral *Pocillopora damicornis* along an environmental gradient. Estuar Coast Shelf S 83: 186–196.
54. Fagoonee I, Wilson HB, Hassell MP, Turner JR (1999) The dynamics of zooxanthellae populations: a long-term study in the field. Science 283: 844–845.
55. Ferrier-Pagès C, Schoelzke V, Jaubert J, Muscatine L, Hoegh-Guldberg O (2001) Response of a scleractinian coral, *Stylophora pistillata*, to iron and nitrate enrichment. J Exp Mar Biol Ecol 259: 249–261.
56. Stimson JS (1987) Location, quantity and rate of change in quantity of lipids in tissue of Hawaiian hermatypic corals. Bull Mar Sc 41: 889–904.
57. Barnes DJ, Blackstock J (1973) Estimation of lipids in marine animals and tissues: detailed investigation of the sulphophospho- vanillin method for 'total' lipids. J Exp Mar Biol Ecol 112: 103–118.
58. Harland AD, Navarro JC, Spencer Davies P, Fixter LM (1993) Lipids of some Caribbean and Red Sea corals: total lipid, wax esters, triglycerides and fatty acids. Mar Biol 117: 113–117.

59. Patton JS, Abraham S, Benson AA (1977) Lipogenesis in the intact coral *Pocillopora capitata* and its isolated zooxanthellae: evidence for a light-driven carbon cycle between symbiont and host. Mar Biol 44: 235–247.

60. Oku H, Yamashiro H, Onaga K, Sakai K, Iwasaki H (2003) Seasonal changes in the content and composition of lipids in the coral *Goniastrea aspera*. Coral Reefs 22: 83–85.

61. McCowan DM, Pratchett MS, Paley AS, Seeley M, Baird AH (2011) A comparison of two methods of obtaining densities of zooxanthellae in *Acropora millepora*. Galaxea Journal of Coral Reef Studies 13: 29–34.

62. Hughes TP (1994) Catastrophes, phase-shifts, and large-scale degradation of a Caribbean coral reef. Science 265: 1547–1551.

63. Hall VR (1997) Interspecific differences in the regeneration of artificial injuries on scleractinian corals. J Exp Mar Biol Ecol 212: 9–23.

64. Done TJ (1988) Simulation of recovery of pre-disturbance size structure in populations of *Porites* spp. damaged by the crown of thorns starfish *Acanthaster planci*. Mar Biol 100: 51–61.

65. Ward S (1995) Two patterns of energy allocation for growth, reproduction and lipid storage in the scleractinian coral *Pocillopora damicornis*. Coral Reefs 14: 87–90.

66. Stimson JS (1997) The annual cycle of density of zooxanthellae in the tissues of field and laboratory-held *Pocillopora damicornis* (Linnaeus). J Exp Mar Biol Ecol 214: 35–48.

67. Bonaldo RM, Welsh JQ, Bellwood DR (2012) Spatial and temporal variation in coral predation by parrotfishes on the GBR: evidence from an inshore reef. Coral Reefs 31: 263–272.

68. Rotjan RD, Lewis SM (2009) Predators selectively graze reproductive structures in a clonal marine organism. Mar Biol 156: 569–577.

69. Motta PJ (1988) Functional morphology of the feeding apparatus of ten species of Pacific butterflyfishes (Perciformes, Chaetodontidae): An ecomorphological approach. Environ Biol Fish 22: 39–67.

70. Cole A, Pratchett MS, Jones GP (2009) Effects of coral bleaching on the feeding response of two species of coral-feeding fish. J Exp Mar Biol Ecol 373: 11–15.

71. Pisapia C, Cole AJ, Pratchett MS (2012) Changing feeding preferences of butterflyfishes following coral bleaching. Proc 12th Int Coral Reef Sym 13: 1–5.

72. Reigl B, Branch GM (1995) Effects of sediment on the energy budgets of four scleractinian (Bourne 1900) and five alcyonacean (Lamouroux 1816) corals. J Exp Mar Biol Ecol 186: 259–275.

73. Madin JS, Dell AI, Madin EMP, Nash MC (2013) Spatial variation in mechanical properties of coral reef substrate and implications for coral colony integrity. Coral Reefs 32: 173–179.

74. Mass T, Brickner I, Hendy E, Genin A (2011) Enduring physiological and reproductive benefits of enhanced flow for a stony coral. Limnol Oceanogr 56: 2176–2188.

75. Muscatine L, Falkowski PG, Dubinsky Z, Cook PA, McCloskey LR (1989) The effect of external nutrient resources on the population dynamics of zooxanthellae in a reef coral. Proc R Soc Lond B 236: 311–324.

76. Strickland D (2010) Variations in coral condition within the hydrodynamic regime at Sandy Bay, Ningaloo Reef, Western Australia. Thesis University Western Australia.

77. Rowan R, Knowlton N, Baker AC, Jara J (1997) Landscape ecology of algal symbionts creates variation in episodes of coral bleaching. Nature 388: 265–269.

78. Jones RJ (1997) Zooxanthellae loss as a bioassay for assessing stress in corals. Mar Ecol Prog Ser 149: 163–171.

79. Cooper T, Lai M, Ulstrup KE, Saunders SM, Flematti GR, et al. (2011) Symbiodinium genotypic and environmental controls on lipids in reef building corals. PLOSONE 6(5): e20434.

80. Crossland CJ, Barnes DJ, Borowitzka MA (1980) Diurnal lipid and mucus production in the staghorn coral *Acropora acuminata*. Mar Biol 60: 81–90.

81. Hoogenboom M, Rodolfo-Metalpa R, Ferrier-Pagès C (2010) Co-variation between autotrophy and heterotrophy in the Mediterranean coral *Cladocora caespitosa*. J Exper Biol 213: 2399–2409.

82. Richmond RH (1987) Energetics, competency, and long-distance dispersal of planula larvae of the coral *Pocillopora damicornis*. Mar Biol 93: 527–533.

83. Pernet V, Gavino V, Gavino G, Anctil M (2002) Variations of lipid and fatty acid contents during the reproductive cycle of the anthozoan *Renilla koellikeri*. J Comp Physiol B 172: 455–465.

84. Leuzinger S, Anthony K, Willis B (2003) Reproductive energy investment in corals: scaling with module size. Oecologia 136: 524–531.

85. Yamashiro H, Oku H, Onaga K, Iwasaki H, Takara K (2001) Coral tumors store reduced level of lipids. J Exp Mar Biol Ecol 265: 171–179.

86. Bak RPM, Meesters EH (1998) Coral population structure: the hidden information of colony-size frequency distribution. Mar Ecol Prog Ser 162: 301–306.

87. Nugues MM, Roberts CM (2003) Partial mortality in massive reef corals as an indicator of sediment stress on coral reefs. Mar Pollut Bull 46: 314–323.

88. Graham JE, van Woesik R (2013) The effects of partial mortality on the fecundity of three common Caribbean corals. Mar Biol 160: 2561–2565.

89. Cumming (2002) Tissue injury predicts colony decline in reef-building corals. Mar Ecol Prog Ser 242: 131–141.

90. Marshall PA, Baird AH (2000) Bleaching of corals on the Great Barrier Reef: differential susceptibilities among taxa. Coral Reefs 19: 155–163.

91. Linares C, Pratchett MS, Coker DJ (2011) Recolonisation of *Acropora hyacinthus* following climate-induced coral bleaching on the Great Barrier Reef. Mar Ecol Prog Ser 438: 97–104.

92. Bachok Z, Mfilinge P, Tsuchiya M (2006) Characterization of fatty acid composition in healthy and bleached corals from Okinawa, Japan. Coral Reefs 25: 545–554.

15

An Experimental Test of whether the Defensive Phenotype of an Aphid Facultative Symbiont Can Respond to Selection within a Host Lineage

Ailsa H. C. McLean*, H. Charles J. Godfray

Department of Zoology, University of Oxford, Oxford, United Kingdom

Abstract

An experiment was conducted to test whether parasitoid resistance within a single clonal line of pea aphid (*Acyrthosiphon pisum*) might increase after exposure to the parasitoid wasp *Aphidius ervi*. Any change in resistance was expected to occur through an increase in the density of protective symbiotic bacteria rather than genetic change within the aphid or the bacterial symbiont. Six aphid lineages were exposed to high parasitoid attack rates over nine generations, each line being propagated from individuals that had survived attack; a further six lineages were maintained without parasitoids as a control. At the end of the experiment the strength of resistance of aphids from treatment and control lines were compared. No differences in resistance were found.

Editor: Nicholas J. Mills, University of California, Berkeley, United States of America

Funding: This study was partially funded by a UK Natural Environment Research Council (http://www.nerc.ac.uk/) grant NE/G017638/1 to HCJG. AHCM was funded by the University of Oxford. The funders had no role in study design, data collection and analysis, decision to publish, or preparation of the manuscript.

Competing Interests: The authors have declared that no competing interests exist.

* Email: ailsa.mclean@zoo.ox.ac.uk

Introduction

Animals are not as well defended against parasites and diseases as is physiologically possible and additive genetic variation in resistance is frequently observed [1,2]. The level of resistance exhibited is likely to reflect a number of different trade-offs between immune function and other aspects of the organism's life-history including development time [3], competitive ability [4] and reproductive output [5]. Artificially strengthening selection for resistance in an experimental setting can prove useful in understanding how this variation arises and what determines the patterns of resistance observed in the field [3,4].

Parasitoids and their insect hosts provide excellent systems for studying the evolution of resistance because their development is intimately intertwined, with survival of one dependent upon the death of the other [6]. In aphids, a major component of the defensive response against endoparasitoids is provided by facultative endosymbiotic bacteria, in particular *Hamiltonella defensa* [7,8]. The ability to confer resistance requires *Hamiltonella* to be infected by particular strains of a bacteriophage, termed APSE (<u>A</u>cyrthosiphon <u>p</u>isum <u>S</u>econdary <u>E</u>ndosymbiont [9]), which encode a number of toxin genes whose products are thought to play a role in attacking the developing parasitoid [10,11]. In other respects, aphids appear to have a reduced immune system relative to most insects [12,13], but whether this is a cause or a consequence of their association with symbiotic bacteria is unknown. Nevertheless, aphids may possess considerable intrinsic as well as symbiont-conferred resistance to parasitoid wasps [14].

The braconid parasitoid wasp, *Aphidius ervi*, has been shown to evolve higher virulence (defined as successful parasitism) when populations were maintained on a line of aphids infected with a protective strain of *Hamiltonella* [15]. In a separate study, adults of *A. ervi* were found to adapt to parasitize aphids with symbiont-associated resistance by laying multiple eggs in a single individual (self-superparasitism) [16]. Only one adult parasitoid can emerge successfully from a single parasitized aphid, but it is thought that laying several eggs can dilute and hence overwhelm host defences. Given the potential for parasitoids to evolve rapidly in response to host defences, it would be of interest to know whether aphid defences can evolve to respond to greater parasitoid challenge.

Aphid populations could achieve better symbiont-associated resistance through a variety of routes. First, they might acquire more highly protective bacteria. The facultative symbiont *Hamiltonella* is transmitted maternally with almost perfect fidelity in the asexual phases of aphid reproduction, and is also inherited during the sexual generation [17]. It can also be transmitted horizontally between individuals of the same [18–20] or different [21] species. Exactly how horizontal transfer occurs is not currently understood (there is experimental evidence that parasitoids are capable of transmitting *Hamiltonella* via oviposition [22]) but acquisition of novel symbionts by whatever route could improve aphid resistance. Second, aphids may gain more protective phages. Again, little is known about how the protective phage moves between *Hamiltonella* strains, although there is evidence for considerable variation between phages in the resistance they confer [9–11] and the *Hamiltonella* genome shows evidence of extensive horizontal transfer of mobile elements [23].

Third, parasitoid attack may select for aphid clones that have higher resistance leading to a change in mean population phenotype. Finally, and our focus here, improved resistance may occur within clonal lineages of aphid by changes in the abundance or genetic composition of the symbiont or phage.

The degree of symbiont-conferred resistance in pea aphids appears to be proportional to the number of *Hamiltonella* cells present. Black bean aphids (*Aphis fabae*) in their first and second instars, which contain fewer *Hamiltonella* cells, are vulnerable to parasitoids, even when they carry a protective strain of *Hamiltonella* [24]. We have also observed that individuals from one pea aphid clone (subsequently used in this study) gain little protection from their symbionts in the first and second instars (83% versus 90% parasitism; N = 116, 92) but are moderately protected from the third instar onwards (53% versus 81% parasitism; N = 98, 76). The toxin genes involved in resistance appear to be constitutively expressed rather than induced [25] and so the increase in resistance as the aphid grows may be a simple reflection of the growing number of bacteria. If there is heritable variation in cell number amongst clonal aphids then parasitism might select for greater bacterial cell densities and hence resistance. Although higher symbiont numbers have previously been associated with fitness costs [26,27], an increase might still be beneficial if the symbionts are protective and the risk of parasitoid attack is high.

We conducted an experiment to test whether parasitoid resistance within a single clonal line of pea aphids (*Acyrthosiphon pisum*) might increase after exposure to the wasp *A. ervi* for nine successive generations. Over this timescale, we expected any change in resistance to occur because of increased densities of protective phage rather than genetic change within the aphid, bacterial symbiont, or the phage it carries. We hypothesised that the density of protective symbionts might increase through growth in bacterial numbers or through an increase in the fraction of bacteria carrying multiple copies of the protective APSE phage. Loss of phage occurs infrequently but regularly in certain aphid lines [9,26] which suggests phage dynamics may lead to variation in copy number upon which natural selection can operate.

Methods

Ethics statement

The pea aphid clone used in our experiment was collected from *Lotus pedunculatus* growing on private land in Berkshire, UK in 2003 (grid reference SU 976 851), with permission from the landowner. Any future researchers wishing to use the same location should contact the owner of the land at the time of their study in order to gain similar permission to collect.

Experimental organisms

A line of aphids was established from a single female and maintained in culture in the laboratory on broad bean (*Vicia faba*). The taxon *A. pisum* consists of a complex amalgam of host plant-associated races [28] but all are able to feed on cultivated *V. faba* [29,30]. Diagnostic PCR was used to confirm that this clone carries the symbiont *Hamiltonella defensa*, but no other known secondary symbionts of pea aphids (see Henry *et al.* [20] for details of primers and PCR conditions used). This clone was chosen because preliminary experiments had shown it to be partially resistant to the parasitoid *A. ervi* and for resistance to increase with age. When carrying *Hamiltonella* about 50% of aphids survive parasitoid attack but this drops to less than 20% in sub-lines from which the symbiont has been removed using antibiotics (for details of antibiotic curing protocol, see McLean *et al.* [31]).

Aphids were maintained in culture and in the experiments at 20°C with a 16:8 h light:dark cycle, and kept in 9 cm Petri dishes containing a single leaf of *V. faba* with the petiole inserted in 2% agar gel to keep it fresh. Aphids were transferred to a fresh leaf once a week. The *A. ervi* parasitoid wasps used were taken from an inbred stock maintained in the laboratory at 20°C and a 16:8 h light:dark cycle for over five years, and reared on a highly susceptible pea aphid clone which lacks any described secondary endosymbionts. Female parasitoids were less than one week old when used in the experiment; all had been exposed to males and so were presumed to be mated, and had been allowed prior experience of oviposition on aphids lacking secondary endosymbionts.

Experimental design

Twelve replicate lineages of aphids were set up, six of which were exposed to parasitoid attack for nine generations, the others acting as controls. All lineages originated from a single asexual adult aphid which had been placed on a leaf of *V. faba* in a Petri dish and allowed to reproduce for 48 hours. Twelve of the offspring were removed and used to initiate the replicate lines. Once adults, these 12 aphids were placed on *V. faba* leaves for 24 hours and their offspring kept and used in the first round of parasitoid exposure.

To initiate the exposure treatment, six adult aphids were separated and allowed to reproduce for 48 hours. Four days later, the third instar offspring were exposed to parasitoid wasps. Offspring from individual females were placed together in a Petri dish (N = 12) without any leaf material and a single female *A. ervi* introduced. The dishes were observed for up to two hours, and individual aphids removed immediately after the wasp had attacked them. We know from previous experiments in which we had exposed aphids to parasitoids and immediately dissected them that a parasitoid egg can be found in >80% of individuals.

The parasitized aphids were then kept on fresh leaves of *V. faba* for 11 days, by which time surviving aphids had begun to reproduce while those that had succumbed to the wasp's attack had become "mummies" containing parasitoid pupae. From two to eight surviving aphids from each of the six experimental lines were then removed to individual 9 cm Petri dishes with fresh leaves and allowed to reproduce for 48 hours to initiate the next generation. This procedure was repeated for eight further generations. The six control lines of aphids were maintained in exactly the same way except for exposure to parasitoids.

Up to 12 Petri dishes could be watched simultaneously and exposure to parasitoids was therefore carried out in three separate blocks for each generation, conducted sequentially on the same day. The only alteration made to the protocol during the experiment was to change the parasitoid exposure arena from a 9 cm Petri dish to a 5 cm Petri dish after the third generation. This was done to increase the likelihood that the wasp located every aphid and that every oviposition event was observed.

Final resistance assay

After nine generations of exposure to parasitoid attack the resistance of experimental lines was compared with the controls. Over the course of the experiment, two of the experimental lines became extinct, leaving four experimental lines for the final assay.

To avoid complications involving any maternal effects all aphid lines were maintained for a tenth generation without exposure to parasitoids before the final resistance assay. Ten females were then taken from each of the four remaining experimental lines and placed in dishes to reproduce for 48 hours. Ten females were also taken from four randomly chosen control lines, giving a total of 80 dishes. When the offspring had reached the third instar, 12 aphids were taken from each dish and exposed to a single parasitoid female in a 5 cm Petri dish for two hours. All wasps were allowed access to an equal number of aphids for the same length of time. The exposed aphids were kept for 12 days and the number of parasitized mummies, surviving aphids and dead aphids recorded. The number of wasps hatching successfully from the mummies was also observed over the subsequent five days.

We compared rates of parasitism and successful wasp emergence in the two treatment groups using generalized mixed modelling techniques, implemented using packages 'lme4' [32] and 'car' [33] in R version 3.0.2 [34]. In each case, we used a binomial distribution, with an individual level error term included to account for any overdispersion in proportion data.

Results

There was no significant difference between exposed and non-exposed treatments ($\chi^2 = 0.115$, d.f. = 1, P = 0.735; Fig. 1a); the average level of successful parasitism was 49.8%. This is very similar to the results of our preliminary assessment of resistance in this aphid clone before the experiment began, which found a mean of 53.0% successful parasitism. Likewise, the rate of successful emergence of parasitoids from the mummies was unaffected by treatment ($\chi^2 = 0.432$, d.f. = 1, P = 0.511; Fig. 1b), with emergence at 72.0% in the non-exposed group and 76.6% in the exposed lines.

Discussion

We set out to discover whether an aphid clone that displays incomplete but significant symbiont-conferred resistance to parasitoids was able to respond to high levels of parasitism by developing increased resistance over a number of generations. We thought this most likely to occur through an increase in the numbers of the symbiont itself, or of the toxin-encoding phages which infect the symbiont and are ultimately responsible for the resistance phenotype. We find no evidence that the aphid clone was able to respond to parasitism by developing increased resistance by any mechanism: there was no difference in the level of symbiont-conferred resistance in the lines that had or had not been exposed to parasitoids.

Our results suggest that there is little scope for short-to medium-term adaptive change in resistance within a particular aphid clone without the horizontal transfer of symbionts. This conclusion must obviously be made with some caveats. First, we used only one clone of pea aphid and the single *Hamiltonella* strain with which it had been collected in the field. It is of course possible that different aphid clones would have displayed a different response, although defence against parasitoids in pea aphids seems largely [8], if not entirely [14,35], to be driven by the properties of the symbiont strain. Given the lack of any phenotypic change in our study, we did not go on to assess using qPCR whether either bacterial symbiont or bacteriophage titre had changed over the course of the experiment. However, it is possible that a different aphid genotype or a different symbiont strain might have provided greater heritable variation in the number of bacteria per aphid, or the number of phage per bacterium, upon which selection could have acted. Second, symbiont associations may be more plastic immediately after they have come together through horizontal transfer (as observed experimentally by Russell & Moran [36]) and the natural association we chose may have been too stable to provide variation for selection. Finally, we used a single, highly inbred, line of a single parasitoid species. Symbiont-wasp genotype × genotype interactions have been observed [37] and the use of a

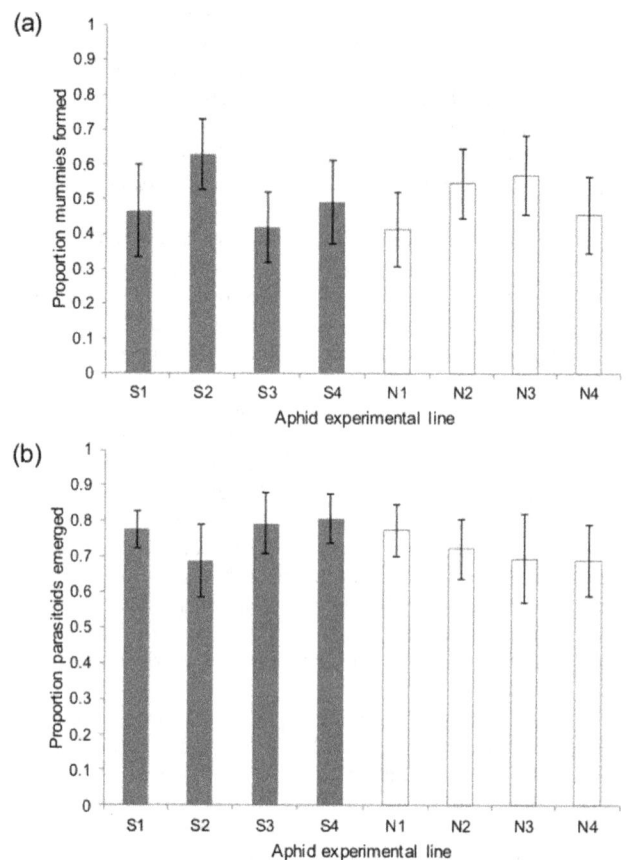

Figure 1. Parasitoid success rates in aphids from exposed and control treatment lines. (a) Proportion of mummies formed for the exposed treatment lines (in grey) and the control lines (in white); (b) Proportion of successful adult emergence of parasitoids from the exposed treatment lines (in grey) and the control lines (in white). No significant differences were observed between or within treatments. Error bars denote standard errors of the mean.

different (perhaps less virulent) strain or species may have led to a response being observed.

Parasitoids bred on partially resistant clones of aphid have been demonstrated to evolve increased virulence [15,38] and to respond to symbiont presence by adjusting their oviposition behaviour [16]. The absence of short-term responses in aphids is not surprising, given their asexual reproduction within a season. Pea aphids in northern Europe have only a single sexual generation per year, and even this may be dispensed with in warmer climates. As their parasitoids can have multiple generations a year, aphids would appear to be at a disadvantage in any coevolutionary interaction. However, their main defence against parasitoids aside from defensive symbionts may be escape in time and space. Aphid colonies can grow very fast because of asexual reproduction and the viviparous telescoping of generations that this allows. They may thus be able to reach a large colony size and produce many dispersive winged aphids before the majority of their natural enemies have increased to levels at which they can cause significant mortality. The type of intra-colony selection imposed in our study may thus occur rarely in the field. Interestingly, there is evidence that winged aphids are produced earlier in the presence of parasitoids [39].

The majority of work on symbiont-conferred resistance against parasitoids has, as in this study, focussed on laboratory studies.

Consequently, little is known about how symbionts affect interactions between aphids and their natural enemies in the field. *Aphidius ervi* shows additive genetic variation in virulence [15,40] and has the capacity to evolve improved performance on different aphid species in the laboratory [41]. However, there is no evidence of genetic differentiation amongst *A. ervi* that attack different pea aphid populations on different host plants, even when these host plant-associated populations differ markedly in resistance [42,43]. If the lack of flexibility in clonal responses to parasitism that we observe is typical, then inter-clonal selection would be expected to promote resistance only on plants where this was most advantageous. The fact that pea aphid populations adapted to different host plants differ so markedly in the defensive symbionts they carry [20,44] suggests that the selection pressure for defence must vary between host plants. Understanding the nature of this variation, and whether the symbionts influence the complex tritrophic interactions that are known to exist between aphids, plants and parasitoids, will be an interesting avenue for future research.

Supporting Information

Table S1 Raw data used for analysis. 'Name' indicates the aphid line (see graphs). Where emergence is marked N/A, there were either no mummies, or an accident prevented the wasps from being assessed following emergence (two cases). Percentage parasitism was counted using the number of mummies divided by the number of live and number of mummified aphids. Dead aphids are ambiguous and so were excluded from the analysis; however, the data are included here for completeness.

Acknowledgments

We are grateful to Lee Henry and Enric Frago and for helpful discussion and to Julia Ferrari for permission to use the aphid clone. The Natural Environment Research Council (UK) funded part of this work (NE/G017638/1). AHCM was funded by the University of Oxford.

Author Contributions

Conceived and designed the experiments: AHCM HCJG. Performed the experiments: AHCM. Analyzed the data: AHCM. Wrote the paper: AHCM HCJG.

References

1. Sheldon BC, Verhulst S (1996) Ecological immunology: Costly parasite defences and trade-offs in evolutionary ecology. Trends in Ecology & Evolution 11: 317–321.
2. Sandland GJ, Minchella DJ (2003) Costs of immune defense: an enigma wrapped in an environmental cloak? Trends in Parasitology 19: 571–574.
3. Boots M, Begon M (1993) Trade-Offs with Resistance to a Granulosis Virus in the Indian Meal Moth, Examined by a Laboratory Evolution Experiment. Functional Ecology 7: 528–534.
4. Kraaijeveld AR, Godfray HCJ (1997) Trade-off between parasitoid resistance and larval competitive ability in *Drosophila melanogaster*. Nature 389: 278–280.
5. Siva-Jothy MT, Tsubaki Y, Hooper RE (1998) Decreased immune response as a proximate cost of copulation and oviposition in a damselfly. Physiological Entomology 23: 274–277.
6. Godfray HCJ (1994) Parasitoids: Behavioural and Evolutionary Ecology. Princeton, NJ: Princeton University Press.
7. Oliver KM, Russell JA, Moran NA, Hunter MS (2003) Facultative bacterial symbionts in aphids confer resistance to parasitic wasps. Proceedings of the National Academy of Sciences of the United States of America 100: 1803–1807.
8. Oliver KM, Moran NA, Hunter MS (2005) Variation in resistance to parasitism in aphids is due to symbionts not host genotype. Proceedings of the National Academy of Sciences of the United States of America 102: 12795–12800.
9. Oliver KM, Degnan PH, Hunter MS, Moran NA (2009) Bacteriophages encode factors required for protection in a symbiotic mutualism. Science 325: 992–994.
10. Degnan PH, Moran NA (2008) Diverse Phage-Encoded Toxins in a Protective Insect Endosymbiont. Applied and Environmental Microbiology 74: 6782–6791.
11. Degnan PH, Moran NA (2008) Evolutionary genetics of a defensive facultative symbiont of insects: exchange of toxin-encoding bacteriophage. Molecular Ecology 17: 916–929.
12. Gerardo NM, Altincicek B, Anselme C, Atamian H, Barribeau SM, et al. (2010) Immunity and other defenses in pea aphids, *Acyrthosiphon pisum*. Genome Biology 11: 16.
13. Laughton AM, Garcia JR, Altincicek B, Strand MR, Gerardo NM (2011) Characterisation of immune responses in the pea aphid, *Acyrthosiphon pisum*. Journal of Insect Physiology 57: 830–839.
14. Martinez AJ, Ritter SG, Doremus MR, Russell JA, Oliver KM (2014) Aphid-encoded variability in susceptibility to a parasitoid. BMC Evolutionary Biology 14: 127.
15. Dion E, Zele F, Simon JC, Outreman Y (2011) Rapid evolution of parasitoids when faced with the symbiont-mediated resistance of their hosts. Journal of Evolutionary Biology 24: 741–750.
16. Oliver KM, Noge K, Huang EM, Campos JM, Becerra JX, et al. (2012) Parasitic wasp responses to symbiont-based defense in aphids. BMC Biology 10: 11.
17. Moran NA, Dunbar HE (2006) Sexual acquisition of beneficial symbionts in aphids. Proceedings of the National Academy of Sciences of the United States of America 103: 12803–12806.
18. Sandström JP, Russell JA, White JP, Moran NA (2001) Independent origins and horizontal transfer of bacterial symbionts of aphids. Molecular Ecology 10: 217–228.
19. Russell JA, Weldon S, Smith AH, Kim KL, Hu Y, et al. (2013) Uncovering symbiont-driven genetic diversity across North American pea aphids. Molecular Ecology 22: 2045–2059.

20. Henry Lee M, Peccoud J, Simon J-C, Hadfield Jarrod D, Maiden Martin JC, et al. (2013) Horizontally Transmitted Symbionts and Host Colonization of Ecological Niches. Current Biology 23: 1713–1717.
21. Russell JA, Latorre A, Sabater-Munoz B, Moya A, Moran NA (2003) Side-stepping secondary symbionts: widespread horizontal transfer across and beyond the Aphidoidea. Molecular Ecology 12: 1061–1075.
22. Gehrer L, Vorburger C (2012) Parasitoids as vectors of facultative bacterial endosymbionts in aphids. Biology Letters 8: 613–615.
23. Degnan PH, Yu Y, Sisneros N, Wing RA, Moran NA (2009) *Hamiltonella defensa*, genome evolution of protective bacterial endosymbiont from pathogenic ancestors. Proceedings of the National Academy of Sciences of the United States of America 106: 9063–9068.
24. Schmid M, Sieber R, Zimmermann YS, Vorburger C (2012) Development, specificity and sublethal effects of symbiont-conferred resistance to parasitoids in aphids. Functional Ecology 26: 207–215.
25. Moran NA, Tran P, Gerardo NM (2005) Symbiosis and insect diversification: An ancient symbiont of sap-feeding insects from the bacterial phylum Bacteroidetes. Applied and Environmental Microbiology 71: 8802–8810.
26. Weldon SR, Strand MR, Oliver KM (2013) Phage loss and the breakdown of a defensive symbiosis in aphids. Proceedings of the Royal Society B: Biological Sciences 280.
27. Oliver KM, Moran NA, Hunter MS (2006) Costs and benefits of a superinfection of facultative symbionts in aphids. Proceedings of the Royal Society B-Biological Sciences 273: 1273–1280.
28. Peccoud J, Ollivier A, Plantegenest M, Simon JC (2009) A continuum of genetic divergence from sympatric host races to species in the pea aphid complex. Proceedings of the National Academy of Sciences of the United States of America 106: 7495–7500.
29. Sandstrom J (1996) Temporal changes in host adaptation in the pea aphid, *Acyrthosiphon pisum*. Ecological Entomology 21: 56–62.
30. Ferrari J, Via S, Godfray HCJ (2008) Population differentiation and genetic variation in performance on eight hosts in the pea aphid complex. Evolution 62: 2508–2524.
31. McLean AHC, van Asch M, Ferrari J, Godfray HCJ (2011) Effects of bacterial secondary symbionts on host plant use in pea aphids. Proceedings of the Royal Society B-Biological Sciences 278: 760–766.
32. Bates D, Maechler M, Bolker B, Walker S (2014) lme4: Linear mixed-effects models using Eigen and S4. R package version 1.0–6.
33. Fox J, Weisberg S (2011) An {R} Companion to Applied Regression, Second Edition. Thousand Oaks, CA: Sage.
34. R Development Core Team (2013) R: a language and environment for statistical computing. Vienna, Austria: R Foundation for Statistical Computing. Available: http://www.r-project.org/. Accessed 2014 Oct 26.
35. Parker BJ, Garcia JR, Gerardo NM (2014) Genetic variation in resistance and fecundity tolerance in a natural host-pathogen interaction. Evolution 68: 2421–2429.
36. Russell JA, Moran NA (2005) Horizontal transfer of bacterial symbionts: Heritability and fitness effects in a novel aphid host. Applied and Environmental Microbiology 71: 7987–7994.
37. Vorburger C, Sandrock C, Gouskov A, Castaneda LE, Ferrari J (2009) Genotypic variation and the role of defensive endosymbionts in an all-parthenogenetic host-parasitoid interaction. Evolution 63: 1439–1450.

38. Rouchet R, Vorburger C (2012) Strong specificity in the interaction between parasitoids and symbiont-protected hosts. Journal of Evolutionary Biology 25: 2369–2375.

39. Weisser WW, Braendle C, Minoretti N (1999) Predator-induced morphological shift in the pea aphid. Proceedings of the Royal Society of London Series B-Biological Sciences 266: 1175–1181.

40. Henter HJ (1995) The potential for coevolution in a host-parasitoid system. 2. Genetic variation within a population of wasps in the ability to parasitize an aphid host. Evolution 49: 439–445.

41. Henry LM, Roitberg BD, Gillespie DR (2008) Host-range evolution in *Aphidius* parasitoids: Fidelity, virulence and fitness trade-offs on an ancestral host. Evolution 62: 689–699.

42. Hufbauer RA (2001) Pea aphid-parasitoid interactions: Have parasitoids adapted to differential resistance? Ecology 82: 717–725.

43. Bilodeau E, Simon JC, Guay JF, Turgeon J, Cloutier C (2013) Does variation in host plant association and symbiont infection of pea aphid populations induce genetic and behaviour differentiation of its main parasitoid, *Aphidius ervi*? Evolutionary Ecology 27: 165–184.

44. Ferrari J, West JA, Via S, Godfray HCJ (2012) Population genetic structure and secondary symbionts in host-associated populations of the pea aphid complex. Evolution 66: 375–390.

16

Membrane Lipid Remodelling of *Meconopsis racemosa* after Its Introduction into Lowlands from an Alpine Environment

Guowei Zheng[1,3,9], **Bo Tian**[2,9], **Weiqi Li**[1,3]*

1 Key Laboratory for Plant Diversity and Biogeography of East Asia, Kunming Institute of Botany, Chinese Academy of Sciences, Kunming, Yunnan, People's Republic of China, **2** Key Laboratory of Tropical Plant Resource and Sustainable Use, Xishuangbanna Tropical Botanical Garden, Chinese Academy of Sciences, Kunming, People's Republic of China, **3** Plant Germplasm and Genomics Center, Germplasm Bank of Wild Species, Kunming Institute of Botany, Chinese Academy of Sciences, Kunming, Yunnan, People's Republic of China

Abstract

Membrane lipids, which determine the integrity and fluidity of membranes, are sensitive to environmental changes. The influence of stresses, such as cold and phosphorus deficiency, on lipid metabolism is well established. However, little is known about how plant lipid profiles change in response to environmental changes during introduction, especially when plants are transferred from extreme conditions to moderate ones. Using a lipidomics approach, we profiled the changes in glycerolipid molecules upon the introduction of the alpine ornamental species *Meconopsis racemosa* from the alpine region of Northwest Yunnan to the lowlands of Kunming, China. We found that the ratios of digalactosyldiacylglycerol/monogalactosyldiacylglycerol (DGDG/MGDG) and phosphatidylcholine/phosphatidylethanolamine (PC/PE) remained unchanged. Introduction of *M. racemosa* from an alpine environment to a lowland environment results in two major effects. The first is a decline in the level of plastidic lipids, especially galactolipids. The second, which concerns a decrease of the double-bond index (DBI) and could make the membrane more gel-like, is a response to high temperatures. Changes in the lipidome after *M. racemosa* was introduced to a lowland environment were the reverse of those that occur when plants are exposed to phosphorus deficiency or cold stress.

Editor: Ing-Feng Chang, National Taiwan University, Taiwan

Funding: The study was supported by grants from NSFC 30670474 & 30870571, West Light Foundation of the Chinese Academy of Sciences (CAS), Germplasm Bank of Wild Species, the CAS Innovation Program of Kunming Institute (540806321211) and the "100 Talents Program, CAS". The study was also supported by grants from (NSFC 31371661). The funders had no role in study design, data collection and analysis, decision to publish, or preparation of the manuscript.

Competing Interests: The authors have declared that no competing interests exist.

* Email: weiqili@mail.kib.ac.cn

9 These authors contributed equally to this work.

Introduction

Plant introduction and acclimatization, the products of which now account in large part for many of our foods and ornamental species, played a critical role in the emergence of civilisation [1]. During the process of introduction, plants are transferred from their native environments to artificial ones where resources are usually plentiful and the plants can avoid stresses, such as freezing, drought, nutrient deprivation, and infection with pathogens [1,2]. The major obstacle to successful introduction is whether plants can adapt to dramatic changes in their environment. For example, the introduction of alpine plants to a lowland environment for the purpose of preservation and sustainable use is very difficult, because few plants can overcome considerable changes in temperature, irradiation, water conditions, and even nutrition [3–5]. Most studies of plant adaptation to environmental changes have focused on the adaption to stresses in which environments shift from optimum to aderverse conditions [6,7]. The mechanisms that plants use to adapt to moderate environments after their transfer from extreme environments are not fully understood. Given that the responses of an organism to two opposite stimuli

are often not simply the direct inverses of each other, understanding how plants adapt to the transfer from alpine to lowland conditions is an issue of biological significance and commercial importance.

Plants can adapt to environmental changes by adjustments at the morphological, physiological, biochemical, and molecular levels [6–8]. Membranes are integral to the structure and function of all cells; maintenance of the integrity and fluidity of membranes is of fundamental importance if plants are to survive environmental changes [9–11]. Glycerolipids are the major constituents of membranes. Lipid remodeling, through adjustment of the composition, unsaturation (represented by the double-bond index, DBI) and the acyl chain lengths (ACL) of their constituent fatty acids, is one of the most important ways that plants use to maintain the function of membranes upon exposure to fluctuating environmental conditions. Plants tend to synthesise additional galactolipids to replace phospholipids under conditions of phosphate deficiency [12–14], but to increase the proportion of phospholipids in response to low temperatures [15,16]. Membrane lipids, such as DGDG and phosphatidylcholine (PC), have relatively large polar head groups that tend to form membranes with the lamellar phase (Lα), which can enhance the stability of the membrane under various stresses. In contrast, monogalactosyldia-cylglycerol (MGDG) and phosphatidylethanolamine (PE)—which have relatively small head groups—show a higher propensity for transition to the non-bilayer H_{II}-type structures. The increased ratio of DGDG/MGDG and PC/PE could enhance the stability of the membrane under temperature and dehydration stresses [17–20].

Changes of the DBI and ACL of membrane glycerolipids that enable the fluidity of membranes to be adjusted are other important responses of membranes to stress, especially that caused by temperature extremes. Low temperature results in a 31% increase in the degree of unsaturation of fatty acids [21]. In contrast, the degree of unsaturation of fatty acids in plants decreases following exposure to high temperatures. For example, the DBI of *Arabidopsis* plants grown at 36°C was 39% lower than that of plants grown at 17°C [22]. Alternatively, whereas longer-chain fatty acids can make the membrane environment more gel-like, shorter chains help to maintain the fluid state of membranes [23]. As such, in bacteria, it is common to see a decrease in the average length of fatty acyl chains as the growth temperature decreases [24].

In alpine-scree ecosystems of the Baima Snow Mountain in Northwest China, the daytime temperature exceeds 35°C, whereas the temperature at night can drop below freezing [5]; in addition, the level of available phosphorus is very low (1.3 ppm) [25]. *Meconopsis racemosa*, a member of the Papaveraceae, is a native of the alpine scree of the Baima Snow Mountain. It is a well-known horticultural plant that bears beautiful flowers. However, the genetic resources of this plant are threatened by habitat destruction. Its introduction into other areas outside of its native habitat is thus one of the most important approaches to protect this ornamental plant. A factor that might limit the introduction of *M. racemosa* into other environments is its poor photosynthetic performance at high temperatures (>30°C) [26,27]. The bio-chemical responses of *M. racemosa* upon its transfer outside of its native habitat remain unknown, especially in terms of membrane lipid profiles.

The process of introducing *M. racemosa* from an extreme alpine environment to a moderate lowland environment involves exposure to substantially different environmental conditions, including considerable changes in temperature and the availability of soil nutrients. This process is different from the study of stress response where plants are transferred from moderate to extreme environments. It remains to be established whether the patterns of change of lipid profiles following transfer from extreme to moderate conditions are the reverse of those that occur upon exposure to the stress caused by transfer from moderate to extreme conditions.

Plant lipidomics based on ESI-MS/MS is a useful method to study the responses of hundreds of lipid molecular species to various environmental stresses [5,20]. The purposes of the current study were to use lipidomics to characterise the changes in *M. racemosa* lipids that occur during its introduction from alpine scree to the lowlands of Kunming, and to compare the lipid remodeling observed when plants are transferred from extreme to moderate environments with those that occur in response to specific environmental stresses.

Materials and Methods

Study site and plant materials

The study site comprised alpine scree at an altitude of 4,560 m on the Baima Snow Mountain in the Hengduan Mountains, a mountain range in Southwest China in the Southeast Qinghai-Tibet Plateau. The collection of samples were permitted by The Forestry Department of Yunnan Province. In this ecosystem, the solar radiation is very strong, and the temperature changes dramatically from about 32°C in the daytime to 3°C at night [5]. In the lowland of Kunming where *M. racemosa* introduced, the altitude is 1,900 m, and the average temperature in the greenhouse during July is approximately 25/15°C (day/night).

M. racemosa Maxim (Papaveraceaehttp://zh.wikipedia.org/w/index.php?title = Arctomecon&action = edit&redlink = 1) is an herb with a height of 20–50 cm, which grows on stony slopes at an altitude of 3,000–4,600 m. It has sharp spines on its leaves and stems, and the stem is branched with many blue or purple flowers. Its seeds are oblong (1–2 mm long on their longest axis) and ripen around September (http://www.efloras.org/florataxon.aspx?flora_id=2&taxon_id=242331771). It is not an endangered or protected species. Field sampling was performed at the alpine study site (28° 23.265′ N, 099° 01.260′ E, 4502 m alt.) on a randomly chosen day in July 2005. The leaves were transferred immediately into 3 mL of isopropanol with 0.01% butylated hydroxytoluene in a boiling water bath for at least 15 min at an altitude of 4,502 m. In the laboratory experiment, seeds of *M. racemosa* Maxim were surface-sterilised; then, after five days incubation at 4°C, the seeds were sown in soil, germinated, and grown in a greenhouse. After three months of growth, lipids were extracted from leaves immediately after their excision.

Lipid extraction, ESI-MS/MS analysis and data processing

Lipid extraction, ESI-MS/MS analysis, and quantification were performed as described previously, with minor modifications [5,20]. To inhibit lipolytic activity, harvested leaves were transferred immediately into 3 mL of isopropanol with 0.01% butylated hydroxytoluene in a boiling water bath (field experi-ment) or a 75°C water bath (laboratory experiment). The leaves were extracted three times with a chloroform:methanol (2:1) mixture, with 12 h of agitation each time. The remaining plant tissue was dried overnight at 105°C and weighed to determine the dry weights of the plants. Lipid samples were analysed using a triple quadrupole MS/MS equipped for ESI. Data processing was performed as described previously [20]. The lipids in each plant were quantified by comparison with two internal standards for the class of lipid studied. Five replicates of each treatment for each plant were analysed. The Q-test was performed on the total amount of lipid in each class, and data from discordant samples were removed [20]. The data were subjected to one-way analysis of variance (ANOVA) with SPSS 13.0. Statistical significance was tested by Fisher's least significant difference (LSD) method. DBI was calculated using the following formula: DBI = $(\sum[N \times mol\%$ lipid])/100, where N is the number of double bonds in each lipid molecule [5]. ACL was calculated using the following formula: ACL = $(\sum[n \times mol\%$ lipid])/100, where n is the number of acyl carbons in each lipid molecule.

Results

Major environmental changes associated with the introduction of *M. racemosa* from alpine scree to a lowland environment

The microclimate at the alpine study site where *M. racemosa* grows was investigated during July. This mainly involved examining the variations in temperature (Fig. 1), solar radiation, and UV radiation within a 24-h period on a clear day. *M. racemosa* growing on alpine scree experienced a dramatic change in temperature over the course of a day, from 33°C in the daytime to 5°C at night. On the day of sampling, the average air temperature was 14.6°C (Fig. 1). Compared with alpine scree, the temperature in the greenhouse in Kunming was about 25/15°C (day/night), with an average of 20°C. The soil in the alpine scree was previously shown to be deficient in nitrogen, potassium and phosphorus, with phosphorus available at a level of only 1.3 ppm, which is very low [25]; in contrast, in the lowlands in Kunming where *M. racemosa* was introduced, the plants were watered with 1/4 Hoagland solution once every month, and the level of phosphorus was sufficient to support normal growth of *M. racemosa*. Considering the differences in microclimate between the alpine scree of Baima Snow Mountain and the lowlands where the greenhouse is situated, this introduction of *M. racemosa* could be described as a process in which plants were introduced from an extreme environment to a moderate one. As one major morphological change associated with this introduction, it was found that leaves of *M. racemosa* grown in the greenhouse were larger than those from the alpine scree (data not shown).

Significant changes in the levels of membrane glycerolipid molecules occurred in *M. racemosa* after its introduction into the Kunming lowlands

An ESI-MS/MS-based lipidomics approach [5,20] that can identify and quantify 11 classes of 130 molecular lipid species was used to examine the patterns of lipid changes in *M. racemosa* during its introduction into a lowland environment. The 11 classes comprise two classes of galactolipids, six classes of phospholipids (Fig.2), and three classes of lysophospholipids (lysoPLs) (Fig.3). The observation that *M. racemosa* contains only 36:6 MGDG

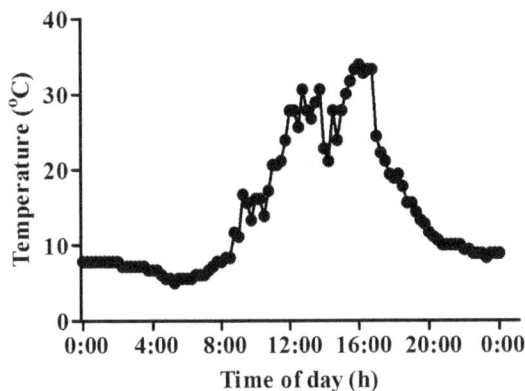

40 ─

Figure 1. Variation in air temperature with time of day in alpine screes of the Hengduan Mountains in July. The temperature was recorded every 30 min on a clear day in July.

molecules (Fig.1) indicates that it is an 18:3 plant that harbours only the eukaryotic lipid synthesis pathway [28].

Detailed analysis of the lipid profiles indicated that the levels of many lipid molecules, but not PA, changed significantly after *M. racemosa* was introduced from an alpine habitat to the Kunming lowlands (Fig.1). Most lipid classes tended to include lipid molecules with fewer double bonds after *M. racemosa* was introduced to a lower altitude; for example, the level of 36:5 DGDG lipid molecules increased almost six-fold (from 0.76 to 4.43 mol%), whereas that of 36:6 DGDG decreased by 25% (from 29.04 to 21.72 mol%). In addition, the level of 36:5 MGDG increased almost four-fold (from 2.42 to 9.22 mol%), whereas that of 36:6 MGDG decreased by 24% (from 40.10 to 30.63 mol%). Levels of 36:4 PC increased more than five-fold (from 0.48 to 2.67 mol%), those of 36:6 PC decreased by 62% (from 1.98 to 0.75), those of 36:4 PE increased more than three-fold (from 0.49 to 1.59 mol%), and those of 36:6 PE decreased by 73% (from 0.81 to 0.22 mol%). The relative increase of PS was 75% (Table 1); in general, the content of most PS molecules increased dramatically, especially for lipids with long acyl chains, such as 40:2 and 42:2 PS (Fig.1).

The lysoPLs—which include lysoPC, lysoPE and lysoPG—are derived from the hydrolysis of phospholipids at the sn-1 or sn-2 position of the glycerol backbone. Upon exposure to low temperature stresses, lysoPLs usually increase by 5- to 20-fold within hours or even minutes [15,20]. Compared with *M. racemosa* in alpine scree, the same species cultured in the Kunming lowlands did not show significant changes in the levels of lysoPL species, except for some individual molecular species, such as 18:3 lysoPC, 18:3 lysoPE, and 18:2 lysoPG. These results might indicate that *M. racemosa* uses lipid remodelling to acclimate to the environmental changes associated with introduction from an extreme to a moderate environment.

Changes in the composition of lipid classes after introduced from an alpine environment to a lowland area

After the introduction of alpine *M. racemosa* to Kunming, the levels of two galactolipids and three lysoPLs decreased, whereas the contents of six phospholipids increased. The content of DGDG decreased significantly from 36.99% to 31.86%, and the content of MGDG decreased from 44.16% to 42.06% (Table 1). The contents of PC and PE increased 3.18% and 1.61%, respectively. The content of PG also increased, from 5.35% to 7.44% (Table 1). The ratio of galactolipids/phospholipids decreased from 4.59 to 2.84; this constitutes a decrease of 38% (Table 1). These results indicate that *M. racemosa* tends to synthesise more phospholipids after its introduction to a lowland environment. Investigation of the ratios of PC/PE and DGDG/MGDG revealed that the changes of these ratios were not statistically significant (Table 1). This result might suggest that membrane integrity was not disrupted during the introduction of *M. racemosa*, despite the major environmental differences between the two habitats.

Different performance of plastidic and extraplastidic lipids upon introduction into a lowland area

DGDG, MGDG and some PG species are plastidic lipids which mainly located in photosynthetic membrane, whereas PC, PE, PA, PI, PS and other PG species which mainly located in plasma membrane are called extraplastdic lipids. Given that DGDG and MGDG are the most abundant plastidic lipids in the leaves of *M. racemosa*, a decrease in their levels represents a decrease in the total level of plastidic lipids. To further compare lipid changes

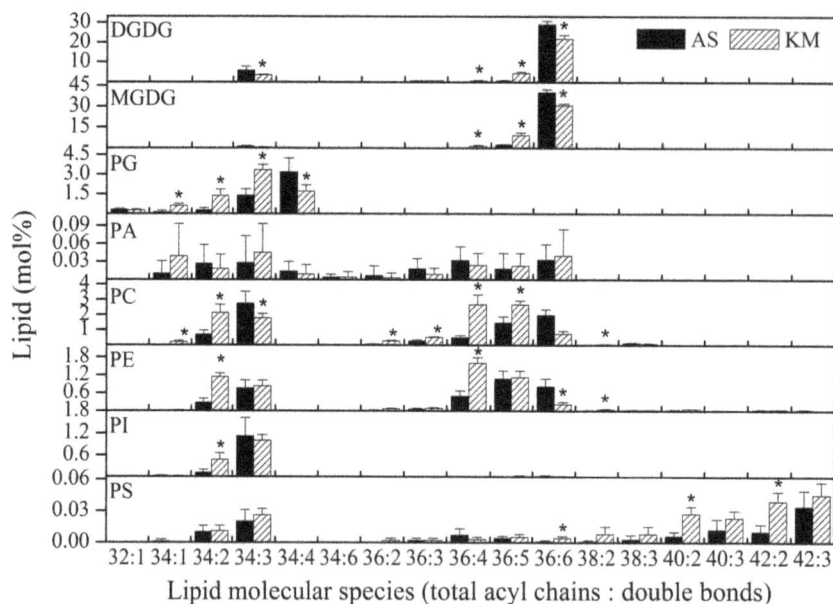

Figure 2. Changes in the molecular species of membrane lipids in *Meconopsis racemosa* **grown in alpine scree (AS) and Kunming (KM).** An asterisk indicates that the value of KM is significantly different from that of AS ($P<0.05$). Values are means ± standard deviation ($n = 4$ or 5).

between plastidic and extraplastidic membranes, the changes in the levels of molecular species of PG were analysed. Whereas 34:4 PG (which harbours a 16:1 acyl chain) is part of the plastidic membrane, both 34:1 and 34:2 PG are extraplastidic lipids. Of the two molecules that correspond to 34:3 PG, one contains a 16:1 acyl chain and is part of the plastidic membrane, whereas the other is extraplastidic [20,29]. Among the five species of PG molecules that were tested, only the level of 34:4 PG decreased considerably, namely, by 45.28%. In contrast, the levels of two plastidic lipids, 34:1 and 34:2, increased more than four-fold (Table 2). The content of 34:3 PG increased almost three-fold, which was a smaller increase than those of the two plastidic lipids.

These results indicate that the levels of extraplastidic lipids increased and plastidic lipids decreased after the introduction of the plant from an alpine environment to a lowland habitat.

Changes in the total degree of unsaturation and acyl chain length after the introduction of *M. racemosa* to the Kunming lowlands area

Changes in the degree of saturation and the lengths of fatty acid chains are very important for cells to modulate the fluidity of their membranes [23]. In this study, DBI was used to indicate the degree of unsaturation of membrane glycerolipids and ACL to

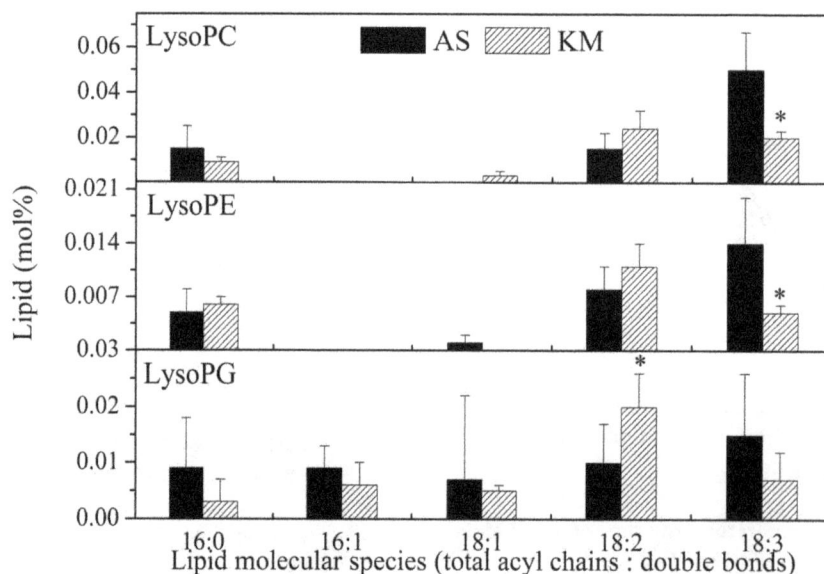

Figure 3. Changes in the molecular species of lysoPLs in *Meconopsis racemosa* **grown in alpine scree (AS) and Kunming (KM).** An asterisk indicates that the value of KM is different from that of AS ($P<0.05$). Values are means ± standard deviation ($n = 4$ or 5).

Table 1. Leaf membrane lipid composition in each head group class and lipid ratios of *M. racemosa* grown in alpine scree (AS) and in Kunming (KM).

Lipid class	mol%		Relative change (%) (%)
	AS	**KM**	
DGDG	36.99±2.93	31.86±1.56*	−13.87
MGDG	44.16±2.49	42.06±1.35	−4.76
PG	5.34±1.55	7.44±1.20*	39.33
PA	0.19±0.10	0.22±0.16	15.79
PC	8.06±1.81	11.24±0.92*	39.45
PE	3.71±0.96	5.32±0.60*	43.40
PI	1.28±0.56	1.54±0.23	20.31
PS	0.12±0.05	0.21±0.04*	75.00
LysoPC	0.08±0.03	0.06±0.01	−25.00
LysoPE	0.03±0.01	0.02±0.00	−33.33
LysoPG	0.05±0.02	0.04±0.01	−20.00
	Lipid ratio		
PC/PE	2.19±0.13	2.13±0.28	−2.74
DGDG/MGDG	0.84±0.07	0.76±0.05	−9.52
Galactolipids/Phospholipids	4.59±1.51	2.84±0.20*	−38.13

The relative change in lipids after introduction of *M. racemosa* to KM is the percentage value for the difference between the values of AS and KM, divided by the value of AS. An asterisk indicates that the value of KM is different from that of AS (*P*<0.05). Values are means ± standard deviation (*n* = 4 or 5).

indicate the lengths of fatty acid chains [5]. For all glycerolipids, the DBI of *M. racemosa* cultured in Kunming was less than that of plants grown in alpine scree, except for DGDG and PA, which showed no significant differences after introduction to the lowland environment. For plants grown in the Kunming lowlands, MGDG, which is one of the predominant lipid constituents of membranes (Table 1), had a DBI of 0.19, which is less than that of plants grown in the alpine scree (Table 3). The DBI of PE decreased by 0.72 after introduction of the plant, with values of 4.31 and 3.59 for the alpine scree and Kunming, respectively. Furthermore, the DBI of total lipids in the alpine scree was 5.30, whereas that for Kunming was 4.95 (Table 3). Except for PS, the ACL of most glycerolipids was not affected by the introduction of *M. racemosa* from an alpine environment to a lowland habitat. The ACL of PS increased from 38.52 to 39.45 after introduction to the lowlands (Table 3). These results suggest that *M. racemosa* changes the fluidity of its membranes to adapt to the environmental changes associated with its introduction into the Kunming

lowlands, and that it does this primarily through adjusting of the level of unsaturation of its membrane lipids.

Discussion

Given that membrane glycerolipids are a major component of membranes, the roles of their remodelling under stressful conditions, such as extreme temperatures and phosphorus starvation, have been one focus of research [5,17,30–33]. However, the consequences of introducing alpine plants to lower altitudes differ from those associated with stress in that the plants experience environmental changes from extreme to moderate conditions. In this study, we profiled the molecular species of lipid in alpine *M. racemosa* during its introduction into a lowland area, and calculated the DBI and ACL of plant membrane glycerolipids. Our results indicated that the ratio of bilayer-stabilising lipids to nonbilayer lipids and ACL of glycerolipids were maintained. There were two major changes with respect to lipid remodelling

Table 2. Levels of PG molecular species in leaves of *M. racemosa* plants.

PG species	mol%		Relative change (%)
	AS	**KM**	
32:1	0.30±0.08	0.26±0.07	−13.33
34:1	0.13±0.12	0.61±0.16*	369.23
34:2	0.27±0.18	1.37±0.49*	407.41
34:3	1.38±0.51	3.34±0.40*	142.02
34:4	3.18±1.05	1.74±0.45*	−45.28

The relative change in PG species after introduction of *M. racemosa* to KM is the percentage value for the difference between the values of AS and KM, divided by the value of AS. An asterisk indicates that the value of KM is significantly different from that of AS (*P*<0.05). Values are means ± standard deviation (*n* = 4 or 5).

Table 3. DBI and acyl chain length of membrane lipids of *M. racemosa* after its introduction from an alpine habitat to a lowland habitat.

Lipid class	Growth Site	DBI	Acyl chain length
DGDG	AS	5.41±0.13	35.68±0.08
	KM	5.34±0.05	35.76±0.02
MGDG	AS	5.85±0.03	35.95±0.02
	KM	5.66±0.07*	35.97±0.01
PG	AS	3.31±0.15	33.86±0.06
	KM	2.77±0.09*	33.90±0.02
PA	AS	3.71±0.96	35.12±0.42
	KM	3.82±0.60	35.13±0.40
PC	AS	4.10±0.06	35.24±0.05
	KM	3.67±0.08*	35.32±0.04*
PE	AS	4.31±0.21	35.67±0.09
	KM	3.59±0.06*	35.42±0.05*
PI	AS	2.93±0.07	34.04±0.03
	KM	2.72±0.14*	34.07±0.03
PS	AS	2.89±0.08	38.52±0.72
	KM	2.69±0.04*	39.45±0.38*
Total	AS	5.30±0.03	35.62±0.02
	KM	4.95±0.06*	35.60±0.03

DBI = $(\sum[N \times \text{mol\% lipid}])/100$, where N is the number of double bonds in each lipid molecule. ACL was calculated using the following formula: ACL = $(\sum[n \times \text{mol\% lipid}])/100$, where n is the number of acyl carbons in each lipid molecule. An asterisk indicates that the value of KM is different from that of AS ($P< 0.05$). Values are means ± standard deviation ($n = 5$).

after *M. racemosa* was introduced into the lowland habitat. The first effect was a significant increase in the levels of phospholipids. The second was a significant decrease in the degree of unsaturation of most glycerolipids. These results might suggest that *M. racemosa* might adjust the composition of different lipids classes and the degree of unsaturation of glycerolipids to adapt to environmental changes after its introduction from an alpine environment to a lowland habitat.

Plastidic lipids, which contain unusually high content of trienoic fatty acids, are the main component of photosystems I and II [34]. Galactolipids, which are the major component of plastidic lipids, harbour more trienoic fatty acids than other membrane phospholipids; for example, the lipid molecules of 36:6 DGDG that were tested here harbour two 18:3 fatty acids [20]. The content of trienoic fatty acids in plants is closely related to their photosynthetic performance at temperature extremes. Plants indigenous to cold areas tend to have a higher content of trienoic fatty acids, and the photosynthesis of plants that have relatively high levels of chloroplast trienoic fatty acids is more sensitive to heat treatment than those with low levels of chloroplast trienoic fatty acids [35]. Although *M. racemosa* maintained a high level of trienoic fatty acids in the galactolipids fraction when grown in the alpine environment, the level of trienoic fatty acids declined after the species was introduced into the lowland environment (Fig. 2). This might contribute to the poor photosynthetic performance of *M. racemosa* at 30°C compared with that at 20°C [26]. The lower level of trienoic fatty acid after introduction to lowland conditions might be an adaptation to the higher temperature in the lowland habitat than in the alpine habitat.

Both DGDG and PC have relatively large head groups, and tend to form bilayer lipid phase. By contrast, MGDG and PE have small head groups, involved in the formation of a nonbilayer lipid

phase [17,20]. Adjusting the molar ratio of these lipids is one of the most important ways that plants use to respond to stresses. The molar ratio of PC/PE decreased in plants subjected to cold and dehydration stresses [19,20], and an increase in the molar ratio of DGDG to MGDG enhances the stability of thylakoid membranes at high temperatures [18]. Notwithstanding the considerable difference in the ambient temperature experienced by *M. racemosa* after its introduction from an alpine to a lowland environment, there were no statistically significant differences in either PC/PE or DGDG/MGDG ratios between *M. racemosa* grown in an alpine habitat and those grown in a lowland environment (Table 1). The difference between these observations and those for plants subjected to environmental stress might suggest that membranes of *M. racemosa* might retain their integrity after introduction into a new environment to ensure adaption to major environmental changes.

The ratio of galactolipids to phospholipids was lower when *M. racemosa* was grown in the lowland habitat than when it was grown in the alpine environment (Table 1), and the content of DGDG decreased significantly after introduction to the lowland habitat (Table 1). Plants replace phospholipids with nonphosphorous galactolipids and transport them to extraplastidic membranes under phosphate starvation, which is very important for various physiological processes [13,14,31]. Whereas the level of phosphorus availability in Baima Snow Mountain (1.3 ppm) is very low [25], the level of phosphorus in the lowland area into which *M. racemosa* was introduced is sufficient to support its growth. The conversion of existing phospholipids to DGDG (recycling pathway) and the synthesis of new DGDG via the Kennedy (*de novo*) pathway are two routes that enable the remodelling of membrane lipids under conditions of phosphate starvation [32]. Alpine-grown *M. racemosa* might synthesise more

DGDG to replace phospholipids in order to ensure the efficient use of the limited amount of available phosphorus. However, during its long-term evolution, this plant might have adapted to the limited availability of phosphorus in alpine areas by deriving DGDG exclusively via the Kennedy pathway. After it was introduced from an extreme to a moderate environment, where the level of phosphorus is sufficient, *M. racemosa* selectively synthesised more extraplastidic lipids and less plastidic lipids. The way in which lipids were remodelled after *M. racemosa* was moved from phosphorus-deficient to phosphorus-sufficient conditions was the reverse of the process observed in plants exposed to phosphorus starvation.

The DBI of membrane lipids is sensitive to temperature changes [23]; it tends to increase in order to maintain the fluid state of membrane under low-temperature treatment, and to decrease to make membranes more gel-like in order to cope with heat treatment [21,22,36]. The DBI of membrane lipids of *M. racemosa* grown under alpine conditions decreased after its introduction to a lowland environment (Table 3). The average temperature at the alpine scree in July was previously reported to be 7.4°C [25], and it was 20°C under the greenhouse conditions used in our study. It was thus proposed that the high degree of membrane saturation would help *M. racemosa* to cope with the low temperature of the alpine scree, and the increased DBI in lowland-grown *M. racemosa* was a response to the relatively high temperature in that habitat.

The present study revealed substantial changes in the lipid profiles of *M. racemosa* as they adapted to a lowland environment. The decreases in the level of galactolipids and DBI were mainly responses to phosphorus-sufficient and the relatively high temperature conditions, which are the opposite of the responses of lipids to phosphorus starvation and cold stresses. This suggests that alpine plants have the genetic potential to adapt to lowland environments through adjusting their membrane lipid composition. However, many issues on this topic remain unresolved. For example, further analysis is needed to elucidate the relationship between the low level of plastidic lipids in lowland plants and their reduced capacity for photosynthesis compared with plants grown at higher altitudes.

Acknowledgments

The authors would like to thank Dr. Shibao Zhang for providing plant material and for constructive advices regarding growth of the plants. Lipid analysis was performed at Kansas Lipidomics Research Centre.

Author Contributions

Conceived and designed the experiments: WQL. Performed the experiments: GWZ BT. Analyzed the data: GWZ BT WQL. Contributed reagents/materials/analysis tools: WQL. Contributed to the writing of the manuscript: GWZ BT WQL.

References

1. Diamond J (2002) Evolution, consequences and future of plant and animal domestication. Nature 418: 700–707.
2. García-Palacios P, Milla R, Delgado-Baquerizo M, Martín-Robles N, Álvaro-Sánchez M, et al. (2013) Side-effects of plant domestication: ecosystem impacts of changes in litter quality. New Phytol 198: 504–513.
3. Körner C (1999) Alpine Plant Life: Functional Plant Ecology of High Mountain Ecosystems. New York, USA: Springer Verlag.
4. Lütz C (2012) Plants in Alpine Regions: Cell Physiology of Adaption and Survival Strategies. New York, USA: SpringerWien.
5. Zheng G, Tian B, Zhang F, Tao F, Li W (2011) Plant adaptation to frequent alterations between high and low temperatures: remodelling of membrane lipids and maintenance of unsaturation levels. Plant Cell Environ 34: 1431–1442.
6. Prasch CM, Sonnewald U (2013) Simultaneous application of heat, drought and virus to Arabidopsis thaliana plants reveals significant shifts in signaling networks. Plant Physiol 162: 1849–1866.
7. Penfield S (2008) Temperature perception and signal transduction in plants. New Phytol 179: 615–628.
8. Wahid A, Gelani S, Ashraf M, Foolad MR (2007) Heat tolerantce in plants: An overview. Environ Exp Bot 61: 199–223.
9. Levitt J (1980) Responses of Plants to Environmental Stresses New York, USA: Academic Press.
10. Wallis JG, Browse J (2002) Mutants of Arabidopsis reveal many roles for membrane lipids. Prog Lipid Res 41: 254–278.
11. Welti R, Shah J, Li W, Li M, Chen J, et al. (2007) Plant lipidomics: Discerning biological function by profiling plant complex lipids using mass spectrometry. Front Biosci 12: 2494–2506.
12. Andersson MX, Stridh MH, Larsson KE, Liljenberg C, Sandelius AS (2003) Phosphate-deficient oat replaces a major portion of the plasma membrane phospholipids with the galactolipid digalactosyldiacylglycerol. FEBS Lett 537: 128–132.
13. Härtel H, Dörmann P, Benning C (2000) DGD1-independent biosynthesis of extraplastidic galactolipids after phosphate deprivation in Arabidopsis. Proc Natl Acad Sci USA 97: 10649–10654.
14. Murphy DJ (2005) Plant lipids: biology, utilisation, and manipulation: Blackwell Pub.
15. Li W, Wang R, Li M, Li L, Wang C, et al. (2008) Differential degradation of extraplastidic and plastidic lipids during freezing and post-freezing recovery in Arabidopsis thaliana. J Biol Chem 283: 461–468.
16. Uemura M, Joseph RA, Steponkus PL (1995) Cold acclimation of Arabidopsis thaliana (Effect on plasma membrane lipid composition and freeze-induced lesions). Plant Physiol 109: 15–30

17. Moellering ER, Muthan B, Benning C (2010) Freezing tolerance in plants requires lipid remodeling at the outer chloroplast membrane. Science 330: 226–228.
18. Chen J, Burke JJ, Xin Z, Xu C, Velten J (2006) Characterization of the Arabidopsis thermo sensitive mutant atts02 reveals an important role for galactolipids in thermotolerance. Plant Cell Environ 29: 1437–1448.
19. Hazei JR, Williams EE (1990) The role of alterations in membrane lipid compositon in enabling physiological adaptation of organisms to their physical environment. Prog Lipid Res 29: 167–227.
20. Welti R, Li W, Li M, Sang Y, Biesiada H, et al. (2002) Profiling membrane lipids in plant stress responses. Role of phospholipase D alpha in freezing-induced lipid changes in Arabidopsis. J Biol Chem 277: 31994–32002.
21. Bakht J, Bano A, Dominy P (2006) The role of abscisic acid and low temperature in chickpea (Cicer arietinum) cold tolerance. II. Effects on plasma membrane structure and function. J Exp Bot 57: 3707–3715.
22. Falcone D, Ogas J, Somerville C (2004) Regulation of membrane fatty acid composition by temperature in mutants of Arabidopsis with alterations in membrane lipid composition. BMC Plant Bio 4: 17.
23. Chintalapati S, Kiran MD, Shivaji S (2004) Role of membrane lipid fatty acids in cold adaptation. Cell Mol Biol 50: 631–642.
24. Denich TJ, Beaudette LA, Lee H, Trevors JT (2003) Effect of selected environmental and physico-chemical factors on bacterial cytoplasmic membranes. J Microbiol Meth 52: 149–182.
25. Li H (2003) Baima Snow Mountain National Nature Reserve. Yunnan, China: The Nationalities Publishing House of Yunnan, 42–58.
26. Zhang S-B (2010) Temperature acclimation of photosynthesis in Meconopsis horridula var. racemosa Prain. Bot Stud 51: 457–464.
27. Zhang S-B, Hu H (2008) Photosynthetic adaptation of Meconopsis integrifolia Franch. and M. horridula var. racemosa Prain. Bot Stud 49: 225–233.
28. Buchanan B, Gruissem W (2002) Biochemistry & Molecular Biology of Plants. Rockville, MD, USA: John Wiley & Sons.
29. Jia Y, Tao F, Li W (2013) Lipid profiling demonstrates that suppressing Arabidopsis phospholipase Dδ retards ABA-promoted leaf senescence by attenuating lipid degradation. PLoS ONE 8: e65687.
30. Burgos A, Szymanski J, Seiwert B, Degenkolbe T, Hannah MA, et al. (2011) Analysis of short-term changes in the Arabidopsis thaliana glycerolipidome in response to temperature and light. Plant J 66: 656–668.
31. Jouhet J, Maréchal E, Baldan B, Bligny R, Joyard J, et al. (2004) Phosphate deprivation induces transfer of DGDG galactolipid from chloroplast to mitochondria. J Cell Biol 167: 863–874.
32. Nakamura Y (2013) Phosphate starvation and membrane lipid remodeling in seed plants. Prog Lipid Res 52: 43–50.

33. Nakamura Y, Koizumi R, Shui G, Shimojima M, Wenk MR, et al. (2009) Arabidopsis lipins mediate eukaryotic pathway of lipid metabolism and cope critically with phosphate starvation. Proc Natl Acad Sci USA 106: 20978–20983.
34. Dörmann P, Benning C (2002) Galactolipids rule in seed plants. Trends Plant Sci 7: 112–118.
35. Murakami Y, Tsuyama M, Kobayashi Y, Kodama H, Iba K (2000) Trienoic fatty acids and plant tolerance of high temperature. Science 287: 476–479.
36. Sakai A, Larcher W (1987) Frost survival of plants: responses and adaptation to freezing stress. New York, USA: Springer Verlag, 124–131.

17

Characterization of Arbuscular Mycorrhizal Fungus Communities of *Aquilaria crassna* and *Tectona grandis* Roots and Soils in Thailand Plantations

Amornrat Chaiyasen[1], J. Peter W. Young[2], Neung Teaumroong[3], Paiboolya Gavinlertvatana[4], Saisamorn Lumyong[1]*

1 Department of Biology, Faculty of Science, Chiang Mai University, Chiang Mai, Thailand, 2 Department of Biology, University of York, York, United Kingdom, 3 Schoool of Biotechnology, Institute of Agricultural Technology, Suranaree University of Technology, Nakhon Ratchasima, Thailand, 4 Thai Orchid Labs Co. Ltd., Khannayao, Bangkok, Thailand

Abstract

Aquilaria crassna Pierre ex Lec. and *Tectona grandis* Linn.f. are sources of resin-suffused agarwood and teak timber, respectively. This study investigated arbuscular mycorrhizal (AM) fungus community structure in roots and rhizosphere soils of *A. crassna* and *T. grandis* from plantations in Thailand to understand whether AM fungal communities present in roots and rhizosphere soils vary with host plant species and study sites. Terminal restriction fragment length polymorphism complemented with clone libraries revealed that AM fungal community composition in *A. crassna* and *T. grandis* were similar. A total of 38 distinct terminal restriction fragments (TRFs) were found, 31 of which were shared between *A. crassna* and *T. grandis*. AM fungal communities in *T. grandis* samples from different sites were similar, as were those in *A. crassna*. The estimated average minimum numbers of AM fungal taxa per sample in roots and soils of *T. grandis* were at least 1.89 vs. 2.55, respectively, and those of *A. crassna* were 2.85 vs. 2.33 respectively. The TRFs were attributed to Claroideoglomeraceae, Diversisporaceae, Gigasporaceae and Glomeraceae. The Glomeraceae were found to be common in all study sites. Specific AM taxa in roots and soils of *T. grandis* and *A. crassna* were not affected by host plant species and sample source (root vs. soil) but affected by collecting site. Future inoculum production and utilization efforts can be directed toward the identified symbiotic associates of these valuable tree species to enhance reforestation efforts.

Editor: Zhengguang Zhang, Nanjing Agricultural University, China

Funding: This research was supported by the Thailand Research Fund for Research-Team Promotion Grant (RTA5580007) and the Commission of Higher Education for National Research University (A1) (http://www.trf.or.th/). AC was funded by the Thailand Research Fund; The Royal Golden Jubilee PhD Program (PHD/0150/2550: http://rgj.trf.or.th/indexth.asp). The funders had no role in study design, data collection and analysis, decision to publish, or preparation of the manuscript.

Competing Interests: Paiboolya Gavinlertvatana is employed by a commercial company (Thai Orchid Labs Co. Ltd).

* Email: saisamorn.l@cmu.ac.th

Introduction

Tropical forests are disappearing at the rate of 13.5 million hectares each year owing to logging, burning and clearing for agriculture and shifting cultivation [1]. At present, managed woodlands are required for timber and non-timber products in many countries. *Aquilaria crassna* Pierre ex Lec. (agarwood) and *Tectona grandis* Linn.f. (teak) are perennial plants that are used extensively to provide aromatic resin-infused wood products [2] and good quality teak wood products [3], respectively. The depletion of wild trees from indiscriminate cutting of *Aquilaria* species has resulted in the trees being listed and protected as endangered species. All *Aquilaria* species were listed in Appendix II of the Convention on International Trade in Endangered Species of Wild Fauna and Flora in 2005 [4]; however, a number of countries have outstanding reservations regarding that listing. Plantlets of *A. crassna* and *T. grandis* are produced in Thailand

for domestic and foreign markets such as Jamaica, Guatemala, Mozambique, Sri Lanka, Indonesia, Laos, Malaysia, and Australia. Most *T. grandis* plantations in Thailand are planted in the northern provinces such as Chiang Mai, Chaing Rai and Phetchabun, while *A. crassna* plantations are mostly in eastern (Rayong, Trat and Chanthaburi provinces) and central (Nakhon-Nayok) Thailand.

Arbuscular mycorrhizal fungi (AMF) are soil fungi in the phylum Glomeromycota [5] that are mutualistically associated with roots of a wide spectrum of tropical and temperate tree species [6]. AM fungi have major effects on plant growth such as enhance the nutrient uptake by plant roots (especially phosphorus), particularly in low fertility soils [7,8], protected plant against drought stress [9,10], protect plant from soil-borne plant pathogenic infection [11], and improve soil aggregate stability through the action of mycelia and glomalin [12,13,14]. AMF inocula applied to plantlets and plant seedlings increased growth during early tree establishment in the field [2,15,16]. AM fungi have been used to inoculate and enhance growth of *T. grandis* [3,17] and *Aquilaria* spp. [2,18] prior to planting out. Therefore, studying the AM fungal communities of these plants in the field should aid plantation establishment and reforestation efforts. Information about the diversity of AM fungi associated with both plants has been reported mostly from natural forests in India [19,20,21,22,23] and only in *T. grandis* from Thailand [24]. These studies characterized communities based upon spore morphology. However, there are no reports of AM fungal communities of either tree using molecular tools. Identification of AM fungi based on spore morphology inevitably has some limitations, e.g. omission of AM fungi that did not produce spores during the sampling period and inability to identify the AM fungi within the roots.

PCR-based methods have been widely used in AM fungal community studies. Various studies have designed sets of specific primers for AM fungi [25,26,27] to facilitate rapid detection and identification directly from field-grown plant roots. Previously, Terminal restriction fragment length polymorphism (T-RFLP) has been used to study the AM fungi community in roots of arable crops [28], perennial herbs [29], herbaceous flowering plants [30], grass species [31,32], grass species with herbaceous flowering plants [33,34], and temperate deciduous trees [35]. Populations of AM fungi have been well studied in a number of ecosystems around the world, but there is scant information available for tropical forests and plantations of tropical and sub-tropical species.

This study provides the first molecular community analysis of AM fungi associated with field-collected roots and rhizosphere soils of the tropical trees *A. crassna* and *T. grandis*, and is part of a long term goal of optimizing AM fungus inoculation strategies to enhance reforestation efforts with these trees. It also provides an early insight into the biodiversity of AM fungi in Thailand to test the hypothesis that differences in AM fungal communities present in the roots and rhizosphere soils are determined by collecting sites, host plant species, and local environmental factors.

Materials and Methods

Ethics Statement

No specific permits were required to carry out research in the plantations: Chiang Mai (99°15' E, 18°58' N), Chiang Rai (99°29' E/19°14' N), Nakhon-Nayok (101°16' E, 14°9' N), Phetchabun (100°47' E, 16°2' N) and Thai Orchids Lab Ltd. (101°7' E, 14°16' N). The field studies did not involve endangered or protected species in Thailand. *Aquilaria crassna* is defined to be the forbidden forest item in only the forest area as the Forest Act. Therefore, the *A. crassna* planting and deforestation in the land of ownership is legal. All *A. crassna* samples were obtained from privately-owned plantations and are therefore not subject to the restrictions of the Forest Act of Thailand. Permission to sample the *T. grandis* and *A. crassna* were granted by the landowner.

Study sites and sampling

Rhizosphere soils and roots were sampled from plantations of *T. grandis* and *A. crassna* in four provinces of Thailand (Table 1). Two sampling sites were located in Chiang Mai and Chiang Rai provinces in the northern region. These sites are monocultures of *T. grandis* planted at 2 m spacings and left to grow naturally with accumulated leaf litter and negligible understory perennial plants. Only roots attached to the main roots of *T. grandis* were sampled. At the sites in the central region; Nakhon-Nayok and Thai Orchids Lab Ltd., Nakhon-Nayok province, and in the northern region; Phetchabun province, *T. grandis* and *A. crassna* were

planted alternately 2 m apart at Thai Orchids Lab Ltd. and Phetchabun. At both sites, weeds were controlled by ploughing and herbicide treatment. Thus, both species were planted without any above-ground vegetation, while in Nakhon-Nayok site, *A. crassna* was left to grow naturally. Paired soil and root samples from each plant species were randomly collected from 3 locations per site at 0–15 cm depth within 50 m^2 and taken to the laboratory. All collections were carried out in July 2010. Root fragments were washed free of soil and air dried on tissue paper. Root fragments and soil samples were stored frozen at $-20°C$ until further analysis.

Soil analyses

Soil pH and electrical conductivity (EC) were determined in a 1:1 soil: water slurry by direct measurement with pH-meter (Waterproof EC Testr, EUTECH instruments). Available phosphorus was measured employing the Bray II method [36]. Total inorganic nitrogen, exchangeable potassium and soil organic carbon were quantified following the methods of soil analysis outlined in Sparks et al. [37].

Molecular analysis

Three replicate rhizosphere soil and root samples from each plant species were used to represent each site of collection. DNA was extracted from rhizosphere soils and roots using the PowerSoil DNA isolation kit (MoBio Laboratories, CA) and Nucleospin Plant II (Macherey-Nagel GmbH & Co. KG, Düren), respectively according to the manufacturers' protocols. DNAs were amplified separately by nested PCR and then 20 µl of each of the three replicates from each sampling site were pooled and purified before restriction digestion [38]. The first round of AMF-specific PCR amplification was performed using the unlabelled primers AML1 and AML2 [26] with 30 cycles. In this first PCR, 40 µl reactions were carried out and each mixture contained 10 pmol of each primer, 1 unit of Taq polymerase (Promega) and 25 mM of each dNTP (Invitrogen) in manufacturer's reaction buffer (Promega). PCR was performed on a PTC100 thermocycler (MJ Research) with an initial denaturation at 94°C for 15 min, followed by 30 cycles of denaturation at 94°C for 30 s, annealing at 57°C for 45 s, extension at 72°C for 45 s, followed by a final extension of 72°C for 5 min. PCR products were visualized on a 1% agarose gel containing 0.1× SybrSafe (Invitrogen). The second round primers, 0.5 unit of Taq polymerase (Promega) and 20 pmol of HEX-labeled NS31 and FAM-labeled AML3 were added directly into 24 µl of each resulting product. Second-round PCR was conducted with 5 additional cycles using the same PCR conditions as the first PCR. The PCR products were purified using the QIAquick PCR purification kit (Qiagen). The purified PCR products were digested separately with the selected restriction enzymes, HinfI, Hsp92II and MboI (Promega) [31,39] for 3 h at 37°C. Digested products were purified as mentioned above. Terminal restriction fragment (TRF) sizes from each sample were determined using the ABI PRISM 3130 Genetic Analyzer System (Applied Biosystems) with GeneScan LIZ-600 (Applied Biosystems) as internal size standards. The GeneMapper software (Applied Biosystems) was used for the analysis of fragment data. To reduce data noise, only fragments containing intensity above a baseline threshold (50 fluorescence units) were recorded. Relative peak heights were calculated and fragments with an average relative abundance <5% were excluded from further analysis.

Screening and DNA sequence analysis

The remainders of the first PCR products were combined and purified using the QIAquick PCR purification kit (Qiagen).

Table 1. Chemical characteristic of soils (mean value ± SEM) in wet season (July 2010) which soils and roots were sampled.

Study plot	Soil pH[a]	Electrical conductivity[a]	Soil organic carbon (%)[a]	Total inorganic N (g kg^{-1} soil)[a]	Available P (mg kg^{-1} soil)[a]	Exchangeable K (mg kg^{-1} soil)[a]	Agricultural management
Phetchabun (PB)	5.77±0.25ab	0.18±0.02a	4.26±0.45ab	222±15a	156±72ab	449±163a	plowing, organic fertilizer, herbicide
Thai Orchid Lab (TO)	6.68± 0.28a	0.14±0.01a	4.15±1.48ab	188±57a	370±158a	296±81a	plowing, organic fertilizer, herbicide
Chiang Mai (CM)	5.70± 0.26bc	0.11±0.05a	6.10 ±0.87a	210±68a	171±133ab	284±10a	No management
Chiang Rai (CR)	5.23±0.11c	0.10±0.04a	2.83±0.25b	140±17a	24±3b	243±28a	No management
Nakhon Nayok (NN)	6.19±0.14ab	0.21±0.07a	3.08±0.66b	200±8a	149±63ab	347±78a	No management

^{a}Means of three observations. Means in the same column followed by the same letter are not significantly different ($\alpha = 0.05$).

Purified DNA was cloned into pGEM-T Easy Vector (Promega) and transformed into *Escherichia coli* JM109. One hundred transformants were selected randomly and their insertion checked by PCR using the same primers, AML1 and AML2. The amplified DNAs were digested by the restriction enzymes HinfI and Hsp92II separately. One clone of each RFLP type was screened and sequenced using sequencing primers SP6 and T7 on an ABI PRISM 3130 Genetic Analyzer System (Applied Biosystems). Sequences were trimmed to the NS31-AML3 region and virtually digested with the restriction enzymes HinfI, Hsp92II, and MboI using an online restriction mapping website (RestrictionMapper).

Phylogenetic analysis

Phylogenetic analysis was carried out on the sequences obtained in this study and those corresponding to the closest matches from Genbank, as well as sequences from cultured AMF taxa including representatives of the major groups of Glomeromycota from GenBank. All sequences obtained from this study were aligned by ClustalX using the BioEdit sequence alignment editor [40] along with 28 AMF sequences from GenBank. The aligned SSU rRNA dataset was trimmed to 450 bp by excluding the terminal primer sequences. A neighbour-joining (NJ) phylogeny was constructed using PAUP*4b10 [41] with the Kimura 2-parameter model and 1000 bootstraps. The nucleotide sequences of the clones retrieved in this study have been deposited in GenBank (accession numbers JQ8643324-JQ864355).

Statistical analysis

The total number of TRFs was used as an AM fungal community diversity measurement [31]. The main and interaction effects of collecting sites, host plant species and sample source (root vs. soil) on number of TRFs using three restriction enzymes were tested with two-way factorial ANOVA using SPSS 11.5 for Windows (SPSS Inc., Chicago, IL, USA). Jaccard similarity coefficients were calculated for the T-RFLP patterns of root and soil samples of both plants, which were clustered by the unweighted pair-group average (UPGMA) algorithm with 1000 bootstrap replicates to obtain confidence estimates. These calculations were performed using FreeTree [42] and the results displayed using TreeView [43].

Results

Soil analyses and correlation with TRFs

Chemical characteristics of soil varied among sites (Table 1). Soil pH values ranged from 5.23 to 6.68. No significant different was observed in electrical conductivity, exchangeable potassium, and total inorganic nitrogen. Available phosphorus in soils tended to be highest at the Thai Orchid Lab site (370 mg kg^{-1}soil) and differed significantly from the Chiang Rai site (24 mg kg^{-1}soil). Soil organic carbon was highest at the Chiang Mai site (6.10%) and differed significantly from the Chiang Rai and Nakhon Nayok sites. Pearson correlation analysis between the soil properties measured and TRFs showed that TRFs were positively correlated with available phosphorus, organic matter, and pH (Table S1).

AM fungal community of root and soil samples from *T. grandis* and *A. crassna*

The total number of different TRFs was used as a measure of AM fungal community diversity. Thirty eight TRFs were found in total for the AML3 (FAM-labelled) primer, while the NS31 (HEX-labelled) primer identified 30 TRFs. Since the AML3 primer revealed many more TRFs than the NS31 primer, only the AML3 fragments were used. Overall, in the roots and soils of *T. grandis* and *A. crassna*, we found 13 different AML3 TRFs after digestion with HinfI, 14 after digestion with Hsp92II and 11 after digestion with MboI. The mean number of TRFs in *T. grandis* root and soil samples was 5.67 and 7.67, respectively when the TRF data of the three enzymes were pooled (Figure 1). It is possible to estimate the minimum average number of AM fungi colonizing the *T. grandis* root samples by dividing the average number of TRFs by 3 (three enzymes and one labeled end) [31]. Thus, there were on average at least 1.89 fungal taxa colonizing each *T. grandis* root sample and 2.55 fungal taxa in surrounding soils, respectively. The values for *A. crassna* were at least 2.85 fungal taxa per root sample and 2.33 fungal taxa in surrounding soils. The mean number of TRFs per sample was not significantly affected by source of samples (root and soil) ($F = 0.159$, $P = 0.693$) and host plant ($F = 3.452$, $P = 0.074$) (Table S2), but there was a statistically significant effect of collecting sites ($F = 42.77$, $P<0.01$), and a significant interaction among those three factors (Table S2). The cluster analysis of TRF patterns in roots (R-) and rhizosphere soils (S-) of *A. crassna* and *T. grandis*, based on Jaccard similarities, showed no significant grouping of samples by sites and source of samples (root and soil) (Figure 2a). This suggested that the AM fungal community in roots and rhizosphere soils was almost independent in *A. crassna* (A) and *T. grandis* (T) plots. Some TRF patterns in roots and rhizosphere soils that were collected from the same site were similar, e.g. R-CRT versus S-CRT and R-TOA versus R-TOT. Combining roots and rhizosphere soils of each plant by sampling site (CM: Chiang Mai, CR: Chiang Rai, NN: Nakhon-Nayok, PB: Phetchabun and TO: Thai Orchids Lab) indicated a tendency for *T. grandis* plots to be grouped together (PBT, CMT and TOT) as well as some *A. crassna* plot samples (PBA and TOA) (Figure 2b). This suggests that the AM fungal community associated with each tree species was more similar across plots than were communities for different trees species at the same location. The response for CRT and NNA, however, does not support this.

Occurrence of AM fungi in soils and roots of both plants

Nearly all of the distinct TRFs (31 out of 38) were found in both host plant species (Figure 3). There were some differences in AM fungal communities between *T. grandis* and *A. crassna* because the TRF 329c (TRFs are identified by their relative mobility and a code indicating the restriction enzyme that generated them: a: MboI, b: HinfI and c: Hsp92II) was not found in *T. grandis*, while 5 TRFs (135c, 141b, 158c, 176b, and 435b) were not found in *A.*

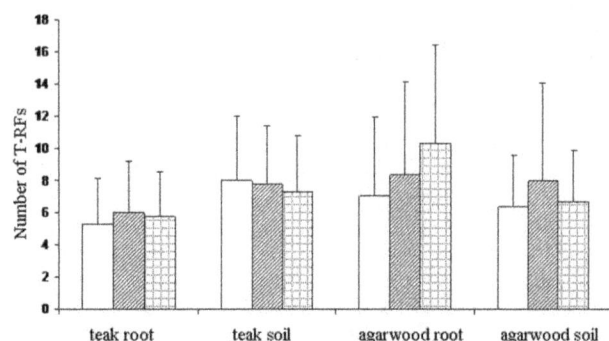

Figure 1. Effects of host plant, *Aquilaria crassna* (agarwood) and *Tectona grandis* (teak), and source of samples (root and soil) on mean number of terminal restriction fragments (TRFs) per sample using three restriction enzymes *MboI* (open bars), *HinfI* (hatched bars) and *Hsp92II* (cross-hatched bars). Values are mean ± SEM (n = 4 for teak and n = 3 for agarwood).

crassna. Comparison of the population in roots and soils of *T. grandis* (Fig 3a) showed that 6 TRFs (135c, 158c, 176b, 181c, 435b and 438b) were found only in roots, while 141b and 281a were only found in soils. In *A. crassna* (Figure 3b), TRFs 176c, 181c and 438b were only found in root samples.

Sequence and phylogenetic analysis

Clones were selected for sequencing on the basis of HinfI and Hsp92II RFLP typing. DNA sequences of 32 selected clones were determined, 7 clones from *A. crassna* and 25 clones from *T. grandis*. Predicted TRFs from the 32 virtually digested clone sequences were compared to observed TRFs from all three restriction enzymes (Table S3). A difference in size of up to 7 nucleotides was accepted as a match, because migration in capillary electrophoresis is sequence-specific, so that mobility (in rmu) is only approximately equivalent to sequence length (in bp). All predicted TRFs were observed, and the great majority of the observed TRFs were represented in the cloned sequences.

Our phylogenetic analysis was based on the new classification of Krüger et al. [44]. The 32 clone sequences were aligned with 23 sequences identified as closely related reference sequences in GenBank and a phylogenetic tree was constructed using the 18S rRNA gene sequences of *Paraglomus occultum* (GenBank accessions AJ276081 and JN687477) as outgroup. This indicated the presence of five AM fungal clades belonging to the families Claroideoglomeraceae, Diversisporaceae, Gigasporaceae, and Glomeraceae (Figure 4), the most frequent sequences corresponding to Glomeraceae. The subclusters contained close matches to taxa previously identified by Singh et al. [22] based on spore morphology of AM fungi in rhizosphere soils of *T. grandis*: TR1-16, TR1-43, TS4-4, AR5-7 and TS6-1 are close to *Rhizophagus intraradices* or *R. irregularis*, while TR1-27 is close to *Redeckera fulvum*. Clone sequences TS4-9 and TS4-32 are similar to *Diversispora aurantia*, while TR3-R10 is probably *Gigaspora margarita*. When sequence data are compared with individual TRFs (Table S3 and Figure 4), it is clear that individual TRFs cannot be used to identify sequence type, because many different species may generate a TRF of the same size. For example, the FAM fragment at 164b could equally well be from *G. indicum*, *Re. fulvum* or *Claroideoglomus etunicatum*.

Discussion

This study examined the AM fungal communities of *A. crassna* and *T. grandis* plantations in Thailand. The estimated numbers of AM fungal taxa in roots and soils of *T. grandis* seedlings were 1.89 and 2.55 respectively, while in roots and soils of *A. crassna* there were 2.85 and 2.33 respectively. The AM fungal diversity was low compared with other plants. Using similar methodologies and definitions, Vandenkoornhuyse et al. [31] reported an average of 6.1 AM fungal taxa colonizing grass roots in a temperate seminatural grassland system, and 5.5 AM fungal taxa were found colonizing each *Solidago virgaurea* L. seedling root sample in low-Arctic meadow habitat [29].

Previous studies quantified the AM fungal diversity in rhizospheres of *T. grandis* and *A. crassna* mainly based on spore morphology and aimed to select efficient AM fungal isolates for growth enhancement. For example, Singh et al. [22] found an average of nine species per 100 g dry soil in a Jhum fallow site at which *T. grandis* was the dominant tree species, and most species belonging to the genus Glomus. Tamuli and Boruah [21] studied the AM fungi association of agarwood (*Aquilaria malaccensis*) plantations in Jorhat District of the Brahmaputra Valley, India. They found that the genus *Glomus* was dominant; among these

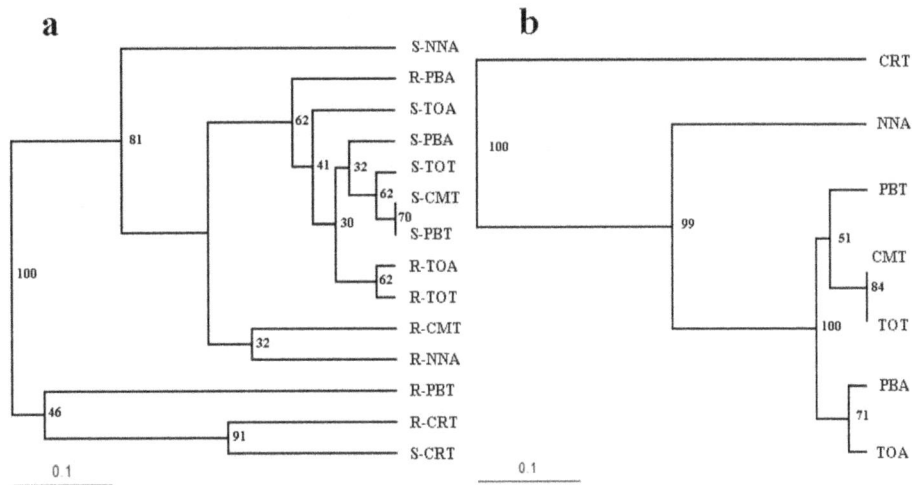

Figure 2. Cluster analysis of terminal restriction fragment length polymorphism patterns from AM fungal communities associated with *Aquilaria crassna* **(A) and** *Tectona grandis* **(T); a) TRFs patterns in roots (R-) and rhizosphere soils (S-) and b) TRFs patterns in five sites (CM: Chiang Mai, CR: Chiang Rai, NN: Nakhon-Nayok, PB: Phetchabun and TO: Thai Orchid Labs).** The unweighted pair-group average (UPGMA) algorithm was used to cluster patterns based on Jaccard similarities. Percentage values based on 1000 bootstrap replicates are given at each node.

G. fasciculatum (now known as *Rhizophagus fasciculatus*; [45]) was the most dominant followed by *G. aggregatum*. We are not aware of any information on the diversity of AM fungi on *A.* *crassna*. According to previous studies, we also found that most sequences belonged to the family Glomeraceae that includes *Glomus* and *Rhizophagus*. This result is consistent with previously

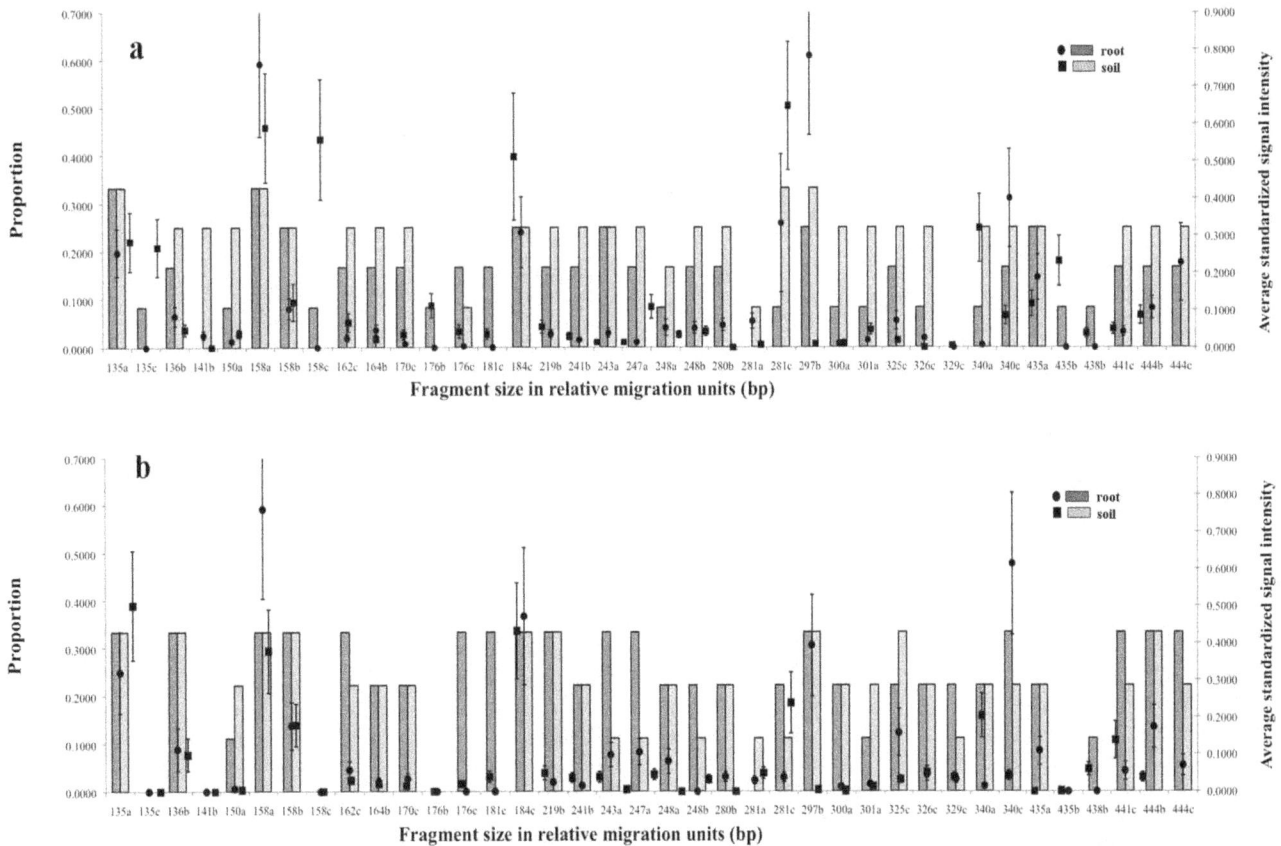

Figure 3. Occurrence of TRFs from roots and soils in (a) *Tectona grandis* **and (b)** *Aquilaria crassna.* Bars indicate the proportion of samples that yielded each TRF; dots indicate the average intensity of that fragment (± SEM) in those samples. The letters indicate the restriction enzyme involved in each fragment size, a: *MboI*, b: *HinfI* and c: *Hsp92II*.

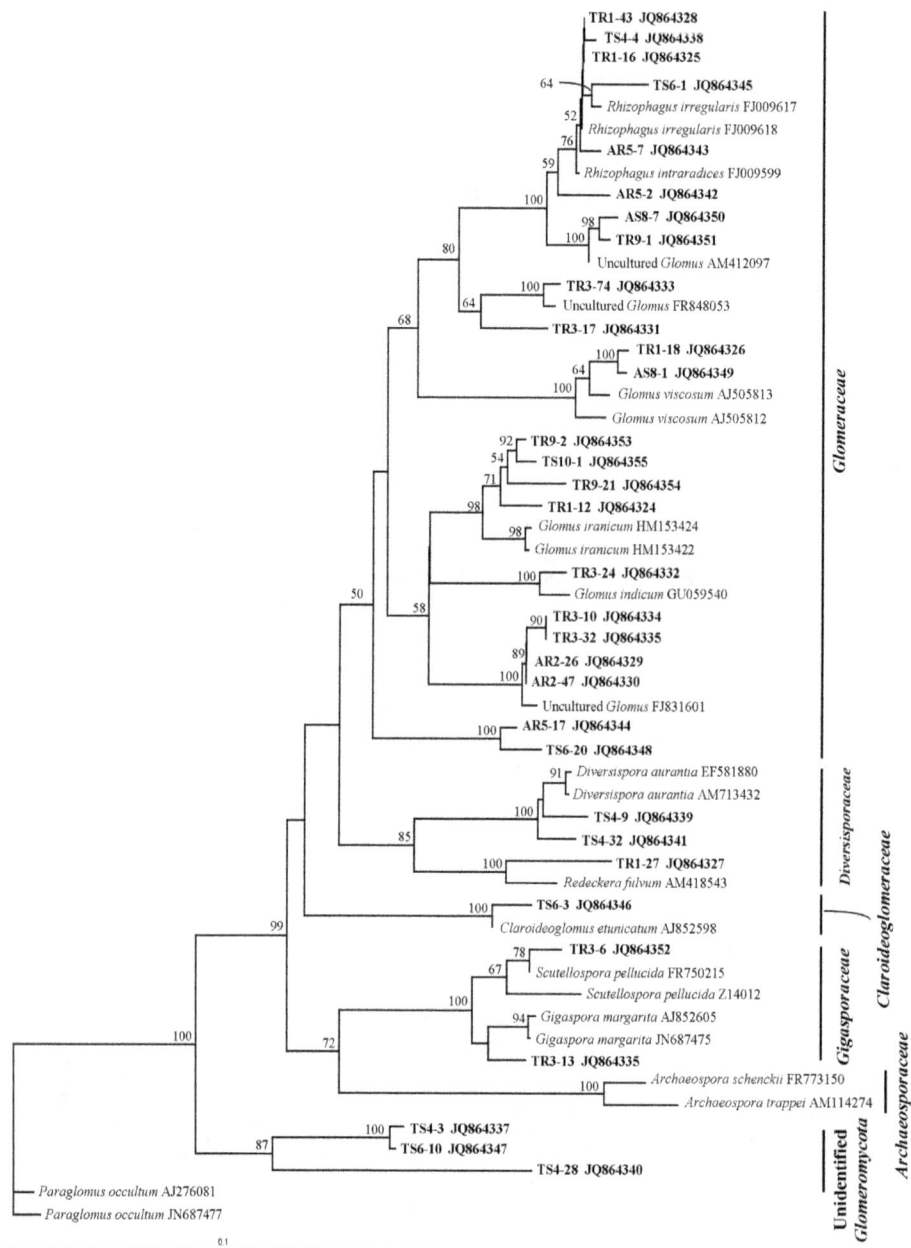

Figure 4. Neighbour-joining (NJ) phylogenetic tree of partial small subunit rRNA gene. Phylogeny was constructed using the region from NS31 to AML3. The percentage support values are based on 1000 bootstraps.

published phylogenies [29,39,46]. The dominance of this family suggests that they able to survive under various agricultural conditions such as soil disturbance from plowing and cultivation and pesticide usage like that used here in the Phetchabun and Nakhon-Nayok sites. Those conditions may be unfavorable for other AM fungi. One possible reason why *Glomus* species have the ability to survive in a disturbed system is related to differences in propagation strategies [29]. Glomeraceae are capable of colonizing via fragments of mycelium, mycorrhizal root pieces, and spores, while Gigasporaceae are only capable of propagation via spores because they do not produce intra-radical vesicles: lipid-rich storage structures which allow for re-growth of hyphae from previously colonized root pieces [46,47,48,49]. This difference can explain the dominance of the Glomeraceae over Gigasporaceae

members in an environment with repetitive agricultural disturbance. Oehl et al. [50] revealed a clear seasonal and successional AMF sporulation dynamics and implied that different life strategies of different ecological AMF groups could be defined on the basis of diverging temporal sporulation dynamics.

This study shows that the choice of restriction enzymes (HinfI, Hsp92II, MboI) did not significantly affect AM fungal diversity found per sample. Using a combination of those three restriction enzymes could detect possible species of AM fungi in the samples, even if they resulted in similar-sized fragments. HinfI and Hsp92II were chosen in this study because they showed the highest polymorphism of cleavage sites at the extremities of the amplified DNA fragment [31]. Mummey and Rillig [39] and Wolfe et al. [51] also found that HinfI and MboI can separate different closely-related species of AM fungi identified from phylogenetic analyses. For example, *R. irregularis* and *R. intraradices* are closely related species that group in the same clade (Figure 4). Six clone sequences (TR1-16, TR1-43, TS4-4, AR5-2, AR5-7and TS6-1) that were related to both species were not completely separated using phylogenetic analysis, but virtual digesting with those three enzymes did separate them by using the combination of restriction pattern of each enzyme (Table S3). Clone sequences TR1-16, 1-43, and 6-1 grouped with *R. irregularis* and TS4-4, 5-2, 5-7 grouped with *R. intraradices*.

Some TRFs were only found in roots or only in soils, suggesting that some AM fungi may be rare in soil but produce fungal structures in roots that are rich enough for T-RFLP detection, while some were found only as spores in soils and did not colonize roots. While the majority of TRFs were associated with both *T. grandis* and *A. crassna*, some TRFs were associated with just one plant (i.e. 135c, 141b, 158c, 176b, 329c and 435b). In clustering analysis, samples from each plant species were grouped together even if they were collected from different sites. *A. crassna* samples seemed to group together, but since many AMF taxa were shared by both trees, *A. crassna* shared some AM fungal community patterns with *T. grandis* (Figure 2). Statistical analysis revealed significant effects of collecting sites and the interaction between collecting sites, host plant species and source of samples on TRFs (Table S2). Thus, specific AM taxa in roots and soils of *T. grandis* and *A. crassna* were affected by site but not affected by host plant species and source of samples (root and soil). This is in accordance with the observation of Bever et al. [52] that the host-dependence of the relative growth rates of fungal populations may play an important role in the maintenance of fungal species diversity. Previously, it has been reported that neighboring plants may have a significant impact on the AM fungal colonization and community composition of AM fungi in plant roots [34]. Although *T. grandis* at the Chiang Mai site had other *T. grandis* as closest neighbors with some negligible understory perennial plants, and at the other two sites the closest neighbors were *A. crassna*, the cluster analysis did not reveal any effect of this difference in neighbors. AM fungal community patterns in CMT were grouped with PBT and TOT sites in which weeds were controlled by agricultural management.

In conclusion, we demonstrated here that AM fungal community patterns in rhizosphere soils and roots of *T. grandis* and *A. crassna* were similar even if they were collected from different sites. AM fungal communities of *T. grandis* samples from different sites were similar, as were those in *A. crassna* samples. We also found that most sequences represented Glomeraceae, including *Glomus* spp. and *Rhizophagus* spp. Virtual digestion of sequences using the target sequences of the restriction enzymes HinfI, Hsp92II and MboI yielded expected fragments that mostly matched observed TRFs, linking possible AM fungal species to each TRF. Specific AM taxa in roots and soils of *A. crassna* and *T. grandis* were affected by site but were not affected by host plant species and source of samples (root and soil). Although the T-RFLP technique can provide important information about the AM fungal diversity associated with plant species of interest, trap cultures and cultured spores from the field site are still important in order to assess the ability of the AM fungi to enhance the growth of the plants, and to provide effective candidates for inoculum production targeted for these economically important tree species.

Supporting Information

Table S1 Correlation matrix of soil factors and terminal restriction fragments (TRFs) of study areas in wet season (July 2010) which soils were sampled.

Table S2 Summary of two-way analysis of variance for main and interaction effects of host plants (*Aquilaria crassna* and *Tectona grandis*), sites, and source of samples (root and soil) on AM fungal community diversity measured as the number of different TRFs per sample. Significant P-values are shown in bold.

Table S3 Clone sequences and TRFs derived from roots and rhizosphere soils of *T. grandis* and *A. crassna*. Values in bold indicate TRFs that match the sizes of virtual digest fragments (with differences ranging from 0 to 7 bp).

Author Contributions

Conceived and designed the experiments: SL JY. Performed the experiments: AC JY. Analyzed the data: AC JY. Contributed reagents/materials/analysis tools: JY NT PG SL. Wrote the paper: AC JY SL. Collected samples: PG.

References

1. Kobayashi S (2004) Landscape rehabilitation of degraded tropical forest ecosystems, Case study of the CIFOR/Japan project in Indonesia and Peru. Forest Ecol Manag 201: 13–22.
2. Turjaman M, Tamai Y, Santoso E, Osaki M, Tawaraya K (2006) Arbuscular mycorrhizal fungi increased early growth of two nontimber forest product species Dyera polyphylla and Aquilaria filaria under greenhouse conditions. Mycorrhiza 16: 459–464.
3. Rajan SK, Reddy BJD, Bagyaraj DJ (2000) Screening of arbuscular mycorrhizal fungi for their symbiotic efficiency with Tectona grandis. Forest Ecol Manag 126: 91–95.
4. CITES (2004) Amendments to Appendices I and II of the Convention on International Trade in Endangered Species of Wild Flora and Fauna (CITES). Thirteenth meeting of the conference of the parties, Bangkok, Thailand, 3–14 October, pp. 1–9.
5. Schüßler A, Schwarzott D, Walker C (2001) A new fungal phylum, the Glomeromycota: phylogeny and evolution. Mycol Res 105 (12): 1413–1421.
6. Habte M (2000) Mycorrhizal fungi and plant nutrition. In: Silva JA, Uchida R, editors. Plant Nutrient Management in Hawaii's Soils, Approaches for Tropical and Subtropical Agriculture. College of Tropical Agriculture and Human Resources, Manoa: University of Hawaii. pp. 127–131.
7. Brundrett MC, Bougher N, Dell B, Grove T, Malajczuk N (1996) Working with mycorrhizas in forestry and agriculture. Australian Centre for International Agricultural Research Monograph 32, Canberra.
8. Smith S, Read D (2008) Mycorrhizal Symbiosis. New York: Academic Press.
9. Porcel R, Ruiz-Lozano JM (2004) Arbuscular mycorrhizal influence on leaf water potential, solute accumulation, and oxidative stress in soybean plants subjected to drought stress. J Exp Bot 55: 1743–1750.
10. Doubková P, Vlasáková E, Sudová R (2013) Arbuscular mycorrhizal symbiosis alleviates drought stress imposed on Knautia arvensis plants in serpentine soil. Plant Soil 370: 149–161.
11. Azcón-Aguilar C, Barea JM (1996) Arbuscular mycorrhizas and biological control of soil-borne plant pathogens – an overview of the mechanisms involved. Mycorrhiza 6: 457–464.
12. Wright SF, Upadhyaya A (1998) A survey of soils for aggregate stability and glomalin, a glycoprotein produced by hyphae of arbuscular mycorrhizal fungi. Plant Soil 198: 97–107.
13. Rillig MC (2004) Arbuscular mycorrhizae, glomalin, and soil aggregation. Can J Soil Sci 84: 355–363.
14. Rillig MC, Ramsey PW, Morris S, Paul EA (2003) Glomalin, an arbuscular-mycorrhizal fungal soil protein, responds to land-use change. Plant Soil 253: 293–299.
15. Habte M, Miyasaka SC, Matsuyama DT (2001) Arbuscular mycorrhizal fungi improve early forest-tree establishment. In: Horst WJ et al., editors. Plant nutrition–Food Security and Sustainability of Agro-ecosystems. Dordrecht: Kluwer Academic Publishers. pp. 644–645.
16. Urgiles N, Loján P, Aguirre N, Blaschke H, Günter S, et al. (2009) Application of mycorrhizal roots improves growth of tropical tree seedlings in the nursery: a

step towards reforestation with native species in the Andes of Ecuador. New Forest 38: 229–239.

17. Swaminathan C, Srinivasan VM (2006) Influence of microbial inoculants on seedling production in teak (*Tectona grandis* L.f.). J Sustain Forest 22 (3): 63–76.

18. Tabin T, Arunachalam A, Shrivastava K, Arunachalam K (2009) Effect of arbuscular mycorrhizal fungi on damping-off disease in *Aquilaria agallocha* Roxb. seedlings. Trop Ecol 50 (2): 243–248.

19. Thapar HS, Khan SN (1988) Seasonal frequency of *Endogone* spores in new-forest soils. In: Khosla PK, Sehgal RN, editors. Trends in Tree Sciences. Solan: Indian Society of Tree Scientists. pp.161–162.

20. Kanakadurga VV, Manoharachary D, Rama RP (1990) Occurrence of endomycorrhizal fungi in teak. In: Bagyaraj DJ, Manjunath A, editors. Mycorrhizal symbiosis and plant growth.Bangalore: University of Agricultural Sciences. pp.17.

21. Tamuli P, Boruah P (2002) Vesicular-abuscular mycorrhizal (VAM) association of agarwood tree in Jorhat District of the Brahmatputra Valley. Indian For 128 (9): 991–994.

22. Singh SS, Tiwari SC, Dkhar MS (2003) Species diversity of vesicular-arbuscular mycorrhizal (VAM) fungi in Jhum fallow and natural forest soils of Arunachal Pradesh, north eastern India. Trop Ecol 44 (2): 207–215.

23. Dhar PP, Mridha MAU (2012) Arbuscular mycorrhizal associations in different forest tree species of Hazarikhil forest of Chittagong, Bangladesh. J For Res 23 (1): 115–122.

24. Ramanwong K (1998) Species diversity of vesicular-arbuscular mycorrhizal fungi of teak (*Tectona grandis* Linn.f.) and their effects on growth of teak seedlings. M.S. Thesis, Kasetsart University, Chatuchak, Bangkok.

25. Helgason T, Daniell TJ, Husband R, Fitter AH, Young JPW (1998) Ploughing up the wood-wide web? Nature 394: 431.

26. Lee J, Lee S, Young JPW (2008) Improved PCR primers for the detection and identification of arbuscular mycorrhizal fungi. FEMS Microbiol Ecol 65: 339–349.

27. Krüger M, Stockinger H, Krüger C, Schüßler A (2009) DNA-based species level detection of Glomeromycota: one PCR primer set for all arbuscular mycorrhizal fungi. New Phytol 183: 212–223.

28. Daniell TJ, Husband R, Fitter AH, Young JPW (2001) Molecular diversity of arbuscular mycorrhizal fungi colonising arable crops. FEMS Microbiol Ecol 36: 203–209.

29. Pietikäinen A, Kytöviita MM, Husband R, Young JPW (2007) Diversity and persistence of arbuscular mycorrhizas in a low-Arctic meadow habitat. New Phytol 176: 691–698.

30. Mummey DL, Rillig MC (2006) The invasive plant species Centaurea maculosa alters arbuscular mycorrhizal fungal communities in the field. Plant Soil 288: 81–90.

31. Vandenkoornhuyse P, Ridgway KP, Watson IJ, Duck M, Fitter AH, et al. (2003) Co-existing grass species have distinctive arbuscular mycorrhizal communities. Mol Ecol 12: 3085–3095.

32. Johnson D, Vandenkoornhuyse PJ, Leake JR, Gilbert L, Booth RE, et al. (2004) Plant communities affect arbuscular mycorrhizal fungal diversity and community composition in grassland microcosms. New Phytol 161: 503–515.

33. van der Heijden MGA, Wiemken A, Sanders IR (2003) Different arbuscular mycorrhizal fungi alter coexistence and resource distribution between co-occurring plant. New Phytol 157: 569–578.

34. Mummey DL, Rillig MC, Holben WE (2005) Neighbouring plant influences on arbuscular mycorrhizal fungal community composition as assessed by T-RFLP analysis. Plant Soil 271: 83–90.

35. Barto EK, Antunes PM, Stinson K, Koch AM, Klironomos JN, et al. (2011) Differences in arbuscular mycorrhizal fungal communities associated with sugar maple seedlings in and outside of invaded garlic mustard forest patches. Biol Invasions 13: 2755–2762.

36. Houba VJG, Van Der Lee JJ, Novozamsky I, Wallinga J (1988) Soil and Plant analysis. Part 5: Soil Analysis Procedure. Agricultural University, Wageningen.

37. Sparks DL, Page AL, Helmke PA, Loeppert RH, Soltanpour PN, et al. (1996) Methods of Soil Analysis. Part 3. Chemical Methods. Madison, Wisconsin: Soil Science Society of America

38. Render C, Weißhuhn K, Kellner H, Buscot F (2006) Rationalizing molecular analysis of field-collected roots for assessing diversity of arbuscular mycorrhizal fungi: to pool, or not to pool, that is the question. Mycorrhiza 16: 525–531.

39. Mummey DL, Rillig MC (2007) Evaluation of LSU rRNA-gene PCR primers for analysis of arbuscular mycorrizal fungal communities via terminal restriction fragment length polymorphism analysis. J Microbiol Meth 70: 200–204.

40. Hall T (1999) BioEdit: a user-friendly biological sequence alignment editor and analysis program for Windows95/98/NT. Nucl Acid S 41: 95–98.

41. Swofford DL (2002) PAUP* Phylogenetic analysis using parsimony (*and other methods). version4b10. Sunderland. MA: Sinauer Associates

42. Hampl V, Pavlícek A, Flegr J (2001) Construction and bootstrap analysis of DNA fingerprinting-based phylogenetic trees with a freeware program Free-Tree: Application to trichomonad parasites. Int J Syst Evol Micr 51: 731–735.

43. Page RDM (1996) TreeView: An application to display phylogenetic trees on personal computers. Comput Appl Biosci 12: 357−358.

44. Krüger M, Krüger C, Walker C, Stockinger H, Schüßler A (2012) Phylogenetic reference data for systematics and phylotaxonomy of arbuscular mycorrhizal fungi from phylum to species-level. New Phytol 193: 970–984.

45. Schüßler A, Walker C (2010) The Glomeromycota: A species list with new families and new genera (Libraries at the Royal Botanic Garden Edinburgh, Edinburgh, UK; The Royal Botanic Garden Kew, Kew, UK; Botanische Staatssammlung Munich, Munich, Germany; and Oregon State University, Corvallis, Oregon, pp.1–56.

46. Helgason T, Fitter AH, Young JPW (1999) Molecular diversity of arbuscular mycorrhizal fungi colonising *Hyacinthoides non-scripta* (bluebell) in a seminat-ural woodland. Mol Ecol 8: 659–666.

47. Gazey C, Abbott KK, Robson AD (1993) VA mycorrhizal spores from 3 species of *Acaulospora*—germination, longevity and hyphal growth. Mycol. Res. 97: 785–790.

48. INVAM Newsletter 3 (1993) Properties of infective propagules at the suborder level (*Glomineae* versus *Gigasporineae*). West Virginia University Web Services. Available: http//invam.caf.wvu.edu/articles/propagules.htm. Accessed 9 July 2012.

49. Brundrett M, Abbott LK, Jasper DA (1999) Glomalean mycorrhizal fungi from tropical Australia. I. Comparison of the effectiveness and specificity of different isolation procedures. Mycorrhiza 8: 305–314.

50. Oehl F, Sieverding E, Ineichen K, Mäder P, Wiemken A, et al. (2009) Distinct sporulation dynamics of arbuscular mycorrhizal fungal communities from different agroecosystems in long-term microcosms. Agr Ecosyst Environ 134: 257–268.

51. Wolfe BE, Mummey DL, Rillig MC, Klironomos JN (2007) Small-scale spatial heterogeneity of arbuscular mycorrhizal fungal abundance and community composition in a wetland plant community. Mycorrhiza 17: 175–183.

52. Bever JD, Morton JB, Antonovics J, Schultz PA (1996) Host-dependent sporulation and species diversity of arbuscular mycorrhizal fungi in a mown grassland. J Ecol 84: 71–82.

18

Phactr3/Scapinin, a Member of Protein Phosphatase 1 and Actin Regulator (Phactr) Family, Interacts with the Plasma Membrane via Basic and Hydrophobic Residues in the N-Terminus

Akihiro Itoh[1]♎, Atsushi Uchiyama[1]♎, Shunichiro Taniguchi[2], Junji Sagara[1]*

1 Department of Biomedical Laboratory Sciences, Health Sciences, Shinshu University, Matsumoto, Japan, **2** Department of Molecular Oncology, Medical Sciences, Shinshu University Graduate School of Medicine, Matsumoto, Japan

Abstract

Proteins that belong to the protein phosphatase 1 and actin regulator (phactr) family are involved in cell motility and morphogenesis. However, the mechanisms that regulate the actin cytoskeleton are poorly understood. We have previously shown that phactr3, also known as scapinin, localizes to the plasma membrane, including lamellipodia and membrane ruffles. In the present study, experiments using deletion and point mutants showed that the basic and hydrophobic residues in the N-terminus play crucial roles in the localization to the plasma membrane. A BH analysis (http://helixweb.nih.gov/bhsearch) is a program developed to identify membrane-binding domains that comprise basic and hydrophobic residues in membrane proteins. We applied this program to phactr3. The results of the BH plot analysis agreed with the experimentally determined region that is responsible for the localization of phactr3 to the plasma membrane. *In vitro* experiments showed that the N-terminal itself binds to liposomes and acidic phospholipids. In addition, we showed that the interaction with the plasma membrane via the N-terminal membrane-binding sequence is required for phactr3-induced morphological changes in Cos7 cells. The membrane-binding sequence in the N-terminus is highly conserved in all members of the phactr family. Our findings may provide a molecular basis for understanding the mechanisms that allow phactr proteins to regulate cell morphogenesis.

Editor: Eugene A. Permyakov, Russian Academy of Sciences, Institute for Biological Instrumentation, Russian Federation

Funding: This work was supported by a Grant-in aid 16570156 to JS from the Minister of Education, Science and Culture, Japan. The funders had no role in study design, data collection and analysis, decision to publish, or preparation of the manuscript.

Competing Interests: The authors have declared that no competing interests exist.

* Email: sagaraj@shinshu-u.ac.jp

♎ These authors contributed equally to this work.

Introduction

Protein phosphatase 1 (PP1) and actin regulatory (Phactr) proteins are a family that comprises four members in humans and other vertebrates; phactr1–4 [1]. The phactr gene is present in worms and insects but not in protozoa. Accumulating evidence indicates the involvement of phactr proteins in human diseases such as myocardial infarction, Parkinson's disease, and cancers [2–4].

Each phactr protein contains four G-actin-binding RPEL motifs, including an N-terminal motif and a C-terminal triple RPEL repeat. The C-terminal triple RPEL repeat is adjacent to the PP1-binding domain. RPEL motifs are also found in the regulatory domains of myocardin-related transcription factor (MRTF) transcriptional coactivators where they control subcellular localization and activity by sensing signal-induced changes in the G-actin concentration [6,7]. Subcellular localization of phactr1 that was similar to that of MRTF is controlled by RPEL motifs. Phactr1 exhibits nuclear accumulation in response to serum-induced G-actin depletion [8]. However, there is no evidence for the serum-induced nuclear accumulation of phactr proteins other than phactr1 [8,9].

Phactr proteins are involved in cell migration both *in vitro* and *in vivo*, and it is believed that phactr is a novel protein family that regulates cytoskeleton dynamics [9–14]. However, the mechanisms that regulate actin cytoskeleton dynamics are poorly understood. It has been reported that G-actin and PP1 competitively bind to the C-terminal region and the formation of the phactr–PP1 complex is inhibited by an increase in the cytoplasmic G-actin concentration, which is induced by extracellular signals such as serum. The current hypothesis suggests that the phactr–PP1 complex is controlled by the changes in the cytoplasmic G-actin concentration, which regulate the actin cytoskeleton dynamics by modulating the phosphorylation status of actin regulatory protein(s) [8,15]. This suggests that phactr proteins regulate both the PP1 activity and subcellular localization by sensing the cytoplasmic actin concentration through RPEL motifs.

The catalytic subunit of PP1 interacts with noncatalytic subunits that determine the activity, substrate specificity, and subcellular localization of the phosphatase [16]. PP1 can dephosphorylate cofilin and myosin [17,18]. The actin filament-severing activity of cofilin, which stimulates the treadmill-like movement of the actin cytoskeleton in the lamellipodia and filopodia, is controlled by its phosphorylation status, and the force-generating activity of myosin is controlled by the phosphorylation status of myosin itself. In this context, several studies have shown that the phactr–PP1complex modulates the phosphorylation status of cofilin or myosin, and therefore regulating actin cytoskeleton dynamics [8,12,15].

Phactr3 was originally named as the nuclear scaffold-associated PP1-inhibiting protein (scapinin), which is found in the nuclear insoluble fraction of the leukemia cell line HL-60 [19]. However, phactr3 is distributed to the plasma membrane in adherent cells, and it enhances cell migration [9]. In the present study, we explored the domains that direct the membrane localization of phactr3 on the basis of deletion and point mutation experiments. We identified a membrane-targeting domain that comprises basic and hydrophobic residues in the N-terminus. This indicates that phactr3 is a membrane-associated PP1 and an actin regulator. Our findings provide a molecular basis for understanding the mechanisms that allow phactr3 to regulate membrane-cytoskeleton dynamics. The amino acid sequences of the N-terminal membrane-targeting domain are highly conserved in the phactr protein family.

Materials and Methods

Ethics

This study was carried out in strict accordance with the recommendations in the Guide for the Care and Use of Laboratory Aimals of Shinshu University. The protocol was approved by the Shinshu University Ethics Committee for animal care, handing, termination (Permit Number: 230051). All surgery was performed under Ketamine/xylazine anesthesia, and all efforts were made to minimize suffering.

GFP-Phactr3 mutants

The full-length phactr3 cDNA (NP_001186435, 518 aa) was obtained by RT-PCR using an RT primer 5′-AATCTC-TATGGCCTGTGGAA-3′, a reverse primer 5′-TCTCTATGGCCTGTGGAATCT-3′, a forward primer 5′-CTGGATGAGATGGACCAAACG-3′, a template poly A+ RNA of HL-60 cells, and a high fidelity RNA PCR kit (Takara, Japan). It was subcloned into the SmaI site in pKF18k (Takara, Japan). pEGFP-c2-phactr3 was produced by inserting the BamHI/EcoRI fragment of pKF18k-phactr3 into BglII/EcoRI sites of pEGFP-c2 (Clontech) [10]. A mutant with a deleted N-terminal region (ΔNt) (Fig. 1C) was generated by PCR-based mutagenesis using a reverse primer 5′-CATCTCATCCAGGGG-GATCT-3′ and a forward primer 5′-GCGCTGGAGAAGAA-GATGGC-3′. Expand High-Fidelity DNA polymerase (Roche Molecular Biochemicals) was used for the PCR analysis. Deletion mutants of Nt, PP1-binding domain (ΔPP1), ΔRx3/ΔPP1, and

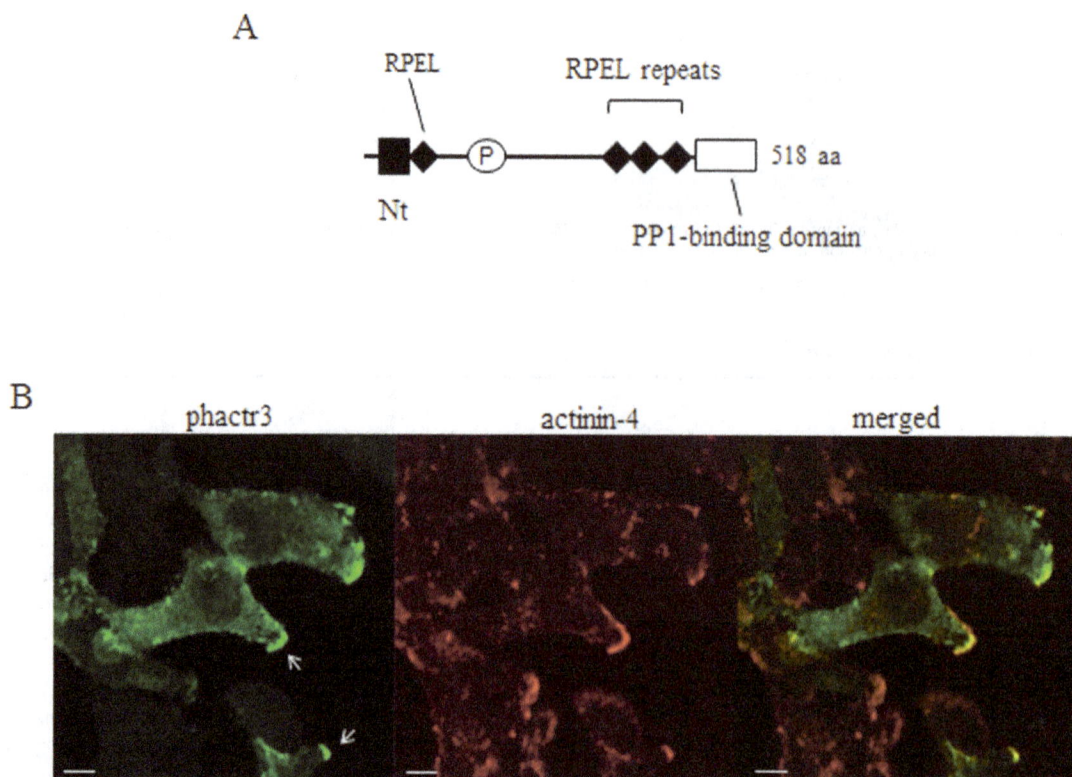

A

B

Figure 1. The distribution patterns of phactr3 in cells. (A) Primary structure of phactr3 (NP_001186435, 518 aa). RPEL, G-actin binding motif, and PP1-binding domain (PP1) are indicated. P, a region with a proline-rich sequence. The N-terminal region (Nt), RPEL motifs, and PP1-binding domain are highly conserved in the phactr protein family. (B) Localization of phactr3 in HeLa cells. We have previously established a HeLa cell line where phactr3 expression was induced by tetracycline [5]. The HeLa cells were cultured with tetracycline (0.5 μg/ml) for 24 h and immunostained with anti-phactr3 monoclonal antibody. The distribution patterns of phactr3 were compared with that of α-actinin 4, which is known to localize to the lamellipodia and membrane ruffles [20]. Bars, 20 μm.

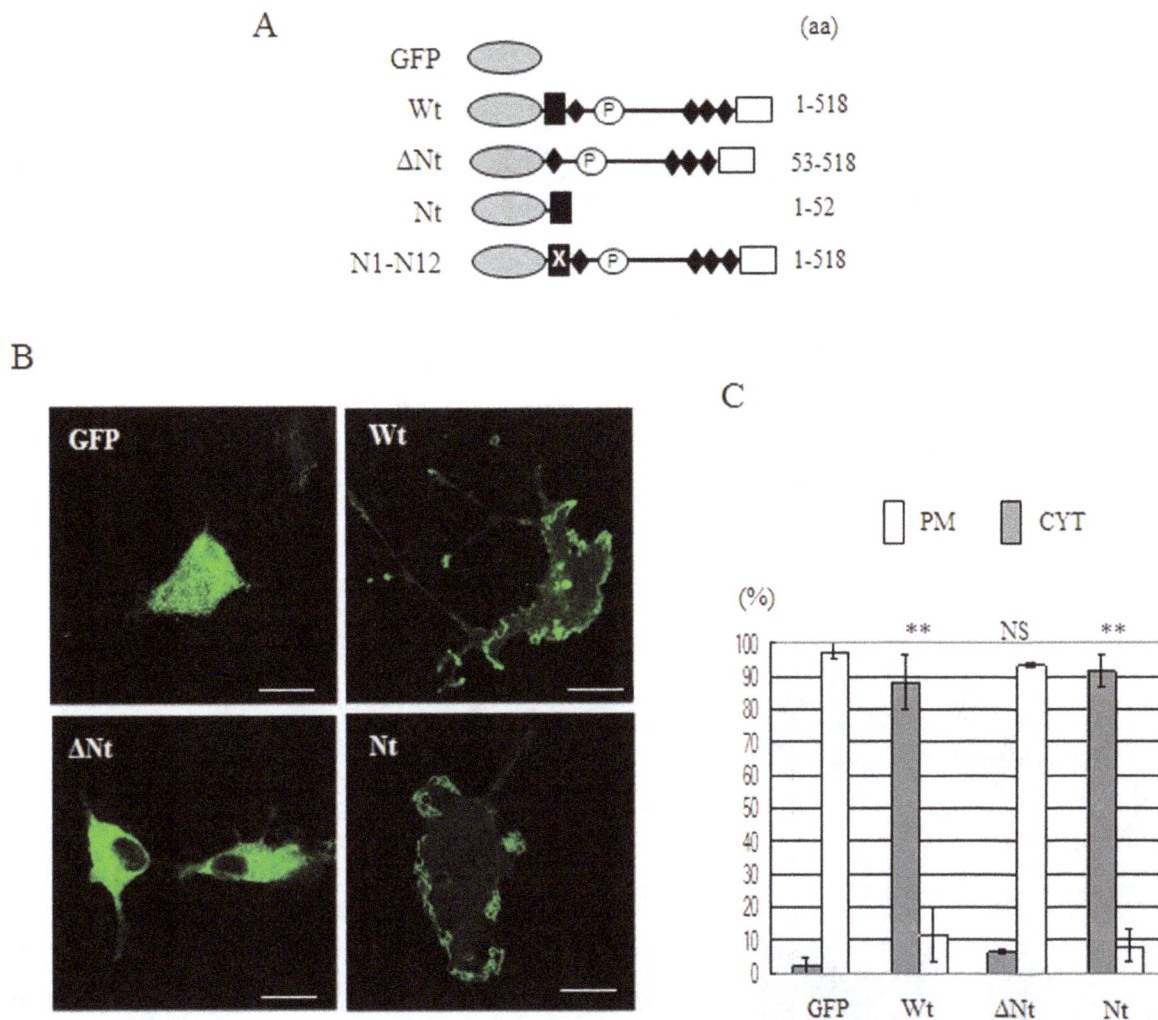

Figure 2. The N-terminal region (Nt) is responsible for the distribution of phactr3 in the plasma membrane. (A) Deletion mutants of GFP-phactr3 used in this study. (B) GFP and GFP-phactr3 constructs were expressed in Cos7 cells and their subcellular distributions were observed by confocal laser scanning microscopy. Bars, 20 µm. (C) The numbers of cells with predominantly plasma membrane (PM) distributions and cytoplasm (CYT) distributions were counted by fluorescence microscopy. The mean and standard error (SE) were calculated on the basis of three independent experiments (shown in the figure as mean ± SE). The asterisks in the histograms indicate significant differences between Wt and each mutant (Student's t-test; **$P < 0.005$). NS, not significant.

NtR (Fig. 1C) were generated by introducing stop codon sequences to terminate protein synthesis at targeted positions using the QuikChange XL site-directed mutagenesis kit (Agilent Technologies). Primers were designed according to the cDNA sequence data of phactr3 (AB098521). The mutations in pEGFP-c2-phactr3 were confirmed by sequencing. Alanine substitution mutants of N1–12 were also generated using the QuikChange XL site-directed mutagenesis kit.

Cell Culture and Transfection

Cos7 cells (ATCC CRL-1651) were cultured in Dulbecco's modified minimal essential medium (DMEM) containing 8% fetal bovine serum, 100 units/ml penicillin, and 100 µg/ml streptomycin with 5% CO_2 and at 37°C. We used FuGENE 6 (Roche Molecular Biochemicals) as the transfection reagent to introduce the plasmid DNA into Cos7 cells. One day before transfection, 0.45 ml of cell suspensions (3×10^5 cells/ml) were mounted on a silicone-coated two-well glass slide (Matsunami, Japan). Subsequently, 0.5 µg of plasmid DNA (pEGFP-phactr3), its mutants,

and 2 µl of FuGENE 6 were mixed in 50 µl of DMEM (antibiotic free) and then allowed to stand for 15 min at room temperature. Twenty microliters of the FuGENE6/DNA mixture was added to each well before cell culture. At 24–48 h post-transfection, 45 µl of formaldehyde (36%–38%) was added to the cell culture to fix the cell morphology. In some cases, the nuclei were stained with Heochst 33342 (1 µg/ml). Fluorescent cells were observed using a BX60-34-FLB-1 fluorescence microscope (Olympus) or an LSM 5 EXCITER confocal laser scanning microscope (Carl Zeiss MicroImaging Inc.). The subcellular localizations of GFP-phactr3 and its mutants were assessed on the basis of three independent experiments.

Expression of phactr3 in HeLa cells and immunostaining

We established a HeLa cell line using a Tet-ON vector (Invitrogen Life Technology) [9]. Phactr3 expression was induced by the addition of tetracycline at a final concentration of 0.5 µg/ml. After 24 h, the cells were fixed with formaldehyde (3.7%) and immunostained with anti-phactr3 monoclonal antibody [5],

A

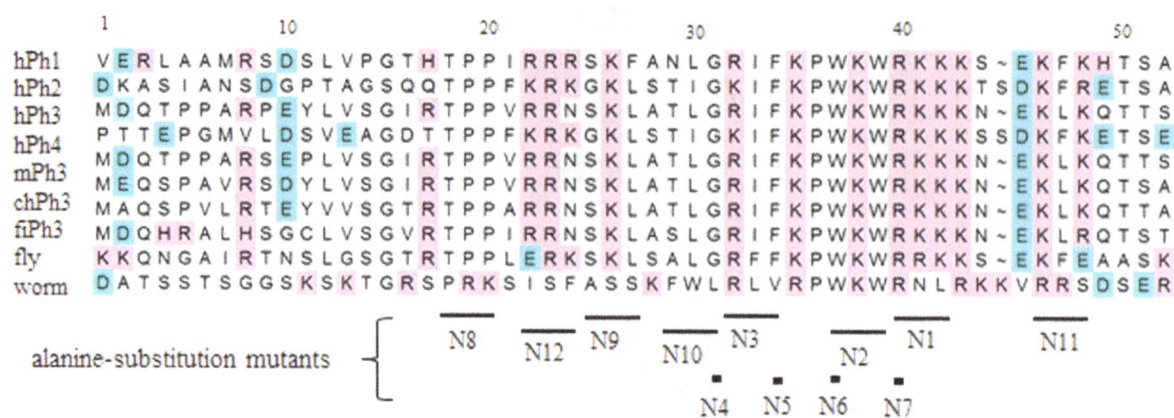

B

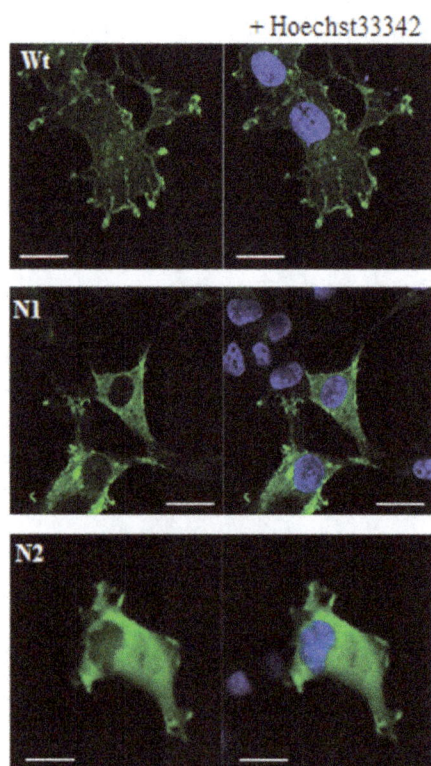

C

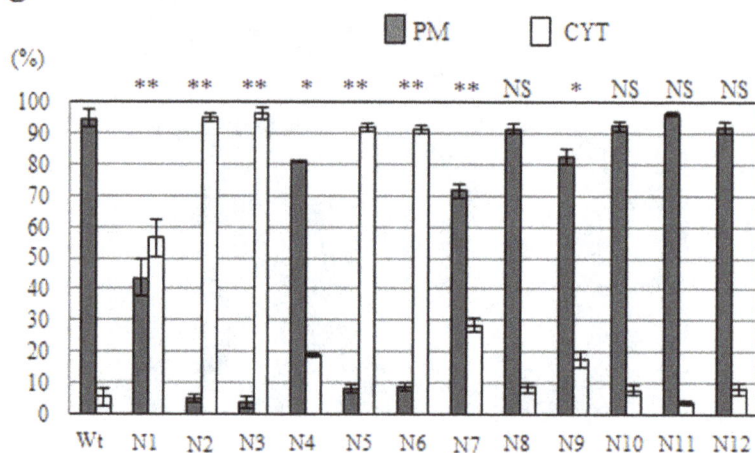

D

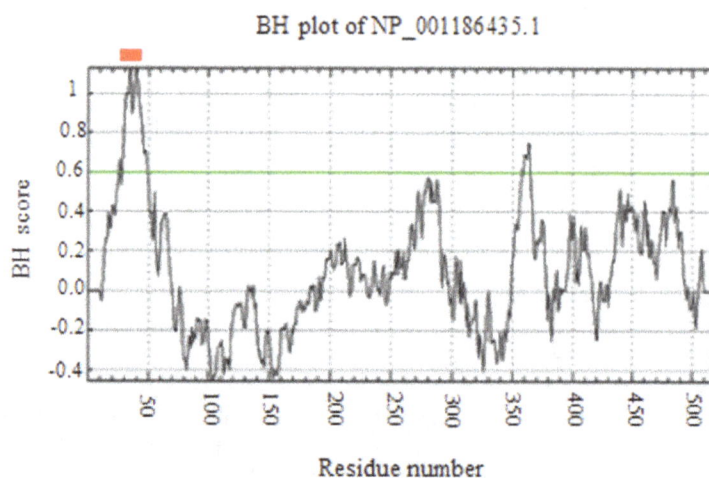

Figure 3. Basic and hydrophobic residues in the Nt are responsible for the distribution in the plasma membrane. (A) Alignment of the Nt amino acid (aa) sequences of phactr. Human phactr 1–4 (hPh1–4), mouse phactr3 (mouse), chicken phactr3 (bird), zebra fish phactr3 (fish), *Drosophila melanogaster* phactr (fly), and *Caenorhabditis elegans* phactr (worm) were aligned. N1–12 indicate substituted with alanine residues. (B) The distribution patterns of GFP-phactr3 Wt, N1, and N2. Cos7 cells expressing each construct were photographed using confocal laser microscopy at

24 h after transfection. The nucleus was stained with Hoechst 33342. Bars, 20 µm. (C) Amino acid residues in the indicated Nt (A) were substituted with alanine residues and expressed in Cos7 cells. The numbers of cells with predominantly plasma membrane (PM) distributions and cytoplasm (CYT) distributions were counted by fluorescence microscopy [(denoted as mean ± standard error (SE)]. The asterisks in the histograms indicate significant differences between Wt and each mutant (Student's t-test; *$P<0.05$, **$P<0.005$). NS, not significant. (D) The BH plot analysis of the membrane interaction domains [24] detected a potent membrane interaction domain in Nt. Red bar indicates the experimentally deduced region responsible for the distribution in the plasma membrane.

followed by FITC-conjugated anti-mouse antibody. Actinin-4 rabbit antibody (rabbit) was generously provided by Dr Yamada T. (National Cancer Center Research Institute, Japan) [20].

GST-Nt fusion protein

Nt was generated as a fusion protein with glutathione transferase (GST) in *Escherichia coli*. The Nt cDNA was amplified by PCR using a forward primer (5'-CCCGAATTCGATGA-GATGGACCAAACGCCCCCG-3') and a reverse primer (5'-CCCCTCGAGCTACGTTGTCTGCTTCAGTTTTTCGT-3'). The forward primer included an *Eco*RI cleavage site

(underlined above), whereas the reverse primer included an *Xho*I cleavage site (CTCGAG) and a stop codon (underlined above). The PCR product was double-digested with *Eco*RI and *Xho*I and integrated between the *Eco*RI and *Xho*I sites of pGEX 4T-1. The BL21 strain of *E. coli* was used to reduce the proteolytic products of a GST fusion protein. GST-Nt was generated by induction with 0.5 mM isopropyl β-D-1-thiogalactopyranoside for 4 h at 25°C and purified with glutathione-Sepharose 4B beads (GE Healthcare).

A

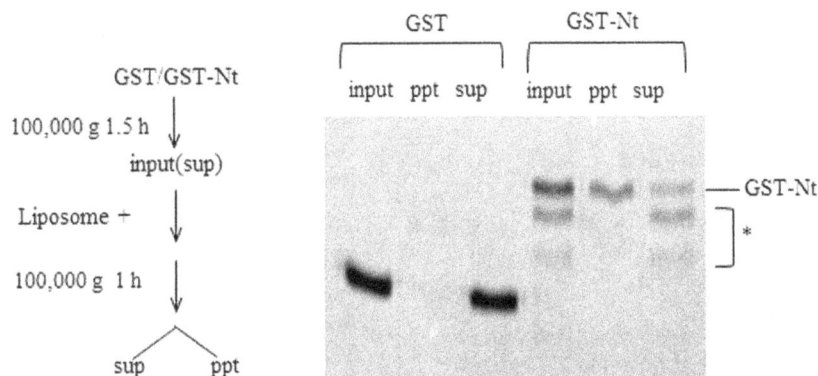

B

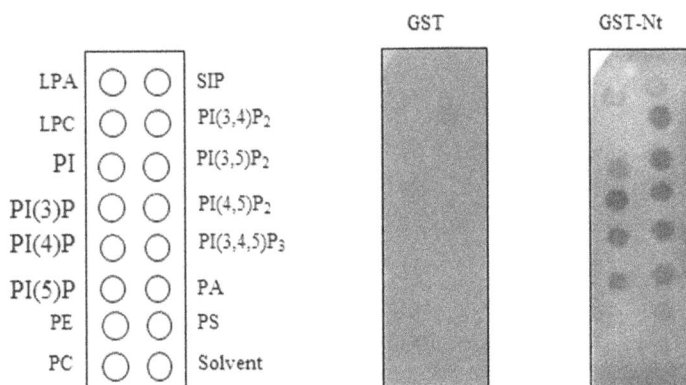

Figure 4. The interaction between the Nt and lipid bilayers. (A) Liposome co-sedimentation assay. Nt was produced as a fusion protein with glutathione transferase (GST-Nt) in the BL21 strain of *Escherichia coli* and purified with glutathione-Sepharose 4B beads. GST was used as a control. Diagram showing the liposome sedimentation assay. GST and and GST-Nt were centrifuged before use to sediment insoluble proteins and mixed with liposomes (input). After centrifugation, the liposome-bound (ppt) and liposome-unbound (sup) fractions were recovered. Aliquots of each fraction were separated on an SDS-PAGE and stained with Coomassie brilliant blue. (B) Lipid binding assay using the spot array method. Lipid-spotted membrane strips were incubated with GST or GST-Nt and the lipid-bound proteins were detected using anti-GST antibody. LPA, lysophosphatidic acid; PI, phosphatidylinositol; PI(3)P, phosphatidylinositol-(3)-phosphate; PI(4)P, phosphatidylinositol-(4)-phosphate; PI(5), phosphatidylinositol-(5)-phosphate; PE, phosphatidylethanolamine, PC, phosphatidylcholine; SIP, sphingosine-1-phosphate; PI(3,4)P₂; phosphatidylinositol-(3,4)-bisphosphate; PI(3,5)P₂, phosphatidylinositol-(3,5)-bisphosphate; PI(3,4,5)P₃, phosphatidylinositol-(3,4,5)-trisphosphate; PA, phosphatidic acid; PS, phosphatidylserine.

Liposome co-sedimentation assay

All surgery was performed under Ketamine/xylazine anesthesia, and all efforts were made to minimize suffering. Rat brains were homogenized in a methanol/chloroform mixture that included acetic acid, and the total lipid fraction was extracted and dried with nitrogen gas [21]. The dried lipids were hydrated in a binding buffer (25 mM HEPES-NaOH, pH 7.4, 100 mM NaCl, and 0.5 mM EDTA) at a final concentration of 2 mg/ml by shaking for 60 min at $37°C$. The hydrated lipid sample was then sonicated in a bath sonicator to generate liposomes. A liposome co-sedimentation assay was performed [22]. The GST-Nt and GST solutions were centrifuged for 90 min at $100,000 \times g$ at $4°C$ to remove any aggregates. The protein concentrations were measured using a BCA protein assay kit (Pierce) and diluted to 100 µg/ml. Subsequently, 250 µl of the liposome and protein solutions, respectively, were mixed and incubated for 20 min at room temperature. The protein/liposome mixture was centrifuged for 60 min at $100,000 \times g$ at room temperature to yield the supernatant (sup) and precipitate (ppt). ppt was resuspended in 500 µl of binding buffer. Aliquots of sup and ppt were subjected to SDS-PAGE and stained with Coomassie brilliant blue.

Lipid-binding assay with Membrane Lipid Strips

PIP strips and Membrane Lipid Strips (Echelon Biosciences Inc.) were used in this study. The membrane strips were prespotted with various lipids, and lipid-binding assays were performed [23]. 100 moles of each of the following lipids were spotted: lysophosphatidic acid (LPA), lysophosphocholine (LPC), phosphatidylinositol (PI), phosphatidylinositol-(3)-phosphate (PI(3)P), phosphatidylinositol-(4)-phosphate (PI(4)P), phosphatidylinositol-(5)-phosphate (PI(5)P), phosphatidylethanolamine (PE), phosphatidylcholine (PC), sphingosine-1-phosphate (SIP), phosphatidylinositol-(3,4)-bisphosphate $(PI(3,4)P_2)$, phosphatidylinositol-(3,5)-bisphosphate $(PI(3,5)P_2)$, phosphatidylinositol-(4,5)-bisphosphate $(PI(4,5)P_2)$, phosphatidylinositol-(3,4,5)-trisphosphate $(PI(3,4,5)P_3)$, phosphatidic acid (PA), and phosphatidylserine (PS). The strips were blocked with 3% bovine serum albumin in PBS-T [PBS (−) containing 0.02% Tween 20] and then incubated with GST-Nt or GST at a protein concentration of 0.5 µg/ml. GST-Nt and GST were detected using rabbit anti-GST antibody (MBL, Japan) and horseradish peroxidase-conjugated anti-rabbit immunoglobulin (the second antibody). Finally, the strips were immersed in a solution of Immobilon Western Chemiluminescent HRP Substrate (Merck Millipore) and exposed to X-ray films.

Results

Phactr3 localizes to the plasma membrane through its N-terminal region (Nt)

The entire sequence of Nt (1–52 aa), the functional roles of which are unknown, is encoded by exon2 of the *phactr3* gene (Fig. 1A). In HeLa cells, phactr3 was distributed throughout the cells, but it was frequently localized to the plasma membrane including the lamellipodia and membrane ruffles (Fig. 1B). Figure 1B compares its distribution pattern with that of α-actinin 4, which localizes to the lamellipodia and membrane ruffles [20]. In the present study, we introduced deletions and point mutations into GFP-phactr3 (Fig. 1) to determine the domain responsible for localization to the plasma membrane. GFP-phactr3 constructs were expressed in Cos7 cells and their distribution patterns were observed by fluorescent microscopy and confocal laser scanning microscopy. GFP-Phactr3-Wt was distributed in the plasma membrane including the lamellipodia and membrane ruffles [9]. The distribution pattern of GFP-phactr3-Wt in Cos7 cells was

essentially the same as that of phactr3 (no tag) in HeLa cells (Fig. 1B and 2B).

ΔNt greatly impaired localization to the plasma membrane. However, Nt could be localized to the plasma membrane by itself (Fig. 2). These results demonstrate that the localization of phactr3 to the plasma membrane requires Nt.

In addition to the lamellipodia and membrane ruffles, GFP-phactr3-Wt was often distributed in long/slender cytoplasmic extensions in Cos7 cells where it was concentrated at the tips of these extensions (Fig. 2B). We also analyzed these cytoplasmic extensions using mutants (see below).

N-terminal basic and hydrophobic residues are crucial for membrane targeting

To determine the N-terminal amino acid (aa) sequence required for the interaction with the plasma membrane, we substituted the aa residues in Nt with alanine (N1–12; Fig. 3A) and observed subcellular distributions. For example, the N1 (RKK>AAA) and N2 (WKW>AAA) mutants were predominantly distributed in the cytoplasm (Fig. 3B). We counted the number of cells that predominantly exhibited plasma membrane distributions or cytoplasm distributions using a fluorescent microscopy. In the triplet mutants N1, N2, N3, and N9, and the point mutants N4, N5, N6, and N7, the number of cells that localized to the plasma membrane was greatly or moderately reduced (Fig. 3C). These mutation experiments demonstrated that aa sequence of GRIFKPWKWRKK (31–42 aa) in Nt is crucial for localization to the plasma membrane. In addition, the peptide of SKL (25–27 aa) may be involved in membrane targeting.

A BH plot analysis (http://helixweb.nih.gov/bhsearch) is a program developed to identify or predict unstructured membrane-binding domains that comprise basic and hydrophobic residues in membrane proteins [24]. Therefore, we applied this program to phactr3 (HP_001186435, 518 aa). The BH plot analysis detected a peak score in Nt (Fig. 3D). The region with a BH score of >0.8 ranged from 29 to 45 aa, and the region with a BH score of >0.6 ranged from 25 to 49 aa. The results of the BH plot analysis agreed with the experimentally determined region that is responsible for the localization of phactr3 to the plasma membrane. Furthermore, the BH plot analysis indicated that a wide range of N-terminal polypeptides, including GRIFKPWKWRKK (31–42 aa), may be involved in the membrane targeting of phactr3. However, we could not exclude the possibility of contributions by other regions. A weak membrane-targeting signal is present within the C-terminal RPEL repeats (Fig. 3D).

Interaction between Nt and lipid layers

The following two possible molecular mechanisms may allow Nt to interact with the plasma membrane: direct or indirect binding (binding via other membrane proteins) with lipid bilayers. The BH search results indicated that Nt interacts directly with the lipid bilayers [24].

We examined whether Nt directly binds to lipid bilayers using a liposome co-sedimentation assay. First, the insoluble proteins were precipitated by centrifugation. Subsequently, GST-Nt (1–52 aa) or GST (control) were mixed with liposomes and separated into liposome (ppt) and soluble (sup) fractions by centrifugation (Fig. 4A). GST-Nt sedimented with the liposomes, whereas GST did not. It was notable that the cleaved products did not sediment with the liposomes. Although we used the BL21 strain of *E. coli* to express GST-Nt and to reduce proteolytic products, the GST-Nt samples always included cleaved products (star-shaped products).

Therefore, it appears that Nt is highly sensitive to cleavage by the endogenous bacterial protease.

Subsequently, we examined the species of lipids that interact with Nt using lipid spot array filters (Fig. 4B). This assay showed that Nt predominantly bound to acidic phospholipids but not with neutral phospholipids such as phosphatidylcholine and phosphatidylethanolamine. Nt did not bind to cholesterol and phosphatidylglycerol (data not shown).

Stimulation of the formation of long/slender cytoplasmic extensions in Cos7 cells

GFP-phactr3-Wt was often distributed in long/slender cytoplasmic extensions in Cos7 cells, where it was frequently concentrated at the tips of these extensions (Fig. 2B and 5A). We have previously shown that phactr3-expressing Hela cells frequently exhibited elongated shapes that were rarely seen in the parental Hela cells [9]. However, the long/slender extensions were not seen in HeLa cells that expressed phactr3 (no tag) and GFP-phactr3-Wt [9,19]; therefore, indicating that the formation of these structures is cell type-dependent. Long/slender cytoplasmic extensions were observed in Cos7 cells that expressed GFP, although the number of cytoplasmic extensions was less than that with GFP-phactr3-Wt. We compared the number of cytoplasmic extensions in Cos7 cells that expressed wild-type or mutant phactr3 at 24 h after transfection (Fig. 5B). The number of cytoplasmic extensions was reduced in Cos7 cells that expressed N1 and N2 (mutants with incomplete membrane-targeting sequences), but it was not reduced in Cos7 cells that expressed N8 and N11 (mutants with the complete membrane-targeting sequence). In addition, the N10 and N12 mutants with complete membrane-targeting sequences stimulated the formation of long/slender cytoplasmic extensions, whereas the N3, N5, N6, and N7 mutants with incomplete membrane-targeting sequences failed to stimulate the formation of long/slender cytoplasmic extensions (data not shown). Nuclear morphologies that are characteristic to apoptosis or necrosis were not seen in Cos7 cells that expressed wild-type or mutants of phactr3. These results indicate that the integrity of the membrane-targeting sequence in the N-terminus is required to stimulate the formation of long/slender cytoplasmic extensions in Cos7 cells. It should be noted that GFP-Nt did not stimulate the formation of these cytoplasmic extensions; therefore, implying that Nt is sufficient for membrane targeting whereas it is insufficient for stimulating the formation of long/slender cytoplasmic extensions in Cos7 cells.

Discussion

The phactr protein family is considered to be involved with cell migration and morphogenesis by modulating the actin cytoskeleton, but their regulatory mechanisms are poorly understood [8–10,12]. Our experiments with deletion mutants demonstrated that Nt is required for the localization of phactr3 to the plasma membrane (Fig. 2).

Phactr3 lacks fatty acid-attachment sites (e.g., palmitoylation, prenylation, and myristorylation sites) and lipid-binding domains with defined ternary structures (e.g., a PH domain). The mutation experiments showed that the N-terminal peptide (31–42 aa), which comprises basic and hydrophobic residues, is critical for localization to the plasma membrane (Fig. 3C). The BH plot analysis, a program that identifies membrane-targeting domains containing unstructured clusters of basic and hydrophobic residues in membrane proteins, supported this result (Fig. 3D). The BH plot analysis also indicated that the N-terminal sequence can directly interact with lipid bilayers without any requirement for other

membrane proteins. We also performed an *in vitro* analysis and showed that Nt interacts with the liposome without the involvement of other membrane proteins (Fig. 4A); therefore, demonstrating the direct interaction between phactr3 and lipid bilayers.

The region with a BH score of >0.8 ranged from 29 to 45 aa, and the region with a BH score of >0.6 ranged from 25 to 49 aa. This indicates that although the 31–42 aa sequence in Nt is the most critical for interactions, other aa residues adjacent to the critical sequence may be involved in membrane interactions. Analyses using synthetic peptides are necessary to determine the precise ranges of the membrane-interacting domains.

Point mutations of phenylalanine (N5) or tryptophan (N6) in the membrane interaction sequence resulted in impaired localization to the plasma membrane. This indicates that these hydrophobic residues play critical roles in the interactions with the plasma membrane possibly by inserting their aromatic head groups into lipid bilayers. Analyses using lipid blot array filters demonstrated that Nt of phactr3 binds to a broad range of acidic phospholipids (Fig. 4B). It is well known that the clusters of basic residues in membrane proteins are involved in electrostatic interactions with the negative surface charge on the inner leaflet of the plasma membrane [25–27]. The aa sequence of GRIFKPWKWRKK (31–42 aa) was shown to be critical for localization to the plasma membrane and it includes six basic residues (Fig. 3A). Nt with a BH score of >0.6 included 10 basic residues.

Some actin regulatory proteins, such as N-WASP, WAVE, profilin, and cofilin, selectively bind to specific species of phosphoinositides such as phosphatidylinositol-(4,5)-bisphosphate and phosphatidylinositol-(3,4,5)-trisphosphate [28–29]. The selective recognition of phospholipids facilitates the controlled (spatial and temporal) recruitment of these actin regulatory proteins to specific membrane domains. However, Nt of phactr3 nonselectively binds to negatively charged phospholipids. This nonselective recognition mode may allow the constitutive recruitment of phactr3 to the plasma membrane and/or intracellular membrane [27].

The experimentally determined membrane-targeting sequence is highly conserved in phactr proteins from insects to humans and partly conserved in that from worms (Fig. 3A). The BH plot analysis showed that the highly or partly conserved sequence of all members of the phactr family is probably the membrane-binding site. Huet et al., [15] demonstrated that phactr4 localizes to the plasma membrane through its Nt (1–111 aa). They measured the Forster resonance energy transfer between GFP-Phactr4 and mCherry-CAAX (a plasma membrane probe) to assess the interaction between phactr4 and the plasma membrane. Their conclusion appears to be in agreement with our results, although they did not characterize the membrane interaction sequence. Therefore, it is likely that the N-terminal conserved sequence acts as the membrane-targeting sequence in all members of the phactr family.

GFP-phactr3-Wt stimulated the formation of long/slender cytoplasmic extensions in Cos7 cells (Fig. 2B and 5). GFP-phactr3 mutants with incomplete membrane-targeting sequences failed to stimulate the formation of these cytoplasmic extensions, whereas GFP-phactr3 mutants with the complete membrane-targeting sequence resulted in morphological changes that were similar to those with GFP-phactr3-Wt. GFP-Nt did not increase the number of long/slender cytoplasmic extensions (Fig. 5); therefore, implying that Nt is sufficient for membrane targeting but that it is insufficient to stimulate the formation of long/slender cytoplasmic extensions in Cos7 cells (Fig. 5). Therefore, the interactions with PP1 and actin via the C-terminal region and the interaction with

A

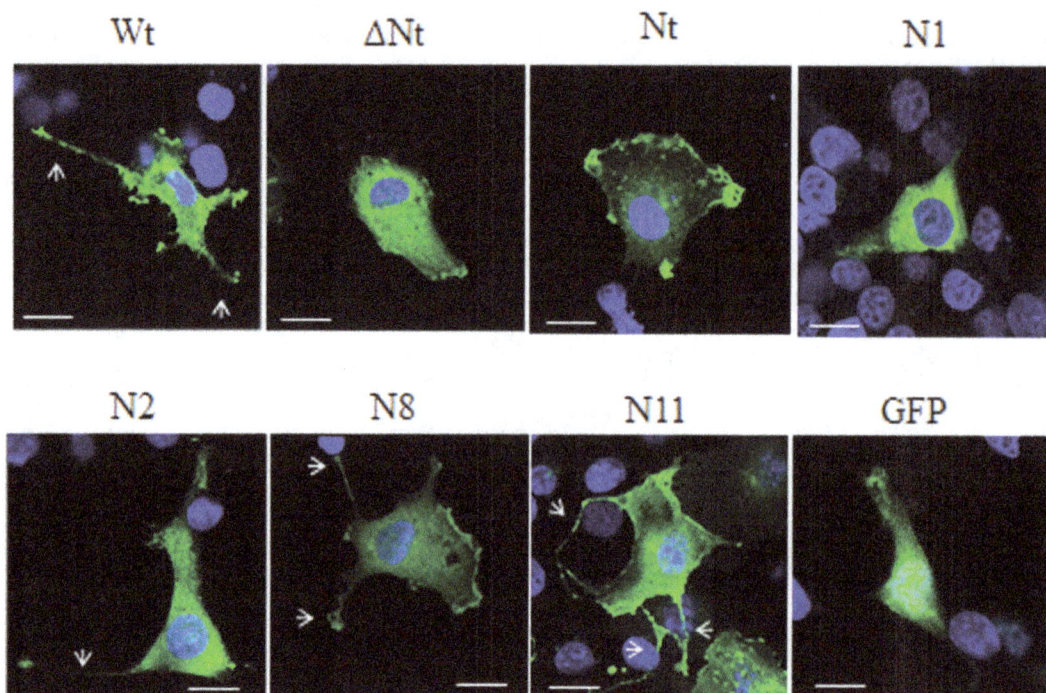

B

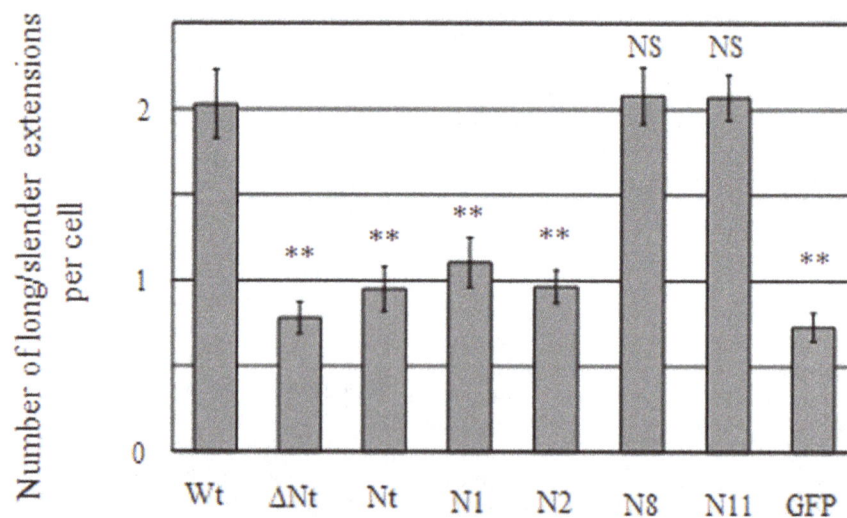

Figure 5. Long/slender cytoplasmic extensions induced by the expression of GFP-phactr3 and its mutants in Cos 7 cells. (A) Distribution patterns of GFP-phactr3 Wt and mutants in Cos7 cells. Cos7 cells expressing Wt or each mutant were imaged using confocal laser microscopy at 24 h after transfection. The nucleus was stained with Hoechst 33342. The arrows show long/slender cytoplasmic extensions. (B) The number of long/slender extensions in Cos7 cells expressing each mutant. Each plasmid was transfected into Cos7 cells and the number of long/slender extensions per cell was counted by fluorescent microscopy at 24 h after transfection. Mean $\pm$ standard error (SE). The asterisks in the histograms indicate significant differences between Wt and each mutant (Student's t-test; $**P<0.005$). NS, not significant.

the plasma membrane via Nt may be necessary for inducing morphological changes in Cos7 cells.

G-actin and PP1 competitively bind to the C-terminal regions of phactr proteins and the cytoplasmic G-actin concentration is determined by RPEL motifs, which control the formation of the phactr-PP1 complex [8,15]. Previous studies suggest that the phactr-PP1complex modulates the phosphorylation status of cofilin or myosin; therefore, regulating actin cytoskeleton dynamics [8,12,15]. Overall, these results suggest that phactr protein is a membrane-associated PP1 regulator, which is itself regulated by signal-induced changes in the cytoplasmic G-actin levels. phactr1 modulates lamellipodial dynamics in human endothelial cells, which are regulated by dynamic interactions between the plasma membrane and actin cytoskeleton [13]. Therefore, it is likely that the phactr-PP1 complex modulates the phosphorylation status of actin cytoskeleton regulators such as cofilin and myosin, which associate with the plasma membrane.

Acknowledgments

We thank Drs. Shigeaki Hida and Naoki Itano for their kind assistance.

Author Contributions

Conceived and designed the experiments: AI ST JS. Performed the experiments: AI AU JS. Analyzed the data: AI AU JS. Contributed reagents/materials/analysis tools: ST JS. Contributed to the writing of the manuscript: AI AU JS.

References

1. Allen PB, Greenfield AT, Svenningsson P, Haspeslagh DC, Greengard P (2004) Phactrs 1–4: A family of protein phosphatase 1 and actin regulatory proteins. Proc. Natl. Acad. Sci. U.S.A. 101: 7187–7192. doi: 10.1073/pnas.0401673101.
2. Myocardial infraction Consortium (2009) Genome-wide association of early-onset myocardial infarction with single nucleotide polymorphisms and copy number variants. Nat. Genet. 41: 334–341. doi:10.1038/ng.327.
3. Wider C, Lincoln SJ, Heckman MG, Diehl NN, Stone JT, et al. (2009) Phactr2 and Parkinson's disease. Neurosci Lett 453: 9–11. doi: 10.1016/j.neulet.2009.02.009.
4. Solimini NL, Liang AC, Xu C, Pavlova NN, Xu Q, et al. (2013) STOP gene Phactr4 is a tumor suppressor. Proc Natl Acad Sci USA 110: E407–414. doi: 10.1073/pnas.1221385110.
5. Mouilleron S, Wiezlak M, O'Reilly N, Treisman R, McDonald NQ (2012) Structures of the Phactr1 RPEL domain and RPEL motif complexes with G-actin reveal the molecular basis for actin binding cooperativity. Structure 20: 1960–1970. doi: 10.1016/j.str.2012.08.031.
6. Miralles F, Posern G, Zaromytidou AU, Treisman R (2003) Actin dynamics control SRF activity by regulation of its coactivator MAL. Cell 113: 329–342. doi: 10.1016/S0092-8674(03)00278-2.
7. Vartiainen MK, Guettler S, Larijani B, Treisman R (2007) Nuclear actin regulates dynamic subcellular localization and activity of the SRF cofactor MAL. Science 316: 1749–1752. doi: 10.1126/science.1141084.
8. Wiezlak M, Diring J, Abella J, Mouilleron S, Way M, et al. (2012) G-actin regulates the shuttling and PP1 binding of the RPEL protein Phactr1 to control actomyosin assembly. J Cell Sci 125: 5860–5872. doi: 10.1242/jcs.112078.
9. Sagara J, T. Arata T, Taniguchi S (2009) Scapinin, the protein phosphatase 1 binding protein, enhances cell spreading and motility by interacting with the actin cytoskeleton. PLoS One 4: e4247. doi:10.1371/journal.pone.0004247.
10. Favot L, Gillingwater M, Scott C, Kemp PR (2005) Overexpression of a family of RPEL proteins modifies cell shape. FEBS Lett 579: 100–104. doi: 10.1016/j.febslet.2004.11.054.
11. Farghaian H, Chen Y, Fu AW, Fu AK, Ip JP, et al. (2011) Scapinin-induced inhibition of axon elongation is attenuated by phosphorylation and translocation to the cytoplasm. J Biol Chem 286: 19724–19734. doi: 10.1074/jbc.M110.205781.
12. Zhang Y, Kim TH, Niswander L (2012) Phactr4 regulates directional migration of enteric neural crest through PP1, integrin signaling, and cofilin activity. Genes Dev 26: 69–81. doi: 10.1101/gad.179283.111.
13. Allain B, Jarray R, Borriello L, Leforban B, Dufou S, et al. (2012) Neuropilin-1 regulates a new VEGF-induced gene, Phactr-1, which controls tubulogenesis and modulates lamellipodial dynamics in human endothelial cells. Cell Signal 24: 214–223. doi: 10.1016/j.cellsig.2011.09.003.
14. Fils-Aimé N, Dai M, Guo J, El-Mousawi M, Kahramangil B, et al. (2013) MicroRNA-584 and the protein phosphatase and actin regulator 1 (PHACTR1), a new signaling route through which transforming growth factor-β Mediates the migration and actin dynamics of breast cancer cells. J Biol Chem 288: 11807–11823. doi: 10.1074/jbc.M112.430934.
15. Huet G, Rajakylä EK, Viita T, Skarp KP, Crivaro M, et al. (2013) Actin-regulated feedback loop based on Phactr4, PP1 and cofilin maintains the actin monomer pool. J Cell Sci 126: 497–507. doi: 10.1242/jcs.113241.
16. Bollen M, Stalmans W (1992) The structure, role, and regulation of type 1 protein phosphatases, Crit. Rev. Biochem. Mol Biol 27: 227–281. doi: 10.3109/10409239209082564.
17. Alessi D, MacDougall LK, Sola MM, Ikebe M, Cohen P (1992) The control of protein phosphatase-1 by targetting subunits. The major myosin phosphatase in avian smooth muscle is a novel form of protein phosphatase-1. Eur J Biochem 210: 1023–1035. doi: 10.1111/j.1432-1033.1992.tb17508.x.
18. Ambach A, Saunus J, Konstandin M, Wesselborg S, Meuer SC, et al. (2000) The serine phosphatases PP1 and PP2A associate with and activate the actin-binding protein cofilin in human T lymphocytes. Eur J Immunol 30: 3422–3431. doi: 10.1002/1521-4141(2000012)30: 12<3422::AID-IMMU3422>3.0.CO; 2-J.
19. Sagara J, Higuchi T, Hattori Y, Moriya M, Sarvotham H, et al. (2003) Scapinin, a putative protein phosphatase-1 regulatory subunit associated with the nuclear nonchromatin structure. J Biol Chem 278: 45611–45619. doi: 10.1074/jbc.M305227200.
20. Honda K, Yamada T, Endo R, Ino Y, Gotoh M, et al. (1998) Actinin-4, a novel actin-bundling protein associated with cell motility and cancer invasion. J Cell Biol 140: 1383–1393. doi: 10.1083/jcb.140.6.1383.
21. Bligh EG, Dyer WJ (1959) A rapid method of total lipid extraction and purification. Can J Biochem Physiol 37: 911–917. doi: 10.1139/o59-099.
22. Patki V, Virbasius J, Lane WS, Toh BH, Shpetner HS, et al. (1997) Identification of an early endosomal protein regulated by phosphatidylinositol 3-kinase. Proc Natl Acad Sci USA 94: 7326–7330.
23. Kavran JM, Klein DE, Lee A, Falasca M, Isakoff SJ, et al. (1998) Specificity and promiscuity in phosphoinositide binding by pleckstrin homology domains. J Biol Chem 273: 30497–30508. doi: 10.1074/jbc.273.46.30497.
24. Brzeska H, Guag J, Remmert K, Chacko S, Korn ED (2010) An experimentally based computer search identifies unstructured membrane-binding sites in proteins: application to class I myosins, PAKS, and CARMIL. J Biol Chem 285: 5738–5747. doi: 10.1074/jbc.M109.066910.
25. Niggli V (2001) Structural properties of lipid-binding sites in cytoskeletal proteins. Trends Biochem Sci 26: 604–611. doi: 10.1016/S0968-0004(01)01927-2.
26. McLaughlin S, Murray D (2005) Plasma membrane phosphoinositide organization by protein electrostatics. Nature 438: 605–611. doi:10.1038/nature04398.
27. Moravcevic K, Oxley CL, Lemmon MA (2012) Conditional peripheral membrane proteins: facing up to limited specificity. Structure 20: 15–27. doi: 10.1016/j.str.2011.11.012.
28. Yin HL, Janmey PA (2003) Phosphoinositide regulation of the actin cytoskeleton. Annu Rev Physiol 65: 761–789. doi: 10.1146/annurev.physiol.65.092101.142517.
29. Takenawa T, Suetsugu S (2007) The WASP-WAVE protein netwaok: connecting the membrane to the cytoskeleton. Nat Rev Mol Cell Biol 8: 37–48. doi:10.1038/nrm2069.

19

Genetic Analysis of Loop Sequences in the Let-7 Gene Family Reveal a Relationship between Loop Evolution and Multiple IsomiRs

Tingming Liang[1], Chen Yang[1], Ping Li[1], Chang Liu[1], Li Guo[2]*

1 Jiangsu Key Laboratory for Molecular and Medical Biotechnology, College of Life Science, Nanjing Normal University, Nanjing, 210023, China, **2** Department of Epidemiology and Biostatistics, School of Public Health, Nanjing Medical University, Nanjing, 211166, China

Abstract

While mature miRNAs have been widely studied, the terminal loop sequences are rarely examined despite regulating both primary and mature miRNA functions. Herein, we attempted to understand the evolutionary pattern of loop sequences by analyzing loops in the let-7 gene family. Compared to the stable miRNA length distributions seen in most metazoans, higher metazoan species exhibit a longer length distribution. Examination of these loop sequence length distributions, in addition to phylogenetic tree construction, implicated loop sequences as the main evolutionary drivers in miRNA genes. Moreover, loops from relevant clustered miRNA gene families showed varying length distributions and higher levels of nucleotide divergence, even between homologous pre-miRNA loops. Furthermore, we found that specific nucleotides were dominantly distributed in the 5' and 3' terminal loop ends, which may contribute to the relatively precise cleavage that leads to a stable isomiR expression profile. Overall, this study provides further insight into miRNA processing and maturation and further enriches our understanding of miRNA biogenesis.

Editor: John J. Rossi, Beckman Research Institute of the City of Hope, United States of America

Funding: This work was supported by National Basic Research Program of China (973 Program) (No. 2012CB947600 and 2013CB911600), the National Natural Science Foundation of China (No. 31171137, 31271261, 31401009 and 61301251), the Program for New Century Excellent Talents in University by the Chinese Ministry of Education (No. NCET-11- 0990), the Program for the Top Young Talents by the Organization Department of the CPC Central Committee, the Collaborative Innovation Center for Cardiovascular Disease Translational Medicine (Nanjing Medical University), the Research Fund for the Doctoral Program of Higher Education of China (No. 20133234120009), the National Natural Science Foundation of Jiangsu (No. BK20130885), the Natural Science Foundation of the Jiangsu Higher Education Institutions (No. 12KJB360001 and 13KJB330003), NSFC for Talents Training in Basic Science (J1103507), and the Priority Academic Program Development of Jiangsu Higher Education Institutions (PAPD). The funders had no role in study design, data collection and analysis, decision to publish, or preparation of the manuscript.

Competing Interests: The authors have declared that no competing interests exist.

* Email: gl8008@163.com

Introduction

MicroRNAs (miRNAs), a class of small non-coding RNA, are widely studied as crucial regulatory molecules able to modulate broad regulatory networks at the post-transcriptional levels [1,2]. miRNA is generated from primary miRNA (pri-miRNA) and precursor miRNA (pre-miRNA) in animals [3], with the pre-miRNA presenting a stable stem-loop structure. Both of the two arm products, miR-#-5p and miR-#-3p, have been reported to form mature and functional miRNAs [4–8], with the loop sequences connecting miR-#-5p and miR-#-3p in the stem-loop structure typically degraded during miRNA biogenesis. These transitory intermediates are rarely examined due to an unseen direct role in the miRNA regulation process. Instead, most studies have focused on the functional and regulatory roles of miRNAs during miRNA-mRNA recognition as it relates to expression or translational repression. However, indeed, both pri-miRNA and pre-miRNA may contribute to the regulatory process [9]. Specifically, the loop nucleotides may tune and alter miRNA activity, controlling the processing precision during the miRNA maturation process [9].

Recently, the potential impact of hairpin loops has been examined, with short hairpin RNA (shRNA) loops possibly influencing effectivity [10–12] and alternative loop conformations in miRNAs potentially effecting expression [13,14]. The loop sequence may also affect Dicer recognition and possibly specificity, thus affect miRNA cleavage during the maturation process [11,15,16]. As mentioned above, the loop sequence in pre-miRNAs has an important role in regulating the activities and specificities of related molecules that can facilitate Drosha and Dicer, with a mutation in the loop possibly affecting miRNA processing [17,18]. Furthermore, miRNAs have been a key focal point because a miRNA locus can yield multiple sequences have diverse 5' and/or 3' ends and expression levels [19–25]. These miRNA variants, or physiological miRNA isoforms (isomiRs), may be mainly generated by imprecise Drosha and Dicer cleavage during pre-miRNA processing [22]. Additionally, the loop sequences may also impact pre-miRNA processing by affecting Drosha and/or Dicer directly or indirectly. Even though pre-miRNA loop sequences have been shown to directly or indirectly impact Drosha and Dicer during miRNA processing, few studies have performed an evolutionary analysis of the loop sequences,

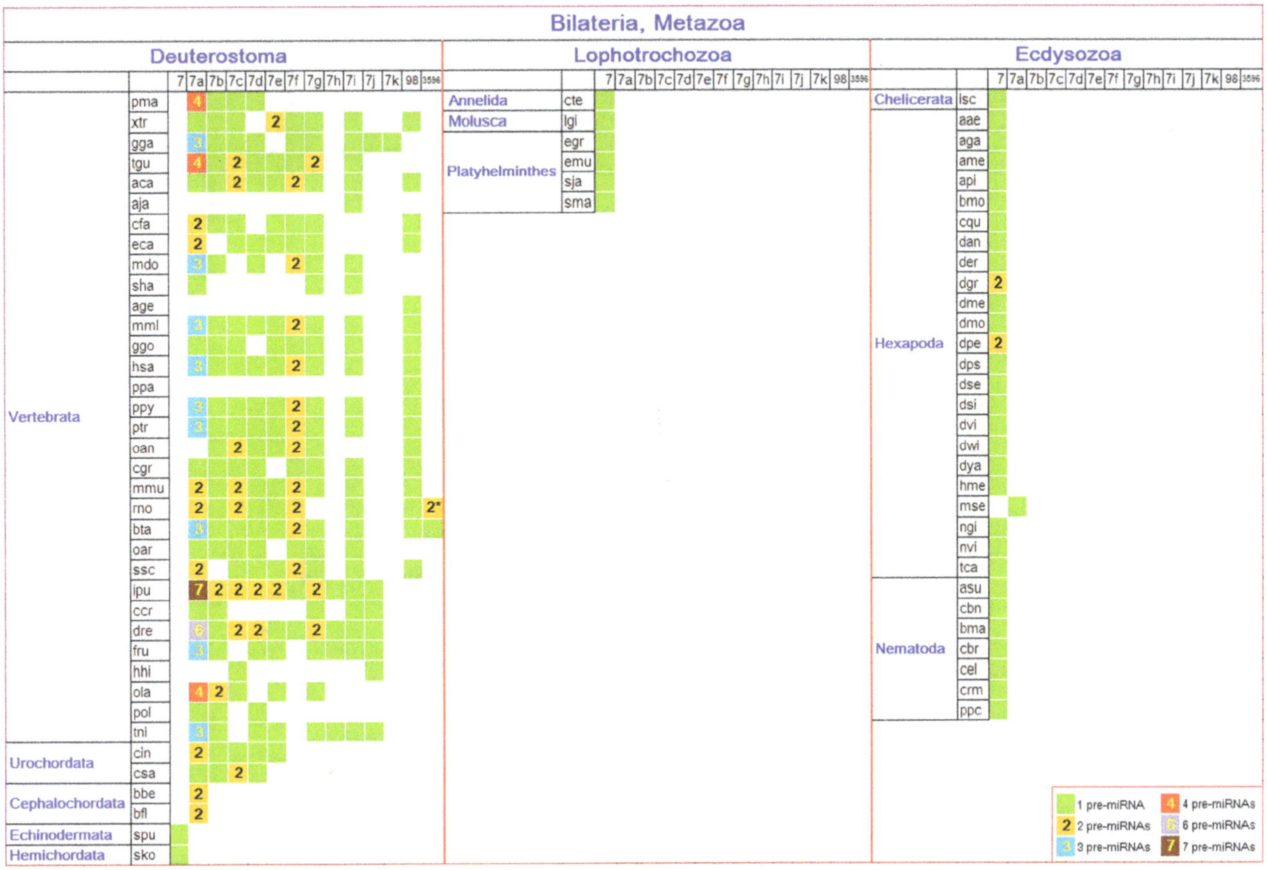

Figure 1. The distribution of the let-7 family among metazoan animals. The let-7 gene was named and homologous miRNAs were further found and named. These miRNA names were consistent with the current annotations in the miRBase database. "2*" indicates that there are two genes, rno-mir-3596b and rno-mir-3596d, and not two pre-miRNAs for the mature miRNA.

especially one focused on homology and clustered miRNAs across different animal species.

Herein, we attempt to perform an evolutionary analysis of the loop sequences in let-7 and locate related miRNAs across different animal species. The let-7 gene family has been widely detected in metazoans, with its associated miRNAs thought to have co-expanded with the HOX gene clusters [26]. A series of homologous miRNAs can be found in a miRNA gene family, including multicopy pre-miRNAs from specific animal species. Simultaneously, some members can be located in a cluster with other related miRNAs, with miRNAs within a close proximity on a chromosome being co-transcribed as a polycistronic transcript [27–30]. Therefore, a classical miRNA gene family and related miRNAs are typically selected to track and reveal the evolutionary patterns of the ever present yet ignored loop sequences. Thus, this study will examine the roles of loops during miRNA maturation and processing.

Materials and Methods

MiRNA members in the let-7 gene family, to include homologous miRNAs and miRNAs from different animal species, were obtained from the miRBase database (http://www.mirbase. org/cgi-bin/mirna_summary.pl?fam=MIPF0000002). According to the location distributions, some were located within a gene cluster and all pre-miRNA and miRNA let-7 sequences and other related miRNAs were simultaneously collected.

pre-miRNA loop sequences were collected, including miR-#-5p and miR-#-3p sequences. If the miR-#-5p or miR-#-3p sequences were not annotated in the miRBase database, the complementary antisense miRNA consensus sequences was collected following alignment with the Clustal X 2.0 software [31]. Despite the sequence diversity of miRNAs among different animal species, miRNAs are phylogenetically conserved, particularly in the functional seed sequences (nucleotides 2–8). miRNAs can also vary in their length distributions, which can be attributed to being derived from various 3' ends. Additionally, multiple isomiRs have been widely detected from miRNA loci, which indicates that flexible 5' and 3' ends mediate these loop sequences. The present study mainly focused on the core loop sequences. The 5' and 3' ends of loop sequences are cleavage sites of Dicer, with previous studies showing that imprecise cleavage by Drosha and Dicer can lead isomiRs generation via biased cleavage [24,32,33]. Based on this phenomenon, in addition to biased cleavage, the continuous 5, 3 and 1 nucleotides in the 5' and 3' ends of the loop sequences were specifically analyzed. To understand the effect of changes in the pre-miRNA loop sequences, the minimum free energy of relevant pre-miRNA sequences were predicted using the RNAfold webserver [34].

Phylogenetic trees were constructed using pre-miRNAs and loop sequences were constructed based on the Neighbor-Joining (NJ) method using the MEGA 6.06 software [35]. Evolutionary relationships among loop sequences within the hsa-let-7 gene family were reconstructed using SplitsTree 4.10 [36]. Sequence

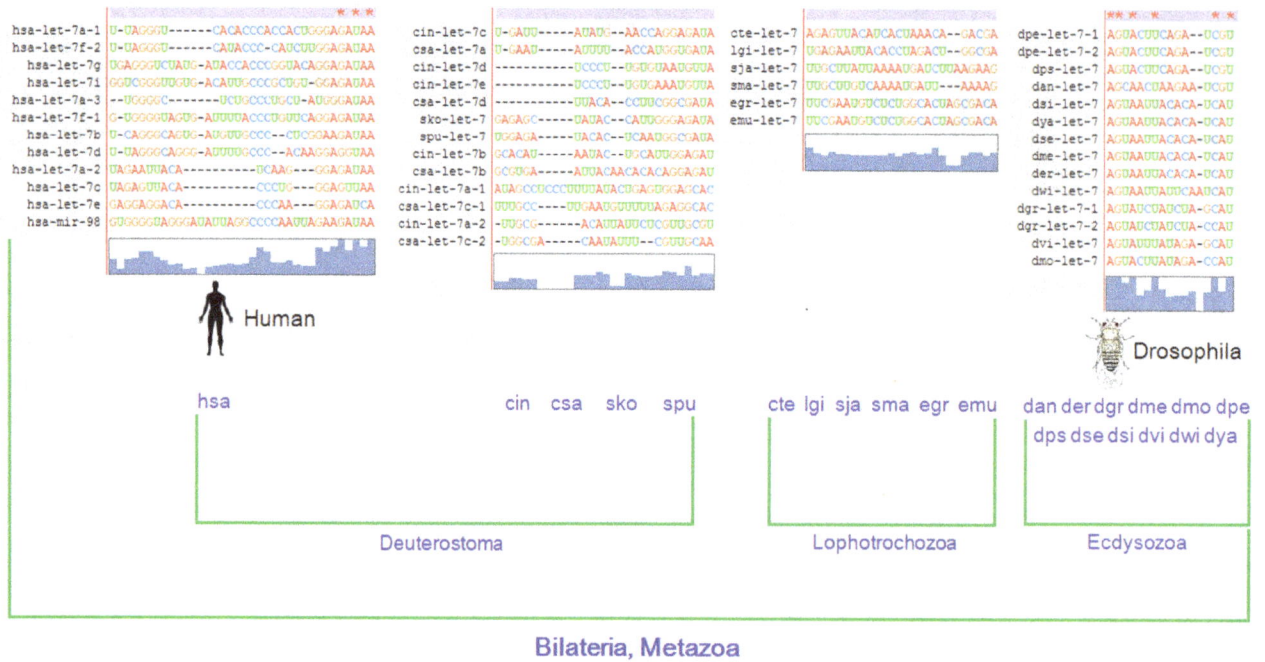

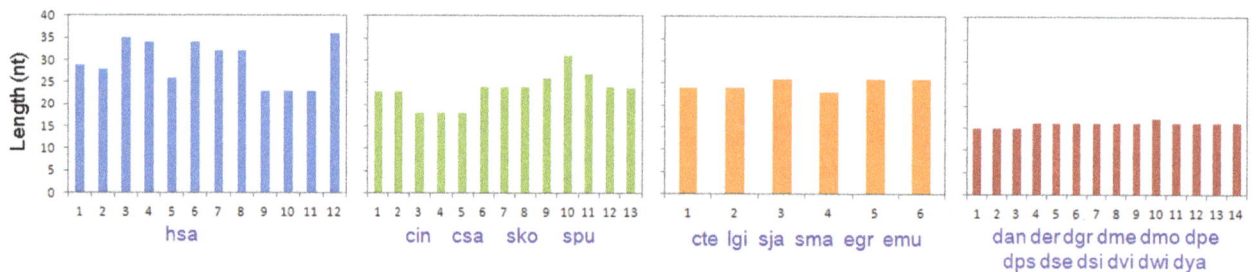

Figure 2. Sequence diversity and length distribution of the loop sequences among different animal species. (aae, *Aedes aegypti*; aca, *Anolis carolinensis*; aga, *Anopheles gambiae*; age, *Ateles geoffroyi*; aja, *Artibeus jamaicensis*; ame, *Apis mellifera*; api, *Acyrthosiphon pisum*; asu, *Ascaris suum*; bbe, *Branchiostoma belcheri*; bfl, *Branchiostoma floridae*; bma, *Brugia malayi*; bmo, *Bombyx mori*; bta, *Bos taurus*; cbn, *Caenorhabditis brenneri*; cbr, *Caenorhabditis briggsae*; ccr, *Cyprinus carpio*; cel, *Caenorhabditis elegans*; cfa, *Canis familiaris*; cgr, *Cricetulus griseus*; cin, *Ciona intestinalis*; cqu, *Culex quinquefasciatus*; crm, *Caenorhabditis remanei*; csa, *Ciona savignyi*; cte, *Capitella teleta*; dan, *Drosophila ananassae*; der, *Drosophila erecta*; dgr, *Drosophila grimshawi*; dme, *Drosophila melanogaster*; dmo, *Drosophila mojavensis*; dpe, *Drosophila persimilis*; dps, *Drosophila pseudoobscura*; dre, *Danio rerio*; dse, *Drosophila sechellia*; dsi, *Drosophila simulans*; dvi, *Drosophila virilis*; dwi, *Drosophila willistoni*; dya, *Drosophila yakuba*; eca, *Equus caballus*; egr, *Echinococcus granulosus*; emu, *Echinococcus multilocularis*; fru, *Fugu rubripes*; gga, *Gallus gallus*; ggo, *Gorilla gorilla*; hhi, *Hippoglossus hippoglossus*; hme, *Heliconius melpomene*; hsa, *Homo sapiens*; ipu, *Ictalurus punctatus*; isc, *Ixodes scapularis*; lgi, *Lottia gigantean*; mdo, *Monodelphis domestica*; mml, *Macaca mulatta*; mmu, *Mus musculus*; mse, *Manduca sexta*; ngi, *Nasonia giraulti*; nvi, *Nasonia vitripennis*; oan, *Ornithorhynchus anatinus*; oar, *Ovis aries*; ola, *Oryzias latipes*; pma, *Petromyzon marinus*; pol, *Paralichthys olivaceus*; ppa, *Pan paniscus*; ppc, *Pristionchus pacificus*; ppy, *Pongo pygmaeus*; ptr, *Pan troglodytes*; rno, *Rattus norvegicus*; sha, *Sarcophilus harrisii*; sja, *Schistosoma japonicum*; sko, *Saccoglossus kowalevskii*; sma, *Schistosoma mansoni*; spu, *Strongylocentrotus purpuratus*; ssc, *Sus scrofa*; tca, *Tribolium castaneum*; tgu, *Taeniopygia guttata*; tni, *Tetraodon nigroviridis*; xtr, *Xenopus tropicalis*).

logos of relevant loop sequences across homologous miRNAs or clustered miRNAs were analyzed using the WebLogo program (http://weblogo.berkeley.edu/logo.cgi) [37]. An un-paired t test was used to estimate length distribution differences between the two groups and the chi-square (χ^2) test was performed to estimate differences in loop nucleotide composition. Length distribution differences among multiple groups were estimated using ANOVA analysis ($P<0.05$), followed by a q test if a significant P-values was obtained. Adjusted P-values were obtained based on pairwise comparison. All the relevant statistical analysis was performed using the Stata software (Version 11.0).

Results

Overview of the distribution and sequence characteristics of let-7 loop sequences

Let-7 gene family members have been found in 75 metazoan animal species (Figure 1). Multiple homologous miRNA genes, some being multicopy pre-miRNAs, were more prevalent in vertebrates and urochordates, while single miRNA genes were more common in other metazoan species (Figure 1). Among these miRNAs, let-7a showed higher pre-miRNA multicopy numbers (1–7) relative to other homologous miRNAs. Although these multicopy pre-miRNAs could generate the same let-7a-5p

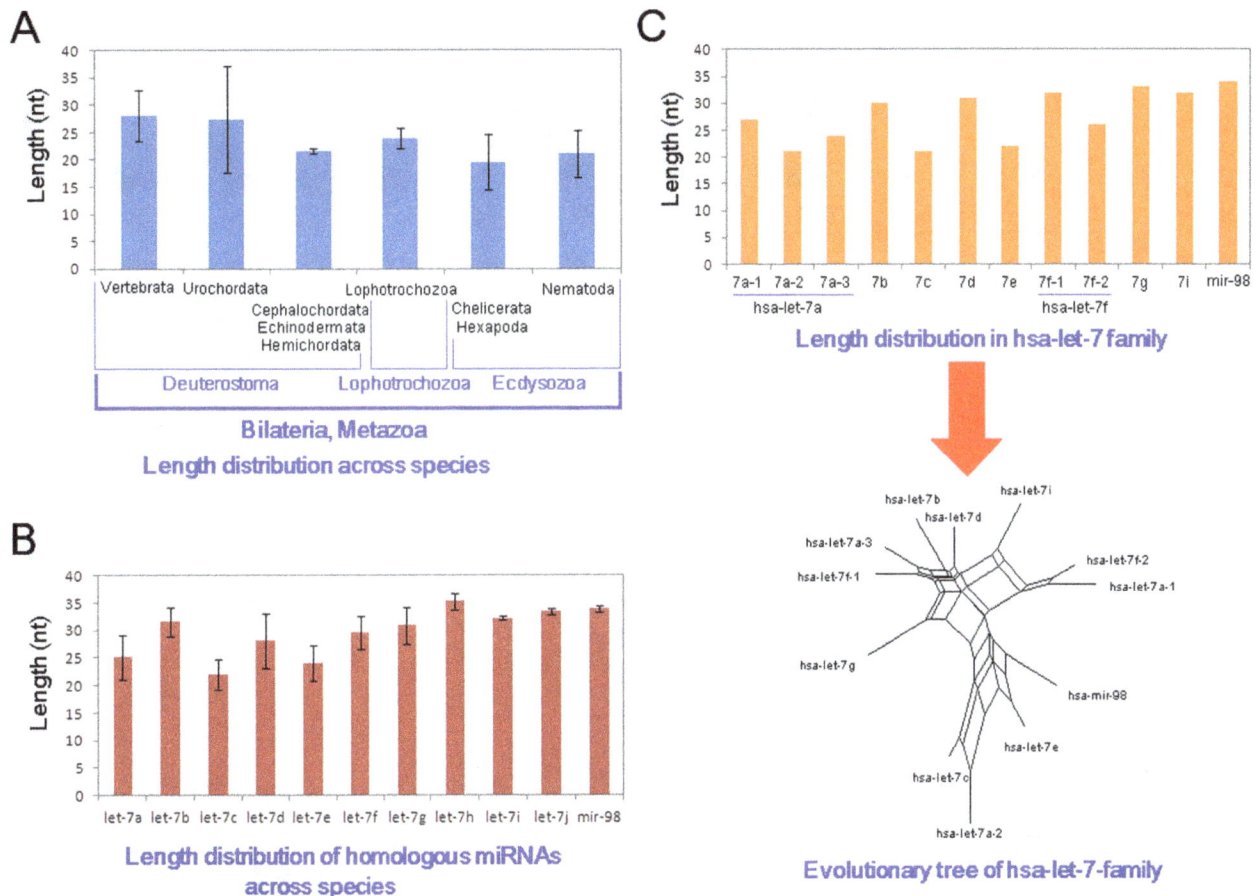

Figure 3. Length distributions of loops and their relevant evolutionary tree. (A) Length distribution of the let-7 loop among different animal species; (B) length distribution of loops in homologous miRNAs within the let-7 family across different animal species; and (C) length distribution of homologous hsa-let-7 loops and relevant evolutionary tree.

sequences, their loop sequences showed larger genetic distances (Figure S1 in File S1). Phylogenetic tree construction showed that the pre-miRNA multicopy loop sequences may be divided into two clusters based on loop and pre-miRNA sequences, with distribution differences noted (Figure S1 in File S1).

Although loop sequences were believed to not be well-conserved relative to the miR-#-5p and miR-#-3psequences, loop sequences in Drosophila were found to be highly conserved (Figure 2A). Loop sequences among different species or from homologous miRNAs or multicopy pre-miRNAs showed higher levels of sequence divergence. Nucleotide insertions/deletions were common in the loop sequences, even between homologous miRNAs in the same species or between multicopy pre-miRNAs. Interestingly, varying length distributions between loop sequences from different animal species were noted (Figure 2B, $P_{adj}<0.05$, except between Urochordata and Lophotrochozoa, Table S1 in File S1).

Based on the whole let-7 population, changes in length distribution across different species were found to be significant ($F = 19.14$, $P<0.0001$, Figure 3A). The longest average loop length was found in vertebrates (28.13 ± 4.66 nts), followed by those in urochordates (27.36 ± 9.78 nts), which were longer loop sequences than those found in other metazoans (Figure 3A). Each member in the let-7 gene family showed an inconsistent length distribution (Figure 3), despite being homologous miRNAs with higher sequence similarity. Specifically, let-7a (25.12 ± 3.99 nts), let-7c (22.03 ± 2.81 nts) and let-7e (24.17 ± 3.21 nts) had shorter

length distributions than other homologous miRNAs (from 28.12 ± 4.87 nts to 35.25 ± 1.50 nts) ($F = 39.97$, $P<0.0001$, Figure 3B). To further examine potential relationships between length distributions and evolutionary patterns, hsa-let-7 family loop sequences were analyzed. Loop sequences from multicopy pre-miRNAs showed larger divergence lengths, with these relevant loop sequences located in different evolutionary clusters (Figure 3C and Figure S1C in File S1). Larger genetic distances were noted in homologous miRNAs, with shorter loops from the let-7a-2, let-7c and let-7e genes located in a single cluster (Figure 3C and Figure S1C in File S1).

Nucleotide characteristics in loop sequence

According to the 10 nucleotides collected from the 5′ and 3′ ends of the loop sequences, guanine was the most prevalent (40.99%), followed by adenine (29.09%), while the cytosine presence was very low (5.32%) (Table S2 in File S1). Furthermore, if 3 continuous nucleotides were selected from either the 5′ or 3′ ends, no significant difference in nucleotide composition could be detected between the ends ($\chi^2 = 6.20$, $P = 0.102$). However, significant differences could be found between the two ends if only one terminal nucleotide was analyzed ($\chi^2 = 63.90$, $P<0.0001$). Uracil and guanine were the dominant nucleotides at the 5′ terminal ends (61.11% and 28.95%), while uracil was the most dominant nucleotide at the 3′ terminal ends (81.58%) (Figure 4, Table S2 in File S1 and Figure S2A in File S1). Guanine

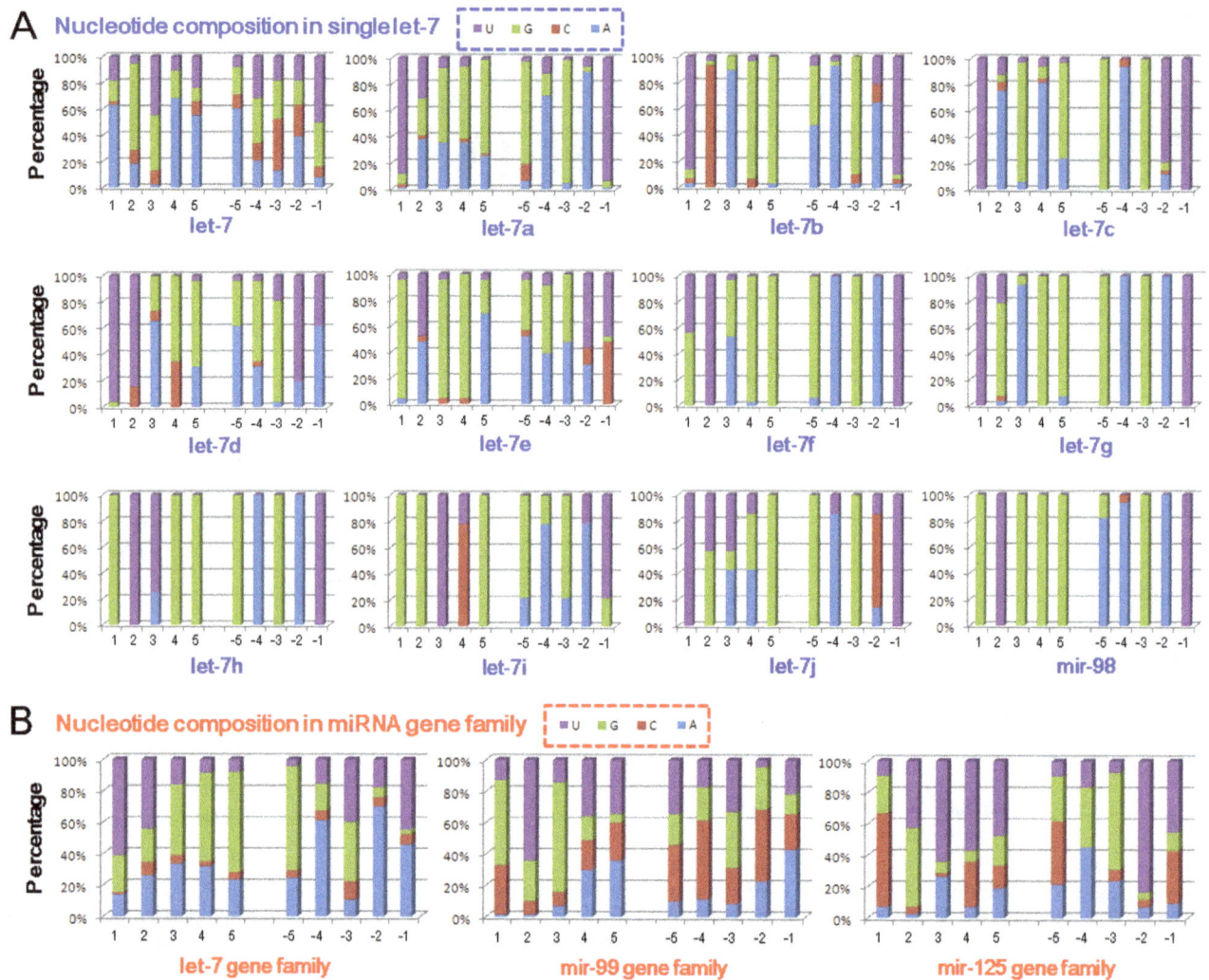

Figure 4. Nucleotide composition of the 5 continuous nucleotides at the 5′ and 3′ ends of the loop sequence. (A) Loop nucleotide composition in single let-7 genes (including let-7, let-7a, let-7b, etc.), with rare let-7 genes, such as let-7k and mir-3596, were not analyzed. (B) Loop nucleotide composition of relevant clustered miRNA gene families.

was present at an elevated rate, appearing at especially high frequencies in positions 4 and 5 at the 5′ ends and positions −3 and −5 at the 3′ ends (Table S2 in File S1), with a dominance also noted in the middle of the loop (Figure 5 and Figure S2A in File S1).

As a larger miRNA gene family, loop sequences showed larger nucleotide divergence than let-7-5p and let-7-3p (Figure 5A and Figure 5C). Let-7-5p acts as a canonical miRNA and is well conserved across animal species, particularly in the seed sequences. Among these homologous let-7 sequences, their loops showed varying levels of sequence similarity when examining single miRNA genes across animal species (Figure 5A and Figure S2A in File S1). While mature miRNAs, including miR-#-5p and miR-#-3p, are well conserved (especially for canonical mature miRNA) (Figure S2B in File S1), the loops commonly contain diverged nucleotides (Figure S2 in File S1) relative to the terminal regions that are relatively conserved.

Loop sequences in related miRNAs clustered with let-7

Based on the close physical distance between let-7 and other miRNAs, relevant miRNAs clustered with let-7 were collected

from different species. Similarly to the let-7 gene family, loops in related miRNA gene family were not as conserved (Figure 5B and Figure 5C). Interestingly, let-7 loops showed longer length distributions among their miRNA cluster (let-7 loops: 95% CI, 25.28–27.00 nts; clustered loops: 95% CI, 14.50–16.40 nts; $t = 16.14$, $P < 0.0001$), with a similar relation seen among the human let-7 genes and its clustered miRNAs (Table S3 in File S1). Based on the 5 continuous nucleotides extending from the 5′ and 3′ ends, significant difference could be detected between the three physically distant relevant miRNA gene families ($\chi^2 = 301.41$, $P < 0.0001$). Most miRNA gene families showed divergence in the dominant nucleotides between 5′ and 3′ ends (Table S2 in File S1). Curiously, we found that the let-7 cluster showed inconsistent dominant nucleotides at the 3′ terminal across the let-7 gene family. Furthermore, diverse dominant nucleotides and nucleotide compositions could be detected between relevant alternative miRNA gene families (Table S2 in File S1).

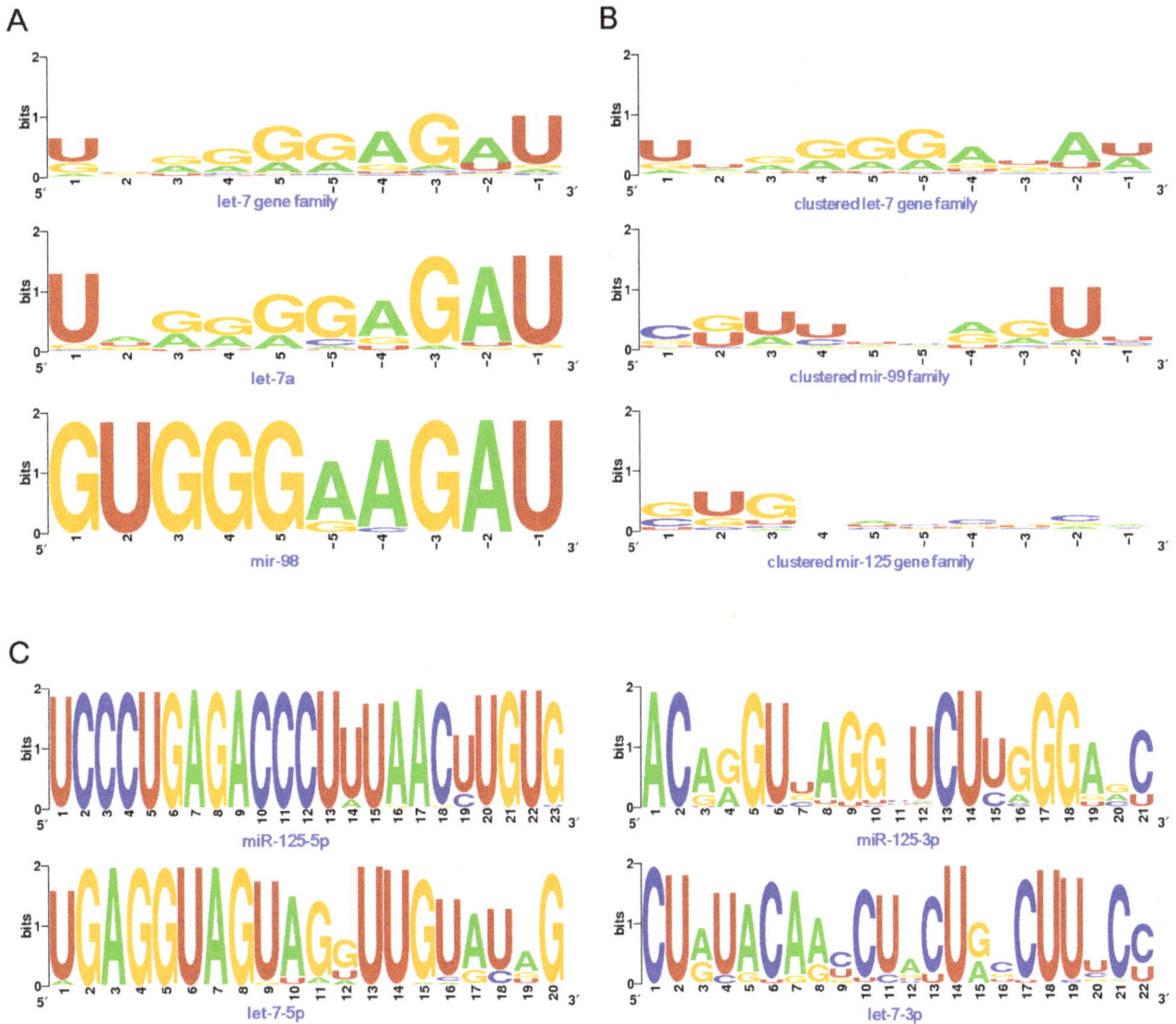

Figure 5. Sequence logo of loop sequences and miRNA sequences. (A) Sequence logo of all of the let-7 loop sequences, including the single let-7a and mir-98 loop sequences; (B) sequence logo of loop sequences from relevant clustered miRNA gene families; and (C) sequence logo of relevant mature miRNAs (including miR-#-5p and miR-#-3p).

Discussion

miRNAs are generated from pre-miRNA through Dicer recognition and cleavage of the terminal loop [38], with the loop connecting to miR-#-5p and miR-#-3p, which may contribute to the hairpin activity [39]. As a class of small non-coding RNAs, the negative regulatory RNAs have been widely studied, but little focus has been placed on the loop sequences. Indeed, the loops, including the loop structure and sequence, may tune and alter miRNA activity [9], which may further affect miRNA expression [13,14] by affecting Dicer recognition and cleavage during miRNA maturation [11,15]. The potential roles of loop sequences during miRNA biogenesis have been studied, especially for occurrence of multiple isomiRs. These isomiRs are mainly derived from imprecise and alternative cleavage by Drosha and Dicer during miRNA processing and maturation [19–24]. These loop sequences may provide more information for miRNA biogenesis, especially based on the analysis of loop sequences across different

animal species. In the present study, let-7 gene family loop sequences and other relevant miRNAs were analyzed and clustered to elucidate evolutionary and functional roles relating to miRNA biogenesis.

Multicopy miRNA genes may be located on different chromosomes and are mainly derived from historical duplication events [40–43]. Although the same mature miRNAs can be derived from these multicopy pre-miRNAs, the arms and loop sequences can be involved in larger divergence events, particularly the loops (Figure S1 in File S1) [7]. These loops may show diverse genetic distances relative to other homologous miRNA genes, and are always grouped in different clusters (Figure 3C and Figure S1 in File S1). Based on the phylogenetic relationships of pre-miRNAs and loops, similar distributions suggest that loop sequences dominate nucleotide divergence and evolutionary patterns in pre-miRNAs (Figure S1 in File S1). While both miR-#-5p and miR-#-3p sequences are conserved, loop sequences exhibit divergence through both varied nucleotides (including insertion/deletion) and lengths (Figure 2, Figure 5 and Figure S2 in File S1). In drosophila, the loop sequences and mature miRNAs are well conserved as well, with varied nucleotides and insertion/deletions found in the middle region of the loop (Figure 2A). The two terminal ends of the loops are relative conserved, which may contribute to stem-loop structure and cleavage by Dicer. These results suggest rapid evolution of the loops that further drives the evolution of miRNA genes.

Interestingly, we found changed length distributions of the loop sequences across different animal species and among different homologous miRNAs (Figure 2, Figure 3, Table S1 in File S1 and Table S3 in File S1). In higher metazoan species, the let-7 loops tend to be longer than those seen in lower species, with varying length distributions also seen different homologous miRNAs (Figure 3B and 3C). Clustered miRNAs tend to have similar length distribution, which implicates that the loop lengths may be affected by evolutionary relationships (Figure 3 and Figure S1 in File S1). Additionally, longer loop sequences may be an evolutionary trend in let-7 gene family, which may be of importance during miRNA biogenesis. Specifically, loop length may influence the stem-loop structure and stability (Table S4 in File S1), with longer loop sequences providing a possibility to dominate the evolution of miRNA genes across different animal species and homologous miRNAs within a specific species. These length variances further increase loop sequence diversity, which contributes to the evolutionary divergence between different miRNA genes, including homologous miRNAs (especially for multicopy pre-miRNAs). Indeed, although varied loop sequences exist, we still found the potential nucleotide characteristics in the 5′ and 3′ ends of the loops (Figure 4, Figure 5, Figure S2 in File S1 and Table S2 in File S1). Dominant nucleotides, such as uracil and guanine, are present at higher frequencies at the terminal ends, with these biased nucleotide compositions possibly influencing Dicer cleavage and contributing to the phenomenon of multiple isomiRs. While many isomiRs can be produced from a miRNA loci, only several isomiRs are dominantly expressed [21,24,33].

Some let-7 sequences are located in gene clusters with homologous miRNAs or other miRNAs, such as with the mir-125 and mir-99 gene families. Loops of clustered miRNA gene families may have different lengths, suggesting the length difference of loops from different miRNA genes (Table S3 in File S1). Significant differences in 5′ and 3′terminal nucleotide compositions are noted among these clustered miRNAs, with both uracil and guanine dominating in the two terminus ends (Table S2 in File S1). Compared to the stable length distributions of miRNAs, rapid loop sequence evolution can drive miRNA gene

evolution and may further affect isomiR expression profiles during pre-miRNA processing [11,15]. Expression analysis indicates stable isomiR expression profiles [21,25], even across different tissues and animal species, suggesting stable Drosha and Dicer cleavage during miRNA processing and maturation. Furthermore, loop sequences also show higher levels of nucleotide divergence between homologous miRNA genes, especially between multicopy pre-miRNAs (Figure 2 and Figure S1 in File S1). These findings suggest that nucleotide divergence does not influence cleavage precision, but that nucleotide bias in the 5′ and 3′ ends of the loop potentially influences miRNA maturation.

Herein, canonical miRNA loop sequences were collected, with variations in the loops potentially based on the phenomenon of multiple isomiRs, with the canonical miRNA sequences not necessarily the most dominant sequence. However, according to length distributions and terminal end nucleotide compositions, it can be concluded that loop sequences tend to be longer in higher animal species, with rapid evolution of the loop sequences further driving miRNA gene evolution. Stable isomiR expression profiles indicate relative cleavage, which may be closely related to the dominant nucleotide distributions. This study further enriches the understanding of miRNA biogenesis as it relates to loop sequences across different animal species and among homologous miRNAs, particularly considering the phenomenon of multiple isomiRs.

Supporting Information

File S1   Figure S1. Examples of neighbor-joining tree of the loop and pre-miRNA sequences in the let-7 gene family. Figure S2. Sequence logo of loop sequences and miRNA sequences. (A) Sequence logo of loop sequences from single let-7 genes and (B) sequence logo of relevant mature miRNAs (including miR-#-5p and miR-#-3p). **Table S1. Loop sequence length distribution from Figure 2B. Table S2. Nucleotide composition and frequency. Table S3. 95% confidence interval (CI) and difference between length distributions of let-7 and related clustered miRNAs. Table S4. Effect on minimum free energy (MFE) when changing loop sequence lengths in hsa-let-7a-1.**

Author Contributions

Conceived and designed the experiments: TML CL LG. Performed the experiments: TML CL LG. Analyzed the data: TML CY PL. Contributed reagents/materials/analysis tools: CY PL CL. Contributed to the writing of the manuscript: TML LG.

References

1. Lai EC (2003) microRNAs: runts of the genome assert themselves. Curr Biol 13: R925–936.
2. Flynt AS, Lai EC (2008) Biological principles of microRNA-mediated regulation: shared themes amid diversity. Nature Reviews Genetics 9: 831–842.
3. Lee Y, Jeon K, Lee JT, Kim S, Kim VN (2002) MicroRNA maturation: stepwise processing and subcellular localization. Embo Journal 21: 4663–4670.
4. Okamura K, Phillips MD, Tyler DM, Duan H, Chou YT, et al. (2008) The regulatory activity of microRNA star species has substantial influence on microRNA and 3′ UTR evolution. Nature Structural & Molecular Biology 15: 354–363.
5. Czech B, Zhou R, Erlich Y, Brennecke J, Binari R, et al. (2009) Hierarchical rules for Argonaute loading in Drosophila. Mol Cell 36: 445–456.
6. Okamura K, Liu N, Lai EC (2009) Distinct mechanisms for microRNA strand selection by Drosophila Argonautes. Mol Cell 36: 431–444.
7. Guo L, Lu Z (2010) The Fate of miRNA* Strand through Evolutionary Analysis: Implication for Degradation As Merely Carrier Strand or Potential Regulatory Molecule? Plos One 5: e11387.
8. Jagadeeswaran G, Zheng Y, Sumathipala N, Jiang HB, Arrese EL, et al. (2010) Deep sequencing of small RNA libraries reveals dynamic regulation of conserved

and novel microRNAs and microRNA-stars during silkworm development. BMC Genomics 11: 52.
9. Yue SB, Trujillo RD, Tang Y, O'Gorman WE, Chen CZ (2011) Loop nucleotides control primary and mature miRNA function in target recognition and repression. RNA Biol 8: 1115–1123.
10. Brummelkamp TR, Bernards R, Agami R (2002) A system for stable expression of short interfering RNAs in mammalian cells. Science 296: 550–553.
11. Hinton TM, Wise TG, Cottee PA, Doran TJ (2008) Native microRNA loop sequences can improve short hairpin RNA processing for virus gene silencing in animal cells. J RNAi Gene Silencing 4: 295–301.
12. Vlassov AV, Korba B, Farrar K, Mukerjee S, Seyhan AA, et al. (2007) shRNAs targeting hepatitis C: effects of sequence and structural features, and comparision with siRNA. Oligonucleotides 17: 223–236.
13. Zeng Y, Cullen BR (2004) Structural requirements for pre-microRNA binding and nuclear export by Exportin 5. Nucleic Acids Res 32: 4776–4785.
14. Boudreau RL, Martins I, Davidson BL (2009) Artificial microRNAs as siRNA shuttles: improved safety as compared to shRNAs in vitro and in vivo. Mol Ther 17: 169–175.

15. McManus MT, Petersen CP, Haines BB, Chen J, Sharp PA (2002) Gene silencing using micro-RNA designed hairpins. Rna 8: 842–850.

16. Vermeulen A, Behlen L, Reynolds A, Wolfson A, Marshall WS, et al. (2005) The contributions of dsRNA structure to Dicer specificity and efficiency. Rna 11: 674–682.

17. Liu G, Min H, Yue S, Chen CZ (2008) Pre-miRNA loop nucleotides control the distinct activities of mir-181a-1 and mir-181c in early T cell development. PLoS One 3: e3592.

18. Michlewski G, Guil S, Semple CA, Caceres JF (2008) Posttranscriptional Regulation of miRNAs Harboring Conserved Terminal Loops. Molecular Cell 32: 383–393.

19. Landgraf P, Rusu M, Sheridan R, Sewer A, Iovino N, et al. (2007) A mammalian microRNA expression atlas based on small RNA library sequencing. Cell 129: 1401–1414.

20. Morin RD, Aksay G, Dolgosheina E, Ebhardt HA, Magrini V, et al. (2008) Comparative analysis of the small RNA transcriptomes of Pinus contorta and Oryza sativa. Genome Research 18: 571–584.

21. Guo L, Yang Q, Lu J, Li H, Ge Q, et al. (2011) A Comprehensive Survey of miRNA Repertoire and 3′ Addition Events in the Placentas of Patients with Pre-eclampsia from High-throughput Sequencing. PLoS ONE 6: e21072.

22. Neilsen CT, Goodall GJ, Bracken CP (2012) IsomiRs - the overlooked repertoire in the dynamic microRNAome. Trends Genet 28: 544–549.

23. Lee LW, Zhang S, Etheridge A, Ma L, Martin D, et al. (2010) Complexity of the microRNA repertoire revealed by next generation sequencing. Rna-a Publication of the Rna Society 16: 2170–2180.

24. Guo L, Chen F (2014) A Challenge for miRNA: Multiple IsomiRs in miRNAomics. Gene 544: 1–7.

25. Burroughs AM, Ando Y, de Hoon MJL, Tomaru Y, Nishibu T, et al. (2010) A comprehensive survey of 3′ animal miRNA modification events and a possible role for 3′ adenylation in modulating miRNA targeting effectiveness. Genome Research 20: 1398–1410.

26. Tanzer A, Stadler PF (2004) Molecular evolution of a microRNA cluster. Journal of Molecular Biology 339: 327–335.

27. Lagos-Quintana M, Rauhut R, Meyer J, Borkhardt A, Tuschl T (2003) New microRNAs from mouse and human. Rna-a Publication of the Rna Society 9: 175–179.

28. Lim LP, Glasner ME, Yekta S, Burge CB, Bartel DP (2003) Vertebrate MicroRNA genes. Science 299: 1540–1540.

29. Kim VN, Nam JW (2006) Genomics of microRNA. Trends in Genetics 22: 165–173.

30. Xu JZ, Wong CW (2008) A computational screen for mouse signaling pathways targeted by microRNA clusters. Rna-a Publication of the Rna Society 14: 1276–1283.

31. Larkin MA, Blackshields G, Brown NP, Chenna R, McGettigan PA, et al. (2007) Clustal W and clustal X version 2.0. Bioinformatics 23: 2947–2948.

32. Guo L, Li H, Lu J, Yang Q, Ge Q, et al. (2012) Tracking miRNA precursor metabolic products and processing sites through completely analyzing high-throughput sequencing data. Mol Biol Rep 39: 2031–2038.

33. Guo L, Chen F, Lu Z (2013) Multiple IsomiRs and Diversity of miRNA Sequences Unveil Evolutionary Roles and Functional Relationships Across Animals. MicroRNA and Non-Coding RNA: Technology, Developments and Applications. 127–144.

34. Lorenz R, Bernhart SH, Honer Zu Siederdissen C, Tafer H, Flamm C, et al. (2011) ViennaRNA Package 2.0. Algorithms Mol Biol 6: 26.

35. Tamura K, Stecher G, Peterson D, Filipski A, Kumar S (2013) MEGA6: Molecular Evolutionary Genetics Analysis version 6.0. Mol Biol Evol 30: 2725–2729.

36. Huson DH (1998) SplitsTree: analyzing and visualizing evolutionary data. Bioinformatics 14: 68–73.

37. Crooks GE, Hon G, Chandonia JM, Brenner SE (2004) WebLogo: a sequence logo generator. Genome Res 14: 1188–1190.

38. Tsutsumi A, Kawamata T, Izumi N, Seitz H, Tomari Y (2011) Recognition of the pre-miRNA structure by Drosophila Dicer-1. Nat Struct Mol Biol 18: 1153–1158.

39. Schopman NC, Liu YP, Konstantinova P, ter Brake O, Berkhout B (2010) Optimization of shRNA inhibitors by variation of the terminal loop sequence. Antiviral Res 86: 204–211.

40. Heimberg AM, Sempere LF, Moy VN, Donoghue PCJ, Peterson KJ (2008) MicroRNAs and the advent of vertebrate morphological complexity. Proceedings of the National Academy of Sciences of the United States of America 105: 2946–2950.

41. Hertel J, Lindemeyer M, Missal K, Fried C, Tanzer A, et al. (2006) The expansion of the metazoan microRNA repertoire. BMC Genomics 7: 25.

42. Zhang R, Peng Y, Wang W, Su B (2007) Rapid evolution of an X-linked microRNA cluster in primates. Genome Research 17: 612–617.

43. Guo L, Sun BL, Sang F, Wang W, Lu ZH (2009) Haplotype Distribution and Evolutionary Pattern of miR-17 and miR-124 Families Based on Population Analysis. PLoS ONE 4: e7944.

20

Long-Term Reduction of T-Cell Intracellular Antigens Reveals a Transcriptome Associated with Extracellular Matrix and Cell Adhesion Components

Mario Núñez, Carmen Sánchez-Jiménez, José Alcalde, José M. Izquierdo*

Centro de Biología Molecular 'Severo Ochoa', Consejo Superior de Investigaciones Científicas, Universidad Autónoma de Madrid (CSIC/UAM), Madrid, Spain

Abstract

Knockdown of T-cell intracellular antigens TIA1 and TIAR contributes to a cellular phenotype characterised by uncontrolled proliferation and tumorigenesis. Massive-scale poly(A+) RNA sequencing of TIA1 or TIAR-knocked down HeLa cells reveals transcriptome signatures comprising genes and functional categories potentially able to modulate several aspects of membrane dynamics associated with extracellular matrix and focal/cell adhesion events. The transcriptomic heterogeneity is the result of differentially expressed genes and RNA isoforms generated by alternative splicing and/or promoter usage. These results suggest a role for TIA proteins in the regulation and/or modulation of cellular homeostasis related to focal/cell adhesion, extracellular matrix and membrane and cytoskeleton dynamics.

Editor: Didier Auboeuf, INSERM, France

Funding: This work was supported by grants from Ministerio de Economía y Competitividad-FEDER (BFU2008-00354 and BFU2011-29653). The CBMSO receives an institutional grant from Fundación Ramón Areces, Spain. The funders had no role in study design, data collection and analysis, decision to publish, or preparation of the manuscript.

Competing Interests: The authors have declared that no competing interests exist.

* Email: jmizquierdo@cbm.uam.es

Introduction

T-cell intracellular antigen 1 (TIA1) and TIA1 related/like (TIAR/TIAL1) are two proteins that impact several molecular aspects of RNA metabolism at different transcriptional and post-transcriptional layers of gene expression [1–4]. In the nucleus, TIA proteins regulate and/or modulate DNA-dependent transcription by interacting with DNA and RNA polymerase II [5–8]. Also, they facilitate splicing of pre-mRNAs (around 10–20% of splicing events in human genome) *via* improving the selection of constitutive and atypical 5′ splice sites through shortening the time available for definition of an exon by enhancing recognition of the 5′ splice sites [9–12]. In the cytoplasm, they regulate and/or modulate localization, stability and/or translation of human mRNAs by binding to the 5′ and/or 3′ untranslatable regions [13–25]. Therefore, these multifunctional proteins impress prevalently on the molecular and cellular biology of specific RNAs and proteins, altering their lives and destinies in response to environmental cues and challenges.

TIA proteins appear to have a pleiotropic role in the control of cell physiology. For example, they have been shown to play an important role during embryogenesis. Accordingly, mice lacking TIA1 and TIAR die before embryonic day 7, indicating that one or both proteins must be properly expressed for normal early embryonic development. Indeed, mice lacking TIA1 or TIAR, or ectopically over-expressing TIAR, show higher rates of embryonic lethality [17,26–28]. Further, TIA regulators are known to target genes with relevant biological associations with cell networks involving complex responses such as death/survival, proliferation/differentiation, inflammation, adaptation to environmental stress, viral infections and tumorigenesis [1,2,13–28].

Although the relevance of TIA proteins in key cellular processes involving, for example, inflammation and the stress responses, are well established, their roles on proliferation/differentiation events and survival/cell death responses in patho-physiological settings are not completely known. To assess the potential long-term regulatory roles of TIA proteins in cellular responses, we used an RNA interference strategy to stably down-regulate TIA1/TIAR expression together with genome-wide profiling analysis, to identify genes and processes involved in cell phenotypes regulated and/or modulated by TIA proteins. Our findings suggest that TIA proteins regulate and/or modulate membrane dynamics linked to extracellular matrix and focal/cell adhesion components.

Materials and Methods

Cell cultures and immunofluorescence analysis

Adherent HeLa cell lines, silenced for expression of TIA1, TIAR and HuR, or control cells, were constructed by stable transfection of corresponding short hairpin RNAs (shRNAs) (Fig. S1). Cell lines were maintained under standard conditions and analyzed by confocal microscopy [23–25].

RNA purification

Total RNA was purified with TRIzol Reagent (Invitrogen). RNA quality was assessed using the Agilent 2100 Bioanalyzer.

Library preparation and sequencing

cDNA libraries were prepared with Illumina's mRNA-Seq Sample Prep kit following the manufacturer's protocol. Each library was run on one RNASeq Multiplexed 75-bp paired-end sequence using the Illumina Genome Analyzer (GAIIx), facilitated by the Madrid Science Park.

Primary processing of Illumina RNA-seq reads

RNA-seq reads were obtained using Bustard (Illumina Pipeline version 1.3). Reads were quality-filtered using the standard Illumina process. Three sequence files were generated in FASTQ format; each file corresponded to the HeLa cell line from which the RNA originated [29]. The total number of reads and additional metrical data are shown in Fig. S2. The sequence data have been deposited in the NCBI Gene Expression Omnibus database (http://www.ncbi.nlm.nih.gov/geo/info/linking.html) and are accessible through the GEO Series accession number GSE46516.

Mapping of RNA-seq reads using TopHat

Reads were processed and aligned to the University of California, Santa Cruz, reference human genome (UCSC, build hg19) using the TopHat tool [30].

Transcript assembly and abundance estimation using Cufflinks

The aligned read files were processed using the Cufflinks software suite [31]. Cufflinks uses the normalized RNA-seq fragment counts to measure the relative abundance of transcripts. The unit of measurements is fragments per kilobase of exon per million fragments mapped (FPKM). Confidence intervals for FPKM estimates were calculated using Bayesian inference [32].

Comparison of reference annotation and differential expression testing using Cuffcompare and Cuffdiff

Once all short read sequences were assembled with Cufflinks, the output. GTF files were sent to Cuffcompare along with a reference. GTF annotation file downloaded from the Ensembl database. This classified each transcript as known or novel. Cuffcompare produces a combined. GTF file which is passed to the Cuffdiff tool with the original alignment (.SAM) files produced by TopHat. We used Cuffdiff to perform two pairwise comparisons of expression, splicing and promoter use between control, TIA1 or TIAR-silenced samples.

Visualization of mapped reads

Mapping results were visualized using both the UCSC genome browser [33,34] and the Integrative Genomics Viewer software, available at http://www.broadinstitute.org/igv/. Views of individual genes were generated by uploading coverage.wig files to the UCSC Genome browser as a custom track. Data files were restricted to the chromosome in question due to upload limits imposed by the genome browser. The same method was used to generate coverage plots for all chromosomes.

Genome-wide profiling by microarray analysis

RNA quality checks, amplification, labelling, hybridization with Human Genome U133 Plus 2.0 Array Chips (approximately 55,000 transcripts) (Affymetrix Inc. Santa Clara, CA) and initial data extraction were performed at the Genomic Service Facility at the Centro Nacional de Biotecnología (CNB-CSIC). Comparison of multiple cDNA array images (three independent experiments per each biological condition tested) was carried out by using an average of all of the gene signals on the array (global normalization) to normalize the signal between arrays. The quantified signals were background-corrected (local background subtraction) and normalized using the global Lowess (LOcally WEighted Scatterplot Smoothing) regression algorithm. Local background was corrected by the normexp method with an offset of 50. Background-corrected intensities were transformed to log scale (base 2) and normalized by Lowess for each array [35]. Finally, to obtain similar intensity distribution across all arrays, Lowess-normalized-intensity values were scaled by adjusting their quantiles [36]. After data processing, each probe was tested for changes in expression over replicates using an empirical Bayes moderated t statistic [37]. To control the false discovery rate (FDR), P values were corrected using the method of Benjamini and Hochberg [38]. FIESTA viewer (http://bioinfogp.cnb.csic.es/tools/ FIESTA) was used to visualize all microarray results and to evaluate the numerical thresholds ($-1.5 \geq$ fold change ≥ 1.5; FDR<0.05) applied for selecting differentially expressed genes [39]. Microarray data have been deposited in the NCBI Gene Expression Omnibus database (http://www.ncbi.nlm.nih.gov/geo/info/ linking.html) and are accessible through the GEO Series accession number GSE47664.

Functional analysis of gene lists

The Gene Ontology (GO) and Kyoto Encyclopedia of Genes and Genomes (KEGG) database analyses were conducted using software programmes provided by GenCodis3 (http://genecodis.cnb.csic.es) [40,41]. The functional clustering tools were used to look for functional enrichment for significantly over- and under-expressed genes ($P<0.05$) in control, TIA and/or TIAR-silenced HeLa cell lines.

QPCR analysis

Reverse transcription (RT) reactions and real-time polymerase chain reaction (PCR) was performed at the Genomic PCR Core Facility at Universidad Autónoma de Madrid in the Centre of Molecular Biology *Severo Ochoa*. Analysis was performed on two independent samples in triplicate, including no-template and RT-minus controls. Beta-actin (ACTB) mRNA expression was used as an endogenous reference control. Relative mRNA expression was calculated using the comparative cycle threshold method. The primer pairs used in the analysis are described in Fig. S8. The following mRNAs were quantified: TIA1, T-cell intracellular antigen 1; TIAR, TIA1 related protein; ACTB; beta-actin; CLGN, calmegin; COL1A2, collagen type 1, alpha 2; FAM129A, cell growth inhibiting protein 39; FBN2, fibrillin 2; LGR5, leucine-rich repeat containing G protein-coupled receptor 5; MKX, mohawk homeobox; TWIST1, twist basic helix-loop-helix; AKR1C1-4, aldo-keto reductase family, members C1–C4; CCL2, chemokine (C-C motif) ligand 2; COL4A4, collagen type IV, alpha 4; COL5A3, collagen type V, alpha 3; OSGIN1, oxidative stress induced growth inhibitor 1 and TNFRF11B, tumor necrosis factor receptor superfamily member 11b.

Fluorescence microscopy

Control and TIA-silenced HeLa cells were grown for 24 h on coverslips, washed with phosphate-buffered saline (PBS), fixed in formalin (Sigma) at room temperature for 10 min, washed with PBS, and processed. For immunofluorescence, the coverslips were incubated for 45 min at room temperature with the primary antibodies against TIA1, TIAR, HuR (Santa Cruz Biotechnology) or α-tubulin (Sigma) proteins [23,24]. The samples were then washed with PBS and incubated for 45 min with the corresponding secondary antibodies (Invitrogen). The samples were then

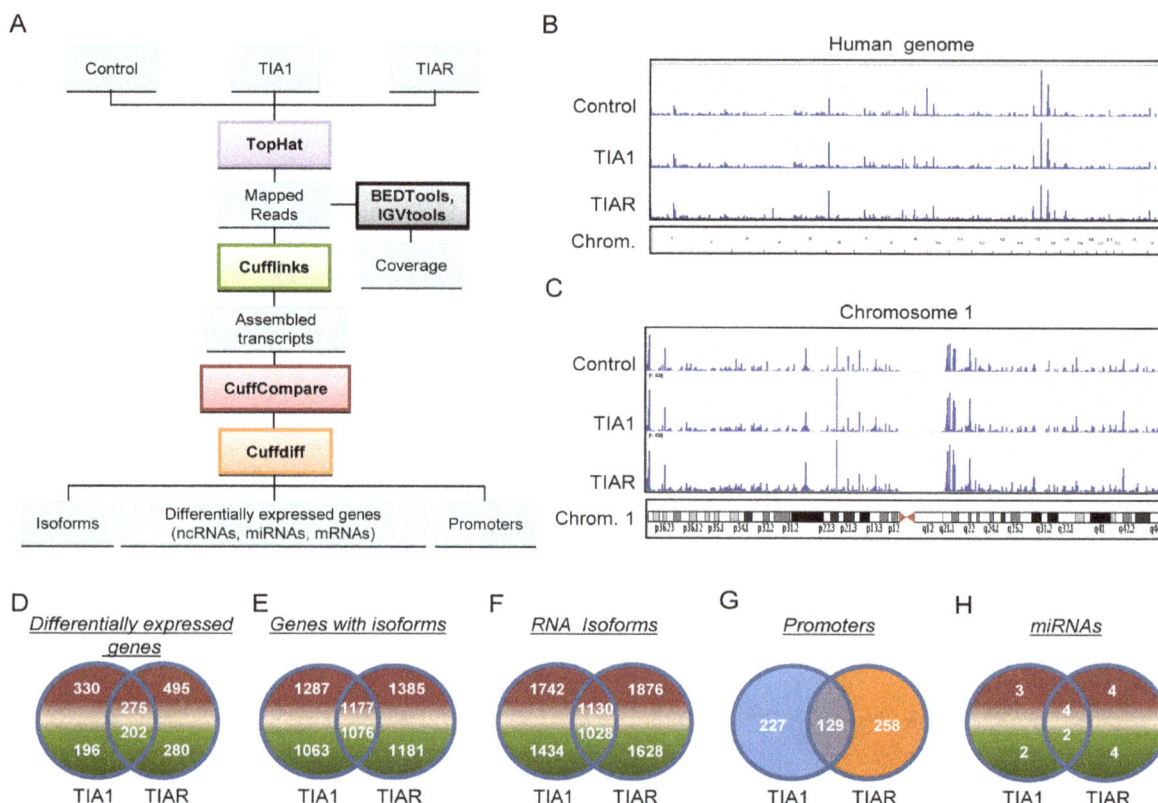

Figure 1. Characterization by massive poly(A$^+$) sequencing of the transcriptomes associated with TIA1 or TIAR-silenced HeLa cells. (**A**) Workflow followed to analyze the data from massive-scale poly(A+) RNA sequencing. (**B** and **C**) Transcriptional profiling in human chromosomes. The RNA-seq read density along the length of the human genome (**B**) and chromosome 1 (**C**) are shown. (**D–H**) Venn diagrams displaying distributions and numbers of differentially-expressed genes (**D**), genes with isoforms (**E**), RNA isoforms (**F**), promoters (**G**) and microRNAs (**H**) in TIA1 or TIAR-silenced HeLa cells. The numbers of genes that were up- (in red) and down- (in green) regulated, as well as those shared, are indicated.

washed in PBS and mounted with Mowiol (Calbiochem). Microscopy was performed with a confocal microscope.

Cell adhesion assays

A cell adhesion assay was carried out as described [42]. Briefly, control and TIA-silenced HeLa cells were exponentially grown in DMEM supplemented with 10% FBS (Sigma). Cells were deprived of serum for 12 h prior to the assay and then were washed three times with serum-free DMEM and grown in DMEM. Cells were detached using 10 mM EDTA and observed under a microscope to confirm complete dissociation (aprox. 10 min.). The cells were then washed twice with DMEM to remove EDTA and were resuspended at 2×10^5 cells/ml in DMEM/0.1% BSA in 12-well plates coated with either BSA (10 µg/ml), rat collagen I (150 µg/ml) or non-coated. Cells (500 µl) were allowed to adhere for 20 min at 37°C. Then, each well was washed (four times) with 300 µl DMEM to eliminate non-adherent cells. After washing, DMEM containing 10% FBS was added and cells were allowed to recover at 37°C for 4 h. Following this, 30 µl of MTT substrate (5 mg/ml) was added to each well and the incubation was continued for an additional 2 h at 37°C. Finally, the MTT-treated cells were lysed in buffer containing 10% SDS and 0.03% HCl and the absorbance was measured at 570 nm on a spectrophotometer. Where indicated, cells were cultured in the presence of DMSO or PMA (100 nM) for 4 h and assayed for MTT activity.

Transfections and luciferase assays

The human COL1A2 promoter construct containing the partial promoter sequence fused to a firefly luciferase (Luc) reporter gene was generated as previously described [43]. This construct was kindly provided by Prof. Miyazahi (Tokyo Medical University, Tokyo, Japan). Control, TIA1 and/or TIAR-silenced HeLa cells were transiently co-transfected with 500 ng of pEGFP-C1 and firefly luciferase reporter plasmids. Cells were incubated at 37°C for 24 h and lysed with 200 µl of cell culture lysis reagent (Promega) and microcentrifugated at 14,000 rpm for 5 minutes at 4°C; supernatant was used to determine firefly luciferase activity in a Monolight 2010 luminometer (Analytical Luminiscence Laboratory). Luciferase activity was expressed as relative light units (RLU) per milligram of protein and normalized to GFP expression determined by immunoblot. Co-transfection experiments were performed in duplicate and the data presented as the means of the ratio RLU/GFP, expressed as fold induction relative to the corresponding control values (means ± standard error of the mean).

Results

RNA-seq coverage

To study the individual contribution of TIA1 and TIAR proteins on a genome-wide basis, we set out to characterize the transcriptomes associated with TIA1 or TIAR-down-regulation in HeLa cells ([25] and Fig. S1). Three transcriptomic libraries were generated from poly(A+) RNAs isolated from control, TIA1 or

TIAR-knocked down HeLa cells, and were analyzed using the Illumina platform. The alignment and annotation of resulting reads, splice junctions and profiles of differentially-expressed RNAs were performed with bioinformatic tools indicated in Fig. 1A [29]. A total of 67.37 M reads were obtained from control, TIA1 and TIAR-silenced cells (Fig. S2). Unique reads were mapped to the *Homo sapiens* genome for each condition. The distribution of mapped reads to different chromosomal features revealed that a significantly large number (46.3%, 46.4% and 47.9%, respectively) of the reads mapped within mRNA coding regions. The remaining reads were mapped within untranslated (UTR) (23.6%, 23.8% and 23.3%, respectively), intronic (11.8%, 11.9% and 11.9%, respectively) and intergenic (18.1%, 17.8% and 16.7%, respectively) regions (Fig. 1B and Fig. S2).

To investigate the level and uniformity of the read coverage against the human genome, we plotted mapped reads of the control, TIA1 and TIAR samples along the human genome as well as for each human chromosome (Fig. S3). The coverage of the RNA-seq analysis on the human genome was extensive, as illustrated by chromosome 1 since this is the largest chromosome in the human karyotype, encoding over 13.6% of all human genes. As expected, no reads mapped to the centromeres but revealed extensive transcriptional activity throughout the chromosomes (Fig. 1B and C and Fig. S3). The distribution of reads within distinct chromosomal regions indicated a very homogeneous scattering among the experimental situations compared (Fig. 1B and C and Fig. S3). Between 91% (corresponding to control sample) and 92% (corresponding to both TIA1 and TIAR samples) of reads aligned to the reference genome in a unique manner. Due to low quality, a small percentage of reads were removed from the analysis prior to mapping to the reference (Fig. S2 and S3).

Analysis of differentially expressed genes (DEGs), genes with isoforms, RNA isoforms, promoters and microRNAs

Cufflinks uses the Cuffdiff algorithm to calculate differential expression at both the gene and transcript level ([29] and Fig. 1A). Differential gene expression for control versus TIA1 or TIAR samples was calculated using the ratio of TIA1 or TIAR versus control FPKM values for every gene. The differential gene expression ratios were tested for statistical significance as described [29]. To detect transcriptional regulation, RNA-seq data was analyzed by Cufflinks. Cufflinks also identifies post-transcriptional regulation by looking for changes in relative abundances of mRNAs spliced from the same primary transcript between control and TIA1 or TIAR-silenced conditions, which is detected as alternative splicing. Thus, Cufflinks discriminates between transcriptional and post-transcriptional processing ([29] and Fig. 1A). Using Venn diagrams, a summary of the number of shared and differential expressed genes (DEGs), genes with isoforms (splicing data filtered by gene name), RNA isoforms (splicing data filtered by NM identifier), promoters and micro(mi)RNAs is shown in Fig. 1D–H and Fig. S4–S6. Collectively, these results indicate that TIA1 and/or TIAR regulate both specific and overlapping aspects of the human transcriptome, suggesting that their functional roles can be redundant, additive and/or independent, in agreement with previous findings [22,24,25,28,44].

Classification and cluster analysis of genes, RNAs and promoters identified by RNA-seq

As a first attempt to understand the functional relevance of differentially expressed up- and down-regulated genes, isoforms and promoters in TIA or TIAR-silenced HeLa cells, Gene Ontology (GO) and Kyoto Encyclopedia of Genes and Genomes (KEGG) database analysis was performed. GO and KEGG analysis were able to identify the main categories of biological processes and pathways of DEGs controlled by TIA1 and TIAR ($P < 0.05$). GO categories related to regulation of DNA transcription, signal transduction, multicellular organismal development, cell adhesion and differentiation, cell proliferation, apoptosis and nervous system were among the enriched categories in both up- and down-regulated genes obtained through silencing of TIA1 or TIAR (Fig. 2A, C, E, and G and Fig. S4–S6). KEGG database analysis, integrating individual components into unified pathways, was used to identify the enrichment of specific pathways in functionally-regulated gene and RNA groups. The results showed that several KEGG pathways were significantly enriched ($P < 0.05$) in both up- and down-regulated genes, including pathways involved in focal adhesion, cancer, axon guidance, extracellular matrix components (EMC)-receptor interaction, MAPK signalling and regulation of actin cytoskeleton (Fig. 2B, D, F, and H and Fig. S4–S6). GO categories related to alternative promoter usage involved biological processes associated with DNA transcription, cell differentiation and cycle, signal transduction and metabolism (Fig. 2I and J and Fig. S4–S6). Collectively, these results suggest that both TIAR and TIA1 modulate specific and overlapping aspects of the transcriptome (DEGs, alternative splicing and promoters) related to the extracellular environment and signal transduction pathways, which could contribute to the cell proliferation and differentiation phenotypes described in TIA1 and/or TIAR-knocked down HeLa cells [19,25].

Microarray analysis and functional categories

To validate genes and regulatory trends found after silencing both TIA proteins, we used genome-wide expression microarrays. This yielded a set of differentially expressed genes in TIA1 and TIAR-silenced HeLa cells: 251 genes were up-regulated and 173 genes were down-regulated by more than 1.5-fold (FDR RP< 0.05) (Fig. 3A and Fig. S7). GO and KEGG database analysis of the up- and down-regulated genes showed significant enrichments in genes associated with signal transduction, cell and focal adhesion, multicellular organismal development and the nervous system (Fig. 3B and C). A comparison of the results obtained from RNA-seq and genome-wide analysis revealed the superior power of massive-RNA sequencing (Fig. 3D), suggesting that our understanding of transcriptional complexity linked to TIA proteins is far from complete.

Validation of RNA-seq and mRNA array-predicted changes by QPCR

To validate previous results on identified genes and their relative expression levels, several up- and down-regulated mRNAs were confirmed by quantitative RT-PCR (QPCR) analysis (Fig. S8). As shown in Fig. 4, the results obtained by quantitative amplification (Fig. 4A) were fully consistent with the data observed by hybridization using microarrays, and the relative fold changes in individual massive RNA sequencing for TIA1 or TIAR (Fig. 4B).

Knockdown of TIA proteins results in increased cell adhesion rates

As shown in Fig. 2, 3 and S4–S7, GO and KEGG analysis suggested a long-term role for TIA1 proteins in molecular events linked with cell adhesion. To validate these results at the functional level, we examined the cell adhesion potential of HeLa cells with reduced TIA1 and TIAR expression. As shown in Fig. S9, the

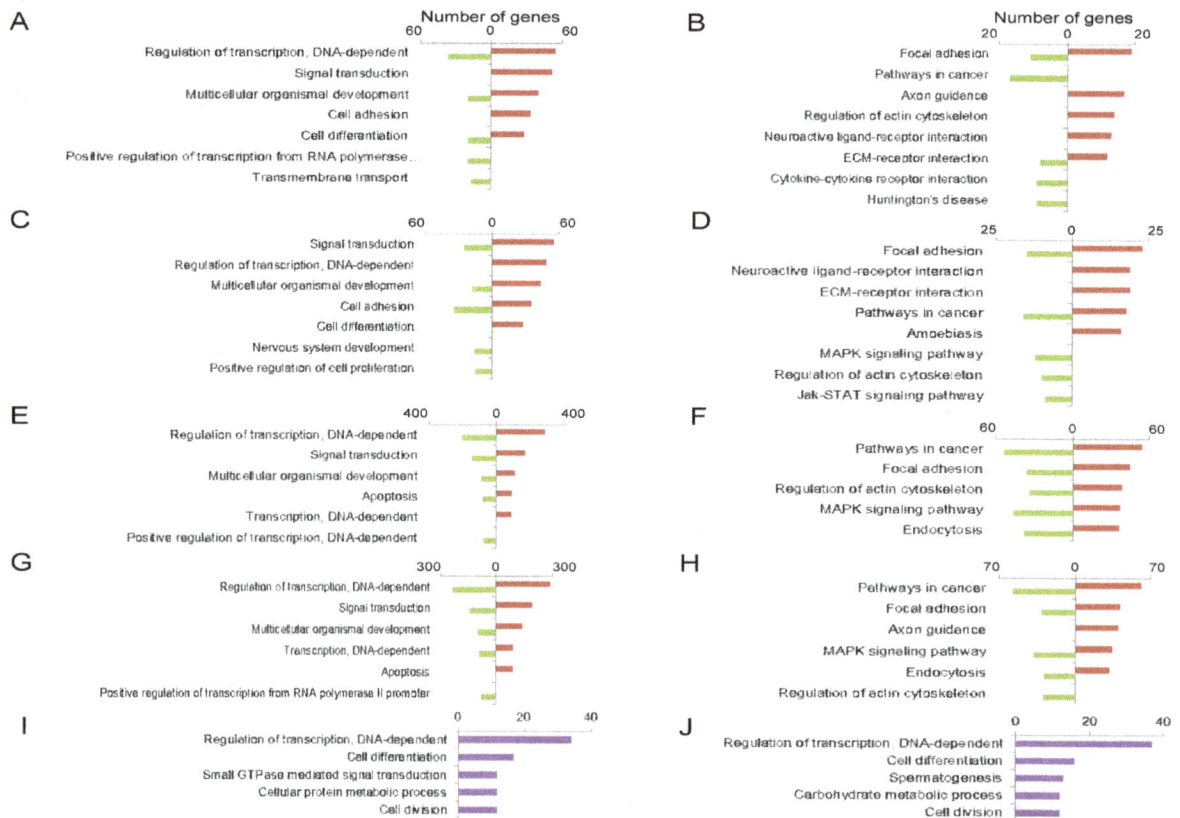

Figure 2. Top-five categories of biological processes and pathways associated with TIA1 or TIAR-silenced HeLa cells. (A–D) Histograms of the distribution of up- (red) and down- (green) regulated genes using the Gene Ontology (GO) (**A** and **C**) and Kyoto Encyclopedia of Genes and Genomes (KEGG) (**B** and **D**) databases ($P<0.05$) in TIA1 (**A** and **B**) or TIAR (**C** and **D**)-silenced HeLa cells. (**E–H**) Histograms of the distribution of up- (red) and down- (green)-regulated RNA isoforms using the GO (**E** and **G**) and KEGG (**F** and **H**) databases ($P<0.05$) in TIA1- (**E** and **F**) or TIAR- (**G** and **H**) silenced HeLa cells. (**I** and **J**) Histograms of the gene distribution of alternative promoter usage in the GO database ($P<0.05$) in TIA1- (**I**) or TIAR- (**J**) silenced HeLa cells.

reduction of TIA1 plus TIAR expression in HeLa cells resulted in an increased cell adhesion capacity compared with control HeLa cells. Indeed, measurement of metabolic activity by methyl thiazolyl tetrazolium (MTT) assay, support a role for TIA proteins in the regulation and/or modulation of cell-substratum adhesion. This capacity, associated with the reduction of the levels of TIA proteins, was stimulated with phorbol 12-myriaste 13-acetate (PMA) in control HeLa cells after 4 hours of incubation (Fig. S9). Furthermore, as shown in Fig. S10, the sustained reduction of TIA1 and/or TIAR could promote a significant and reproducible induction (2- to 16-fold) of a luciferase construct under the control of COL1A2 human promoter sequences [43], suggesting that diminished expression of TIA proteins deregulates the basal transcriptional activity of this promoter. Taken together, these observations may suggest that the gene expression patterns detected in TIA1/TIAR-knocked down HeLa cells might be the result of an overlapping regulation, implying the involvement of several molecular events regulated and/or modulated by these multifunctional proteins on cell adhesion and extracellular matrix components.

Discussion

Polyvalent TIA proteins have pleitropic effects on RNA biology and function as cell sensors to sustain homeostasis and facilitate adaptation, survival or death responses to a great variety of stressing conditions involving environmental and epigenetic

challenges in the short- and long-term [19,22–25,28,44]. To do this, these multifunctional regulators orchestrate transcriptome dynamics associated with transcriptional and/or post-transcriptional regulatory programs, which regulate and/or modulate molecular events and processes with relevant biological implications in cellular phenotypes and behaviours related with apoptosis, inflammation, viral infections, embryogenesis and oncogenesis [13–25]. Through genome-wide expression profiling approaches, we have identified many important genes/RNAs and have made a preliminary attempt to construct the regulatory pathways implicated in TIA-downregulated HeLa cells. Herein, we report that sustained TIA down-regulation has a functional impact on a human transcriptome associated to differentially-expressed genes and RNA isoforms generated by alternative splicing and/or promoter usage which are prevalently linked to gene categories related with focal/cell adhesion, extracellular matrix, membrane and cytoskeleton dynamics.

Cancer cells often exhibit hyperactive signalling pathways activated by membrane components. Cell and focal adhesions and cytoskeleton dynamics provide contact between neighbouring cells or between a cell and the extracellular matrix and, as such, play essential roles in the regulation of migration, proliferation, differentiation and apoptosis, and also tumorigenesis [45]. The reduction of TIA expression results in altered expression of extracellular matrix, membrane and cytoskeleton components and also signal transduction genes [23,24,44,46], thus providing an

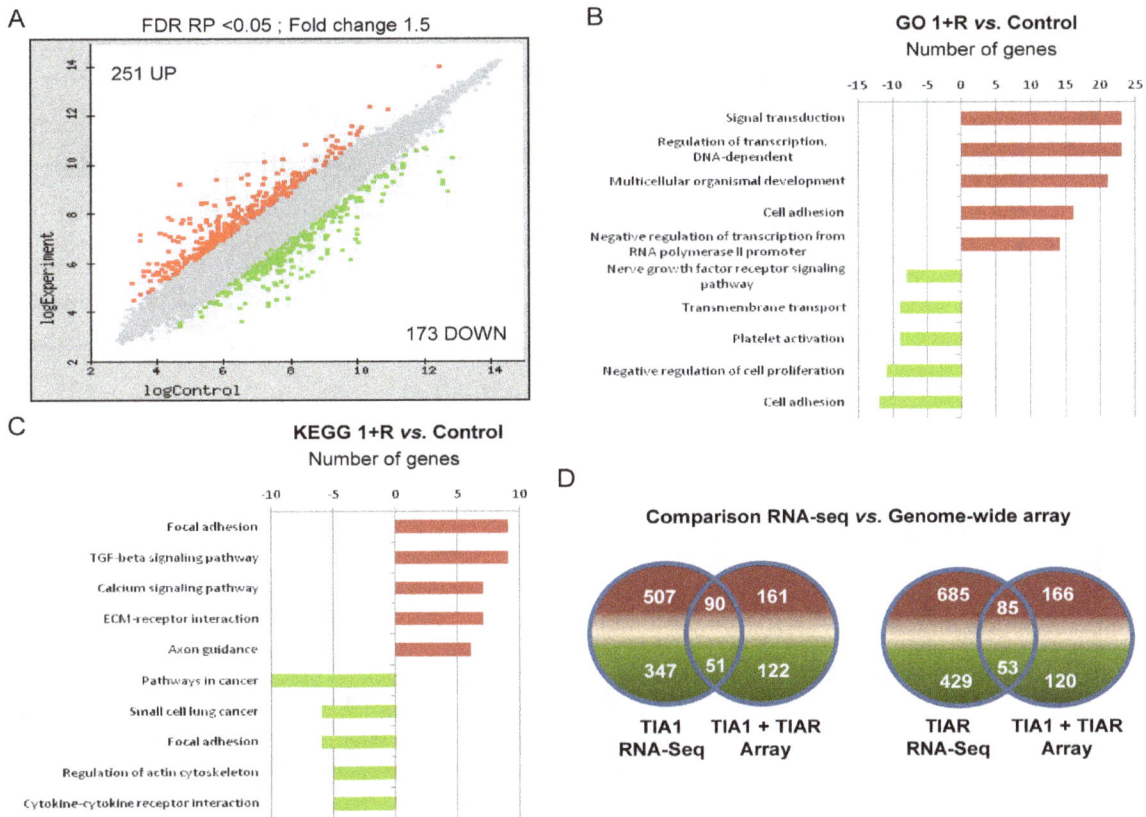

Figure 3. Characterization by genome-wide profiling of the transcriptome associated with TIA-silenced HeLa cells. (A) MA plot representation of the distribution of up- (spots in red) and down- (spots in green) regulated RNAs (−1.5≥ fold-change ≤1.5; FDR<0.05) in TIA1 and TIAR-silenced *versus* control HeLa cells. **(B and C)** Histograms of the distribution of up- (red) and down- (green) regulated genes using the GO **(B)** and KEGG **(C)** databases (*P*<0.05) in TIA1 and TIAR-silenced HeLa cells. **(D)** Comparison between RNA-seq and genome-wide array data. Venn diagrams depicting the numbers of genes that were up- (red) and down- (green) regulated as well as those shared in TIA1 or/and TIAR-knocked down HeLa cells.

additional molecular basis for the observation that sustained TIA down-regulation contributes to the deleterious phenotypes observed in HeLa cells lacking TIA proteins [23–25]. Notably, the down-signature enriched category included the focal and cell adhesion components (as for example, ALCAM, AMBP, CD9, DMKN, FLRT2, FOLR1, ITGA6, L1CAM, NOTCH3, OLR1, PCDHA1, PPP2R5C, RSPO3, TNNC1 and TSPAN8). These components play important roles in cell biology, and phenotypes linked to transformed cancer cells [45,47,48]. Moreover, genes in the more significantly up-regulated network module included collagens (as for example, COL1A1, COL1A2, COL3A1 and COL8A1) and extracellular matrix-related components (as for example, CD36, CFH, CGA, FBN2, IGFBP7, LGALS8, LOX, PDGFC and TNC). These observations strengthen the association of this gene subset with most transformed cell phenotypes and relate it to cell growth and survival responses [47,48]. The effects of TIA expression on extracellular matrix and focal/cell adhesion phenomena would depend on how transcriptome dynamics are integrated in a given cell type and physiological context, and could have significant implications in inflammatory diseases and cancer progression [17,19,22–25,49,50].

TIA proteins are able to bind the same sequence motifs and sites on some human RNAs [12–16], and they can regulate and/or modulate some overlapping aspects of the human transcriptome [12–16]. A previous TIA-iCLIP analysis in HeLa cells showed that the highest enrichment density of cDNAs were seen in introns and

3′-UTRs [12]. These observations are consistent with the role of TIA proteins in pre-mRNA splicing, localization, stability and/or translational regulation, via binding U-, C- and AU-rich sequence elements located at the 5′ spliced sites and the 5′ and 3′-UTRs, respectively [9–21]. It is interesting to note that no overlap was observed between gene targets reported to be regulated by TIA1 at the translational level [13], and genes predicted *in silico* to be regulated by TIA1/TIAR at the alternative pre-mRNA splicing level [51]. This observation suggests that TIA proteins could be regulating and/or modulating distinct subsets of genes at the splicing and translational levels. The TIA-iCLIP data have expanded the total number of post-transcriptional events and associated gene functions that could be predicted to be regulated by TIA proteins. Thus, the estimated frequency of cellular events regulated by TIA1/TIAR via specific-sequences motifs might be approximately 20–30% [3,4,12,51].

Tumorigenesis is caused both by genetic and epigenetic events. The progressive acquisition of mutations in oncogenes or tumor-suppressor genes might act in concert with epigenetic events, such as functional down-regulation of TIA proteins, to give cells a competitive growth advantage. TIA1 and TIAR genes are mutated [52] and down-regulated [18,21,25,44,49,50] in several types of human cancers. In human transformed cells, TIA proteins regulate the transcription, alternative splicing, stability and/or translation of many target genes associated with tumor development and progression, involving the control of cell proliferation,

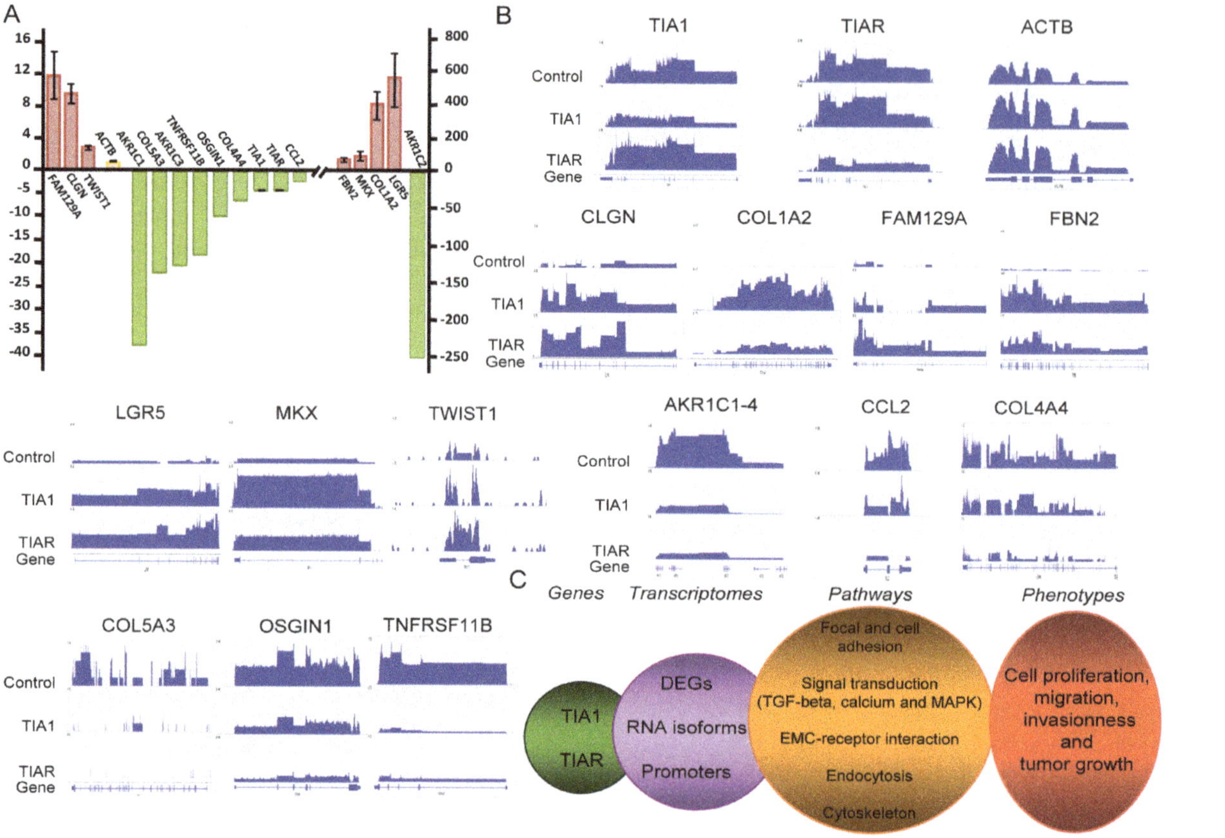

Figure 4. Validation of RNA-seq and microarray predicted changes. (A) Quantification of relative expression levels of indicated genes by qPCR. **(B)** RNA-seq mapping to the UCSC reference genome (hg19) of the genes indicated in control and TIA1 or TIAR-knocked down HeLa cells. A schematic representation of every gene analyzed is shown at the bottom. **(C)** Summary of the molecular and cellular events associated with transcriptome dynamics in TIA-downregulated HeLa cells.

apoptosis, angiogenesis, invasion, metastasis, evasion of immune recognition and metabolic reprogramming [13–25]. Thus, the sustained reduction of TIA proteins could facilitate the acquisition of oncogenic phenotypes [25] characterized by severe alterations in cell proliferation/growth, invasion and morphology, via modulation of several gene expression layers involving changes to global and specific translational rates of cell-cycle G2/M phase transition and DNA replication/repair factors encoding mRNAs [24] as well as actin [23] and tubulin [46] cytoskeletons, together with transcriptome dynamics linked to extracellular matrix, focal/cell adhesion and membrane regulatory events. Thus, prolonged TIA knockdown may lead to cell proliferation and neoplastic growth by simultaneously activating translation of mRNAs that encode proteins involved in cell cycle progression and cell dynamics. These findings also suggest cooperation of transcriptomic and translational programs that underpin prevalent biological effects on TIA-associated gene expression networks. This cellular and molecular scenario could contribute to the different phenotypes observed in TIA-knocked down HeLa cells. Therefore, our observations support the idea that these proteins might function as cell growth and tumor suppressor genes, through the regulation and/or modulation of many RNAs which are targeted by transcriptional and post-transcriptional regulatory events. However, further investigation is required to discern the cellular and molecular mechanisms underlying TIA protein control/modulation of gene expression. Results from these studies could potentially give insights into the differential gene networks

that contribute to cell phenotypes observed in the absence of these regulators, both at short- and long-term, in homeostasis and environmental stress conditions.

Supporting Information

Figure S1 shRNA-mediated knockdown of TIA1 and TIAR in HeLa cells. HeLa cells silenced for expression of TIA1, TIAR or HuR were stained with anti-TIA1, anti-TIAR, anti-HuR and anti-α-tubulin antibodies and were visualized by confocal microscopy, as described [23,24]. The scale bar is 10 μm.

Figure S2 Metrical data of massive-scale poly(A+) RNA sequencing analysis.

Figure S3 Summary of contig distribution on human chromosomes in control, TIA1 and TIAR-silenced HeLa cells. The RNA-seq read density along the lengths of the human chromosomes is illustrated. Each bar represents the log₂ of the frequency reads plotted against chromosome coordinates. RNA-seq data were mapped to the UCSC Human genome build 19. The RNA map corresponding to RNA binding proteins TIA1 and TIAR is included at the bottom. The results were adapted using the TIA1 and TIAR *in vivo* ultraviolet (UV)-crosslinking and immunoprecipitation (iCLIP) analysis provided by the Ule laboratory [12].

Figure S4   List of massive sequencing-predicted genes in TIA1-silenced HeLa cells. Functional clusters based on Gene Ontology (GO) and Kyoto Encyclopedia of Genes and Genomes (KEGG) databases.

Figure S5   List of massive sequencing-predicted genes in TIAR-silenced HeLa cells. Functional clusters based on GO and KEGG databases.

Figure S6   List of shared and contrasting expressed genes and their GO/KEGG analysis from TIA1 and TIAR-silenced HeLa cells.

Figure S7   List of differentially expressed genes in TIA-silenced HeLa cells by microarray analysis. Functional clusters based on GO and KEGG databases.

Figure S8   List of primer pair sequences used in QPCR analysis.

Figure S9   Effect of TIA1 and TIAR knockdown on cell adhesion. Cell adhesion of control and TIA-knocked down HeLa cells was assessed using either plastic, BSA-coated or collagen-coated plates. Cells were seeded and processed as indicated. Thereafter, the number of adhered cells was quantified by measuring the conversion of MTT into DMSO-soluble formazan by living cells, at 570 nm. The represented values were normalized and expressed relative to control values (whose value was fixed arbitrarily to 1 and are mean ± standard error of the mean (SEM) of at least two independent experiments.

Figure S10   Effect of TIA1 and/or TIAR knockdown on transcriptional activation of the COL1A2 gene promoter. Schematic representation of the COL1A2 human gene promoter is shown. *Cis*-acting consensus sequences are represented by boxes. Control and TIA1 and/or TIAR-silenced HeLa cells were transiently cotransfected with the COL1A2 promoter-driven firefly luciferase construct together with a GFP-expressing plasmid (used as a transfection control). The represented values –the ratio between luciferase relative light units (RLU)/GFP expression measured by Western blot– were normalized and expressed relative to the control sample, whose value was fixed arbitrarily to 1, and are mean ± SEM of at least two independent experiments.

Acknowledgments

We wish to thank B Pardo and JM Sierra for critical reading of the manuscript and encouragement. We are grateful for the generosity of J Millán, F Rodríguez-Pascual, H Miyazaki and J Ule, who have provided us with different reagents and advice.

Author Contributions

Conceived and designed the experiments: JMI. Performed the experiments: MN CSJ JA. Analyzed the data: MN CSJ JA JMI. Contributed to the writing of the manuscript: MN CSJ JMI.

References

1. Tian Q, Streuli M, Saito H, Schlossman SF, Anderson P (1991) A polyadenylate binding protein localized to the granules of cytolytic lymphocytes induces DNA fragmentation in target cells. Cell 67: 629–639.
2. Kawakami A, Tian Q, Duan X, Streuli M, Schlossman SF, et al. (1992) Identification and functional characterization of a TIA-1-related nucleolysin. Proc Natl Acad Sci 89: 8681–8685.
3. Barbosa-Morais NL, Irimia M, Pan Q, Xiong HY, Gueroussov S, et al. (2012) The evolutionary landscape of alternative splicing in vertebrate species. Science 338: 1587–1593.
4. Merkin J, Russell C, Chen P, Burge CB (2012) Evolutionary dynamics of gene and isoform regulation in mammalian tissues. Science 338: 1593–1599.
5. Suswam EA, Li YY, Mahtani H, King PH (2005) Novel DNA-binding properties of the RNA-binding protein TIAR. Nucleic Acids Res 33: 4507–4518.
6. McAlinden A, Liang L, Mukudai Y, Imamura T, Sandell LJ (2007) Nuclear protein TIA1 regulates COL2A1 alternative splicing and interacts with precursor mRNA and genomic DNA. J Biol Chem 282: 24444–24454.
7. Das R, Yu J, Zhang Z, Gygi MP, Krainer AR, et al. (2007) SR proteins function in coupling RNAP II transcription to pre-mRNA splicing. Mol Cell 26: 867–881.
8. Kim HS, Wilce MC, Yoga YM, Pendini NR, Gunzburg MJ, et al. (2011) Different modes of interaction by TIAR and HuR with target RNA and DNA. Nucleic Acids Res 39: 1117–1130.
9. Del Gatto-Konczak F, Bourgeois CF, Le Guiner C, Kister L, Gesnel MC, et al. (2000) The RNA-binding protein TIA1 is a novel mammalian splicing regulator acting through intron sequences adjacent to a 5′ splice site. Mol Cell Biol 20: 6287–6299.
10. Förch P, Puig O, Kedersha N, Martínez C, Granneman S, et al. (2000) The apoptosis-promoting factor TIA1 is a regulator of alternative pre-mRNA splicing. Mol Cell 6: 1089–1098.
11. Izquierdo JM, Majós N, Bonnal S, Martínez C, Castelo R, et al. (2005) Regulation of Fas alternative splicing by antagonistic effects of TIA1 and PTB on exon definition. Mol Cell 19: 475–484.
12. Wang Z, Kayikci M, Briese M, Zarnack K, Luscombe NM, et al. (2010) iCLIP predicts the dual splicing effects of TIA-RNA interactions. PLoS Biol 8: e1000530.
13. López de Silanes I, Galbán S, Martindale JL, Yang X, Mazan-Mamczarz K, et al. (2005) Identification and functional outcome of mRNAs associated with RNA-binding protein TIA1. Mol Cell Biol 25: 9520–9531.
14. Mazan-Mamczarz K, Lal A, Martindale JL, Kawai T, Gorospe M (2006) Translational repression by RNA-binding protein TIAR. Mol Cell Biol 26: 2716–2727.
15. Kim HS, Kuwano Y, Zhan M, Pullmann R Jr, Mazan-Mamczarz K, et al. (2007) Elucidation of a C-rich signature motif in target mRNAs of RNA-binding protein TIAR. Mol Cell Biol 27: 6806–6817.
16. Yamasaki S, Stoecklin G, Kedersha N, Simarro M, Anderson P (2007) T-cell intracellular antigen-1 (TIA1)-induced translational silencing promotes the decay of selected mRNAs. J Biol Chem 282: 30070–30077.
17. Piecyk M, Wax S, Beck AR, Kedersha N, Gupta M, et al. (2000) TIA1 is a translational silencer that selectively regulates the expression of TNF-alpha. EMBO J 19: 4154–4163.
18. Liao B, Hu Y, Brewer G (2007) Competitive binding of AUF1 and TIAR to MYC mRNA controls its translation. Nat Struct Mol Biol 14: 511–518.
19. Reyes R, Alcalde J, Izquierdo JM (2009) Depletion of T-cell intracellular antigen (TIA)-proteins promotes cell proliferation. Genome Biol 10: R87.
20. Damgaard CK, Lykke-Andersen J (2011) Translational coregulation of 5′TOP mRNAs by TIA-1 and TIAR. Genes Dev 25: 2057–2068.
21. Gottschald OR, Malec V, Krasteva G, Hasan D, Kamlah F, et al. (2010) TIAR and TIA1 mRNA-binding proteins co-aggregate under conditions of rapid oxygen decline and extreme hypoxia and suppress the HIF-1α pathway. J Mol Cell Biol 2: 345–356.
22. Sánchez-Jiménez C, Carrascoso I, Barrero J, Izquierdo JM (2013) Identification of a set of miRNAs differentially expressed in transiently TIA-depleted HeLa cells by genome-wide profiling. BMC Mol Biol 14: 4.
23. Carrascoso I, Sánchez-Jiménez C, Izquierdo JM (2014) Long-term reduction of T-cell intracellular antigens leads to increased beta-actin expression. Mol Cancer 13: 90.
24. Carrascoso I, Sánchez-Jiménez C, Izquierdo JM (2014) Genome-wide profiling reveals a role for T-cell intracellular antigens TIA1 and TIAR in the control of translational specificity in HeLa cells. Biochem J 461: 43–50.
25. Izquierdo JM, Alcalde J, Carrascoso I, Reyes R, Ludeña MD (2011) Knockdown of T-cell intracellular antigens triggers cell proliferation, invasion and tumor growth. Biochem J 435: 337–344.
26. Beck AR, Miller JJ, Anderson P, Streuli M (1998) RNA-binding protein TIAR is essential for primordial germ cell development. Proc Natl Acad Sci 95: 2331–2336.
27. Kharraz Y, Salmand PA, Camus A, Auriol J, Gueydan C, et al. (2010) Impaired embryonic development in mice overexpressing the RNA-binding protein TIAR. PLoS One 5: e11352.
28. Sánchez-Jiménez C, Izquierdo JM (2013) T-cell intracellular antigen (TIA)-proteins deficiency in murine embryonic fibroblasts alters cell cycle progression and induces autophagy. PLoS One 8: e75127.

29. Twine NA, Janitz K, Wilkins MR, Janitz M (2011) Whole transcriptome sequencing reveals gene expression and splicing differences in brain regions affected by Alzheimer's disease. PLoS One 6: e16266.

30. Trapnell C, Pachter L, Salzberg SL (2009) TopHat: discovering splice junctions with RNA-seq. Bioinformatics 25: 1105–1111.

31. Trapnell C, Williams BA, Pertea G, Mortazavi A, Kwan G, et al. (2010) Transcript assembly and quantification by RNA-seq reveals unannotated transcripts and isoform switching during cell differentiation. Nat Biotechnology 28: 511–515.

32. Jiang H, Wong WH (2009) Statistical inferences for isoform expression in RNA-seq. Bioinformatics 25: 1026–1032.

33. Zweig AS, Karolchik D, Kuhn RM, Haussler D, Kent WJ (2008) UCSC genome browser tutorial. Genomics 92: 75–84.

34. Langmead B, Trapnell C, Pop M, Salzberg SL (2009) Ultrafast and memory-efficient alignment of short DNA sequences to the human genome. Genome Biol 10: R25.

35. Smyth GK, Speed TP (2003) Normalization of cDNA microarray data. Methods 31: 265–273.

36. Bolstad BM, Irizarry RA, Astrand M, Speed TP (2003) A comparison of normalization methods for high density oligonucleotide array data based on bias and variance. Bioinformatic 19: 185–193.

37. Smyth GK (2001) Limma: linear models for microarray data. In: Gentleman R, Carey V, Dudoit S, Irizarry R, Huber W, editors. Bioinformatics and Computational Biology Solutions using R and Bioconductor. Springer: New York. 397–420.

38. Benjamini Y, Drai D, Elmer G, Kafkafi N, Golani I (2001) Controlling the false discovery rate in behavior genetics research. Behav Brain Res 125: 279–284.

39. Oliveros JC (2007) FIESTA@BioinfoGP.An interactive server for analyzing DNA microarray experiments with replicates. FIESTA Viewer New FIESTA Server Available: http://bioinfogp.cnb.csic.es/tools/FIESTA.

40. Carmona-Saez P, Chagoyen M, Tirado F, Carazo JM, Pascual-Montano A (2007) GENECODIS: a web-based tool for finding significant concurrent annotations in gene lists. Genome Biol 8: R3.

41. Nogales-Cadenas R, Carmona-Saez P, Vazquez M, Vicente C, Yang X, et al. (2009) GeneCodis: interpreting gene lists through enrichment analysis and integration of diverse biological information. Nucleic Acids Res 37: W317–322.

42. Chen Y (2011) Cell adhesion assay. Bio-protocol 1: e98.

43. Miyazaki H, Kobayashi R, Ishikawa H, Awano N, Yamagoe S, Miyazaki Y, Matsumoto T (2012) Activation of COL1A2 promoter in human fibroblasts by Escherichia coli. FEMS Immunol Med Microbiol 65: 481–487.

44. Heck MV, Azizov M, Stehning T, Walter M, Kedersha N, et al. (2014) Dysregulated expression of lipid storage and membrane dynamics factors in Tia1 knockout mouse nervous tissue. Neurogenetics 15: 135–144.

45. Watt FM, Huck WT (2013) Role of the extracellular matrix in regulating stem cell fate. Nat Rev Mol Cell Biol 14: 467–473.

46. Li X, Rayman JB, Kandel ER, Derkatch IL (2014) Functional Role of Tia1/Pub1 and Sup35 Prion Domains: Directing Protein Synthesis Machinery to the Tubulin Cytoskeleton. Mol Cell 55: 305–318.

47. Duperret EK, Ridky TW (2013) Focal adhesion complex proteins in epidermis and squamous cell carcinoma. Cell Cycle 12: 3272–3285.

48. González DM, Medici D (2014) Signaling mechanisms of the epithelial-mesenchymal transition. Sci Signal 7: re8.

49. Karalok HM, Karalok E, Saglam O, Torun A, Guzeloglu-Kayisli O, et al. (2014) mRNA-binding Protein TIA-1 Reduces Cytokine Expression in Human Endometrial Stromal Cells and is Down-Regulated in Ectopic Endometrium. J Clin Endocrinol Metab Aug 20: jc20133488.

50. Hamdollah Zadeh MA, Amin EM, Hoareau-Aveilla C, Domingo E, Symonds KE, et al. (2014) Alternative splicing of TIA-1 in human colon cancer regulates VEGF isoform expression, angiogenesis, tumour growth and bevacizumab resistance. Mol Oncol Aug 20. pii: S1574-7891(14)00172-0. doi:10.1016/j.molonc.2014.07.017.

51. Aznarez I, Barash Y, Shai O, He D, Zielenski J, et al. (2008) A systematic analysis of intronic sequences downstream of 5′ splice sites reveals a widespread role for U-rich motifs and TIA1/TIAL1 proteins in alternative splicing regulation. Genome Res 18: 1247–1258.

52. Gonzalez-Pérez A, Pérez-Llamas C, Deu-Pons J, Tamborero D, Schroeder MP, et al. (2013) IntOGen-mutations identifies cancer drivers across tumor types. Nat Methods 10: 1081–1082.

21

Stable Composition of the Nano- and Picoplankton Community during the Ocean Iron Fertilization Experiment LOHAFEX

Stefan Thiele[1☯¶¤a], Christian Wolf[2☯¶], Isabelle Katharina Schulz[2,3¤b], Philipp Assmy[4], Katja Metfies[2], Bernhard M. Fuchs[1]*

1 Department of Molecular Ecology, Max Planck Institute for Marine Microbiology, Bremen, Germany, **2** Department of Polar Biological Oceanography, Division of Bioscience, Alfred Wegener Institute - Helmholtz Centre for Polar and Marine Research, Bremerhaven, Germany, **3** Bremen International Graduate School for Marine Sciences (GLOMAR), MARUM - Center for Marine Environmental Sciences, University of Bremen, Bremen, Germany, **4** Center for Ice, Climate and Ecosystems (ICE), Norwegian Polar Institute, Fram Centre, Tromsø, Norway

Abstract

The iron fertilization experiment LOHAFEX was conducted in a cold-core eddy in the Southern Atlantic Ocean during austral summer. Within a few days after fertilization, a phytoplankton bloom developed dominated by nano- and picoplankton groups. Unlike previously reported for other iron fertilization experiments, a diatom bloom was prevented by iron and silicate co-limitation. We used 18S rRNA gene tag pyrosequencing to investigate the diversity of these morphologically similar cell types within the nano- and picoplankton and microscopically enumerated dominant clades after catalyzed reported deposition fluorescence *in situ* hybridization (CARD-FISH) with specific oligonucleotide probes. In addition to *Phaeocystis*, members of Syndiniales group II, clade 10–11, and the *Micromonas* clades ABC and E made up a major fraction of the tag sequences of the nano- and picoplankton community within the fertilized patch. However, the same clades were also dominant before the bloom and outside the fertilized patch. Furthermore, only little changes in diversity could be observed over the course of the experiment. These results were corroborated by CARD-FISH analysis which confirmed the presence of a stable nano- and picoplankton community dominated by *Phaeocystis* and *Micromonas* during the entire course of the experiment. Interestingly, although Syndiniales dominated the tag sequences, they could hardly be detected by CARD-FISH, possibly due to the intracellular parasitic life style of this clade. The remarkable stability of the nano- and picoplankton community points to a tight coupling of the different trophic levels within the microbial food web during LOHAFEX.

Editor: Rodolfo Paranhos, Instituto de Biologia, Brazil

Funding: This work was funded by the Max Planck Society and the Helmholtz Association (Young Investigator Group PLANKTOSENS, VH-NG-500). The funders had no role in study design, data collection and analysis, decision to publish, or preparation of the manuscript.

Competing Interests: The authors have declared that no competing interests exist.

* Email: bfuchs@mpi-bremen.de

☯ These authors contributed equally to this work.

¶ These authors are shared first authors on this work.

¤a Current address: Stazione Zoologica Anton Dohrn, Naples, Italy
¤b Current address: Red Sea Research Center, King Abdullah University of Science and Technology (KAUST), Thuwal, Kingdom of Saudi-Arabia

Introduction

Phytoplankton blooms occur seasonally in large parts of the oceans. Typically, a spring or upwelling bloom dominated by large diatoms is followed closely by a community dominated by small nanoplankton. However, wide ocean areas exhibit low phytoplankton standing stocks despite perennially high nutrient concentrations. Such high nutrient - low chlorophyll areas (HNLCs) are present in the subarctic and equatorial Pacific Ocean but also in most of the Southern Ocean. John Martin and colleagues postulated in the early 1990-ies that iron availability limits phytoplankton growth in these HNLC areas [1]. In the following years a dozen Lagrangian experiments in iron-limited HNLC waters have shown that phytoplankton blooms can be induced by artificial iron fertilization [2]. In most experiments the iron induced blooms were dominated by large diatoms which stimulated the idea that the export of rapidly sinking diatom aggregates from iron-induced blooms could enhance the strength and efficiency of the biological carbon pump [3]. Additionally, the enhanced primary production in surface waters would lead to an increase of dissolved organic carbon (DOC) and particulate organic carbon (POC), both of which are the basis nutrition for different levels of the microbial loop, in particular the bacterial and archaeal community [4]. To quantify the extent of carbon export of phytoplankton biomass and the impact of the microbial loop in surface waters the Indo-German iron fertilization experiment LOHAFEX ('loha' is Hindi for 'iron'; FEX for Fertilization EXperiment) was conducted in late austral summer of 2009 in a cold core eddy north of the Antarctic Polar Front in the Atlantic sector of the subantarctic Southern Ocean. In previous iron fertilization experiments the abundance and diversity of the large bloom-forming diatoms had been explored in detail [5–8], while smaller Eukarya, ranging from 2–20 μm (nanoplankton) and 0.2–2 μm (picoplankton), have been rarely explored and were treated as "black boxes" in most of the studies so far. Eukaryotic nano- and picoplankton have been observed to dominate blooms after iron-fertilization [9,10], especially in areas with a co-limitation of iron and silicate. For example during the SAGE iron fertilization experiment, haptophytes and prasinophytes accounted for ∼75% of the chlorophyll *a* content [11]. During LOHAFEX mainly *Phaeocystis*-like small flagellated and non-flagellated taxa dominated the bloom upon fertilization and only little export could be measured which was possibly the consequence of the co-limitation of dissolved iron and silica in the fertilized patch [12–14].

With this study we identified and quantified the response of the eukaryotic nano- and picoplankton during the LOHAFEX experiment. Several methods are available for identification and quantification. The most wide-spread method is to count cells in Lugol- or formaldehyde-fixed water samples settled in sedimentation chambers by inverted light microscopy and to quantify total cell numbers based on different size classes and morphologies [15]. The cell numbers of eukaryotic nano- and picoplankton during LOHAFEX have been quantified by this method and are reported in detail in an accompanying study [12]. However with the exception of a few morphologically distinct species, quantification of specific groups of these otherwise featureless small eukaryotes remains problematic. Scanning electron microscopy provides more morphological details due to higher resolution [16], yet it is not suited for high throughput analyses.

Molecular biological tools based on ribosomal RNA genes, like catalysed reporter deposition (CARD-) fluorescence *in situ* hybridization (FISH) [17,18] and tag pyrosequencing [19], provide a stable phylogenetic framework with a resolution superior to that of other molecular methods such as marker pigment analyses [20]. Using the rRNA approach, a wealth of previously unexplored diversity was recently revealed from different ocean areas [21–24]. FISH is well established for the identification and quantification of Bacteria and Archaea in complex environmental samples, and was also successfully applied to investigate eukaryotic nano- and picoplankton communities [22,25]. A combined approach using sequencing and FISH methods is commonly used for the identification of bacterial and archaeal communities [17]. Therefore in this study we aimed at combining methods established for nano- and picoplankton analyses, like light microscopic quantification of Lugol-fixed samples with tag pyrosequencing [19] and FISH [18] to characterize the eukaryotic nano- and picoplankton community composition with higher taxonomic resolution. This combination of methods has also the capacity to tap into yet unknown diversity and to discover novel organisms involved in iron-induced phytoplankton blooms.

Material & Methods

Sampling

The iron fertilization experiment LOHAFEX was conducted during the RV "Polarstern" cruise ANT XXV/3 (12th January to 6th March, 2009) as described previously [13,26]. Briefly, the closed core of a stable cyclonic eddy adjacent to the Antarctic Polar Front in the Atlantic sector of the subantarctic Southern Ocean was fertilized with 2 t of Fe (10 t of $FeSO_4 \times 7 H_2O$) on 27th January. A second fertilization was applied using 2 t of Fe (10 t of $FeSO_4 \times 7 H_2O$) after 18 days (on 14th February). The fertilized patch was monitored for 38 days. As a response to the fertilization, Fv/Fm ratios increased from below 0.3 to above 0.45 and

chlorophyll *a* concentrations increased from 0.5 μg l^{-1} to 1.0–1.2 μg l^{-1} within 14 days [12,13]. The peak chlorophyll value of 1.6 μg l^{-1} was reached at the end of the third week. Both Fv/Fm and chlorophyll values decreased thereafter to values of 0.35 and 0.7 μg l^{-1}, respectively. Samples were taken on day −1 prior to the start of the experiment, on days 5, 9, 14, 18, 22, 24, 33, 36 inside the fertilized patch ("IN" stations) and days 4, 16, 29, 35 (only 20 m) and 38 outside the fertilized patch ("OUT" stations, Figure 1) [12]. Both IN and OUT stations were situated within the eddy. On each day, 190 ml of water from 20 m depth and 40 m depth were fixed with 10 ml acidic Lugol solution (5% final conc. v/v) and stored in brown glass bottles at 4°C in the dark for 1.5–2.5 years until manual counting and 3 years until CARD-FISH analysis. Due to the well mixed water column from surface down to 60–80 m depth (data not shown), both samples (20 and 40 m) were treated as replicates. For DNA extraction 90 l (day −1), 85 l (day 9), 75 l (day 16/OUT), and 67 l (day 18) were sampled at 20 m depth and filtered on 0.2 μm pore size cellulose acetate filters (Sartorius, Göttingen, Germany) after a prefiltration step with a 5 μm. filter. These samples were stored at −80°C.

DNA extraction and tag-pyrosequencing

DNA extraction was done using the E.Z.N.A. SP Plant DNA Kit (Omega Bio-Tek, Norcross, USA). Initially, the filters were incubated in lysis buffer (provided in the kit) at 65°C for 10 min before performing all further steps as described in the manufacturer's instructions. The eluted DNA was stored at −20°C until further analysis. We amplified ∼670 bp fragments of the 18S rRNA gene, containing the highly variable V4-region, using the primer-set 528F (5′-GCGGTAATTCCAGCTCCAA-3′) and 1055R (5′-ACGGCCATGCACCACCACCCAT-3′) modified after Elwood [27] as described by Wolf [28]. Pyrosequencing (single reads, forward direction) was performed on a Genome Sequencer FLX system (Roche, Penzberg, Germany) by GATC Biotech AG (Konstanz, Germany). Raw sequence reads were processed to obtain high quality reads (Table S1). Reads with a length below 300 bp, reads longer than >670 bp and reads with more than one uncertain base (N) were excluded from further analysis. Chimera sequences were excluded using the software UCHIME 4.2 [29]. The high quality reads of all samples were clustered into operational taxonomic units (OTUs) at the 98% identity level using the SILVAngs pipeline (https://www.arb-silva.de/ngs/; [30]) (Table S1). Consensus sequences of each OTU were generated and used for further analyses. The 98% identity level is conservative, but was found suitable to reproduce original eukaryotic diversity [31] and to embrace the error-rate of 454 pyrosequencing. The consensus sequences were aligned and imported into a manually curated reference tree containing 51.553 high quality sequences of Eukarya of the SILVA reference database (release SSU_Ref_119, July 2014) by parsimony criteria using the ARB software suite [30]. Classification was done based on the resulting positioning of the consensus sequence in the tree. The raw sequence data generated in this study has been deposited at GenBank's Short Read Archive (SRA) under the accession number SRA064723.

CARD-FISH

In tests, the margins of many Lugol-fixed cells appeared disrupted, shrunken or shapeless after CARD-FISH and indicated an elevated cell loss. Therefore, an additional fixation step with formaldehyde was introduced to further stabilize the cells for CARD-FISH and to ensure bright signals and stable cell counts. Hundred milliliter of Lugol-fixed sample was incubated for 1 h with formaldehyde (1% final concentration), neutralized with 1 M

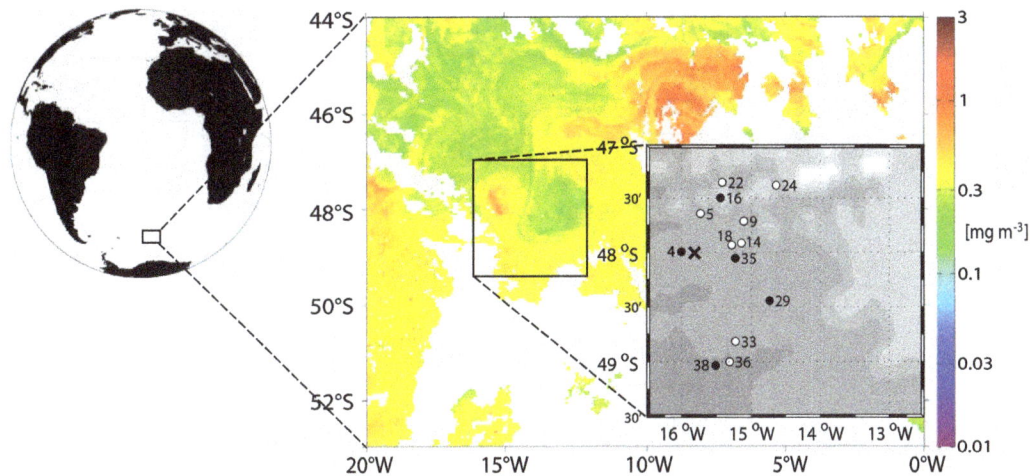

Figure 1. Map and sampling scheme of LOHAFEX. MODIS (Moderate Resolution Imaging Spectroradiometer) satellite image from 14. February 2009 showing chlorophyll *a* concentrations for the Polar Frontal Zone with the LOHAFEX bloom encircled. Stations and experiment days of both the IN (black) and OUT stations (white) are shown in the small map. The X marks day −1 before the iron addition on 27th January. The globe and the inset map were generated with the M_Map package for Matlab (version 7.12.0.635; MathWorks, Natick, MA). The chlorophyll *a* data were downloaded from the NASA website http://oceancolor.gsfc.nasa.gov/.

sodium thiosulfate and filtered onto polycarbonate filters with 0.8 µm pore size after pre-filtration using 20 µm pore size filters (Millipore, Tullagreen, Ireland). Due to limited sample amount, only 25 ml and 70 ml were filtered for samples from day −1 and day 38 (both 20 m depth).

CARD-FISH was done as described previously [32]. Briefly, samples were embedded in 0.1% agarose. A permeabilization step was done with Proteinase K (5 µg/ml) for 15 minutes for hybridizations with the probe PHAEO03 due to the length of 34 bp of this probe. Hybridization and amplification was done on glass slides using 50 ml tubes or in Petri dishes using 700 ml glass chambers as moisture chambers at 46°C. We used 14 horseradish peroxidase (HRP) labeled oligonucleotide probes (Table 1) including the probe NON338 as a negative control. All other probes were chosen according to 454 tag sequencing results. For signal amplification, Alexa488 labeled tyramides were used for all probes. After the CARD procedure samples were stained with DAPI for quantification of total cell numbers.

Cell quantification

For nano- and picoplankton cell quantification, two different methods were used. Quantification of CARD-FISH positive cells was done manually on an Eclipse 50i microscope (Nikon, Amstelveen, Netherland) at 1000x magnification in 50 fields of view (FOV) per sample in duplicates (Table S2). Total cell numbers were counted from the same CARD-FISH preparations using an automated counting routine. A Zeiss AxioImager. Z2 microscope (Zeiss, Jena, Germany) equipped with an automated stage was used to automatically acquire images from the preparations using the software package AxioVision Release 4.7 (Zeiss, Jena, Germany) and the macro [33] MPISYS. Image acquisition comprised an automated focusing routine, an automated sample area definition and a manual image quality assessment [33]. The software takes three images of each field of view along a given track on the sample, one in the DAPI channel (350 nm), one in the FISH channel (488 nm), and one at 594 nm at the main autofluorescence of the cells, caused by various cell components which were not further analyzed (Figure S1). These picture triplets were further processed using the software

ACMEtool 0.76 (an updated version including description is available on www.technobiology.ch). Before processing further, a manual quality check was done for every picture triplet and non-usable triplets were discarded. After this quality control, cells were detected automatically using an algorithm optimized for nano- and picoplankton quantification in the ACMEtool 0.76. Since DAPI signals were often quenched by strong autofluorescence and not all cells were stained by the general eukaryotic probe EUK516, the algorithm combines the green probe signal, the orange autofluorescence of accessory pigments and the green autofluorescence of the cells to define nano- and picoplankton cells (Figure S1). After automatic cell detection, the pictures were again manually evaluated to include cells missed by the evaluation algorithm and then all cells were quantified. For quantification only samples with a minimum of 15 image triplets were considered. Total nano- and picoplankton cell counts were calculated as a mean value from a minimum of 13 CARD-FISH preparations.

Probe design

Two new probes for the subclades I and II of the Syndiniales clade (Table 1) were designed using the probe design function of ARB [30] based on the SILVA ref 108 database [30] including the consensus sequences from tag pyrosequencing. Re-evaluations of the probes were done based on the SILVA ref 119 database from July 2014. Probe SYN-I-1161 had 48 target hits outside the Syndiniales group I (28 in dinoflagelates and 2 in Syndiniales group II, the rest scattered through the Eukarya). Probe SYN-II-675 showed no false-positives outside the Syndiniales group II. Optimal stringency of the probes was tested *in situ* by a series of increasing formamide concentration in the CARD-FISH buffer on a sample from day 38 (20 m depth) (Table 1).

Statistics

The total cell numbers obtained by manual counting using light microscopy and automated counting were compared using linear regressions. Normal distribution of the data was tested using the Kolmogorov-Smirnov test. Normal distributed data were tested using one way ANOVAs including Holm-Sidak comparison and

Table 1. List of oligonucleotides used in this study.

Probe	Target organism	Sequence (5′→3′)	FA (%)[a]	Reference
EUK516	Eukarya	ACCAGACTTGCCCTCC	0	[17]
NON338	Control	ACTCCTACGGGAGGCAGC	35	[45]
PRAS04	Mamiellophyceae	CGTAAGCCCGCTTTGAAC	40	[38]
PRYM02	Haptophyta	GGAATACGAGTGCCCCTGAC	40	[46]
MICRO01	Micromonas pusilla	AATGGAACACCGCCGGCG	40	[38]
PHAEO03	Phaeocystis	GAGTAGCCGCGGTCTCCGG AAAGAAGGCCGCGCC	20	[47]
PELA01	Pelagophyceae	GCAACAATCAATCCCAATC	20	[46]
MAST1A	MAST 1 clade	ATTACCTCGATCCGCAAA	30	[22]
MAST1B	MAST 1 clade	AACGCAAGTCTCCCCGCG	30	[22]
MAST1C	MAST 1 clade	GTGTTCCCTAACCCCGAC	30	[22]
MAST3	MAST 3	ATTACCTTGGCCTCCAAC	30	[48]
MAST4	MAST 4	TACTTCGGTCTGCAAACC	30	[48]
SYN-I-1161	Syndiniales group I	TCCTCGCGTTAGACACGC	20	This study
SYN-II-675	Syndiniales group II	CACCTCTGACGCGTTAAT	20	This study

A Probe-check of PRAS04 on SILVA ref 119 targeted 95% for the class Mamiellophyceae [38] with only one false-positive hit in the Dinophyceae and one in the Chrysophyceae, but no other hits in the Prasinophyceae. Thus, probe PRAS04 is specific only for Mamiellophyceae (Figure S3). Similarly, probe SYN-I-1161 resulted in a 30% coverage of the Syndiniales group I (48 outgroup hits) and SYN-II-675 targeted 42% of Syndiniales group II (no outgroup hits). However, SYN-II-675 targeted 82% of the Syndiniales group II clade 10–11, the main Syndiniales clade during LOHAFEX.
[a]Formamide concentration in the CARD-FISH hybridisation buffer.

not-normal distributed data were tested using ANOVA on ranks. All analyses were done using SigmaStat 3.5 (Statcon, Witzenhausen, Germany).

Results

Community composition

The diversity of eukaryotic nano- and picoplankton was assessed by tag pyrosequencing in the 0.2–5 μm fraction one day before the start of the experiment, during the experiment on days 9 and 18 inside the fertilized patch, and on day 16 outside the fertilized patch (Figure 2). All four samples had a similar composition with respect to the abundant OTUs. The most frequent tags in all samples originated from Alveolata (31–37%), Chlorophyta (24–29%), Haptophyta (19–27%), and Stramenopiles (14–21%). Some of the 22 abundant OTUs (>100 reads/OTU) showed fluctuations in sequence abundance over the course of the experiment, while in general the community was rather stable (Figure 2). The frequency of sequence tags originating from members of the genus *Phaeocystis* (Haptophyta) decreased from 23% (day −1) to 15% (day 18, Figure 2). Among the Mamiellophyceae, a class within the Chlorophyta, the genus *Micromonas* was quite frequent within the induced bloom (17–18%), but were considerably lower at day 16 (9%) outside the fertilized patch. Group E of *Micromonas* was slightly less abundant (4–10%) than *Micromonas* group ABC (5–10%) and both showed lowest abundance outside of the fertilized patch (Figure S2). Furthermore *Bathycoccus* sp. was found abundant (4–5%) in the Mamiellophyceae. Pelagophyceae showed a decrease in sequence abundance from 3.3% at day −1 to 1.0% at day 18 inside the patch, while they were found in highest abundance at the OUT station on day 16, with 4.6% (Figure 2). The most important Alveolata were Syndiniales represented by group I, II, and III. Group I was dominated by clades 1 (∼1%) and clade 4 (0.2–1.7%), while the total sequence abundance of this group never exceeded 3.5% (Figure S3 A). The abundance of

Syndiniales group III ranged around 1% with a minimum of 0.3% at day 16 outside the fertilized patch (Figure S3 A). Members of the Syndiniales group II were dominant inside and outside the bloom, showing relative abundance of 23–26%. The most dominant clade within this group was clade 10–11 with 12–17% relative abundance inside the fertilized patch (Figure S3 B). Clade 5 showed abundance of 1.2–3.1%, while clades 1, 6, 13, 16, 20, and 32 rarely occurred in abundance higher than 1.5% (Figure S3 B). Among the Stramenopiles, the most dominant OTU belonged to the MAST-1 clade (∼2%), and the MAST-3 clade (0.4–2.1%). The clades MAST-2, MAST-4 and MAST-7 were <1%. However, the bulk of OTUs within the Stramenopiles accounted for 4–10% of the sequences.

Nano- and picoplankton cell numbers

Nano- and picoplankton cells were first enumerated on Lugol fixed samples inside and outside the fertilized patch by manual counting using light microscopy [12]. In the mixed surface water layer nano- and picoplankton abundance was quite stable around 1.0×10^4 ml^{-1} $\pm 1.3 \times 10^3$ cells ml^{-1} but increased slightly from $8.9 \times 10^3 \pm 3.7 \times 10^2$ cells ml^{-1} on day 5 to $1.3 \times 10^4 \pm 4.6 \times 10^2$ cells ml^{-1} on day 22 after the second iron addition. Cell numbers remained at this elevated level during the later phase of the experiment (Figure 3 A). Outside the fertilized patch cell numbers were almost identical to inside the patch in the early phase of the experiment and remained rather stable over the course of the experiment ($9.4 \times 10^3 \pm 1.1 \times 10^3$ cells ml^{-1}). Only on the last day 38 of the experiment cell numbers increased to $1.1 \times 10^4 \pm 1.5 \times 10^3$ cells ml^{-1} (Figure 3 B).

In comparison cell counts obtained with automated cell counting after CARD-FISH of Lugol- and formaldehyde fixed samples were by a factor of ∼1.5 (range 1.0–2.5) lower compared to the manual counts. We calculated that during the CARD-FISH procedure an average cell loss of 26% ±11% occurred. However, similar to the manual counts, nano- and picoplankton abundance

		Day	-1	IN 9	18	OUT 16
Haptophyta	*Phaeocystis* sp.					
	Other Prymnesiophyceae					
Chlorophyta	*Bathyococcus* sp.					
	Micromonas group ABC					
	Micromonas group E					
	Other Mamiellophyceae					
Alveolata	Syndiniales group I clade 1					
	Syndiniales group I clade 4					
	Other Syndiniales group I					
	Syndiniales group II clade 1					
	Syndiniales group II clade 5					
	Syndiniales group II clade 6					
	Syndiniales group II clade 10-11					
	Syndiniales group II clade 13					
	Syndiniales group II clade 16					
	Syndiniales group II clade 20					
	Syndiniales group II clade 32					
	Other Syndiniales group II					
	Syndiniales Group III					
	Dinoflagellata					
Stramenopiles	Pelagophyceae					
	MAST-1					
	MAST-2					
	MAST-3					
	MAST-4					
	MAST-7					
	Other Stramenopiles					
	Other Eukarya					

0　　　　　　　　　　　25

Sequence abundance [%]

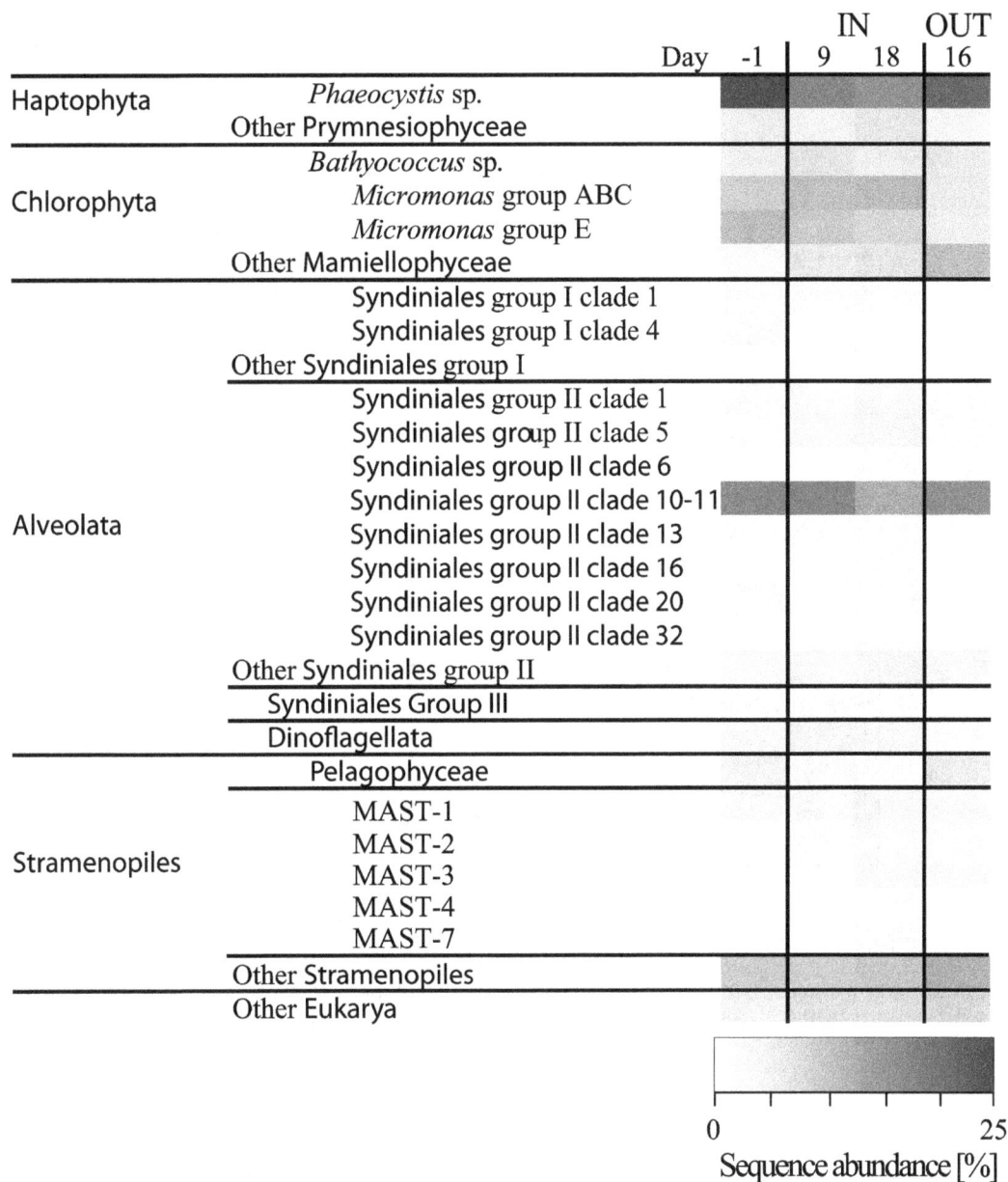

Figure 2. 18S rRNA tag frequency for the most abundant OTUs. An abundant OTU contains >100 sequences at least at one sampling point. Less abundant OTUs were summarized into 'Other [taxon]'.

peaked on day 22 with 9.3×10^3 cells ml^{-1}, but otherwise cell numbers remained rather constant at $6.1 \pm 1.3 \times 10^3$ cells ml^{-1} (Figure 3 A).

Quantification of specific nano- and picoplankton clades

For the detection of nano- and picoplankton cells, the probe EUK516 was used in CARD-FISH, since it targets more than 85% of all Eukarya sequences in the SILVA ref NR 119 rRNA database. On average 60% of the nano- and picoplankton showed a positive signal after CARD-FISH with EUK516 in relation to the counts obtained by automated cell counting (Figure 3A+B). The numbers of EUK516 positive cells were highest on day -1 with 5.8×10^3 cells ml^{-1}, decreasing to 1.9×10^3 cells ml^{-1} on day 9, before a second peak of 4.7×10^3 cells ml^{-1} and 4.8×10^3 cells ml^{-1} on days 22 and 24 inside the fertilized patch (Figure 3 A).

EUK-positive cell numbers were relatively constant outside the patch, but were as high as 7.0×10^3 cells ml^{-1} on day 38, which was significantly different from the comparable IN station on day 36 (p = 0.045) (Figure 3 B).

To investigate the community structure of the nano- and picoplankton, we used CARD-FISH probes with nested specificity for different taxonomic clades based on the tag sequencing data. Within the nano- and picoplankton community inside the fertilized patch, Prymnesiophyceae, mainly from the genus *Phaeocystis* were the main contributors to the nano- and picoplankton community. However, abundance of *Phaeocystis* and other Prymnesiophyceae did not change significantly within the fertilized patch over the course of the experiment (Figure 3A). Values were constant at about 1.0×10^3 cells ml^{-1} for all Prymnesiophyceae and 5.0×10^2 cells ml^{-1} for *Phaeocystis* accounting for about 50% of the Prymnesiophyceae. At the

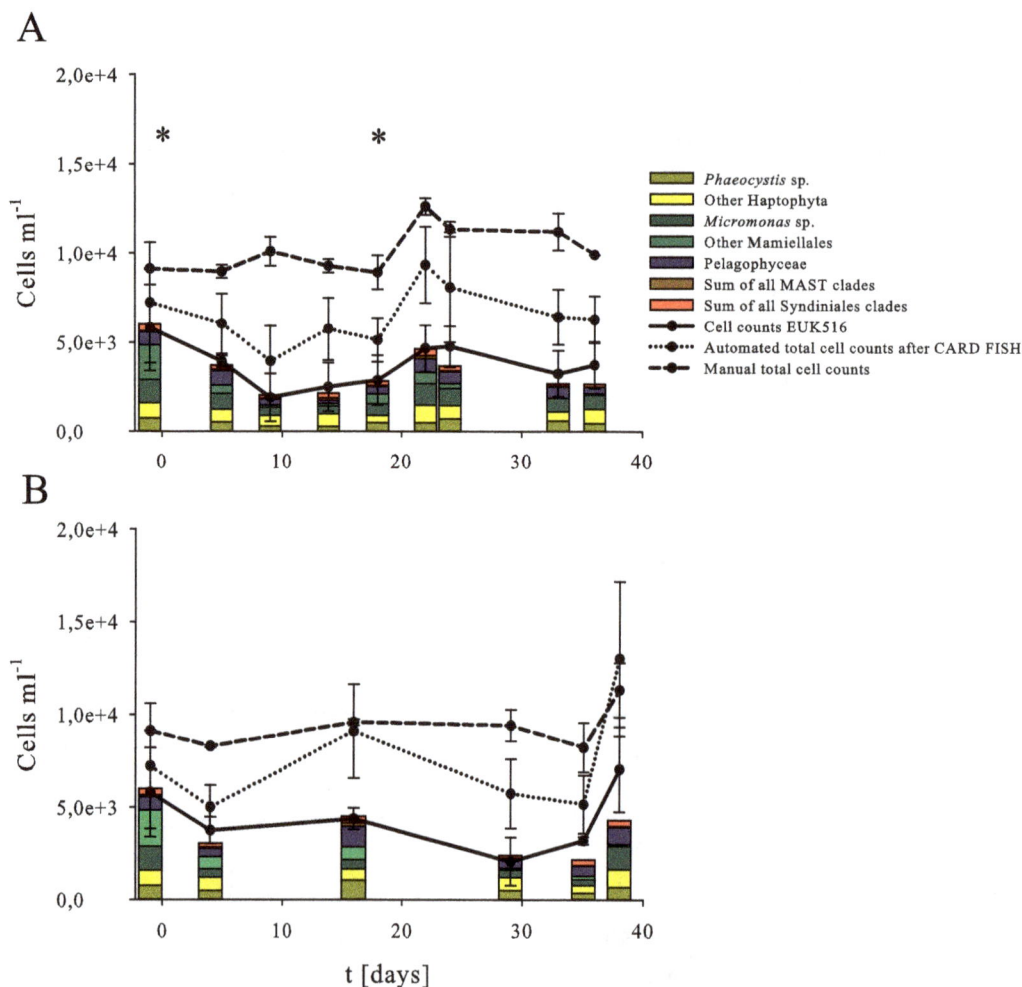

Figure 3. Quantification of the nano- and picoplankton community. Manual total cell counts from Lugol fixed samples (dashed lines), automated total cell counts after CARD-FISH (dotted lines), and cell counts of EUK516 probe (straight line) at IN (A) and OUT (B) stations. Stacked bar charts represent cell numbers of all other probes used in this study. Asterisks mark the iron fertilization events.

OUT station on day 16, higher numbers of *Phaeocystis* were found with 1.1×10^3 cells ml^{-1}, consistent with significantly higher numbers of Prymnesiophyceae (1.5×10^3 cells ml^{-1}) (p = 0.01) (Figure 3B).

Mamiellophyceae, a second dominant class in the tag sequences, showed a higher variation in cell numbers inside the fertilized patch, ranging from 3.2×10^3 cells ml^{-1} to 5.9×10^2 cells ml^{-1}, while cell numbers in the OUT stations remained rather constant and were significantly lower than inside the fertilized patch (p = 0.03). Cell numbers in the dominant subgroup *Micromonas* ranged from 1.3×10^3 cells ml^{-1} to around 4.4×10^2 cells ml^{-1}. On average *Micromonas* accounted for ~72% of the Mamiellophyceae (Figure 3A+B).

Pelagophyceae were also found rather stable in- and outside of the patch with numbers as high as 1.1×10^3 cells ml^{-1} outside the patch on day 16 (Figure 3A+B). Abundance of the group Marine Stramenopiles (MAST) was low and never exceeded 1.7×10^2 cells ml^{-1} during the course of the experiment (Figure 3A+B). Also the numbers of both Syndiniales clades were low and oscillated around 7.7×10^1 cells ml^{-1} for Syndiniales clade I and around 1.8×10^2 cells ml^{-1} for Syndiniales clade II within and outside of the fertilized patch, respectively (Figure 3A+B).

Discussion

A striking outcome of the iron fertilization experiment LOHAFEX was that the phytoplankton standing stocks were dominated by the nano- and picoflagellates, while diatoms never contributed more than 5% [12,13]. A similar response of the plankton community was found during the 15-day SAGE experiment, where diatoms were also co-limited by silicate and consequently picoplankton species dominated the planktonic community [11]. During LOHAFEX the cell numbers of nano- and picoplankton were remarkably stable and showed no large fluctuations during the experiment [12]. This was similar to the response of the bacterioplankton community reported earlier [26].

The diversity of the nano- and picoplankton community using 18S rRNA tag pyrosequencing did not change significantly in the samples analyzed. There were only minor differences between IN and OUT stations and the community composition was highly similar before and after the iron additions. The minor fluctuations found within the bloom could be also attributed to the inconsistencies inherent to every PCR-dependent assay. Consequently, CARD-FISH was used to check for any fluctuations in the major abundant groups of flagellates, which might have been missed by the tag sequencing analysis. We encounter several problems, which are summarized below.

When comparing the cell numbers gained by our automated microscopic cell counting routine with the cell numbers gained by direct light microscopic cell counting [12], we noticed consistently lower cell numbers of the former although both counts were done on the same samples obtained from the same CTD casts during the LOHAFEX experiment. The differences between the cell counts could be due to a number of reasons, maybe also a combination of them. During light microscopic counting, small coccoid cells tend to be underestimated, in contrast, biomass estimations could lead to overestimation of abundance due to cell shrinkage or swelling after fixation. For CARD-FISH, Lugol fixed samples had to be stabilised by additional fixation with formaldehyde [34]. The long storage time in Lugol solution for three years has likely led to cell loss of the more delicate cells. The subsequent filtration step could be an additional source of cell loss [35]. Some of the smaller picoplankton cells might have passed through the pores of 0.8 μm diameter of the polycarbonate filter [36] or cells might have been ruptured during filtration. Several washing steps during CARD-FISH might have led to cell loss, although the samples were embedded in a thin layer of agarose. However, the proportion of clades from tag sequencing and CARD-FISH were highly similar and therefore a preferential loss of a specific group of nano- or picoplankton cells is quite unlikely. However, future studies need to take preservation of these fragile cell types in consideration.

Existing probes were chosen for the dominant groups based on the tag sequencing data and, in the case of the Syndiniales, two new probes were designed. This was necessary to be able to distinguish between the two main Syndiniales groups I and II. Furthermore, Syndiniales probe SYN-II-675 was designed to match with clade 10–11, the most abundant clade during LOHAFEX. The relative abundance of CARD-FISH positive cells corroborated well the relative sequence representations from tag pyrosequencing. However, after CARD-FISH about 30–50% of the nano- and picoplankton cells showed no signal with the general probe EUK516, which was used as a positive control. Most likely this was due to quenching of the probe-conferred fluorescence by elevated autofluorescence of the cells. Alternatively, the accessibility of the ribosomes or even the number of ribosomes in the cells might have been reduced by Lugol fixation and thus no hybridization was possible. Nevertheless the sums for all clade-specific counts were in good agreement with the counts of the EUK516 probe (~90%) and demonstrate that we did not miss a major group of nano- or picoplankton.

Surprisingly during LOHAFEX Pelagophyceae and several MAST clades were found in relatively low abundance, even though ribosomal RNA studies show that members of the class Pelagophyceae (Stramenopiles) were reported frequently as major components of marine nano- and picoplankton communities [22]. Instead, the three most prominent bloom forming clades during the LOHAFEX experiment belonged to the Prymnesiophyceae, Mamiellophyceae, and Alveolata. Both, Prymnesiophyceae and Mamiellophyceae, made up 46–51% of the nano- and picoplankton community. Gomez-Pereira and coworkers found similar numbers of Mamiellophyceae in the same region [37]. The only *Micromonas* species, *M. pusilla*, was often dominating in phytoplankton blooms in the British Channel [38], in Pacific coastal waters and in Arctic waters [23,39], but to our knowledge have been detected only in low numbers in the Southern Ocean so far [28]. During LOHAFEX this species, together with *Phaeocystis*, dominated the iron-induced phytoplankton bloom.

Phaeocystis, a genus of the Prymnesiophyceae, forms large blooms worldwide [40]. During LOHAFEX the bulk of *Phaeocystis* biomass was allocated to solitary cells but formation of colonies attached to diatoms and small free-floating colonies were also observed [12]. The discrepancy between light microscopic ($\sim10^4$ ml^{-1}) and CARD-FISH counts ($\sim10^3$ ml^{-1}) shows the difficulties in counting solely based on morphological features and underpins the necessity to further characterize the clade *Phaeocystis* by molecular tools.

Within the Alveolata, three Syndiniales groups were among the dominant organisms by tag pyrosequencing. Syndiniales were found in the Ross Sea before, though in lower abundance [27]. Most abundant among all OTUs was the Syndiniales group II clade 10–11, while other clades and groups were found only in minor abundance. With the newly developed CARD-FISH probes we could detect Syndiniales only in relatively low abundance, both, inside and outside the fertilized patch. A possible explanation might be that members of the Syndiniales group have been described as endosymbionts and parasites within algae, tintinnids, crustaceans and other Dinophyceae [41], although free-living cells can occur in abundance [42,43]. Cells residing inside these organisms might be inaccessible for large HRP-labeled oligonucleotide probes, although the probe SYN-II-675 targets 82% of the sequences of the dominant Syndiniales group II clade 10–11 in the SILVA database. Syndiniales might have multiple 18S rRNA gene copies per cell, similar to the closely-related group of Dinoflagellata [44], which would partly explain the observed overrepresentation of Syndiniales in the tag sequences.

The taxonomically resolved monitoring of important components of the microbial loop during the LOHAFEX experiment revealed a surprising compositional stability. This stability is most likely caused by silicate limitation of diatoms and the absence of salps in the fertilized waters. While experimentally determined growth rates of diatoms were rather high, the low silicate concentration (<2 μM) were setting a low upper for diatom biomass build-up inside the fertilized patch [2]. It might be speculated, that salp grazing might have exerted a top down control on the nano- and picoplankton community. However, the salp abundance was low, most likely due to predation by *Themisto gaudichaudii* (Smetacek pers. communication). Due to these factors, nano- and picoplankton species were able to maintain high numbers and control the bacterial community [26], though no significant increase or change in diversity was found for the nano- and picoplankton community. It can be speculated that the lack of a pronounced increase by nano- and picoplankton inside the fertilized patch was due to top down control by dinoflagellates, naked ciliates, and tintinnids which were themselves kept in check by high numbers of copepods. During LOHAFEX, the whole planktonic ecosystem, and in particular the microbial loop, seemed to be tightly coupled, resulting in a strong cycling of carbon compounds within the microbial loop, that hence counteracted the efficiency of the biological carbon pump [11,13].

Supporting Information

Figure S1 Picture triplets obtained using the macro MPISYS. Three pictures from the same field of view taken in different channels with excitation light of different wavelength (DAPI: 365 nm, CARD-FISH: 470 nm and autofluorescence: 590 nm), using the probes PRAS04 (Mamiellophyceae) and PHAEO03 (*Phaeocystis*).

Figure S2 18S rRNA-based tree reconstructions of the Syndiniales groups. Tree in Figure S2A shows the different groups within the Syndiniales with a special focus on clades within the group I, while tree in Figure S2B displays clades of the Syndiniales group II. The trees were built using the ARB SILVA ref 119 database [30], calculated using Maximum Likelihood and Neighbour Joining algorithm. The aligned consensus tag sequences were added with parsimony criteria to the trees and percentage of tags falling into the respective clade are given behind the clades. Values in the wedges represent the number of reference sequences. Scale bar represents 5% and 1% estimated base substitution.

Figure S3 18S rRNA-based tree reconstruction of the Mamiellales clades. Values in the wedges represent the number of reference sequences, while values behind the clades show the abundance of LOHAFEX sequences in these clades The tree was build using the ARB SILVA ref 119 database [30], calculated using Neighbour Joining and Maximum Likelihood algorithms. The aligned consensus tag sequences were added with parsimony criteria to the trees and percentage of tags falling into the respective clade are given behind the clades. Values in the wedges represent the number of reference sequences. Scale bar represents 1% estimated base substitution.

Table S1 Summary statistics of pyrosequencing reads. The table also contains values after quality filtering and number of OTUs of the 0.2–5 μm size fraction.

Table S2 Total cell numbers. Results of the quantification of all probes at 20 m (2A) and 40 m (2B) depth. Counts for SYN-I-1161 and SYN-II-675 were not determined at day 29 OUT at 20 m and SYN-II-675 were not determined for day 29 OUT at 40 m (n.d.).

Acknowledgments

We would like to thank the captain and crew of RV Polarstern, the chief scientists of the LOHAFEX project, V. Smetacek and W. Naqvi, and the LOHAFEX scientific party. Furthermore we thank J. Köhler, E. Ruff, and A. Schröer for their help in the lab and P. Yilmaz for assistance with the SILVA taxonomy.

Author Contributions

Conceived and designed the experiments: ST CW IS PA KM BF. Performed the experiments: ST CW IS PA. Analyzed the data: ST CW IS. Contributed reagents/materials/analysis tools: PA KM BF. Wrote the paper: ST CW BF.

References

1. Martin JH (1990) Glacial-interglacial CO_2 change: The iron hypothesis. Paleoceanography 5: 1–13.
2. Boyd PW, Jickells T, Law CS, Blain S, Boyle EA, et al. (2007) Mesoscale Iron Enrichment Experiments 1993–2005: Synthesis and Future Directions. Science 315: 612–617.
3. Ducklow HW, Steinberg DK, Buessler KO (2001) Upper ocean carbon export and the biological pump. Oceanography 14: 50–58.
4. Azam F, Fenchel T, Field J, Gray J, Meyer L, et al. (1983) The ecological role of water column microbes in the sea. Mar Ecol Prog Ser 10: 257–263.
5. Gall MP, Boyd PW, Hall J, Safi KA, Chang H (2001) Phytoplankton processes. Part 1: Community structure during the Southern Ocean Iron RElease Experiment (SOIREE). Deep-Sea Res Pt II 48: 2551–2570.
6. Tsuda A, Kiyosawa H, Kuwata A, Mochizuki M, Shiga N, et al. (2005) Responses of diatoms to iron-enrichment (SEEDS) in the western subarctic Pacific, temporal and spatial comparisons. Prog Oceanograph 64: 189–205.
7. Assmy P, Henjes J, Klaas C, Smetacek V (2007) Mechanisms determining species dominance in a phytoplankton bloom induced by the iron fertilization experiment EisenEx in the Southern Ocean. Deep-Sea Res Pt I 54: 340–362.
8. Smetacek V, Klaas C, Strass VH, Assmy P, Montresor M, et al. (2012) Deep carbon export from a Southern Ocean iron-fertilized diatom bloom. Nature 487: 313–319.
9. Hall JA, Safi K (2001) The impact of in situ Fe fertilisation on the microbial food web in the Southern Ocean. Deep-Sea Res Pt II 48: 2591–2613.
10. Coale KH, Johnson KS, Chavez FP, Buesseler KO, Barber RT, et al. (2004) Southern Ocean iron enrichment experiment: Carbon cycling in high- and low-Si waters. Science 304: 408–414.
11. Peloquin J, Hall J, Safi K, Smith WO Jr, Wright S, et al. (2011) The response of phytoplankton to iron enrichment in Sub-Antarctic HNLCLSi waters: Results from the SAGE experiment. Deep-Sea Res Pt II 58: 808–823.
12. Schulz IK (2013) Mechanisms determining species succession and dominance during an iron-induced phytoplankton bloom in the Southern Ocean (LOHAFEX). Doctoral thesis, University Bremen, Germany. Available: http://nbn-resolving.de/urn:nbn:de:gbv:46-00103521-10. Accessed 15th October 2014.
13. Martin P, van der Loeff MR, Cassar N, Vandromme P, d' Ovidio F, et al. (2013) Iron fertilization enhanced net community production but not downward particle flux during the Southern Ocean iron fertilization experiment LOHAFEX. Global Biogeochem Cy 27: 871–881.
14. Ebersbach F, Assmy P, Martin P, Schulz I, Wolzenburg S, et al. (2014) Particle flux characterisation and sedimentation patterns of protistan plankton during the iron fertilisation experiment LOHAFEX in the Southern Ocean. Deep Sea Research Part I: Oceanographic Research Papers 89: 94–103.
15. Utermöhl H (1958) Zur Vervollkommnung der quantitativen Phytoplankton-Methodik. Mitt int Ver theor angew Limnol 9: 1–38.
16. Vørs N, Buck KR, Chavez FP, Eikrem W, Hansen LE, et al. (1995) Nanoplankton of the equatorial Pacific with emphasis on the heterotrophic protists. Deep-Sea Res Pt II 42: 585–602.
17. Amann RI, Ludwig W, Schleifer KH (1995) Phylogenetic identification and in situ detection of individual microbial cells without cultivation. Microbiol Rev 59: 143–169.
18. Pernthaler A, Pernthaler J, Amann R (2002) Fluorescence in situ hybridization and catalyzed reporter deposition for the identification of marine bacteria. Appl Environ Microbiol 68: 3094–3101.
19. Ronaghi M, Karamohamed S, Pettersson B, Uhlén M, Nyrén P (1996) Real-Time DNA sequencing using detection of pyrophosphate release. Analyt Biochem 242: 84–89.
20. Mackey M, Mackey D, Higgins H, Wright S (1996) CHEMTAX - a program for estimating class abundances from chemical markers: application to HPLC measurements of phytoplankton. Mar Ecol Prog Ser 144: 265–283.
21. Not F, Simon N, Biegala IC, Vaulot D (2002) Application of fluorescent in situ hybridization coupled with tyramide signal amplification (FISH-TSA) to assess eukaryotic picoplankton composition. Aquat Microb Ecol 28: 157–166.
22. Massana R, Terrado R, Forn I, Lovejoy C, Pedrós-Alió C (2006) Distribution and abundance of uncultured heterotrophic flagellates in the world oceans. Environ Microbiol 8: 1515–1522.
23. Kilias E, Wolf C, Nöthig EM, Peeken I, Metfies K (2013) Protist distribution in the Western Fram Strait in summer 2010 based on 454-pyrosequencing of 18S rDNA. J Phycol 49: 996–1010.
24. Unrein F, Gasol JM, Not F, Forn I, Massana R (2014) Mixotrophic haptophytes are key bacterial grazers in oligotrophic coastal waters. ISME J 8: 164–176.
25. Beardsley C, Knittel K, Amann R, Pernthaler J (2005) Quantification and distinction of aplastidic and plastidic marine nanoplankton by fluorescence in situ hybridization. Aquat Microb Ecol 41: 163–169.
26. Thiele S, Fuchs BM, Amann N, Amann R (2012) Microbial community response during the iron fertilization experiment LOHAFEX. Appl Environ Microbiol 78: 8803–8812.
27. Elwood HJ, Olsen GJ, Sogin ML (1985) The small-subunit ribosomal RNA gene sequences from the hypotrichous ciliates Oxytricha nova and Stylonychia pustulata. Mol Biol Evol 2: 399–410.
28. Wolf C, Frickenhaus S, Kilias ES, Peeken I, Metfies K (2013) Regional variability in eukaryotic protist communities in the Amundsen Sea. Antarctic Science 25: 741–751.
29. Edgar RC, Haas BJ, Clemente JC, Quince C, Knight R (2011) UCHIME Improves Sensitivity and Speed of Chimera Detection. Bioinformatics 27: 2194–2200.
30. Quast C, Pruesse E, Yilmaz P, Gerken J, Schweer T, et al. (2012) The SILVA ribosomal RNA gene database project: improved data processing and web-based tools. Nucl Acids Res 41: D590–D596.
31. Behnke A, Engel M, Christen R, Nebel M, Klein RR, et al. (2011) Depicting more accurate pictures of protistan community complexity using pyrosequencing of hypervariable SSU rRNA gene regions. Environ Microbiol 13: 340–349.
32. Thiele S, Fuchs B, Amann R (2011) Identification of microorganisms using the ribosomal RNA approach and fluorescence in situ hybridization. In: Wilderer P, editor. Treatise on Water Science. Oxford: Academic Press, Vol. 3. 171–189.
33. Zeder M, Kohler E, Pernthaler J (2010) Automated quality assessment of autonomously acquired microscopic images of fluorescently stained bacteria. Cytometry A 77: 76–85.
34. Sherr EB, Caron DA, Sherr BF (1993) Staining of heterotrophic protists for visualization via epifluorescence microscopy. In: Kemp PF, Cole JJ, Sherr BF,

Sherr EB, editors. Handbook of Methods in Aquatic Microbial Ecology. Boca Raton, USA: CRC Press. 213–227.

35. Bloem J, Bar-Gilissen MJB, Cappenberg TE (1986) Fixation, counting, and manipulation of heterotrophic Nanoflagellates. Appl Environ Microbiol 52: 1266–1272.

36. Gasol JM, Morn XAG (1999) Effects of filtration on bacterial activity and picoplankton community structure as assessed by flow cytometry. Aquat Microb Ecol 16: 251–264.

37. Gómez-Pereira PR, Kennaway G, Fuchs BM, Tarran GA, Zubkov MV (2013) Flow cytometric identification of Mamiellales clade II in the Southern Atlantic Ocean. FEMS Microbiology Ecology 83: 664–671.

38. Not F, Latasa M, Marie D, Cariou T, Vaulot D, et al. (2004) A single species, *Micromonas pusilla* (Prasinophyceae), dominates the eukaryotic picoplankton in the western English Channel. Appl Environ Microbiol 70: 4064–4072.

39. Balzano S, Marie D, Gourvil P, Vaulot D (2012) Composition of the summer photosynthetic pico and nanoplankton communities in the Beaufort Sea assessed by T-RFLP and sequences of the 18S rRNA gene from flow cytometry sorted samples. ISME J 6: 1480–1498.

40. Schoemann V, Becquevort S, Stefels J, Rousseau V, Lancelot C (2005) *Phaeocystis* blooms in the global ocean and their controlling mechanisms: a review. J Sea Res 53: 43–66.

41. Coats DW, Adam EJ, Gallegos CL, Hedrick S (1996) Parasitism of photosynthetic dinoflagellates in a shallow subestuary of Chesapeake Bay, USA. Aquat Microb Ecol 11: 1–9.

42. Chambouvet A, Morin P, Marie D, Guillou L (2008) Control of toxic marine dinoflagellate blooms by serial parasitic killers. Science 322: 1254–1257.

43. Siano R, Alves-de-Souza C, Foulon E, Bendif EM, Simon N, et al. (2011) Distribution and host diversity of Amoebophryidae parasites across oligotrophic waters of the Mediterranean Sea. Biogeosciences 8: 267–278.

44. Zhu F, Massana R, Not F, Marie D, Vaulot D (2006) Mapping of picoeucaryotes in marine ecosystems with quantitative PCR of the 18S rRNA gene. FEMS Microbiol Ecol 52: 79–92.

45. Wallner G, Amann R, Beisker W (1993) Optimizing fluorescent in situ hybridization with rRNA-targeted oligonucleotide probes for flow cytometric identification of microorganisms. Cytometry 14: 136–143.

46. Simon N, Campbell L, Ornolfsdottir E, Groben R, Guillou L, et al. (2000) Oligonucleotide probes for the identification of three algal groups by dot blot and fluorescent whole-cell hybridization. J Eukaryot Microbiol 47: 76–84.

47. Zingone A, Chretiennot-Dinet M, Lange M, Medlin L (1999) Morphological and genetic characterization of *Phaeocystis cordata* and *P. jahnii* (Prymnesiophyceae), two new species from the Mediterranean Sea. J Phycol: 1322–1337.

48. Massana R, Guillou L, Diez B, Pedrós-Alió C (2002) Unveiling the Organisms behind Novel Eukaryotic Ribosomal DNA Sequences from the Ocean. Appl Environ Microbiol 68: 4554–4558.

22

Dihydroflavonol 4-Reductase Genes Encode Enzymes with Contrasting Substrate Specificity and Show Divergent Gene Expression Profiles in *Fragaria* Species

Silvija Miosic[1], Jana Thill[1], Malvina Milosevic[1], Christian Gosch[1], Sabrina Pober[1], Christian Molitor[2], Shaghef Ejaz[3], Annette Rompel[2], Karl Stich[1], Heidi Halbwirth[1]*

1 Vienna University of Technology, Institute of Chemical Engineering, Vienna, Austria, **2** Institut für Biophysikalische Chemie, Fakultät für Chemie, Universität Wien, Vienna, Austria, **3** Bahauddin Zakariya University, Department of Horticulture, Multan, Pakistan

Abstract

During fruit ripening, strawberries show distinct changes in the flavonoid classes that accumulate, switching from the formation of flavan 3-ols and flavonols in unripe fruits to the accumulation of anthocyanins in the ripe fruits. In the common garden strawberry (*Fragaria×ananassa*) this is accompanied by a distinct switch in the pattern of hydroxylation demonstrated by the almost exclusive accumulation of pelargonidin based pigments. In *Fragaria vesca* the proportion of anthocyanins showing one (pelargonidin) and two (cyanidin) hydroxyl groups within the B-ring is almost equal. We isolated two dihydroflavonol 4-reductase (DFR) cDNA clones from strawberry fruits, which show 82% sequence similarity. The encoded enzymes revealed a high variability in substrate specificity. One enzyme variant did not accept DHK (with one hydroxyl group present in the B-ring), whereas the other strongly preferred DHK as a substrate. This appears to be an uncharacterized DFR variant with novel substrate specificity. Both *DFRs* were expressed in the receptacle and the achenes of both *Fragaria* species and the *DFR2* expression profile showed a pronounced dependence on fruit development, whereas *DFR1* expression remained relatively stable. There were, however, significant differences in their relative rates of expression. The *DFR1/DFR2* expression ratio was much higher in the *Fragaria×ananassa* and enzyme preparations from *F.×ananassa* receptacles showed higher capability to convert DHK than preparations from *F. vesca*. Anthocyanin concentrations in the *F.×ananassa* cultivar were more than twofold higher and the cyanidin:pelargonidin ratio was only 0.05 compared to 0.51 in the *F. vesca* cultivar. The differences in the fruit colour of the two *Fragaria* species can be explained by the higher expression of *DFR1* in *F.×ananassa* as compared to *F. vesca*, a higher enzyme efficiency (K_{cat}/K_m values) of DFR1 combined with the loss of F3'H activity late in fruit development of *F.×ananassa*.

Editor: Stefan Strack, University of Iowa, United States of America

Funding: Investigations were supported by a grant from the Austrian Science Fund (FWF, www.fwf.ac.at): [Projects V18-B03 and P24331-B16]. The funders had no role in study design, data collection and analysis, decision to publish, or preparation of the manuscript.

Competing Interests: The authors have declared that no competing interests exist.

* Email: hhalb@mail.zserv.tuwien.ac.at

Introduction

The strawberry is an appealing plant model for studying flavonoid metabolism during fruit development, as there is not only a change in the flavonoid classes but also in their B-ring hydroxylation patterns. These hydroxylation patterns switch from mainly dihydroxylated flavan 3-ols and flavonols in unripe fruits to monohydroxylated anthocyanins in ripe fruits [1,2,3]. In the *F. vesca* the ratio of anthocyanins possessing one (pelargonidin type) and two (cyanidin type) hydroxyl groups in the B-ring is almost equal, whereas pelargonidin type anthocyanins are particularly prevalent in the *F.×ananassa*. This is frequently reflected in fruit colouration (Figure S1 in File S1) [1,4]. The changes in the hydroxylation patterns can be achieved in two ways: either via the downregulation of *flavonoid 3'-hydroxylase* (*F3'H*) expression in late fruit stages or via the presence of a set of dihydroflavonol 4-reductases (DFR) showing different substrate specificities. Recently we have demonstrated that the differing hydroxylation pattern of

anthocyanins in *F. vesca* and *F.×ananassa* is reflected in the *F3'H* expression pattern. In *F. vesca*, *F3'H* was highly expressed during all developmental stages. This contrasted sharply with a decline in the expression of *F3'H* observed in *F.×ananassa* [1], also reported for *anthocyanidin reductase* and *leucoanthocyanidin reductase*, genes specifically involved in flavan 3-ol formation [5]. To investigate whether DFR substrate specificity could also contribute to the establishment of strawberry fruit anthocyanin hydroxylation patterns, we studied DFR in two species of *Fragaria*.

DFR (EC 1.1.1.219) is an oxidoreductase which catalyzes the NADPH dependent reduction of the keto group in position 4 of dihydroflavonols to produce flavan 3,4-diols (synonym: leucoanthocyanidins), which are the immediate precursors for the formation of anthocyanidins and flavan 3-ols, the building blocks of condensed tannins [6]. DFR competes with flavonol synthase for dihydroflavonols as common substrates and therefore interferes with flavonol formation [7]. DFR thus has a strong influence on the formation of at least 3 classes of flavonoids, anthocyanin pigments, flavanols (which provide protection against herbivore, pests and pathogens), and flavonols (which act as sunscreens) [8,9]. In addition, DFR is unique in the flavonoid pathway, because it can exhibit selectivity for the B-ring hydroxylation pattern of flavonoid substrates. While the DFRs of many plants accept dihydroflavonols possessing one (dihydrokaempferol, DHK), two (dihydroquercetin, DHQ) or three (dihydromyricetin, DHM) hydroxyl groups within the B-ring, specific DFRs have been described from a few plant species that do not convert DHK into the corresponding leucopelargonidin [10,11]. These species do not produce pelargonidin based pigments and therefore lack an orange-red flower colouration. F3'H deficient lines for those species show a white or pale rosy flower colouration. Due to the absence of dihydroxylated precursors the formation of anthocyanins is not possible. A prominent example is petunia (*Petunia hybrida*) [10] where the biotechnological introduction of an non-specific DFR from maize bypassed a gap in the pathway of anthocyanin formation within the F3'H deficient petunia line RL01 [12], resulting in an orangered flower colouration (Figure 1, centre). To probe this phenomenon, an artificial DHK specific DFR was created via site-directed mutagenesis of the DFR from *Gerbera hybrida* with an exchange of an asparagine in position 134 into leucine [10]. This gene is patented for flower colour modification via transgenic approaches.

We report here on the isolation of two DFR cDNA clones from strawberry fruits demonstrating distinct but contrasting substrate specificity. The divergent ratio of expression between these DFRs in two *Fragaria* species contributes to the establishment of the different anthocyanin patterns within these fruits.

Results and Discussion

Cloning of DFR from *Fragaria* species

The NCBI database lists eight *DFR* full-length clones (reference date March 31, 2014) from strawberry all presenting at least 95% sequence similarity with the first isolated *DFR* from *F.×ananassa* (Accession AF029685 [13]). To identify further putative *DFRs* we screened the *F. vesca* genome [14] for the presence of homologues of the well-known *DFR*. With this approach, a *DFR* sequence was found presenting only an 82% sequence identity. The paralogous genes were named *DFR1* (high similarity to Accession AF029685) and *DFR2* (so far unknown). Specific primers were designed for the two *DFR* variant and used for the isolation of cDNA clones from early and late developmental stages of fruits of *F.×ananassa* cv. Elsanta and *F. vesca* cv. Alexandria and cv. Red Wonder. 2–3 allelic variants of both *DFR* variants were isolated from each cultivar (GenBank IDs KC894042-KC894055, Table 1, Figure 2). *DFR1* consisted of 1026 bp with an open reading frame (ORF) of 341 deduced amino acids (*F.×ananassa*) and 999 bp with an ORF of 333 deduced amino acids (*F. vesca*), respectively. *DFR2* of both species consisted of 1050 bp with an ORF of 349 amino acids. *DFR1* and *DFR2* sequences shared 80–83% amino acid sequence identity. The sequence identity of *DFR2* was 97–99% between *F. vesca* and *F.×ananassa* and 98–100% between *F. vesca* cv. Alexandria and cv. Red Wonder. *DFR1* showed 96–98% sequence identity between *F.×ananassa* and *F. vesca*, and 97–100% between *F. vesca* cv. Alexandria and cv. Red Wonder.

The phylogenetic relationship between the *DFRs* of the various *Fragaria* species was analyzed via a neighbor-joining tree that includes further amino acid sequences of *DFRs* accessible in the GenBank (Figure 3). In this tree, the *DFRs* of the Rosaceae family formed a separate cluster. Within this cluster the *DFR2s* from *Fragaria* form a group together with the *DFRs* of rose

(*Rosa×hybrida*, AAX12422), apple (*Malus×domestica*, AAO39816, AAO39817), pear (*Pyrus communis* AAO39818) and hawthorn (*Crataegus monogynae*, AAX1649). *Fragaria DFRs1* and *DFRs2* are clearly revealed in different clusters with *DFR2s* more closely related to the other Rosaceous *DFRs* than to *DFR1*.

Studies with the recombinant enzymes

The cDNA clones were transferred into a pYES expression vector and heterologously expressed in yeast (*Saccharomyces cerevisiae*). All recombinant enzymes demonstrated functional activity catalyzing the NADPH-dependent conversion of dihydroflavonols into leucoanthocyanidins. Control reactions with preparations from yeast cells harbouring the empty vector did not show DFR activity. DFR1s and DFR2s significantly differed in their acceptance of dihydroflavonols. The recombinant DFR2s converted DHQ and DHM to leucocyanidin and leucodelphinidin, but did not accept DHK as a substrate. Recombinant DFR1 s, in contrast, were selective for DHK (Figure S2 in File S1).

The two identified DFR variants were further characterized using the *DFR* pair isolated from ripe fruits of *F.×ananassa* cv. Elsanta. Apart from substrate acceptance, no striking differences were observed in the biochemical or kinetic characteristics of the two recombinant enzymes (Table 2). The highest reaction rates were observed in a weak acidic environment. The kinetic data, particularly the low K_m and the high K_{cat}/K_m value, confirmed the selectivity of DFR1 for DHK. DFR2 had a higher specificity for DHQ than for DHM displaying lower K_m values and higher K_{cat}/K_m values with DHQ as substrate (Table 2). Testing of substrates was performed under optimized conditions for the respective substrate and was within the linear range of the reaction. The selected incubation time and protein concentration ensured that the maximum conversion rate of the best substrate did not exceed 50%. As frequently observed in heterologous expression systems, the substrate selectivity was less obvious, when the protein was present in excess and incubation time was extended beyond the linear range of the reaction. The distinct substrate specificity of the DFRs, however, was confirmed in assays in which DHK and DHQ were simultaneously offered as substrates. When DHK and DHQ were present in equimolar amounts, DFR1 exclusively converted DHK to leucopelargonidin whereas in assays with recombinant DFR2 only the formation of leucocyanidin could be observed (Figure S2 in File S1).

To date, only two variants of DFRs have been reported in the literature: non-specific DFRs, converting all types of dihydroflavonols, and specific DFRs converting only DHQ and DHM. DFR1 from *Fragaria* species represents a third variant of DFR, which prefers DHK. The simultaneous presence of several *DFR* genes has been demonstrated in several plant species [15,16,17,18,19]. It is important to note that gene copies may encode enzymes with different substrate specificities. Drawing conclusions on the substrate specificity of a DFR just from the observed flavonoid hydroxylation pattern should therefore be avoided.

Sequence analysis

DFR1 and DFR2 sequences were analyzed for systematic differences that might be deciding factors for the differing substrate acceptance. The translated amino acid sequences demonstrated the highest divergence at both the N- and C-terminus (Figure 2). DFR2 of both *Fragaria* species had 349 aa and was generally longer than DFR1, which had 341 aa in *F.×ananassa* and only 333 aa in *F. vesca*. Apart from the variable termini, the alignment of the sequences from aa 6–330 (numbering according to DFR1)

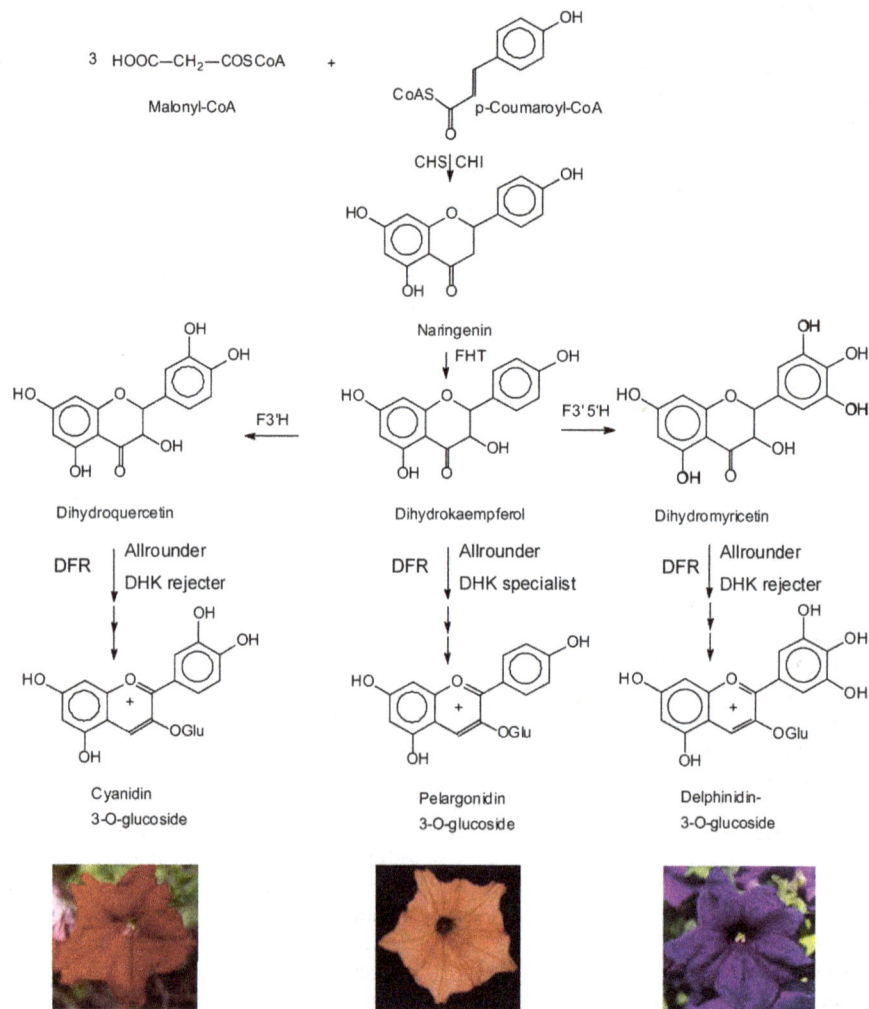

Figure 1. Simplified flavonoid pathway demonstrating the influence of DFR substrate acceptance on the establishment of flower colour.

Table 1. DFR cDNA clones isolated from *Fragaria* species.

DFR variant	Accession No	Species	cultivar	stage
DFR1	KC894042	*F. vesca*	Alexandria	early
DFR1	KC894043	*F. vesca*	Alexandria	late
DFR1	KC894044	*F. vesca*	Alexandria	late
DFR1	KC894045	*F. vesca*	Red Wonder	early
DFR1	KC894046	*F. vesca*	Red Wonder	late
DFR1	KC894047	*F.×ananassa*	Elsanta	early
DFR1	KC894048	*F.×ananassa*	Elsanta	late
DFR2	KC894049	*F. vesca*	Alexandria	early
DFR2	KC894050	*F. vesca*	Alexandria	late
DFR2	KC894051	*F. vesca*	Alexandria	late
DFR2	KC894052	*F. vesca*	Red Wonder	early
DFR2	KC894053	*F. vesca*	Red Wonder	late
DFR2	KC894054	*F.×ananassa*	Elsanta	early
DFR2	KC894055	*F.×ananassa*	Elsanta	late

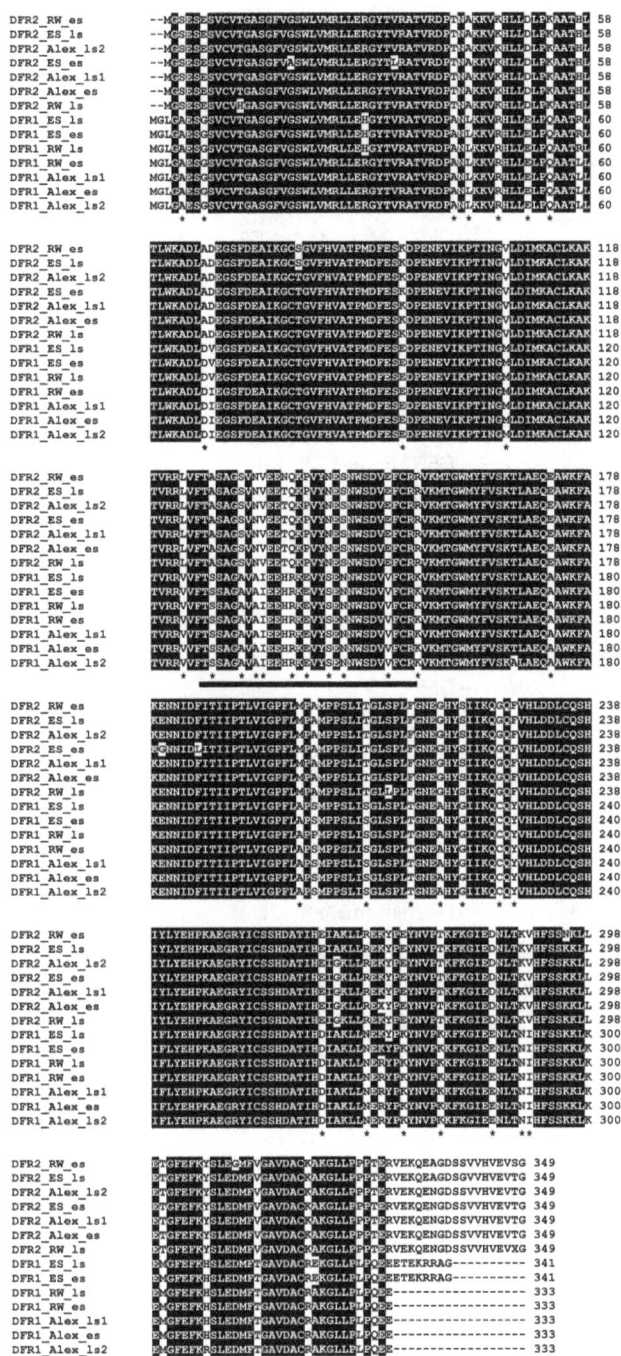

Figure 2. Multiple alignment of the deduced amino acid sequences encoded by DFRs isolated from F. vesca cv Alexandria during early (KC894042, KC894049) and late (KC894043, KC894044, KC894050, KC894051) stages, cv. Red Wonder early (KC894045, KC894052) and late (KC894046, KC894053) stages and F.×ananassa cv. Elsanta early (KC894047, KC894054) and late (KC894048, KC894055) stages. abbreviations: RW: Red Wonder, Alex: Alexandria, ES: Elsanta, es: early stage, ls: late stage.

revealed 36 points of distinct differences. 12 of the differences are located in a region, which was identified by Johnson et al. [10] as being relevant for determining substrate specificity (aa 126–170 in the gerbera DFR corresponding to 128–172 in DFR1). Position 134 in the gerbera DFR is of particular interest in this regard. The

presence of an aspartic acid in the petunia DFR sequence, which contrasts with the more frequently occurring asparagines, was suggested to determine the inability of converting DHK [10]. In addition, the exchange of the asparagine in the gerbera DFR into a non-polar leucine converts the non-specific DFR into a 'DHK specialist' [10]. The crystal structure obtained from a recombinant *Vitis vinifera* DFR confirmed the importance of this position (in this case aa 133). Asparagine 133 coordinates the dihydroflavonol substrate via interaction with the hydroxyl groups in position 3' and 4' [20]. In the DFRs from strawberry, amino acid 133 is an asparagine in DFR2 but an alanine in DFR1. The presence of a non-polar amino acid in the DHK-specialist is in line with [10]. The observed substrate specificity of DFR2, however, cannot be explained by the presence of the asparagine, because this is also found in the non-specific DFR from gerbera.

Relevance of DFR for the anthocyanin hydroxylation pattern in *Fragaria* species

The expression profiles of *DFR1* and *DFR2* were studied in the strawberry fruits. Botanically they are not berries but aggregate fruits, where the so-called 'seeds' are the real *Fragaria* fruits (achenes). The edible part which is commonly referred to as the fruit stems from the receptacle. The polyphenol profile and proteome varies between the tissues [21,22,23]. For this reason, receptacles and achenes were studied separately. The quantitative Real-time PCR data for the *DFRs* were normalized against *actin* (Figure S3 in File S1) and *glyceraldehyde 3-phosphate dehydrogenase* (*GAPDH*) (Figure 4). With both housekeeping genes comparable results were obtained with only slight differences for the early stages of the *F.×ananassa* receptacle (Figure 4). In the receptacle of both species, the profile of *DFR2* expression fluctuated during fruit development while for *DFR1* expression this was less pronounced. In the achenes, the *DFR* expression increased during fruit ripening in both species. In *F. vesca*, *DFR2* expression was drastically higher than the *DFR1* expression in both tissues during all developmental stages and showed two maxima for the receptacle during stage 1 and 3 with a decline in stage 2 (Figure 4). The highest *DFR1* expression in the *F. vesca* fruits was observed in S3 and S4 of the achenes. In *F.×ananassa*, *DFR2* expression was higher in late developmental stages of the fruits, whereas *DFR1* expression remained near stable along the different stages of fruit development and ripening. In the first three stages, *DFR1* expression was higher than *DFR2* expression, in the later stages the ratio reversed, but *DFR1* expression was still significantly higher than in the *F. vesca* receptacle (Figure 4). This was reflected by a differing substrate acceptance of enzyme preparations obtained from strawberry fruits (Figure 5). Preparations from *F. vesca* receptacles demonstrated higher specific activity with DHQ other than with DHK in all the developmental stages of the fruit. Preparations from *F.×ananassa* displayed variable dihydroflavonol preference during fruit development with higher specific activity with DHK as a substrate compared with DHQ in S1-S3, and still lower ones in S4-S6. DHK acceptance of preparations from *F.×ananassa*, however, was always higher than with preparations from *F. vesca*, even in the late developmental stages (Figure 5). Profiles of gene expression and enzyme activities were not always consistent. This was particularly the case in the achenes obtained from S3 and S4 fruits of *F. vesca* with drastically lower enzyme activity observed than what could have been expected from the high *DFR* expression levels. We assume, however, that this could be a problem related to increased levels of polyphenols as well as proteins and lipids in the achenes which might hamper the enzyme activity measurements. In *F. ananassa*, this discrepancy was observed to a much lower extent, which is in

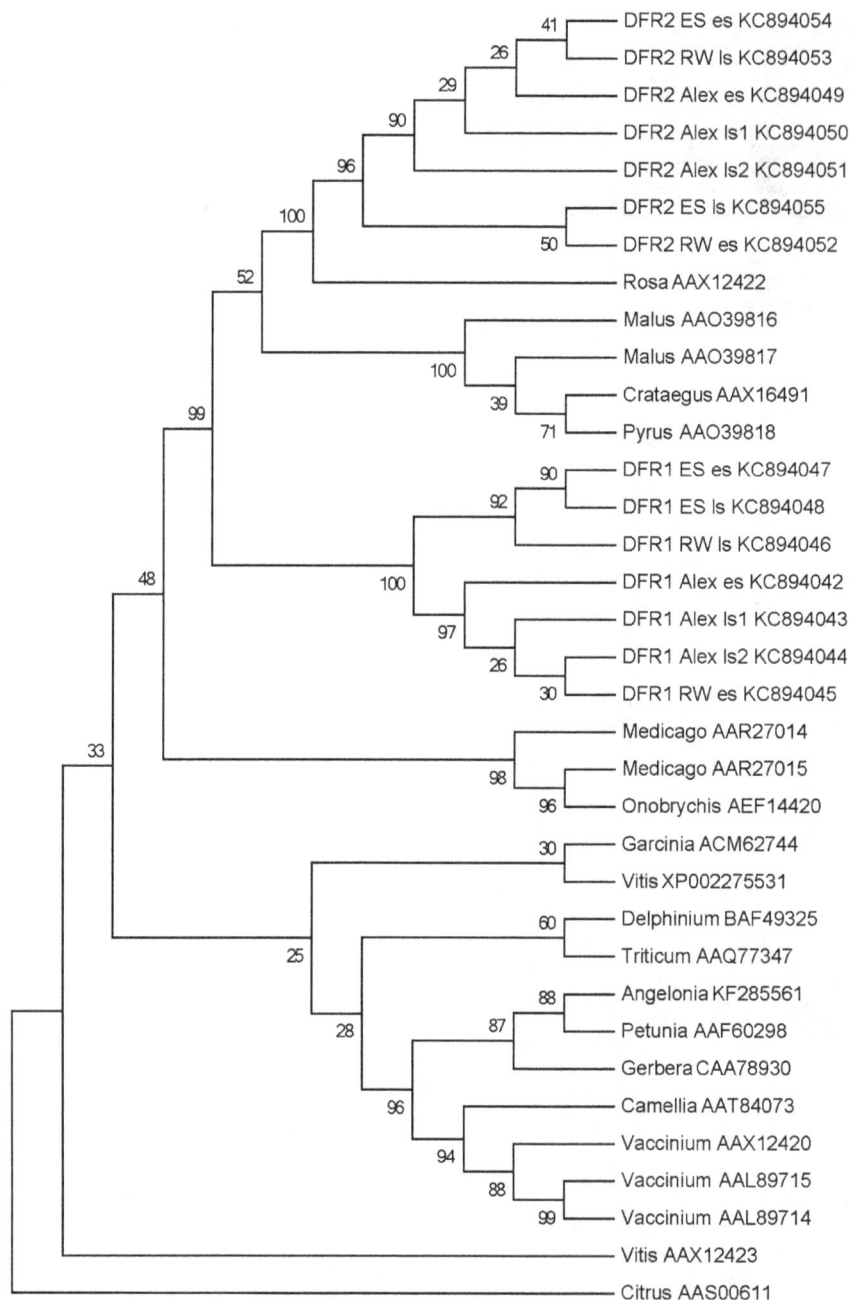

Figure 3. Neighbour-joining tree of *DFR* amino acid sequences of *Fragaria* and various species published in the NCBI GenBank. DFR1 *F. vesca* cv Alexandria, early stage (KC894042), late stage 1 (KC894043), late stage 2 (KC894044), cv Red Wonder early stage (KC894045), late stage (KC894046), *F.×ananassa* cv. Elsanta, early stage (KC894047), late stage (KC894048), DFR2 *F. vesca* cv Alexandria, early stage (KC894049), late stage 1 (KC894050), late stage 2 (KC894051), cv Red Wonder early stage (KC894052), late stage (KC894053), *F.×ananassa* cv. Elsanta, early stage (KC894054), late stage (KC894055), *Angelonia×angustifolia* (KF285561), *Camellia sinensis* (AAT84073), *Citrus sinensis* (AAS00611), *Crataegus monogynae* (AAX16491), *Delphinium belladonna* (BAF49325), *Garcinia mangostane* (ACM62744), *Gerbera hybrida* (CAA78930), *Malus×domestica* (AAO39816, AAO39817), *Medicago truncatula* (AAR27014, AAR27015), *Onobrychis viciifolia* (AEF14420), *Petunia×hybrida* (AAF60298), *Pyrus communis* (AAO39818), *Rosa hybrida* (AAX1242), *Triticum aestivum* (AAQ77347), *Vitis vinifera* (AAX12423, XP002275531), *Vaccinium macrocarpon* (AAL89715, AAL8971, AAX12420). The bootstrap values are indicated next to the relevant nodes (1000 replicates).

line with the fact that wild species frequently show increased levels of polyphenols in comparison to their domesticated counterparts [1]. To contrast with S1 and S2 receptacles of *F.×ananassa DFR* expression was relatively low compared to enzyme activity. Harmonized profiles, however, can not necessarily be expected. The life span of the DFR enzymes in the cell is completely unknown, and observed activity could be a cumulative result of

DFRs produced during different stages, possibly resulting in a shift of maxima between stages. It cannot even be excluded that DFR activities in very early stages result partially from *DFR* expression in the flowering period. In addition to this, post-transcriptional regulation may play a role.

It would appear that DFR does help to determine the pattern of flavonoid hydroxylation in strawberry fruits. In *F. vesca*, a high

Table 2. Characterization of recombinant DFR1 and DFR2 from *Fragaria* × *ananassa* cv. Elsanta obtained from heterologous expression of *DFRs* (KC894048, KC894055) in yeast.

	Recombinant DFR1 (KC894048)	Recombinant DFR2 (KC894055)
pH optimum	6.00^1	$6.25^2/5.75^3$
Temperature optimum [°C]	40	40
Temperature stability [°C]	30	30
Time linearity [min]	25	20
Protein linearity [µg]	20	25
apparent K_{cat} [µmol/kg*s]	11.4^1	$3.1^2/11.2^3$
apparent K_m [µM]	0.40^1	$0.40^2/2.3^3$
K_{cat}/K_m [l/s*kg]	28^1	$7.3^2/4.9^3$

using [1]DHK, [2]DHQ, [3]DHM as substrates.

F3'H expression during all developmental stages continuously promotes the availability of flavonoids with a 3',4'-dihydroxylation pattern including DHQ which is converted into leucocyanidin by the enzyme encoded by the highly expressed *DFR2*. As leucocyanidin is the immediate precursor to the formation of dark-red cyanidin based pigments, this results in a drastically higher cyanidin:pelargonidin ratio in the *F. vesca* fruits (0.51) compared to *F.×ananassa* (0.05). Total anthocyanin concentrations were, however, with 125 mg/kg lower than in *F.×ananassa* (350 mg/kg). We assume, however, that in *F. vesca* an additional factor could be relevant as the observed concentrations of pelargonidin based pigments are higher than would have been expected from the high levels of *DFR2* and *F3'H* transcripts. Measurements of the DFR activity during the four developmental

stages of the *F. vesca* receptacle with DHQ and DHK as substrates confirmed a high DFR activity with a persistent DHQ acceptance in all stages (Figure 5). The relatively high DHK acceptance compared to the low *DFR1* expression level is in line with the higher enzyme efficiency compared to DFR2 (Table 2). In addition, it is possible that F3'H activity in the *F. vesca* fruits is lower than expected from the high transcript levels. However, this could not be verified by enzyme assays as the F3'H is a membrane associated enzyme and is difficult to measure in tissues which are rich in disturbing compounds such as polyphenols, sugars and glucanes [24]. In *F.×ananassa*, the observed transcript levels are in line with the anthocyanin composition of the fruits. Due to the decrease of *F3'H* transcript levels during fruit ripening, a depletion of 3',4'-hydroxylated dihydroflavonols is expected which prevents

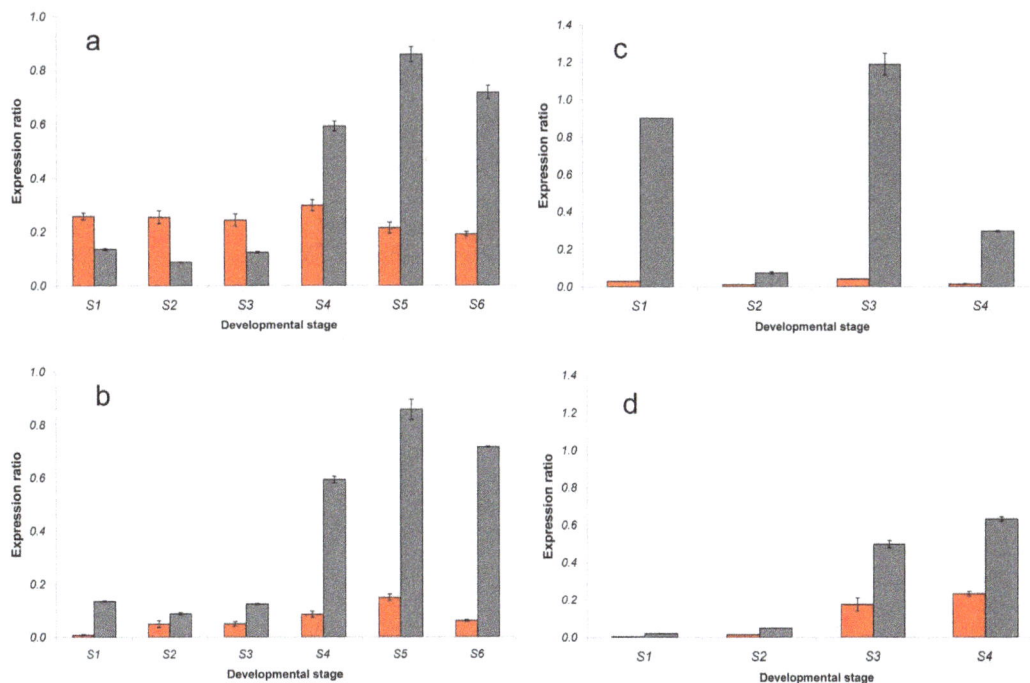

Figure 4. Quantitative expression of *DFR1* and *DFR2* normalized to *glyceraldehyde 3-phosphate dehydrogenase* in receptacle and achenes of *Fragaria* fruits during the different stages of fruit development. a: *F.×ananassa* receptacle, b: *F.×ananassa* achenes, c: *F. vesca* receptacle, d: *F. vesca* achenes. red: *DFR1*, grey: *DFR2*. Data were calculated from three biological replicates with at least two technical replicates for each and with error bars representing the standard deviation.

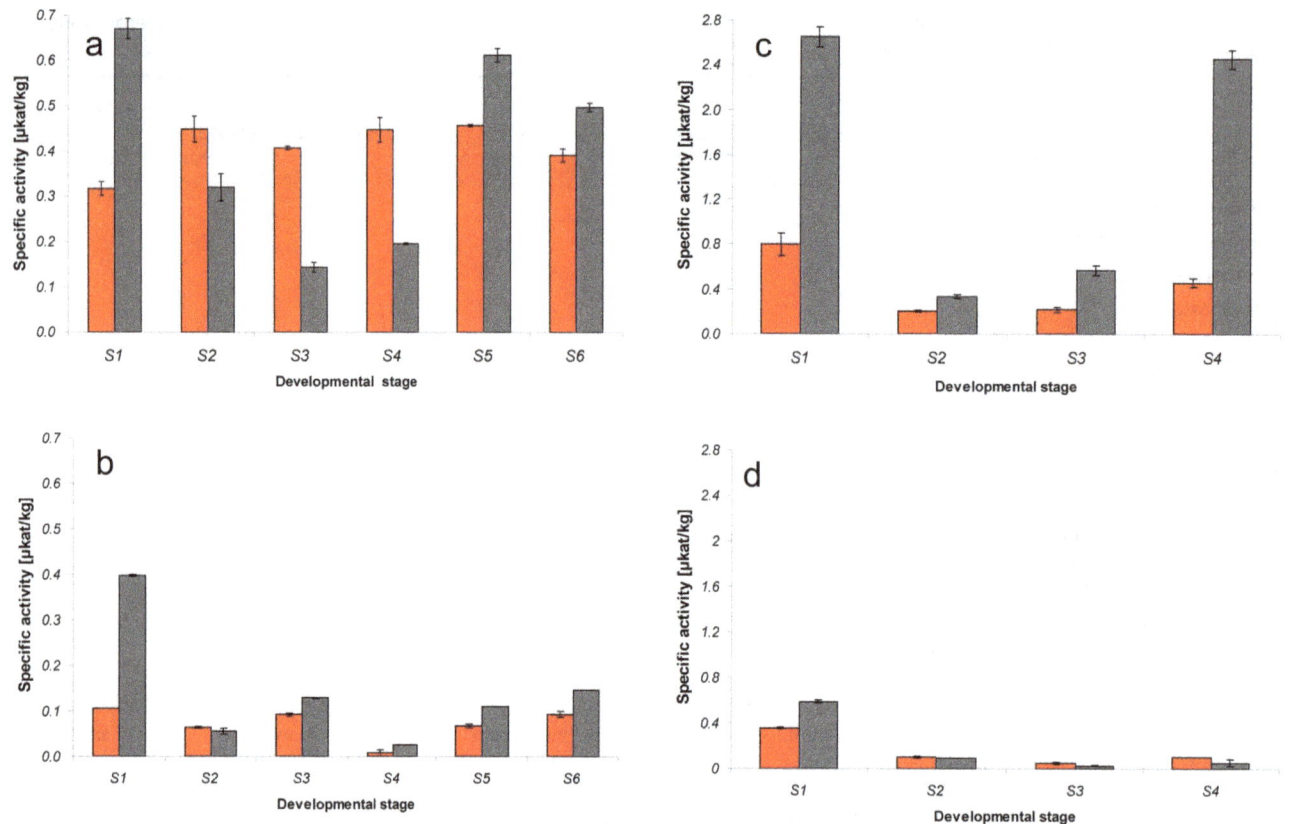

Figure 5. Specific DFR activity [nkat/kg protein] with DHQ and DHK as substrates in receptacle and achenes throughout the different stages of fruit development and ripening in *F.×ananassa* **and** *F. vesca.* a: *F.×ananassa* receptacle, b: *F.×ananassa* achenes, c: *F. vesca* receptacle, d: *F. vesca* achenes. red: *DFR1*, grey: *DFR2*. Data were calculated from three biological replicates and with error bars representing the standard deviation.

the formation of large amounts of cyanidin, despite the high transcript levels of *DFR2*. The relatively high DFR1 expression levels compared to *F. vesca* and the high enzyme efficiency of DFR1 demosntrated by the high K_{cat}/K_m values, however, provides sufficient amounts of precursors to allow the formation of leucopelargonidin as the immediate precursors for pelargonidin formation.

Conclusions

Our studies identified a novel DFR variant in *Fragaria* sp., which demonstrates an unusual DHK preference. The two DFR variants show divergent expression profiles in both of the two species but also with regards to the fruit development. It is likely that these expression profiles are due to differences in the transcript regulation of the two genes. The higher expression rate of DFR1 and the higher enzyme efficiency of DFR1 is an important precondition for the accumulation of pelargonidin based pigments in *F.×ananassa* together with the absence of F3'H expression in the late stages. In *F. vesca*, a high *DFR2* and *F3'H* expression account for the increased levels of cyanidin based pigments. The novel DFR pair derived from a single species will be a perfect model for future studies into the molecular background behind DFR substrate specificity.

Materials and Methods

Plant material

The studies were performed on fruits of *Fragaria×ananassa* cv. Elsanta harvested at the experimental orchard of the Institute of Horticulture and Viticulture (University of Natural Resources and Applied Life Sciences, Vienna, Austria), and *F. vesca* cv. Red Wonder and cv. Alexandria grown at JKI Dresden-Pillnitz, Germany. The fruits were shock-frozen in liquid nitrogen and stored at −80°C. The samples were identical to those recently used for the studies on F3'H in strawberry [25]. Briefly, the six developmental stages of *F.×ananassa* were small-sized (0.7 cm) green fruits (S1), middle-sized (1.5 cm) green fruits (S2), middle-sized (2–2.5 cm) white fruits (S3), middle-sized (2–2.5 cm) turning-stage fruits (S4), middle-sized (2–2.5 cm) late turning-stage fruits (S5), and full-ripe red fruits, 4 cm fruit size (S6). The four developmental stages of *F. vesca* comprised small-sized (0.3 cm) green fruits (S1), middle-sized (0.5 cm) turning-stage fruits (S2), late turning-stage fruits with 0.6 cm fruit size (S3) and full-ripe red fruits of 0.8 cm fruit size (S4) [25]. Due to differences in morphology and fruit development, stages of fruit development between the two species were not defined in a similar way. It is possible, however, to distinguish between unripe fruits (*F. vesca* S1 and *F.×ananassa* S1+2), turning stage fruits (*F. vesca* S2+3 and *F.×ananassa* S4+5) and ripe fruits (*F. vesca* S4 and *F.×ananassa* S6). Receptacle and achenes were separated manually from frozen fruits without defrosting the material. The tissues were analyzed separately for enzyme activity and gene expression.

Chemicals

(2-^{14}C)-Malonyl-coenzyme A (55 mCi/mmol) was purchased from New England Nuclear Corp. GmbH (Vienna, Austria). (^{14}C)-Labeled flavonoids DHK, DHQ, and DHM were synthesized as previously described [26,27] using recombinant F3'5'H from *Sollya heterophylla* and F3'H from *Tagetes erecta*.

Cloning of DFR cDNAs from Fragaria species

mRNA was isolated from strawberry fruits with μMACS mRNA Isolation Kit (Miltenyi Biotec, Germany). cDNA was prepared using the SuperScript II Reverse Transcriptase (Invitrogen, Carlsbad, CA) and the oligo(-dT) anchor primer GAC-CACGCGTATCGATGTCGAC(T)$_{16}$ V. The genome of *F. vesca* was screened for sequences with high similarity to a DFR from *F. x ananassa* (AY695812) using the tools available at www.rosaceae.org.

Specific primers (DFR1_f: ATGGGGTTGGGAGCTGAATC, DFR1_r: TCAACCAGCCCTGCGCTTT, DFR2_f: TGTCAA-GAAACATGGGATCGGAG, DFR2_r: GAAGCTCTCAACA-TACAGAAGATAGA) were designed on the basis of the detected sequences. Proof reading PCR was carried out using the Expand High Fidelity Plus PCR System (Roche, Austria) according to the manufacturer's instructions. PCR products were ligated into the vector pYES2.1/V5-His-TOPO and transformed into *E. coli* strain Top 10 (Invitrogen, Carlsbad, CA) according to the manufacturers instructions. Plasmids were isolated with Wizard Plus SV Minipreps DNA Purification System (Promega, Vienna, Austria) and sequenced by a commercial supplier (Microsynth Austria AG, Vienna, Austria).

Gene expression analysis

The expression of the *DFRs* was quantified by qPCR using a StepOnePlus system (Applied Biosystems, Darmstadt, Germany) and the SybrGreenPCR Master Mix (Applied Biosystems, Vienna, Austria) according to the supplier's instruction. The analysis was carried out in three independent triplicates, and the data was normalized against two control genes, actin and glyceraldehyde 3-phosphate dehydrogenase (*GAPDH*). The relative expression ratio of a target gene was computed applying the equation according to [28]. The efficiency of the PCR-reaction was determined on the basis of standard curves which were obtained by applying different DNA concentrations. The product specificity was confirmed via melt curve analysis and gel electrophoresis. Primers for *DFRs* were designed for the isolated *DFR* sequences (Table S1 in File S1).

Sequence and Phyologenetic Analysis

Multiple alignments were undertaken with Clustal Omega [29,30]. The phylogenetic tree was conducted and bootstrapped with MEGA version 5 [31] using the neighbor-joining method and 1000 replicates.

Anthocyanin determination

For the determination of the anthocyanin content, 10 g of shock-frozen petals were pulverized and mixed with 35 ml 2 M methanolic hydrochloric acid. After shaking for 12 hours at 4°C in an overhead rotator, the suspension was centrifuged for 10 minutes at 19200×g. 10–140 μl of the supernatant was adjusted with 2 M methanolic hydrochloric acid to a final volume of 1000 μl. The absorption at 520 nm was determined on a DU-65 spectrophotometer (Beckman Instruments). The anthocyanin content was calculated as pelargonidin equivalent using a calibration curve obtained with commercially available pelargonidin chloride (Roth, Germany). For acidic hydrolysis of anthocyanins, 20 μl methanolic hydrochloric acid extract were mixed with 180 μl of 4 N HCL and incubated for 60 minutes at 90°C. After cooling for 10 minutes the mixture was centrifuged for 10 minutes at 10000×g. The supernatant was adjusted to 200 μl with 4 N HCL and aliquots were used for HPLC analysis [32] using a Perkin Elmer Series 200 HPLC system equipped with a Perkin Elmer Series 200 diode array detector and Total Chrom Navigator, version 6.3.1 (Perkin Elmer Inc). The column was a BDS Hypersil C18 HPLC column, 5 μm, 250×4.6 mm (Thermo Scientific).

Heterologous expression and protein preparation

The vectors harbouring the *DFR* cDNAs were transformed into yeast strain INVSc1 using the Sc. EasyComp Transformation Kit (Invitrogen, Carlsbad, CA). Preparation of the protein fractions was performed using a modified protocol according to Pompon et al. [33]. Briefly, 250 ml of expression culture was grown in YPGE medium (5 g/l glucose, 10 g/l peptone, 10 g/l yeast extract, 30 ml 100% ethanol) at 28°C and 180 rpm for approximately 6 h, until an OD$_{600}$ of 0.8–1.2 was reached. After addition of 27 ml 20% (w/v) sterile filtered galactose, the culture was incubated at 28°C and 180 rpm for 15 h. Cells were harvested by centrifugation at 4000× g at 4°C for 10 min, with TEK (50 mM Tris/HCl, 1 mM EDTA, 100 mM potassium chloride, pH 7.4), and redissolved in 2.5 ml icecold buffer TES-B* (50 mM Tris/HCl, 1 mM EDTA, 0.6 M sorbitol, 2 mM dithiothreitol, pH 7.4). Disruption of cell walls was achieved via vigourous shaking with glass beads for 30 s every minute during a period of 20 min. Glass beads were removed by centrifugation at 4000×g and 4°C for 10 min. The protein preparation was diluted with 2.5 ml icecold buffer TES-B*, shock frozen in liquid nitrogen and stored at −80°C.

Enzyme preparation

Enzyme preparations from strawberry fruits were obtained by using the protocol of Claudot and Drouet [34] with slight modifications as previously described [24]. To remove low molecular weight compounds, crude enzyme preparations were passed through a gel chromatography column (Sephadex G25, GE Healthcare, Freiburg, Germany). Protein content was determined by a modified Lowry procedure [35] using crystalline bovine serum albumin as a standard.

Enzyme assays

Assays for DFR were performed as described earlier [27,36]. Briefly, the reaction contained in a final volume of 50 μl: 1–10 μl enzyme preparation (2.4–24 μg), 0.048 nmol (^{14}C)-dihydroflavonol, 0.25 nmol NADPH, 44–35 μl 0.1 M KH$_2$PO$_4$/K$_2$HPO$_4$ buffer pH 6.3 containing 0.4% Na ascorbate. The amount of enzyme preparation depended on the recombinant enzyme used and was chosen to ensure a maximum conversion rate with the best substrate at 50%.

Enzyme characterization

All data represents an average of at least three independent experiments. Determination of the pH optimum was carried out as described for the standard DFR assay, but using 0.2 M McIlvaine buffers with pH values between 4.5 and 9.0. Optimal temperature was determined by measuring activities at varying temperatures within 0°C and 60°C. Temperature stability was determined by measuring enzyme activities at 25°C after incubation of the reaction mixture without NADPH at varying temperatures. Kinetic data were calculated from Lineweaver-Burk plots using radiolabeled substrates at varying concentrations.

Supporting Information

File S1   File includes Figures S1–S3 and Table S1. Figure S1: Left: Ripe strawberry fruits of *F. vesca* cv. Red Wonder. Right: *F. ×ananassa* cv. Elsanta. Due to differing magnification factors used, fruit size does not appear at a comparable scale. Figure S2: Radioscan of TLC on cellulose from incubation of recombinant DFR2 (left) and DFR1 (right) in the presence of NADPH offering A: (^{14}C)dihydroquercetin, B: (^{14}C)dihydrokaempferol, C and D: (^{14}C)dihydroquercetin and (^{14}C)dihydrokaempferol in equimolar amounts as substrates. Figure S3: Quantitative expression of *DFR1* and *DFR2* normalized to *actin* in receptacle and achenes of *Fragaria* fruits along the different stages of the fruit development. a: *F. ×ananassa* receptacle, b: *F. ×ananassa* achenes, c: *F. vesca* receptacle, d: *F. vesca* achenes. red: *DFR1*, grey: *DFR2*. Data were calculated from three biological replicates with at least two technical replicates for each and error bars representing the standard deviation. Table S1: List of primers used for quantitative Real-time PCR.

Acknowledgments

The authors kindly acknowledge Klaus Olbricht (Julius Kühn-Institute - Federal Research Centre for Cultivated Plants, Institute for Breeding Research on Horticultural and Fruit Crops, Dresden, Germany) for providing fruits of cv. Red Wonder. We would also like to say thank you to Renate Paltram and Jürgen Greiner for excellent technical assistance. Special thanks to Luke McLaughlin for critically reading the manuscript.

Author Contributions

Conceived and designed the experiments: HH KS AR. Performed the experiments: SM MM SP CM SE CG. Analyzed the data: CM JT HH CG. Contributed to the writing of the manuscript: HH KS SM.

References

1. Thill J, Miosic S, Gotame TP, Mikulic-Petkovsek M, Gosch C, et al. (2013) Differential expression of flavonoid 3′-hydroxylase during fruit development establishes the different B-ring hydroxylation patterns of flavonoids in *Fragaria × ananassa* and *Fragaria vesca* Plant Physiology and Biochemistry 72: 72–78.

2. Halbwirth H, Puhl I, Haas U, Jezik K, Treutter D, et al. (2006) Two-phase flavonoid formation in developing strawberry (*Fragaria× ananassa*) fruit. Journal of Agricultural and Food chemistry 54: 1479–1485.

3. Aharoni A, O'Connell AP (2002) Gene expression analysis of strawberry achene and receptacle maturation using DNA microarrays. Journal of Experimental Botany 53: 2073–2087.

4. Sondheimer E, Karash CB (1956) The major anthocyanin pigments of the wild strawberry (*Fragaria vesca*), Nature 178: 648–649.

5. Almeida JR, D'Amico E, Preuss A, Carbone F, De Vos C, et al. (2007) Characterization of major enzymes and genes involved in flavonoid and proanthocyanidin biosynthesis during fruit development in strawberry *Fragaria×ananassa*. Archives of Biochemistry and Biophysics 465: 61–71.

6. Forkmann G, Heller W (1999) Biosynthesis of flavonoids. In: D Barton, K Nakanishi, O Meth-Cohn, Sankawa U, editors. Comprehensive Natural Products Chemistry: Elsevier Science, Amsterdam. 713–748.

7. Davies KM, Schwinn KE, Deroles SC, Manson DG, Lewis DH, et al. (2003) Enhancing anthocyanin production by altering competition for substrate between flavonol synthase and dihydroflavonol 4-reductase. Euphytica 131: 259–268.

8. Harborne JB (1967) Comparative Biochemistry of Flavonoids. London: Academic Press.

9. Harborne JB, Williams CA (2000) Advances in flavonoid research since 1992. Phytochemistry 55: 481–504.

10. Johnson ET, Ryu S, Yi H, Shin B, Cheong H, et al. (2001) Alteration of a single amino acid changes the substrate specificity of dihydroflavonol 4-reductase. The Plant Journal 25: 325–333.

11. Johnson ET, Yi H, Shin B, Oh B-J, Cheong H, et al. (1999) *Cymbidium hybrida* dihydroflavonol 4-reductase does not efficiently reduce dihydrokaempferol to produce orange pelargonidin-type anthocyanins. The Plant Journal 19: 81–85.

12. Meyer P, Heidmann I, Forkmann G, Saedler H (1987) A new petunia flower colour generated by transformation of a mutant with a maize gene. Nature: 677–678.

13. Moyano E, Portero-Robles I, Medina-Escobar N, Valpuesta V, Muñoz-Blanco J, et al. (1998) A fruit-specific putative dihydroflavonol 4-reductase gene is differentially expressed in strawberry during the ripening process. Plant physiology 117: 711–716.

14. Shulaev V, Sargent DJ, Crowhurst RN, Mockler TC, Folkerts O, et al. (2010) The genome of woodland strawberry (Fragaria vesca). Nature genetics 43: 109–116.

15. Shimada N, Sasaki R, Sato S, Kaneko T, Tabata S, et al. (2005) A comprehensive analysis of six dihydroflavonol 4-reductases encoded by a gene cluster of the *Lotus japonicus* genome. Journal of Experimental Botany 56: 2573–2585.

16. Xie D-Y, Jackson LA, Cooper JD, Ferreira D, Paiva NL (2004) Molecular and biochemical analysis of two cDNA clones encoding dihydroflavonol-4-reductase from *Medicago truncatula*. Plant physiology 134: 979–994.

17. Inagaki Y, Johzuka-Hisatomi Y, Mori T, Takahashi S, Hayakawa Y, et al. (1999) Genomic organization of the genes encoding dihydroflavonol 4-reductase for flower pigmentation in the Japanese and common morning glories. Gene 226: 181–188.

18. Des Marais DL, Rausher MD (2008) Escape from adaptive conflict after duplication in an anthocyanin pathway gene. Nature 454: 762–765.

19. Hua C, Linling L, Shuiyuan C, Fuliang C, Feng X, et al. (2013) Molecular Cloning and Characterization of Three Genes Encoding Dihydroflavonol-4-Reductase from *Ginkgo biloba* in Anthocyanin Biosynthetic Pathway. PloS one 8: e72017.

20. Petit P, Granier T, d'Estaintot BL, Manigand C, Bathany K, et al. (2007) Crystal Structure of Grape Dihydroflavonol 4-Reductase, a Key Enzyme in Flavonoid Biosynthesis. Journal of molecular biology 368: 1345–1357.

21. Aaby K, Skrede G, Wrolstad RE (2005) Phenolic composition and antioxidant activities in flesh and achenes of strawberries (*Fragaria ananassa*). Journal of Agricultural and Food chemistry 53: 4032–4040.

22. Fait A, Hanhineva K, Beleggia R, Dai N, Rogachev I, et al. (2008) Reconfiguration of the achene and receptacle metabolic networks during strawberry fruit development. Plant physiology 148: 730–750.

23. Aragüez I, Cruz-Rus E, Botella MÁ, Medina-Escobar N, Valpuesta V (2013) Proteomic analysis of strawberry achenes reveals active synthesis and recycling of L-ascorbic acid. Journal of Proteomics 83: 160–179.

24. Halbwirth H, Waldner I, Miosic S, Ibanez M, Costa G, et al. (2009) Measuring Flavonoid Enzyme Activities in Tissues of Fruit Species. Journal of Agricultural and Food chemistry 57: 4983–4987.

25. Thill J, Miosic S, Gotame TP, Mikulic-Petkovsek M, Gosch C, et al. (2013) Differential expression of flavonoid 3′-hydroxylase during fruit development establishes the different B-ring hydroxylation patterns of flavonoids in *Fragaria× ananassa* and *Fragaria vesca*. Plant Physiology and Biochemistry.

26. Halbwirth H, Kahl S, Jager W, Reznicek G, Forkmann G, et al. (2006) Synthesis of (^{14}C)-labeled 5-deoxyflavonoids and their application in the study of dihydroflavonol/leucoanthocyanidin interconversion by dihydroflavonol 4-reductase. Plant Science 170: 587–595.

27. Fischer TC, Halbwirth H, Meisel B, Stich K, Forkmann G (2003) Molecular cloning, substrate specificity of the functionally expressed dihydroflavonol 4-reductases from *Malus domestica* and *Pyrus communis* cultivars and the consequences for flavonoid metabolism. Archives of Biochemistry and Biophysics 412: 223–230.

28. Pfaffl MW (2004) Real-time RT-PCR: Neue Ansätze zur exakten mRNA Quantifizierung. BIOspektrum 1: 92–95.

29. McWilliam H, Li W, Uludag M, Squizzato S, Park YM, et al. (2013) Analysis tool web services from the EMBL-EBI. Nucleic Acids Research 41: W597–W600.

30. Sievers F, Wilm A, Dineen D, Gibson TJ, Karplus K, et al. (2011) Fast, scalable generation of high-quality protein multiple sequence alignments using Clustal Omega. Molecular Systems Biology 7.

31. Tamura K, Peterson D, Peterson N, Stecher G, Nei M, et al. (2011) MEGA5: molecular evolutionary genetics analysis using maximum likelihood, evolutionary distance, and maximum parsimony methods. Molecular Biology and Evolution 28: 2731–2739.

32. Chandra A, Rana J, Li Y (2001) Separation, identification, quantification, and method validation of anthocyanins in botanical supplement raw materials by HPLC and HPLC-MS. Journal of Agricultural and Food chemistry 49: 3515–3521.

33. Pompon D, Louerat B, Bronine A, Urban P (1996) [6] Yeast expression of animal and plant P450s in optimized redox environments. Methods in enzymology 272: 51–64.

34. Claudot A-C, Drouet A (1992) Preparation and assay of chalcone synthase from walnut tree tissue. Phytochemistry 31: 3377–3380.

35. Sandermann H, Strominger JL (1972) Purification and properties of C55-isoprenoid alcohol phosphokinase from *Staphylococcus aureus*. Journal of Biological Chemistry 247: 5123–5131.

36. Gosch C, Puhl I, Halbwirth H, Schlangen K, Roemmelt S, et al. (2003) Effect of Prohexadione-Ca on Various Fruit Crops: Flavonoid Com-position and Substrate Specificity of their Dihydroflavonol 4-Reductases. European Journal of Horticultural Science 68: 144–151.

23

On the Normalization of the Minimum Free Energy of RNAs by Sequence Length

Edoardo Trotta*

Institute of Translational Pharmacology, Consiglio Nazionale delle Ricerche (CNR), Roma, Italy

Abstract

The minimum free energy (MFE) of ribonucleic acids (RNAs) increases at an apparent linear rate with sequence length. Simple indices, obtained by dividing the MFE by the number of nucleotides, have been used for a direct comparison of the folding stability of RNAs of various sizes. Although this normalization procedure has been used in several studies, the relationship between normalized MFE and length has not yet been investigated in detail. Here, we demonstrate that the variation of MFE with sequence length is not linear and is significantly biased by the mathematical formula used for the normalization procedure. For this reason, the normalized MFEs strongly decrease as hyperbolic functions of length and produce unreliable results when applied for the comparison of sequences with different sizes. We also propose a simple modification of the normalization formula that corrects the bias enabling the use of the normalized MFE for RNAs longer than 40 nt. Using the new corrected normalized index, we analyzed the folding free energies of different human RNA families showing that most of them present an average MFE density more negative than expected for a typical genomic sequence. Furthermore, we found that a well-defined and restricted range of MFE density characterizes each RNA family, suggesting the use of our corrected normalized index to improve RNA prediction algorithms. Finally, in coding and functional human RNAs the MFE density appears scarcely correlated with sequence length, consistent with a negligible role of thermodynamic stability demands in determining RNA size.

Editor: Danny Barash, Ben-Gurion University, Israel

Funding: The author has no support or funding to report.

Competing Interests: The author has declared that no competing interests exist.

* Email: edoardo.trotta@ift.cnr.it

Introduction

The cell synthesizes various types of RNAs that play distinctive and essential roles in living systems, including coding (mRNA), decoding (tRNA), catalytic (ribozymes), regulatory (e.g., micro-RNA), and structural (e.g., rRNA) functions. The cellular activity of each RNA is normally dependent on the specific structural features of its functional category. This critical role of structure in the function of RNA molecules, together with its difficulty in being determined experimentally [1], have favoured the development of a number of software packages that predict RNA secondary structure. These include computer programs based on minimum free energy (MFE) algorithms. The MFE of an RNA molecule is affected by three properties of nucleotides in the sequence: their number, composition, and arrangement. In fact, longer sequences are on average more stable because they can form more stacking and hydrogen bond interactions, guanine-cytosine (GC)-rich RNAs are typically more stable than adenine-uracil (AU)-rich sequences, and nucleotide order influences the folding structure stability because it determines the number and the extension of loops and double-helix conformations. It has been found that mRNAs and microRNA precursors, unlike other non-coding RNAs, have greater negative MFE than expected given their nucleotide numbers and compositions [2,3]. This led to the observation that free energy can be employed as a criterion for the identification of functional RNAs. However, when the folding energies of different classes of RNA are compared, the dependence of MFE to sequence length can represent a disturbing element. To overcome this obstacle, a new class of free energy indices normalized by sequence length has been proposed. These indices can be conceived as free energy density indicators and were obtained simply by dividing MFE by the number of nucleotides in the sequence [4–9]. A widely used normalized index is the so-called adjusted MFE (AMFE) [9]. AMFE is calculated by dividing MFE by the sequence length and then multiplying the result by 100 to relate the index to a segment of 100 nucleotides. Based on their supposed weak relationship with sequence length, normalized MFEs have been used in a number of published works to compare the free energy among different classes of RNAs. In fact, it has been reported that, after this adjustment, the MFEs of all nucleotide sequences are comparable [9]. Furthermore, it was also reported that length-normalization renders the MFE of hairpins of different lengths comparable [6] and provides an estimate of stability that is not influenced by differences in RNA sequence length [10]. However, even if the length-normalized MFEs have been used in a number of studies, to our knowledge, their relationship with sequence size has not been thoroughly tested and lacks quantitative substantiation. Using simulated sequences, we searched for possible residual components of AMFE associated with length. We found that the suggested procedure for normalizing MFE by length produces unacceptable

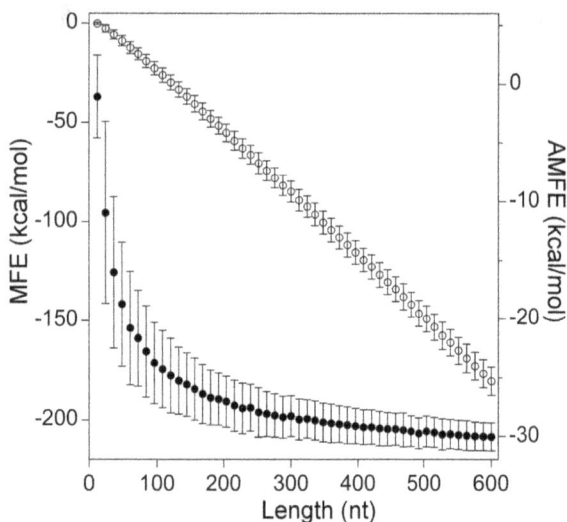

Figure 1. MFE and AMFE versus sequence length. For each sequence length, containing an exact equal frequency of the four nucleotides, 1000 randomly shuffled sequences were simulated. The mean values of the MFE (open circles) and AMFE (closed circles) of the shuffled sequences are plotted versus the sequence length. Vertical bars indicate standard deviations (N = 1000). MFE was computed by RNAfold using default parameters.

results. AMFE is significantly affected by sequence length, leading to substantial errors if the index is used directly to compare the stability of RNA sequences of various lengths. We show that the error is generated by the combined effects of a poor mathematical normalization procedure and a non-perfect linear relationship between MFE and sequence length. To allow the direct comparison of the MFE of differently sized RNAs, here we propose a correction in the normalization procedure that removes the AMFE bias extending its applicability to all RNAs longer than 40 nt. Using the new normalized index, termed MFE density (MFEden), we report the analysis of a set of human coding and functional RNA families.

Results

Comparative software analysis

The most common software programs, employed to predict the secondary RNA structures by MFE algorithms, make use of the so-called nearest-neighbor energy model. This model uses free energy rules based on empirical thermodynamic parameters [11,12] and computes the overall stability of an RNA structure by adding independent contributions of local free energy interactions due to adjacent base pairs and loop regions. In sequences with homogeneous nucleotide arrangements and compositions, the additive and independent nature of the local free energy contributions suggests a linear relationship between computed MFE and sequence length. Normalization by length, obtained by dividing MFE by the number of nucleotides, was introduced to exploit this linear relationship to directly compare the minimum free energies of RNAs of various lengths. To investigate on the relationship of MFE and length-normalized MFE with sequence size, we computed MFE by two of the most common software programs used to predict RNA secondary structure through the free energy minimization approach: Quikfold application, which is incorporated in the Mfold webserver for multiple molecule processing [13,14], and RNAfold, which is included in the

ViennaRNA software package [15,16]. The results obtained from the two programs were very similar, and the differences were irrelevant to the objective of this study. For this reason, we omitted the data from both software programs for each result.

The relationship of length with MFE and normalized-MFE in randomly shuffled sequences containing equal frequencies of A, C, G, and U

The length-normalized index AMFE is computed by using the formula AMFE $= 100*$ MFE/L, where L is the number of nucleotides of the RNA sequence [9]. To determine whether sequence length affects AMFE, as an initial analysis, we generated random sequences of various lengths and equal frequencies of A, C, G, and U. Starting with a set of sequences containing one copy for each different length (from 12 to 600 nt with steps of 12 nt) and exact equal frequencies of the four bases, we generated 1000 sets of randomly shuffled sequences. Then, for each simulated length, represented by 1000 randomized sequences, we computed the mean and standard deviation (SD) values of MFE and AMFE.

As shown by open circles in the graph in Figure 1, the increase of sequence length from 12 to 600 nt causes an apparent linear decrease of the MFE of about -180 kcal/mol at the average rate of -32 kcal/mol every 100 nucleotides. In contrast, indicated by closed circles in the graph in Figure 1, AMFE decreases, by almost 30 kcal/mol, as a hyperbolic function of length, demonstrating that a significant portion of AMFE is correlated with the sequence size. Using the RNA 3.0 (Quickfold) free energy rules [13], we computed the portion of the total minimum free energy associated with the differently classified structural elements. The upper panel in Figure 2 shows the graph of the free energy contributions of the various structural elements versus the sequence length of simulated sequences. As illustrated in the figure, base pair stacking is the most stabilizing element in our simulated sequences by a free energy contribution negatively correlated with length. Loops tend to destabilize minimum folding energy structures by quantities that, distinct from stacking energies, correlate positively with sequence length. Structural elements classified as external loops, which comprise single-stranded nucleotides and base pairs at the end of helices that are not in a loop, are weakly stabilizing and their free energy contribution decreases with length (from -1 to -1.7 kcal/mole).

The normalization by length of the individual MFE contribution from each structural element indicates that stacking and hairpin loop interactions are responsible for almost all AMFE variability associated with sequence length (Figure 2, lower panel).

Comparison of the free energy variability associated with sequence length, nucleotide composition and nucleotide order To evaluate the impact of length to the overall variation of normalized-MFE, we should compare its effects with those generated by varying the order and the composition of nucleotides in the sequences. To this end, we generated 100 sets of randomly shuffled sequences from a set with increasing lengths and GC-contents. The length of shuffled sequences ranged between 20 and 600 nt, with steps of 20 nt. For each length, GC-contents were 20%, 40%, 50%, 60%, and 80%. The results are summarized in Figure 3, where the mean MFEs and the mean AMFEs of each randomly shuffled sequence are plotted versus length and GC-content. As illustrated in Figures 3A and 3B, the average stability of shuffled sequences increases with both length and GC-content. Increasing GC-content at constant length causes a nonlinear decrease of MFE that is more prominent for longer sequences (Figure 3B). From 20% to 80% of GC-content, the folding stability of 20 nt-long RNA increases by -5.4 kcal/mol, whereas that of 600 nt-long sequences increases by -277.0 kcal/mol.

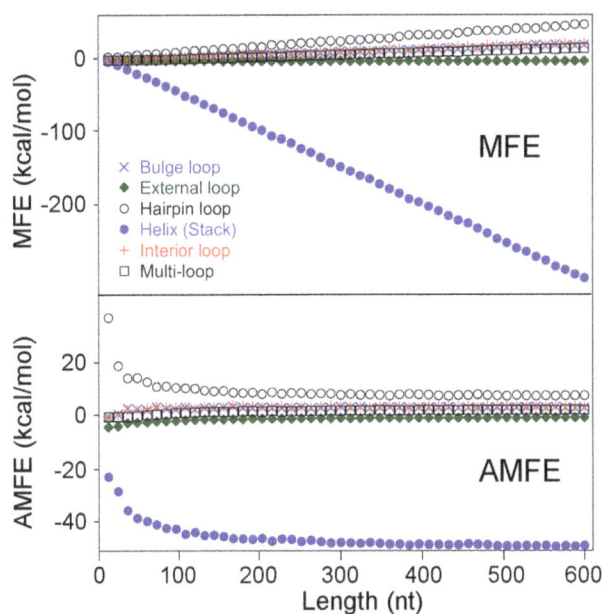

Figure 2. Free energy contributions of RNA structural elements. The free energy contributions of the different structural elements calculated by Quickfold are plotted versus sequence length: external loop (closed diamonds), hairpin loop (open circles), helix (closed circles), bulged loop (X), multi-loop (open squares), and interior loop (plus). The upper panel shows the contributions of structural elements to MFE and the lower panel the contributions to AMFE.

The variation of MFE with length, at constant GC-content, is plotted in Figure 3A. The relationship between MFE and length is apparently linear, and the MFE change rate increases with GC-content. For lengths varying from 20 to 600 nt, MFE changes by about -95.5 kcal/mol in sequences with 20% of GC-content and by -367.2 kcal/mol in sequences with 80% of GC-content.

The MFE variability associated with the nucleotide arrangement in the sequence was quantified by the SD of MFE in randomly shuffled sequences ($N = 100$) at fixed GC-content and length. In the range analyzed ($20\% \leq GC \leq 80\%$ and 20 nt $\leq$ length ≤ 600 nt), SD of MFE varied from 0.45 kcal/mol ($GC = 20\%$, length $= 20$ nt) to 7.2 kcal/mol ($GC = 80\%$, length $= 600$ nt).

The above MFE data were also used to compare the effect of length, nucleotide order, and GC-content on AMFE. Figure 3 illustrates the variations of the mean AMFE with lengths at constant GC-contents (panel C) and with GC-content at constant lengths (panel D). As shown, increasing the sequence length from 20 to 600 nt, at constant GC-content, causes an AMFE change varying from -15.1 kcal/mol – measured in sequences with the lowest GC-content (20%) – to -34.37 kcal/mol for sequences with the highest GC content (80%). The variation of AMFE with GC-content is -26.9 kcal/mol for the shortest sequences (20 nt) and -46.2 kcal/mol for the longest ones (600 nt). Moreover, the SD of AMFE for shuffled sequences at fixed lengths and GC-contents ranges from 0.83 kcal/mol (GC-content $= 20\%$, length $= 480$ nt) to 11.93 kcal/mol (GC-content $= 80\%$, length $= 20$ nt).

Therefore, the results show that sequence length contributes to AMFE of our simulated sequences by an amount ($-15.1 \leq \Delta AMFE \leq -34.4$ kcal/mol) that is comparable to that associated with the variation of nucleotide composition ($-26.9 \leq \Delta AMFE \leq -46.2$ kcal/mol) and with the variability of AMFE produced by a random arrangement of nucleotides in the sequences (0.83 kcal/

mol $\leq SD \leq 11.93$ kcal/mol). This indicates that AMFE, and generally, normalized MFEs, are biased measures of the minimum free energy, tending to decrease significantly with sequence size. These length-dependent differences in AMFE measures raise serious doubts about the validity of the normalization procedure and the reliability of the results obtained using length-normalized MFEs.

Why normalized MFE is not independent of length

We computed the MFE by software tools that apply the nearest-neighbor energy rules to simulate the minimum free energy secondary structure of RNA molecules. According to this model, the free energy of a structure is the result of the sum of independent contributions from various structural elements. All folded structures contain at least one destabilizing loop with a minimum length of three unpaired bases (The Nearest Neighbor Database, NNDB, http://rna.urmc.rochester.edu/NNDB) [17] and at least one base pair. Therefore, regardless of the set of energy parameters used to estimate MFE, negative free energies are not possible for sequences shorter than 5 nt. Accordingly, based on the results of the two different software programs used, the linear fitting of the MFE data versus the sequence size, at constant GC-content, intersects 0 energy axis at lengths higher than 15 nt, depending on base composition. In general, higher fitted lengths at 0 energy are associated with lower GC content. For this reason, in the case of a perfect linear relationship between MFE and length, dividing MFE by the number of nucleotides should result in a new free energy index with a hyperbolic decrease with length: if $MFE = a + b \cdot length$, then $MFE/length = a/length + b$. Although this reason can justify the strong hyperbolic decrease of AMFE with length, this is not the only source of variability of AMFE by length. In fact, as shown by the graph in Figure 4, the residuals from a least-squared linear regression analysis of MFE versus length showed a clear pattern with length, indicating that the assumption of perfect linearity between MFE and length is not valid. In particular, the monotone concave-down curve of the residual plot in Figure 4 indicates that longer sequences tend to be more stable than expected by a linear relationship between MFE and sequence size. Consistent with this, if we translate all the MFE data by a constant amount that shifts its regression line to the origin of the graph, the ratio of the new MFE to length remains significantly dependent on sequence size (data not shown).

A simple correction of the normalization procedure can substantially remove any intrinsic dependence of the MFE on sequence length in RNAs longer than 40 nucleotides

Our results show that the AMFE bias is generated by the combined effect of two causes: the non-perfect linearity of the MFE with sequence length and an inaccurate mathematical procedure that does not take into account that the regression line of the MFE versus the length does not intersect the axes' origin. Here, we introduce a new length-normalized MFE index, termed MFEden, which is computed to reduce the effects of the two causes of AMFE bias:

$$MFEden = 100 * (MFE - MFE_{ref}^{L})/(L - L_0),$$

where L is the length (number of nucleotides) of the analyzed sequence, MFE_{ref}^{L} is the precalculated average MFE computed for a shuffled sample containing L nucleotides and an equimolar ratio of the four nucleotides, and L_0 is a predefined optimal

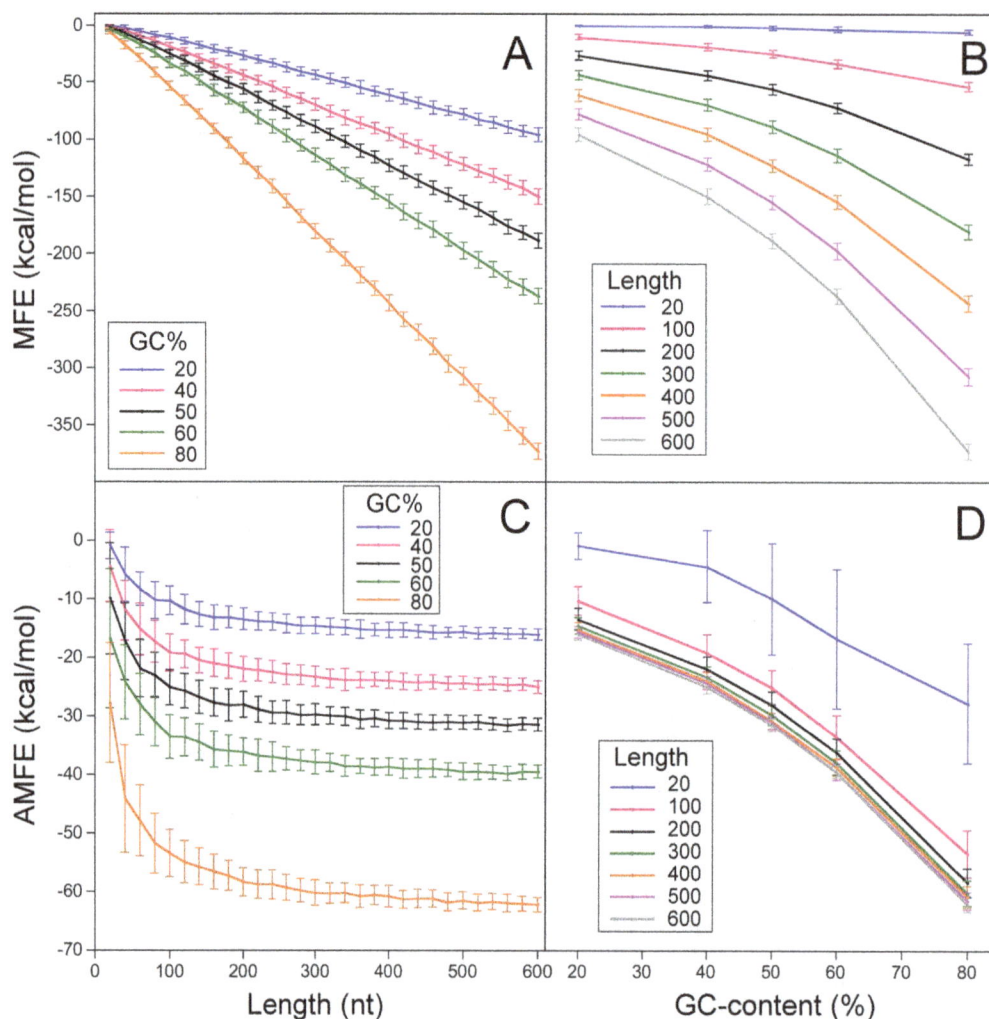

Figure 3. Minimum folding energy of randomly shuffled sequences. MFE (panel A) and AMFE (panel C) versus length at different GC-content: 20%, 40%, 50%, 60%, and 80%. MFE (panel B) and AMFE (panel D) versus GC-content for different sequence lengths: 20 nt, 100 nt, 200 nt, 300 nt, 400 nt, 500 nt, and 600 nt. Vertical bars indicate standard deviations (N = 100).

constant amount that shifts the MFE-versus-length regression line to the origin of the graph. Figure 5 shows the plot of the mean MFEden versus length for shuffled sequences with GC-content of 20%, 40%, 50%, 60% and 80% and for sequence lengths ranging between 40 and 600 nt. The large decrease of AMFE bias in the corrected index MFEden is evident in Figure 6 where the two indices are directly compared. As shown, in the critical range of length, where the bias makes AMFE impractical (between approximately 40 and 300 nt), the MFEden is unaffected by length.

The information content of the MFEden

The MFE of an RNA sequence is determined by the combined contributions of its length, nucleotide content and nucleotide order. MFEden excludes the component of free energy associated with sequence length but includes those related to nucleotide order and composition, also indirectly giving an estimate of their relative contributions. To illustrate the information content of MFEden, here we report an analysis using high confidence sets of two human RNA families: coding sequences (CDSs) and micro RNA precursors (pre-miRNAs) (see Materials and Methods). The panels A and B in Figure 7 show the scatterplot of the MFEden of the

CDSs (red circles) and pre-miRNAs (blue circles) versus the sequence length and the GC-content, respectively. As shown in Figure 7, in agreement with the results previously reported [3], pre-miRNAs are characterized by an MFEden lower than expected according to their nucleotide content. The MFEden of the coding sequences is approximately that expected for our shuffled sequences with a comparable GC-content. Moreover, the MFEden of the coding sequences appears to be scarcely affected by sequence length (Figure 7A), indicating that free energy density, on average, changes little from short to long (<600 nt) CDSs. From the human genomic GC-content, which is approximately 40.9% [18], we estimated the MFEden expected for a typical genomic sequence equal to about 6.2 kcal/mol. This estimated MFEden level is very close to 5.3 kcal/mol, which is the average MFEden that we computed for a sample of 100 genomic sequences, 100 nt-long, randomly chosen inside each human chromosome (2400 sequences in all). In Figure 7B, the estimated level of MFEden for a typical genomic sequence is indicated by an horizontal broken line showing the different nature of the MFE density in CDSs and pre-miRNAs. CDSs on average exhibit more negative MFEden than expected for the genomic GC-content. The folding stability of the CDSs is very close to that expected for

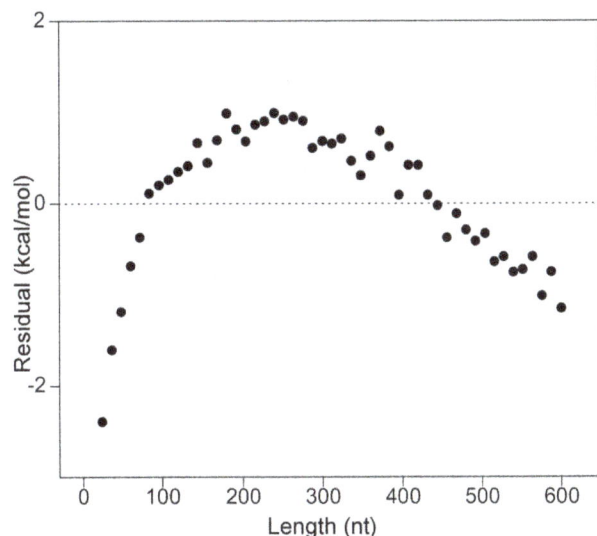

Figure 4. Residual plot from the linear fit of MFE versus length. Residual plot of the linear regression analysis of MFE versus sequence length. The MFE assigned to each length corresponds to the mean value of 1000 shuffled sequences with exact equimolar ratios of A, C, G, and U. Residuals are the differences between the computed MFEs and the corresponding values that are predicted by a linear regression analysis of MFEs with length.

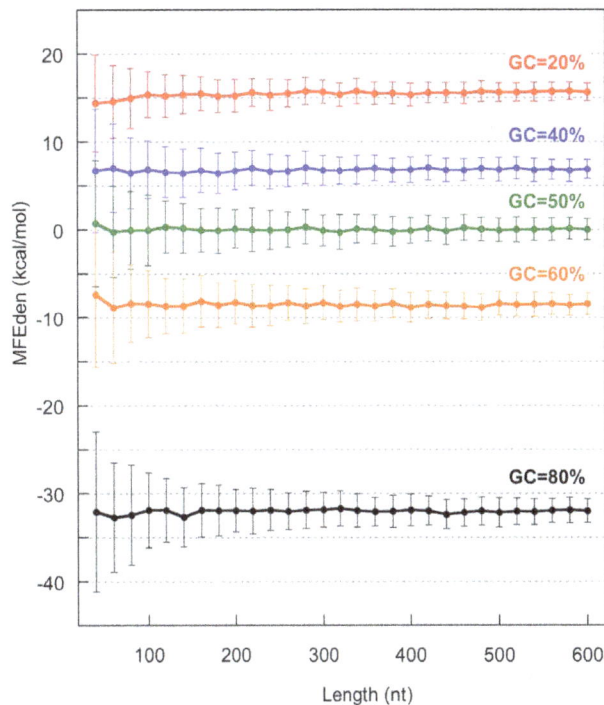

Figure 5. MFEden versus length. Plot of the mean MFEden versus length for shuffled sequences with GC-content of 20%, 40%, 50%, 60% and 80%, and for sequence lengths ranging between 40 and 600 nt with steps of 20 nt. Each point corresponds to the mean value of 100 shufflings. The lines connect MFEden values with the same GC-content. Vertical lines indicate standard deviation (N = 100).

their own GC-content, suggesting a very weak role of nucleotide order in determining their low free energy density. In contrast with CDSs, Figure 7B shows that, although the GC-content also contributes significantly to the high folding stability of pre-miRNAs, for this functional RNA family the nucleotide order plays a dominant role in determining its large stability with respect to a typical genomic sequence.

MFEden analysis of human functional RNAs

Along with CDSs and pre-miRNAs, we analyzed the MFEden of functional RNA sequences ranging between 40 and 600 nt. The datasets used in this study contain the most frequent families of human functional RNAs stored in the Rfam.fasta file of the Rfam database [19] and the sequences of small nucleolar RNAs (snoRNA) H/ACA and C/D box downloaded from the snoRNABase [20].

We found that the average length of each RNA family is not significantly correlated with its average MFEden (Pearson correlation coefficient (R_p) = −0.1531, N = 13, p = 0.6176), indicating that sequence length does not appear to be significantly constrained by folding free energy demands. We roughly estimated the contribution of nucleotide composition to the MFEden of each RNA by the mean MFEden of our shuffled sequences with the corresponding GC-content. The contribution of sequence order was valued by subtracting the estimated contribution of nucleotide composition from the computed MFEden. The average contributions to MFEden of the two sequence properties are, in each RNA family, positively correlated $(R_p = 0.6689, N = 13, p<0.02)$, suggesting that sequence composition and nucleotide order, in contrast with sequence length, concur to determine the level of the thermodynamic stability that characterizes a functional RNA family.

The results of our analysis also show that each RNA family is characterized by a restricted and well-defined combination of MFEden, length and GC-content. As an example, Figure 8

reports the MFEden of signal recognition particle RNAs (SRP RNAs), U6 spliceosomal RNAs (U6 snRNAs), Rous sarcoma virus RNAs (RSV RNAs), and H/ACA box RNAs plotted versus the sequence length (panel A) and the GC-content (panel B). In general, most of the RNA families examined here exhibit an average free energy density more negative than expected for a typical genomic sequence (Figure 9). In particular, SRP and H/ACA box RNAs and pre-miRNAs, exhibit the most negative average free energy density. Only small nuclear ribonucleic acids (snRNA) U4 and U6 and Rous sarcoma virus (RSV) RNAs have an average free energy density equal or slightly more positive than that expected for the genomic sequences. The case of the SRP family sequences stored in the Rfam database is interesting. The MFEden (and MFE) distribution of the SRP RNAs is bimodal, defining two distinct ranges of MFE density that are characterized by a similar range of GC-content (Figure 8). Moreover, surprisingly, the 17 human SRP seed sequences (orange points in Figure 8), which are used as high-quality reference RNAs for predicting SRP sequences stored in the Rfam database, exhibit a GC-content higher than that of the 99% of the SRP sequences in the database (Figure 8).

Discussion

MFE divided by the number of nucleotides is usually defined as length-normalized MFE [2,6–8,21–24]. Strictly speaking, it should mean that, using the normalized MFE indices, the differences in the minimum free energies of RNA molecules can be almost exclusively attributed to their nucleotide order and composition, regardless of their lengths. In fact, the length-normalized index AMFE was introduced by specifying that, after the MFE is

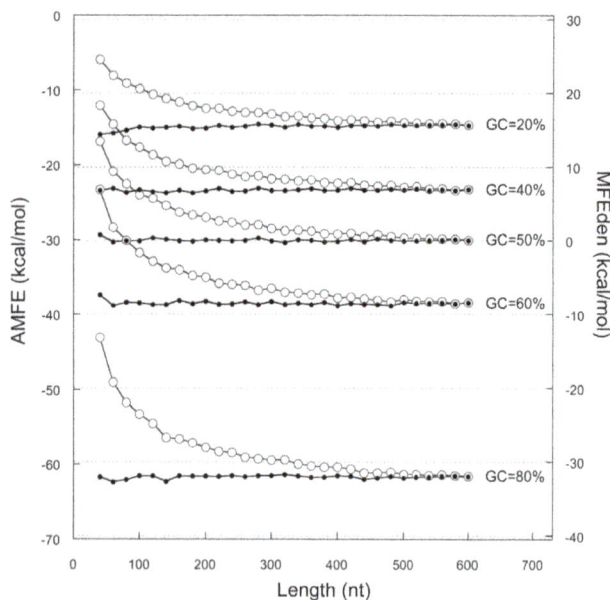

Figure 6. MFEden and AMFE versus length. Comparison of MFEden (black points) and AMFE (grey open circles) for shuffled sequences with GC-content of 20%, 40%, 50%, 60% and 80%, and for sequence lengths ranging between 40 and 600 nt with steps of 20 nt. Each point corresponds to the mean value of 100 shufflings. The lines connect values with the same GC-content.

to analyze the relationship between folding free energy and GC-content in mRNA sequences with different lengths [2].

Normalized MFEs have been employed for various purposes. For example, normalization was used to improve structure prediction by discarding segments whose normalized equilibrium free energies were smaller than a threshold value [7]. Normalized minimum free energy was also used to compare evolutionary relationships between micro-RNA genes and their functions [8], and its usefulness in identifying new non-coding RNAs was compared with other measures [4]. AMFE helped to find thermodynamics differences between nuclear-encoded micro-RNAs localized principally in mitochondria and cytosol [25]. Normalized MFE was also used in the search for distinctive criteria to predicting authentic precursors of microRNAs [22,24], for comparing thermodynamic stability [26], and to improve algorithms for RNA folding predictions [23,27,28].

Despite the significant number of works using normalized MFE, to our knowledge, the linearity of the relationship between MFE and sequence length, as well as the dependence of normalized MFEs on RNA size, has not been thoroughly tested and lacks quantitative substantiation. Here, we show that MFE does not decrease linearly with sequence length, especially in the range of sequences shorter than 100 nt. This deviation from a perfect linear relationship, along with the bias introduced by dividing the MFE by the length of the sequence, makes the normalized MFE of differently sized RNA sequences not directly comparable. In fact, we found that the magnitude of AMFE bias associated with length is comparable to the AMFE variation associated with the GC-content and with the variability produced by the random arrangement of nucleotides in the sequence. We also found that stacking and hairpin loop interactions are responsible for almost all the AMFE bias. The AMFE bias is higher in shorter RNAs and makes the AMFE index unsuitable for sequences shorter than approximately 300 nt. To extend the applicability of normalized MFEs to sequences shorter than 300 nt, we introduce a new index, called MFEden, obtained by a simple correction of the AMFE formula. The new MFEden index extends the applicability of AMFE to RNA longer than 40 nt. This is a big improvement if we

adjusted, sequences with lengths ranging from 60 to 400 nt are comparable based on their MFEs [9]. Accordingly, normalized MFE has been used because MFE values are strongly correlated with length [21], and because it serves as a comparable measure without excessively penalizing the shorter precursor microRNAs or favouring the longer mRNAs [24]. Similarly, it has been reported that normalization renders the MFE of hairpins of different lengths comparable [6], and normalized MFE was used

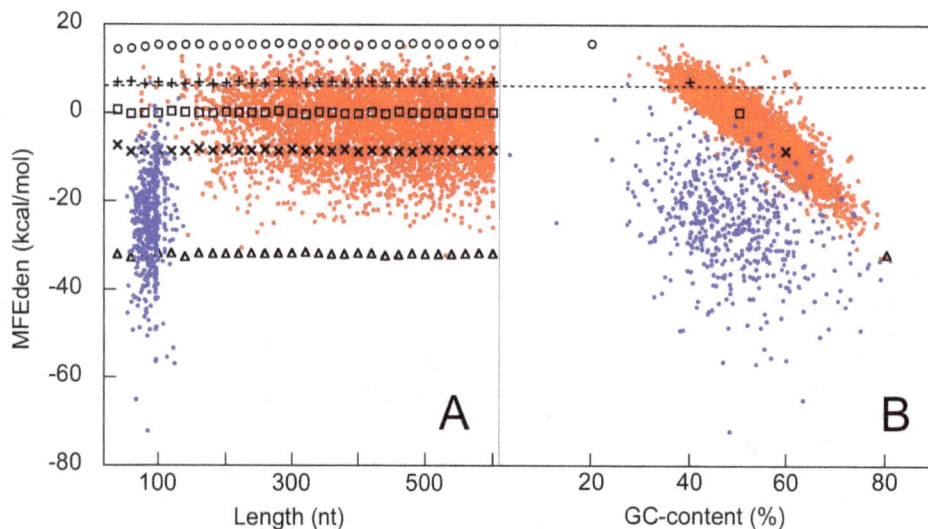

Figure 7. MFEden of human CDSs and pre-miRNA. MFEden of CDSs (red circles) and pre-miRNA (blue circles) are plotted versus sequence length (panel A) and GC-content (panel B). Black symbols indicate the mean MFEden values computed from shuffled sequences: GC-content: 20% (circle), 40% (plus), 50% (square), 60% (×), and 80% (triangle). A horizontal broken line indicates the MFEden level expected for the genomic GC-content.

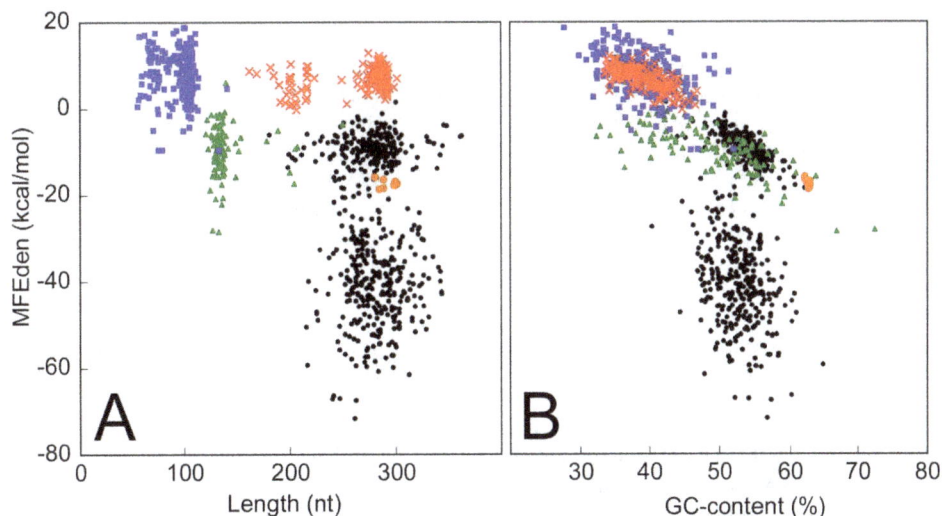

Figure 8. MFEden of human RNA families. The MFEden of the functional RNA families SRP RNAs (black points), U6 snRNAs (blue squares), RSV RNAs (red Xs), and H/ACA box RNAs (green triangles) plotted versus the sequence length (panel A) and the GC-content (panel B). Orange points indicate the 17 human SRP seed sequences of Rfam database.

consider that, of the 2208 functional RNA families stored in the Rfam database, 2023 (92%) have an average length ranging between 40 and 300 nt, and overall, 2104 families (95%) have an average length longer than 40 nt.

The stability of an RNA sequence is determined by the combined contributions of its length, nucleotide content and nucleotide order. In other words, if the local or the overall folding thermodynamic stability is important for the correct functionality of an RNA, it can be reached by acting on these three structural elements. Depending on the specific RNA function, these three

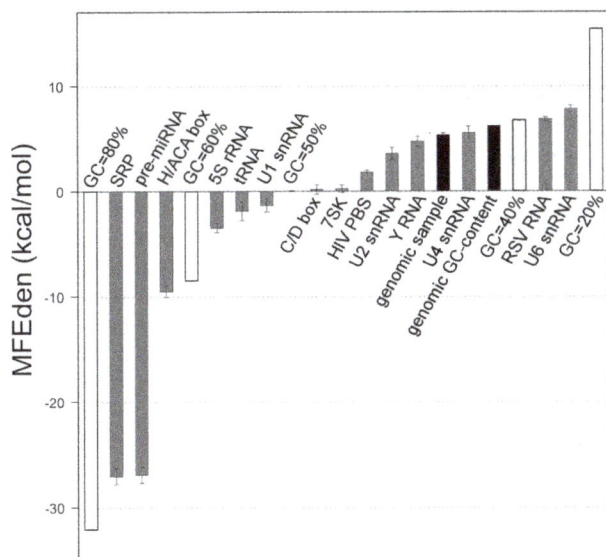

Figure 9. MFEden of 14 human functional RNA families. Bar plot showing the mean MFEden of 14 human functional RNA families (grey bars) compared with the mean MFEden of shuffled sequences with GC-content equal to 20%, 40%, 50%, 60% and 80% (white bars), the mean MFEden of 2400, 100 nt-long, genomic sequences taken at random and the MFEden expected for the genomic GC-content (black bars). The vertical bars indicate the standard errors of the means.

elements could be differently constrained, and the evaluation of their respective contributions to the overall free energy can be useful for their assignment to a functional class. From the perspective of free energy components associated with the three structural elements, the MFE, Z-score [29] and MFEden represent very different indices. The MFE of an RNA includes the free energy components of all three structural elements: sequence length, nucleotide content and nucleotide order. The Z-score represents a different method for quantifying the RNA secondary structure stability [29]. This index measures the distance between the MFE of the analyzed RNA sequence and the average MFE of a number of sequences generated by the random permutation of its nucleotides. The distance is measured in terms of the SD of the permutated sequences and, since the Z-score is a dimensionless index, lacks a direct relationship with the absolute amount of the free energy involved in folding stability. Because the shuffled sequences used as reference contain the same composition and the same number of nucleotides of the analyzed sequence, the Z-score index measures only the component of folding energy associated with the order of nucleotides in the sequence. This important point should be considered when the Z-score of two sequences is compared. In fact, for example, two RNAs with the same length and Z-score can differ significantly in their thermodynamic stability due to different GC-content. In addition, calculating the Z-score, especially for analyzing large RNA families, is laborious and time-consuming because of the sequence randomization procedures and the MFE computation of all simulated sequences. Differently from the MFE and the Z-score, the MFEden excludes the free energy contribution associated with the sequence length but includes the components related to nucleotide order and composition, also, indirectly, providing a rough estimate of their relative contributions. Moreover, the MFEden is measured in free energy units, its computation is not laborious and time-consuming, and it is suitable for large datasets. The MFEden analysis of the human RNA families examined in this work suggests that the GC-content and the nucleotide arrangement generally concur to determine the level of the thermodynamic stability that characterizes a functional RNA family, whereas the sequence length does not appear to be significantly constrained by folding free energy demands. This lack of correlation between the MFEden and the

RNA size suggests that sequence length is scarcely informative about the folding stability demands of an RNA family, and therefore represents a confusing variable when the MFE of different RNA families is compared. For this reason, MFE density appears more informative than MFE about the thermodynamic stability requirements of an RNA family. Accordingly, for example, U6 spliceosomal RNA family is characterized by a significantly high MFEden (Figures 8 and 9) that is consistent with its biological function. Such a low structure stability should facilitate the large conformational changes that U6 RNAs experience during the assembly of the spliceosome [30]. Conversely, pre-miRNA family is characterized by a significantly low MFEden (Figures 7 and 9). This high global structural stability is compatible with the necessity of pre-miRNA to maintain the stem-loop structure that is recognized and cleaved by double-stranded specific nucleases (Dicer family) by a process that is critical for the miRNA biogenesis [31]. The lack of correlation between the MFEden and the RNA size also suggests that the intrinsic higher stability of longer sequences is generally not compensated by a low level of GC-content or a decreased amount of stacking interactions, suggesting that there is not a general optimal level of thermodynamic stability at which every RNA tends. We also found that each RNA family is characterized by a restricted and well-defined combination of MFEden, length and GC-content. Furthermore, pre-miRNA, SRP and ACA_box RNAs exhibit significant negative MFE densities than the other RNAs and the typical genomic sequences. These differences in the MFE density of the RNA families can be exploited to improve the accuracy of sequence filtering for predicting non-coding RNAs.

In conclusion, this work demonstrates that length-normalized indices of MFE are biased measures of folding free energy density and proposes a new index with improved applicability for short RNA sequences. Unlike the Z-score, the new index, termed the MFEden, is simple and not time-consuming to compute, suitable for large datasets, and includes the folding free energy component associated with GC-content. An analysis of the MFEden of real sequences shows the different roles of length, GC-content and nucleotide order in the folding stability of RNA families and suggests the possible use of the MFEden to improve algorithms for predicting new RNAs or for their assignment to a functional class.

Materials and Methods

Data processing and analysis

All data were processed using software programs developed in our laboratory in the C# language that were tested by independent computational tools and manual calculations. Our software also includes programs to randomly shuffle the nucleotides of RNA sequences using the Fisher-Yates algorithm [32] and to read genomic sequences in a random position in the chromosomes. Statistical analysis was performed using STATISTICA (version 8.0, Statsoft, Inc.).

Computation of MFE, AMFE and MFEden

MFE was computed using two programs: RNAfold, included in the ViennaRNA software package version 2.1.5 [15,16]; and Quickfold, from the Mfold web server (http://mfold.rna.albany.edu/?q=DINAMelt/Quickfold) [13,14]. For very short sequences, we found that the MFEs computed by Quickfold (Mfold) were sometimes positive. In these cases, global free energy were set to 0 kcal/mol.

AMFE was calculated by dividing MFE by the sequence length and then multiplying the result by 100 to relate the index to a 100-nucleotides segment: $AMFE = 100 \cdot MFE/length$ [9].

MFEden was computed using the formula $MFEden = 100*(MFE-MFE_{ref}^{L})/(L-L_0)$, where MFE_{ref}^{L} is the expected MFE for a sequence with L nucleotides and equimolar ratios of A, C, G and U nucleotides. The expected MFEs were estimated by the mean MFE of 2000 random shufflings of sequences from a set with increasing lengths: from 40 to 600 nt, in steps of 4 nt. The estimated MFE of intermediate lengths were linearly interpolated (see Dataset S1 for MFEs computed by RNAfold). The optimal L_0 value for MFEs computed by RNAfold was determined empirically equal to 8 nt.

In all simulated sequences, including those with various GC-content, Watson and Crick complementary bases were present at the same frequency: number of As equal to number of Us, and number of Cs equal to number of Gs.

Human RNA sequences

All native sequences analyzed in this work were included in the taxonomic category of Homo Sapiens. RNA coding sequences were downloaded from the consensus CDS database (CCDS) (release 15) (ftp://ftp.ncbi.nlm.nih.gov/pub/CCDS/) [33], which provides high-quality human CDS data. Of the 29064 high-quality sequences downloaded from the CCDS database, we used the 4379 sequences with length included between 40 and 600 nt.

From the miRBase database (ftp://mirbase.org/pub/mirbase) (release 20) we downloaded the set of high-confidence microRNAs [34] which includes 278 human sequences.

The human most frequent families in the Rfam.fasta file stored in the Rfam database [19] (release 11.0) (ftp://ftp.ebi.ac.uk/pub/databases/Rfam/) were used in this study: 5S ribosomal RNAs (ID: RF00001), U1 spliceosomal RNAs (ID: RF00003), U2 spliceosomal RNAs (ID: RF00004), Transfer RNAs (ID: RF00005), U4 spliceosomal RNAs (ID: RF00015), Signal recognition particle RNAs (ID: RF00017), Y RNAs (ID: RF00019), U6 spliceosomal RNAs (ID: RF00026), 7SK RNAs (ID: RF00100) and Rous sarcoma virus RNAs (ID: RF01417).

The sequences of human H/ACA and C/D box small nucleolar RNAs were downloaded from the snoRNABase [20].

Estimate of the MFE density components associated with GC-content and nucleotide order

The expected values of MFEden for a specific GC-content was estimated by a polynomial interpolation of MFEden reference data computed for shuffled sequences with varying levels of GC-content (20%, 40%, 50%, 60% and 80%). The approximate MFEden component associated with GC-content was estimated by subtracting the MFEden expected for the genomic GC-content from the MFEden expected for the GC-content of the sequence analyzed. The estimate of the MFEden associated with nucleotide order was performed by subtracting the MFEden expected for the GC-content from the overall computed MFEden of the analyzed RNA.

Supporting Information

Dataset S1 RNAfold precalculated estimate of MFE expected for RNA sequences with L nucleotides and equimolar ratios of A, C, G and U. The expected MFEs were estimated by the mean MFE of 2000 random shufflings of sequences from a set with increasing lengths: from 40 to 600 nt, in steps of 4 nt. The estimated MFE of intermediate lengths were linearly interpolated.

Author Contributions

Conceived and designed the experiments: ET. Performed the experiments: ET. Analyzed the data: ET. Contributed reagents/materials/analysis tools: ET. Wrote the paper: ET. Wrote algorithms and software code: ET.

References

1. Felden B (2007) RNA structure: experimental analysis. Current Opinion in Microbiology 10: 286–291.
2. Seffens W, Digby D (1999) mRNAs have greater negative folding free energies than shuffled or codon choice randomized sequences. Nucleic Acids Research 27: 1578–1584.
3. Bonnet E, Wuyts J, Rouze P, Van de Peer Y (2004) Evidence that microRNA precursors, unlike other non-coding RNAs, have lower folding free energies than random sequences. Bioinformatics 20: 2911–2917.
4. Freyhult E, Gardner PP, Moulton V (2005) A comparison of RNA folding measures. Bmc Bioinformatics 6.
5. Batuwita R, Palade V (2009) microPred: effective classification of pre-miRNAs for human miRNA gene prediction. Bioinformatics 25: 989–995.
6. Thakur V, Wanchana S, Xu M, Bruskiewich R, Quick WP, et al. (2011) Characterization of statistical features for plant microRNA prediction. BMC Genomics 12: 108.
7. Pervouchine DD, Graber JH, Kasif S (2003) On the normalization of RNA equilibrium free energy to the length of the sequence. Nucleic Acids Res 31: e49.
8. Zhu Y, Skogerbo G, Ning Q, Wang Z, Li B, et al. (2012) Evolutionary relationships between miRNA genes and their activity. BMC Genomics 13: 718.
9. Zhang BH, Pan XP, Cox SB, Cobb GP, Anderson TA (2006) Evidence that miRNAs are different from other RNAs. Cell Mol Life Sci 63: 246–254.
10. Catania F, Lynch M (2010) Evolutionary dynamics of a conserved sequence motif in the ribosomal genes of the ciliate Paramecium. BMC Evol Biol 10: 129.
11. Mathews DH, Sabina J, Zuker M, Turner DH (1999) Expanded sequence dependence of thermodynamic parameters improves prediction of RNA secondary structure. J Mol Biol 288: 911–940.
12. Mathews DH, Disney MD, Childs JL, Schroeder SJ, Zuker M, et al. (2004) Incorporating chemical modification constraints into a dynamic programming algorithm for prediction of RNA secondary structure. Proc Natl Acad Sci U S A 101: 7287–7292.
13. Markham NR, Zuker M (2008) UNAFold: software for nucleic acid folding and hybridization. Methods Mol Biol 453: 3–31.
14. Zuker M (2003) Mfold web server for nucleic acid folding and hybridization prediction. Nucleic Acids Res 31: 3406–3415.
15. Zuker M, Stiegler P (1981) Optimal computer folding of large RNA sequences using thermodynamics and auxiliary information. Nucleic Acids Res 9: 133–148.
16. Lorenz R, Bernhart SH, Honer Zu Siederdissen C, Tafer H, Flamm C, et al. (2011) ViennaRNA Package 2.0. Algorithms Mol Biol 6: 26.
17. Turner DH, Mathews DH (2010) NNDB: the nearest neighbor parameter database for predicting stability of nucleic acid secondary structure. Nucleic Acids Res 38: D280–282.
18. Li XQ, Du D (2014) Variation, evolution, and correlation analysis of C+G content and genome or chromosome size in different kingdoms and phyla. PLoS One 9: e88339.

19. Burge SW, Daub J, Eberhardt R, Tate J, Barquist L, et al. (2013) Rfam 11.0: 10 years of RNA families. Nucleic Acids Res 41: D226–232.
20. Lestrade L, Weber MJ (2006) snoRNA-LBME-db, a comprehensive database of human H/ACA and C/D box snoRNAs. Nucleic Acids Res 34: D158–162.
21. Puzey JR, Kramer EM (2009) Identification of conserved Aquilegia coerulea microRNAs and their targets. Gene 448: 46–56.
22. Zhao DY, Wang Y, Luo D, Shi XH, Wang LP, et al. (2010) PMirP: A pre-microRNA prediction method based on structure-sequence hybrid features. Artificial Intelligence in Medicine 49: 127–132.
23. Spirollari J, Wang JT, Zhang K, Bellofatto V, Park Y, et al. (2009) Predicting consensus structures for RNA alignments via pseudo-energy minimization. Bioinform Biol Insights 3: 51–69.
24. Loong SNK, Mishra SK (2007) Unique folding of precursor microRNAs: Quantitative evidence and implications for de novo identification. Rna-a Publication of the Rna Society 13: 170–187.
25. Bandiera S, Ruberg S, Girard M, Cagnard N, Hanein S, et al. (2011) Nuclear Outsourcing of RNA Interference Components to Human Mitochondria. Plos One 6.
26. Ni M, Shu WJ, Bo XC, Wang SQ, Li SG (2010) Correlation between sequence conservation and structural thermodynamics of microRNA precursors from human, mouse, and chicken genomes. Bmc Evolutionary Biology 10.
27. Alkan C, Karakoc E, Sahinalp SC, Unrau P, Ebhardt HA, et al. (2006) RNA secondary structure prediction via energy density minimization. Research in Computational Molecular Biology, Proceedings 3909: 130–142.
28. Lopes Ide O, Schliep A, de Carvalho AC (2014) The discriminant power of RNA features for pre-miRNA recognition. BMC Bioinformatics 15: 124.
29. Le SY, Maizel JV Jr (1989) A method for assessing the statistical significance of RNA folding. J Theor Biol 138: 495–510.
30. Wolff T, Bindereif A (1993) Conformational changes of U6 RNA during the spliceosome cycle: an intramolecular helix is essential both for initiating the U4-U6 interaction and for the first step of slicing. Genes Dev 7: 1377–1389.
31. Ha M, Kim VN (2014) Regulation of microRNA biogenesis. Nat Rev Mol Cell Biol 15: 509–524.
32. Knuth DE (1997) The art of computer programming, volume 2 (3rd ed.): seminumerical algorithms: Addison-Wesley Longman Publishing Co., Inc. 784 p.
33. Farrell CM, O'Leary NA, Harte RA, Loveland JE, Wilming LG, et al. (2014) Current status and new features of the Consensus Coding Sequence database. Nucleic Acids Res 42: D865–872.
34. Kozomara A, Griffiths-Jones S (2014) miRBase: annotating high confidence microRNAs using deep sequencing data. Nucleic Acids Res 42: D68–73.

24

Evaluation of sgRNA Target Sites for CRISPR-Mediated Repression of *TP53*

Ingrid E. B. Lawhorn, Joshua P. Ferreira, Clifford L. Wang*

Department of Chemical Engineering, Stanford University, Stanford, California, United States of America

Abstract

The CRISPR (clustered regularly interspaced short palindromic repeats) platform has been developed as a general method to direct proteins of interest to gene targets. While the native CRISPR system delivers a nuclease that cleaves and potentially mutates target genes, researchers have recently employed catalytically inactive CRISPR-associated 9 nuclease (dCas9) in order to target and repress genes without DNA cleavage or mutagenesis. With the intent of improving repression efficiency in mammalian cells, researchers have also fused dCas9 with a KRAB repressor domain. Here, we evaluated different genomic sgRNA targeting sites for repression of *TP53*. The sites spanned a 200-kb distance, which included the promoter, transcript sequence, and regions flanking the endogenous human *TP53* gene. We showed that repression up to 86% can be achieved with dCas9 alone (i.e., without use of the KRAB domain) by targeting the complex to sites near the *TP53* transcriptional start site. This work demonstrates that efficient transcriptional repression of endogenous human genes can be achieved by the targeted delivery of dCas9. Yet, the efficiency of repression strongly depends on the choice of the sgRNA target site.

Editor: Robertus A. M. de Bruin, University College London, United Kingdom

Funding: IEBL was funded by the National Science Foundation Graduate Research Fellowship Program http://www.nsfgrfp.org/ and Stanford Graduate Fellowship http://sgf.stanford.edu/. IEBL, JPF, and CLW were funded by the National Institute on Aging (NIH-5R21AG040360-02) http://www.nia.nih.gov/. Data and sorting was performed on instruments in the Stanford Shared FACS Facility obtained using National Institutes of Health S10 Shared Instrumentation Grants S10RR025518-01 and S10RR027431-01 http://dpcpsi.nih.gov/orip/diic/shared_instrumentation. The funders had no role in study design, data collection and analysis, decision to publish, or preparation of the manuscript.

Competing Interests: The authors have declared that no competing interests exist.

* Email: cliff.wang@stanford.edu

Introduction

CRISPR (clustered regularly interspaced short palindromic repeats) serves as an adaptive immune system for many bacteria and archaea [1,2]. In the *Streptococcus pyogenes* type II CRISPR system, CRISPR-associated 9 nuclease (Cas9) binds to an RNA complex of trans-acting RNA (tracrRNA) and CRISPR RNA (crRNA), which guides the complex to DNA sequences complementary to the crRNA. This Cas9-RNA complex can then recognize and cleave the DNA of pathogens or other foreign elements [3–5]. To develop a general, host-independent method for targeting Cas9 and other recombinant proteins to desired DNA sequences, researchers have devised a platform that utilizes a small guide RNA (sgRNA) consisting of the two RNA elements joined by a short hairpin [5]. Targeting is achieved by the base-pairing between a sequence element encoded by the sgRNA and a desired, target DNA sequence. Any sequence can be targeted as long as it lies upstream of a protospacer adjacent motif (PAM) consisting of the sequence NGG (N represents any nucleotide base) [6,7]. Thus far, researchers have employed this technology to achieve targeted genomic insertions, deletions, and mutations in bacteria and eukaryotic cells [8–14].

More recently, researchers have adapted the CRISPR technology to deliver proteins of interest to targeted DNA sequences for applications beyond genome editing. By inactivating the two endonucleolytic domains of Cas9 to create a catalytically "dead"

Cas9 (dCas9) protein [5], researchers have employed the CRISPR gene-targeting platform and conjugated effector domains to achieve transcriptional repression [15–18], activation [18–21], and genomic loci imaging [22]. The CRISPR-mediated repression of gene expression has been termed CRISPR interference (CRISPRi). With CRISPRi, when dCas9 was co-expressed with sgRNA that was complementary to targeted gene, up to 99.7% transcriptional repression was observed in bacteria [15]. Such repression was achieved by targeting the non-template DNA strand of the gene coding region to block or hinder RNA polymerase transcription elongation [23]. Alternatively, repression has been achieved by targeting sites within a gene promoter region to block RNA polymerase and transcription factor binding for transcription initiation.

In contrast to repression of genes in prokaryotes, recent reports of dCas9-mediated repression in mammalian cells have been more modest, with repression reported to be approximately 50%. To improve on the repression reported in mammalian cells, Gilbert *et al.* fused a Krüppel-associated box (KRAB) transcription repressor domain from Kox1 to the C-terminus of dCas9 and multiple sgRNA target sites against the same gene locus were evaluated [16]. While the KRAB effector domain generated up to 93% repression of a reporter gene expression in human cells, the extent of endogenous gene knockdown varied from gene to gene with, in the best case, up to 80% repression of transferrin receptor CD71 at the protein level [16].

The rules for achieving CRISPR-mediated repression in mammalian systems are still not clearly established. When targeting the transcribed regions of a gene, in bacteria it has been reported that targeting the non-template strand leads to greater repression than when targeting the template strand. Based on this finding, it has been previously recommended that only the non-template strand should be targeted for repression in mammalian cells [23], though to our knowledge this has not been explicitly evaluated outside of bacterial systems. There is evidence that the extent of repression can be greater when targeting DNA sequences close to the transcriptional start site (TSS) in bacteria [15] and human embryonic stem cells [18] than sites in the transcribed region. However, to our knowledge a systematic comparison of promoter-proximal and transcribed region sites has not yet been conducted in mammalian cells. Last, while the KRAB effector domain has been to shown to be more effective than dCas9 alone in a select number of genes [16], it is unclear whether the use of the KRAB effector domain is required to achieve strong endogenous repression.

A better understanding of how the choice of dCas9 target sites affects repression of mammalian gene expression would help those who seek to control gene expression for research or synthetic applications [24]. To address these issues, we evaluated the effect of dCas9-mediated transcriptional repression by using sgRNAs complementary to a variety of target sites covering a 200 kb distance: these sites include the promoter, transcribed sequence and regions adjacent to the endogenous TP53 gene locus on both the template and non-template DNA strands. p53, the gene product of TP53, is a transcription factor whose activation by DNA damage or other cellular stresses results in responses that promote cell cycle arrest, apoptosis, or other tumor suppressing processes [25] and has, previously and presently, served as a well-studied model gene for the purpose of studying gene repression [26,27]. We also evaluated long-distance repression achieved by dCas9-KRAB binding in regions flanking the TP53 gene locus. Finally, we evaluated whether repression could be achieved by targeting sites surrounding the transcriptional start site in other genes.

Materials and Methods

Vector design

Target sequences were designed using the protocol recommended in Larson et al. [23]. Briefly, targets of 20–25 nucleotides in length preceding an NGG PAM site were screened for off-target homology using NCBI BLAST and the CRISPR design tool on crispr.mit.edu. Targets were designed to bind to areas of interest within the human TP53 promoter as well as to both the template and non-template strands of TP53 exons 4, 7, and 10. Because the U6 promoter was used for sgRNA expression, all target sequences without a "G" nucleotide in the leading position had one appended to the 5′ end of the target sequence to facilitate expression. Additionally, sgRNA sequences were designed to target sites approximately every 10 kilobases away from the transcriptional start site of TP53 in both the upstream and downstream directions. GFP targets against the non-template strand were created based on the sequences published in Gilbert et al. (GFP1) [16] and Mali et al. (GFP2) [10]. Additional targets were designed to bind to locations 20 to 50 base pairs upstream or 100–120 base pairs downstream of the transcriptional start site (+1) as identified by the DataBase of Transcriptional Start Sites [28,29] for various genes of interest in HEK 293 cells (Sequence Read Archive Accession No. SRA003625) [30]. Scramble1 and Scramble2 "target" controls were designed to have low homology

to the human or murine genome and are used as a non-targeting control to normalize gene expression. The sgRNA sequences are summarized in Tables S1, S2, and S3.

Annealed oligonucleotides containing the sgRNA sequences were ligated into the plasmid pRSET B-U6 sgRNA-term that was digested with BpiI (ThermoScientific, Pittsburgh, PA, USA). This step results in a RNA polymerase III U6 promoter-driven sgRNA expression cassette consisting of a 20–25 nucleotide domain complementary to a target DNA region, a Cas9-binding hairpin, and a transcription terminator. The U6-sgRNA expression construct was then digested with PspOMI and NotI-HF (New England BioLabs, Ipswich, MA, USA) and ligated into pCru5-1.2.1-BFP-IRES-Blast that was digested with NotI-HF to create pCru5-BFP-IRES-Blast-U6 sgRNA. Mammalian codon-optimized Streptococcus pyogenes dCas9 sequence with nuclease-inactivating D10A and H840A substitutions [5] (Addgene plasmid 44246) was fused to three C-terminal NLS (DPKKKRKV) sequences alone or with the KRAB domain sequence. The entire construct was then inserted into pCru5-1.2-mEGFP-IRES-mCherry-F2A-Puro [31] digested with SphI-HF and NsiI (New England BioLabs) to form pCru5-dCas9-3xNLS-IRES-mCherry-F2A-Puro or its variant with the KRAB domain.

Cell culture and transfection

HEK 293 [32] and HEK 293T [33] cells were cultured in DMEM with 10% FBS. Media was supplemented with 1 mM glutamine, 100 U/mL penicillin, and 100 μg/mL streptomycin.

HEK 293 and HEK 293T cells were transfected using polyethylenimine (PEI) transfection. Cells plated in 6-well plates were transfected with 2 μg of the dCas9 expression plasmid and 1.2 μg of the sgRNA expression plasmid per well which can be scaled by surface area for larger or smaller plates. Cells transfected with pSuper or pSuper-p53 plasmids were co-transfected with 0.5 μg pCru5-1.2-GFP-IRES-mCherry-F2A-Puro or pEGFP-N3 for selection or sorting purposes. Multiplexed transfection mixes contained equal amounts of all sgRNA expression plasmids used. 1 day post-transfection, cells were split 50% and selected with 2 μg/mL puromycin.

HEK 293 cells were transduced with retrovirus for stable expression of destabilized GFP. Retrovirus was produced by co-transfecting a pCru5 plasmid and pCL-Ampho into HEK 293T cells using calcium phosphate precipitation. One pCru5 plasmid had a long terminal repeat (LTR) containing a promoter that drives transcription of the neomycin resistance gene and an additional mutated EF-1α promoter (EF-T05) [34] driving expression of destabilized monomeric green fluorescent protein (GFP). The other pCru5 plasmid expressed GFP followed by an internal ribosome entry site (IRES) and a neomycin resistance gene directly off the LTR with a translation initiation sequence variant (1.2) [31]. Virus-containing supernatant was harvested and used to transduce HEK 293 cells. Virus was titered so that transduced cells received a single copy of the vectors. Polybrene (hexadimethrine bromide) was added to cultures at a concentration of 8 μg/mL. 24–48 h post-infection, cells were selected and maintained with 400 μg/mL neomycin.

Flow cytometry

Three days after antibiotic selection, cells were trypsinized, washed once with cold PBS, and resuspended in 1% FBS in PBS. For measurement of GFP knockdown, cells were analyzed on a LSRII (BD Biosciences, San Jose, CA, USA) and gated for mCherry- and TagBFP-positive viable singlet cells. GFP fluorescence on the FITC-A filter was recorded and GFP expression relative to the Scramble1 control was analyzed using FlowJo software (Tree Star, Ashland, OR, USA). For indicated qRT-PCR experiments, samples were sorted on a FACSAria II (BD Biosciences) for TagBFP-positive or GFP-positive viable singlet cells on the Pacific Blue-A filter using the 85 μM nozzle. At least 400,000 cells were sorted per sample.

Quantitative RT-PCR

Cell samples were trypsinized, washed once with cold PBS, and resuspended in RLT buffer supplemented with β-mercaptoethanol before freezing at −80°C for up to a week. Total RNA was isolated using the RNeasy Mini Kit (Qiagen, Valencia, CA, USA) using the manufacturer's directions. RNA was converted to cDNA using the High-Capacity cDNA Reverse Transcription Kit with random primers using the manufacturer's directions (Applied Biosystems, Foster City, CA, USA). All quantitative PCR reactions were done at 15 μL volume using SYBR Green PCR Master Mix (Applied Biosystems). Each independent biological replicate was plated in triplicate and run on the StepOnePlus thermal cycler (Applied Biosystems). Table S4 lists the primer sequences used to measure expression of *TP53* (GenBank Accession No. NM_000546), *WRAP53α* isoform (GenBank Accession No. NM_001143991), *WRAP53* all isoforms (GenBank Accession No. NM_018081), *GAPDH* (GenBank Accession No. NM_001256799), *MAPK1* (GenBank Accession No. NM_138957), *MAPK14* (GenBank Accession No. NM_001315), *PPIB* (GenBank Accession No. NM_000942), and *RB1* (GenBank Accession No. NM_000321).

Expression levels of housekeeping gene *TIMM17B* (GenBank Accession No. NM_001167947) were used for normalization of human gene expression. Relative quantity was calculated by the comparative C_T method ($\Delta\Delta Ct$) using cells transfected with the dCas9 construct and a Scramble sgRNA construct as the control. All reported levels of repression are relative to a Scramble control. Standardization between three independent biological replicates was performed via log transformation, mean centering, and autoscaling described by Willems *et al.* [35]. Relative expression levels from sorted experiments were correlated to their respective target's nucleotide annealing temperature using a linear regression for the target sequence (Figure S4).

Immunoblotting

To quantify p53 abundance, immunoblotting was performed using standard protocols. Endogenous p53 level was detected in lysate of transfected HEK 293T cells using a mouse anti-p53 primary antibody (1C12, #2524, Cell Signaling Technology, Danvers, MA, USA) and a goat anti-mouse IgG antibody conjugated to horseradish peroxidase (HRP). Loading control GAPDH was detected using a rabbit primary antibody conjugated to HRP (14C10, #3683 Cell Signaling Technology, Danvers, MA, USA). HRP activity was measured with the ECL Plus Western Blotting Detection Kit (#RPN2132, GE Healthcare, Piscataway, NJ, USA).

Results

Our goal was to investigate how the location of the sgRNA target site within a gene locus affects dCas9-mediated repression in human cells. We constructed a vector that expressed a human codon-optimized dCas9 protein with three nuclear localization signals (3xNLS). The vector also utilized an internal ribosome entry site (IRES) that allowed bicistronic expression of the dCas9 with mCherry-2A-Puro, which encoded both the mCherry red fluorescent protein and a puromycin resistance gene (Figure S1A). To evaluate the efficiency of the KRAB effector domain, a version

was also created with KRAB fused to the C terminus of the dCas9-3xNLS fusion (Figure S1A). Each sgRNA was expressed from a RNA polymerase III U6 promoter downstream of a cassette expressing blue fluorescent protein (BFP) and a blasticidin resistance gene (Figure S1A). To test the efficacy of our dCas9 fusions, we co-transfected the dCas9 constructs and individual sgRNAs in HEK 293 cells stably expressing green fluorescent protein (GFP). We observed reduction of GFP fluorescence over the non-targeted Scramble1 control in cells expressing a sgRNA targeting GFP and either dCas9 (37% repression) or dCas9-KRAB (60% repression) (Figure S2). As observed previously [16], dCas9-KRAB generated greater GFP repression than dCas9 alone.

In this study, we were interested in evaluating the repression of endogenous genes. Having established that our constructs could repress ectopic GFP expression, we next evaluated the effectiveness of dCas9 and dCas9-KRAB in repressing human *TP53* mRNA expression. To evaluate the effect of target site position on repression, we tested sgRNAs that targeted different sites at or near the *TP53* gene and measured *TP53* mRNA levels by quantitative real-time PCR (qRT-PCR). We hypothesized that the KRAB fusion may have long-distance repressive effects on gene expression since KRAB-mediated promoter silencing has been reported over distances of several tens of kilobases away from the KRAB binding site [36]. To evaluate this hypothesis, we designed sgRNAs to target sites up to 100,000 bp upstream and downstream of the *TP53* transcriptional start site (TSS) in approximately 10,000 bp increments (Figure S1B). Each sgRNA is labeled with a number representing the distance (bp) from the transcriptional start site ("K", one thousand bp) and a "−" or "+" representing upstream or downstream of +1. HEK 293T cells were co-transfected with a dCas9 or dCas9-KRAB expression construct and each sgRNA expression construct, selected with puromycin, and sorted for BFP positive cells. Of the 21 sgRNAs tested, the only appreciable repression of mRNA occurred at a site approximately 36 bp upstream from the *TP53* TSS (85% for dCas9; 90% for dCas9-KRAB; −T36, Figure S1C and Figure S1D). Furthermore, there was little observable additional *TP53* repression with the dCas9-KRAB fusion in comparison to repression with dCas9. Since using multiple sgRNAs has been successful in CRISPR-directed gene activation [19,20], we tested the effect of using multiple sgRNAs for repression by co-transfecting dCas9 or dCas9-KRAB with various combinations of sgRNAs targeting: 10 sites spanning 100,000 bp upstream of *TP53* (−T100K to −T10K), three sites spanning 30,000 bp upstream of *TP53* (−T30K to −T10K), 10 sites spanning 100,000 bp downstream of *TP53* (+T10K to +T100K), three sites spanning 30,000 bp downstream of *TP53* (+T10K to +T30K), and three sites flanking the *TP53* TSS (−T36 +T27 + T110). Again, there was no added transcriptional repression with dCas9-KRAB directed binding as compared to dCas9 (Figure S1E). Moreover, we observed similar repression of *TP53* with the simultaneous targeting of three sites (−T36 +T27 +T100) as with the targeting of the single −T36 site (Figure S1D). This is in contrast to the previously-observed synergistic increase in gene activation induced by use of multiple sgRNA binding sites in conjunction with a dCas9 fused to aVP64 transcription activation domain [20], tetramer of VP16 domains, each capable of recruiting transcriptional machinery to promoter proximal regions. We postulate that this synergy in activation occurred because activation by VP64 can act over long distances [37] thus allowing multiple targeted sites to act together from various distances. In contrast, the repression that we observed may be due to a TSS-proximal mechanism, where additional dCas9 repression factors cannot cooperate significantly to improve the degree of repression. Our results also suggest that fusing KRAB to dCas9 does not significantly improve *TP53* transcriptional repression and that targeting multiple sites concurrently does not necessarily improve the efficiency of repression. However, we cannot rule out the possibility that KRAB may increase repression when targeted to other genes or using alternative sites.

Since dCas9-KRAB had little added benefit over dCas9 for repression of *TP53* and there was significant repression when targeting sites near the TSS, we further explored the positional effect of targeting dCas9 to sites closer to the *TP53* gene itself. 21 sgRNAs were designed to target sequences on the template and non-template strands of the promoter and transcribed region of *TP53* (Table S1 and Figure 1A and B). In bacteria, only sgRNAs binding to the non-template strand could enable repression of a fluorescent protein [15] although that finding has not been replicated in mammalian cells. The sgRNAs were co-expressed with dCas9 in transfected and puromycin-selected HEK 293T cells and directly analyzed by qRT-PCR. As before, the −T36 target site was shown to significantly repress *TP53* expression at the mRNA level ($P<0.01$, Figure 1C) as compared to cells expressing either non-targeted control. Because transfection efficiency varied, we opted to sort for BFP-positive cells after transfection and puromycin selection to enrich for cells expressing both CRISPR constructs. Of the subset of sgRNAs tested, three repressed *TP53* transcriptional expression (Figure 1D), all located within a 200-bp region of the TSS. sgRNA −T36 induced approximately 86% *TP53* mRNA repression, a level comparable to shRNA knock-down with pSuper-p53 construct [26] (Figure S3C). Repression using this target site was also consistently observed at the protein level in unsorted, selected cells (Figure 1E) for three independent experiments. However, when targeting the +T110 and +T111 sites and analyzing protein levels by immunoblotting, not only were the magnitudes of repression lower than −T36 but they were also less reproducible over those same experiments. Additionally, the targeting of sites +T27 and + T110 repressed *TP53* transcription expression. We did not observe significant repression when targeting the template or non-template strands of DNA in the transcribed region. It should be noted that first exon of *TP53* overlaps with exon 1α of *WRAP53* [38,39] which is the first exon of the *WRAP53α* isoform—a known anti-sense transcript of *TP53* (Figure S3A). We observed repression of the *WRAP53α* isoform in cells transfected with dCas9 and the −T36 sgRNA but no similar repression of total *WRAP53* expression from all three isoforms (Figure S3B); we did not see a similar effect with pSuper-p53 shRNA transfection (Figure S3C). Finally, we note that for the many sgRNAs in our study that did not generate appreciable repression of TP53, we cannot rule out that these sgRNAs were somehow incapable of mediating the binding of dCas9 to the DNA.

Observing that target sites −T36 and +T110 were effective at repressing *TP53* transcription, we next designed sgRNAs to target sites at similar distances from the TSS for five additional human genes: *GAPDH*, *PPIB*, *MAPK1*, *MAPK14*, and *RB1* (Figure 2). For each gene, dCas9 was targeted to one site 27 to 52 bp upstream of the TSS on the non-template DNA strand and two sites 81 to 136 bp downstream of TSS on the both the non-template and template DNA strands (Table S3). While it is certainly possible that the KRAB fusion could be effective for repression of these other genes, because we found the dCas9 alone was sufficient for repression of *TP53*, we chose to investigate the whether dCas9 alone could also be effective in repressing other genes. For four of the five genes, modest transcriptional repression was achieved, including over 50% repression in *RB1* with target + R121 (Figure 2F). To explain the variation in repression efficiency,

we hypothesized that the annealing temperature of the target base pair sequence may correlate with the efficiency of repression. Analysis of this parameter and the relative expressions for each target examined in triplicate revealed no correlation between the sgRNA target annealing temperature and mRNA expression (Figure S4). These results indicate modest transcriptional repression can be achieved in a variety of genes by targeting the region proximal to the TSS although there may be additional factors limiting the extent of that knockdown.

Discussion

Prior reports [16,23] recommended targeting either the promoter-proximal region or the transcribed region on the non-template strand for optimal endogenous dCas9 repression in human cells. In the case of *TP53*, while we were unable to achieve significant repression by targeting various sites on either the template or non-template strand of the transcribed region, we were able to repress transcription by targeting sites near or close to the TSS. We report up to 86% repression of *TP53* transcription by using sgRNAs −T36 and +T110, targeting sites 36 bp upstream and 110 bp downstream of the transcriptional start site. Interestingly, the −T36 sgRNA binds to a site that overlaps with previously-reported Myc/Max, USF, and E2F1 binding motifs while +T110 binds to a site overlapping with a Pax binding motif [40]. The −T36 sgRNA is also within 40 bp of the transcription start site, which is the approximate DNA footprint of RNA polymerase II [41]. In support of the notion the dCas9 can interfere with transcription at the DNA level,, recent chromatin immunoprecipitation (ChIP) data by others indicate that sgRNA-mediated dCas9 binding at the transcriptional enhancer of *Nanog* interferes with the binding of several transcription factors in mouse ESCs [42].The CRISPR-mediated repression that we observed is similar to what we were able to achieve using a *TP53* shRNA construct [26] and similar to levels of repression Gilbert *et al.* achieved by targeting sites downstream of the TSS of endogenous *CD71* and *CXCR4* genes [16]. We did not find a clear association between target proximity to the TSS and transcriptional repression, but we were able to achieve modest but significant repression in an additional four of five examined genes.

Interference with transcription initiation provides an alternative to post-transcriptional regulation via RNAi and allows for evaluation of repression at the transcriptional-level. In contrast to gene editing with RNA-directed Cas9, dCas9-mediated repression allows for a reduction in gene expression without modification to the genomic sequence. Thus with dCas9-mediated repression, it should be possible to engineer conditional expression system where repression can be switched "on" or "off" reversibly by other cellular cues or experimental conditions (e.g., addition of a small molecule inducer). In addition, it appears that, depending on the targeted gene, similar levels of transcriptional repression can be achieved without the use of a KRAB effector domain, which, for *TP53* transcription, appeared to have little effect. It is important to note potential limitations in the dCas9-initiated repression. Access of the dCas9-RNA complexes to the target site may be limited or even obstructed by the structure and local environment of the chromatin, epigenetic features, or the presence of regulatory elements [43,44] and thus affect efficiency of repression. Additionally, even if a sequence has few if any off-target binding sites [44–48], some promoter regions may overlap with other genes. We observed concomitant repression of *WRAP53α* isoform mRNA levels (Figure S3B) when targeting *TP53*'s TSS.

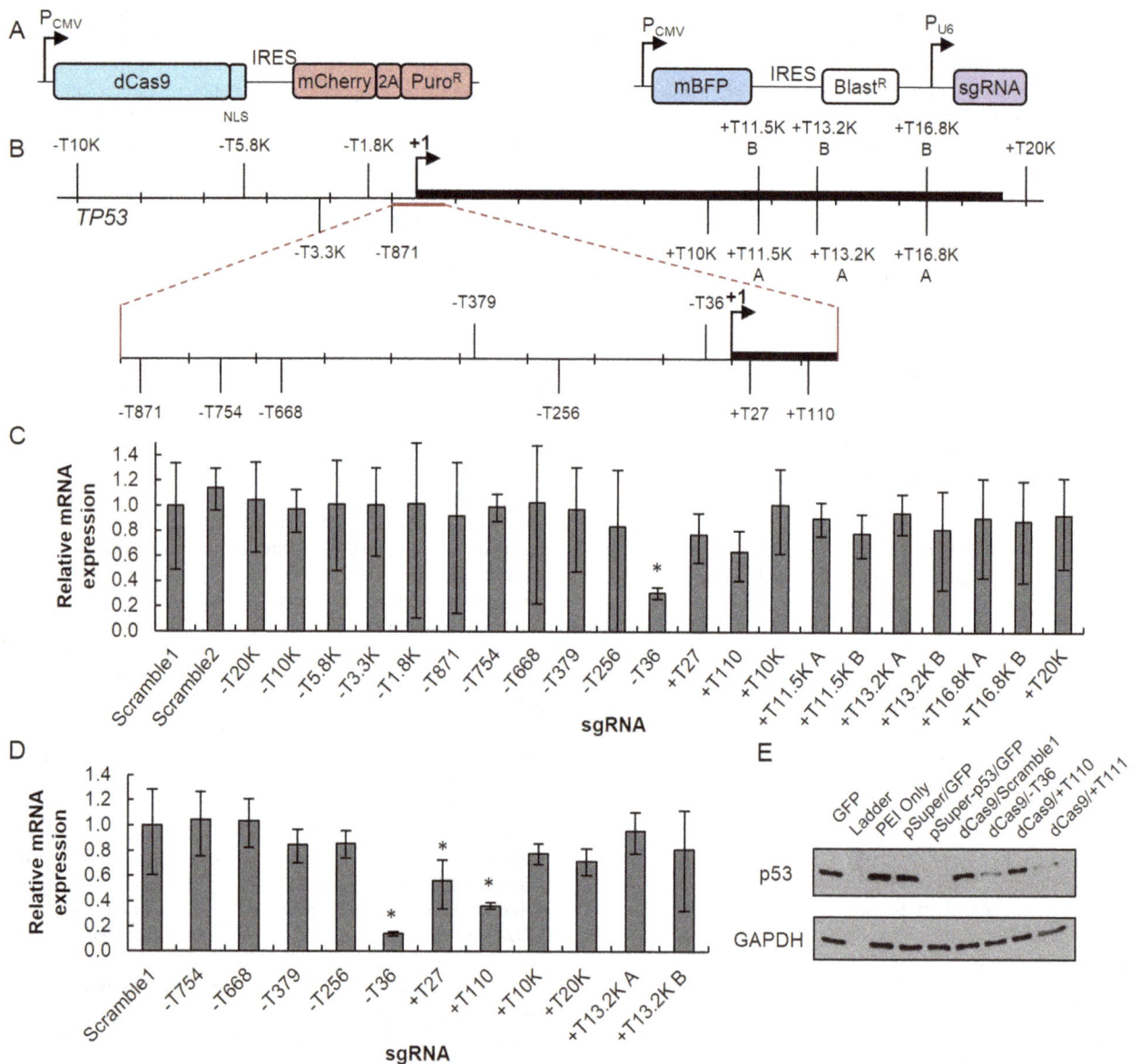

Figure 1. Repression of *TP53* by targeting dCas9 to the transcriptional start site. (A) HEK 293T cells were co-transfected with dCas9 (left) and sgRNA (right) expression plasmids. Codon-optimized dCas9 was fused to three copies of nuclear localization signal (NLS) and was co-expressed with mCherry fluorescent protein. sgRNA plasmid expresses mBFP and sgRNA off separate promoters. P_{CMV}, CMV promoter; 2A, ribosomal slippage site; Puro[R], puromycin resistance gene; IRES, internal ribosome entry site; mBFP, TagBFP fluorescent protein; Blast[R], blasticidin resistance gene; P_{U6}, U6 promoter. (B) Locations of sgRNA binding sites in the *TP53* promoter and transcribed region (thick line). Each sgRNA is numbered by the distance (bp) from the transcriptional start site. "−", upstream of +1; "+", downstream of +1; "K", one thousand bp; +1, transcriptional start site. Labeled sites above and below the transcribed region indicate sgRNAs targeting the template or non-template DNA strands, respectively. (C, D) Relative expression of *TP53* mRNA in cells co-transfected with dCas9 and indicated sgRNA constructs. After three days, cells were (C) directly analyzed by qRT-PCR or (D) sorted for BFP-positive cells, then analyzed by qRT-PCR. Results were normalized, linearly rescaled, and calculated for mean fold change ($n = 3$) $\pm 95\%$ confidence interval, relative to Scramble1 negative control sgRNA. *$P < 0.01$ compared to non-targeted sgRNA control by paired, one-sided t-test. (E) Immunoblot from HEK 293T cell lysate three days post-transfection with shRNA against *TP53*, dCas9 and sgRNAs against *TP53* (e.g., dCas9/−T36), dCas9 and non-targeted sgRNA (dCas9/Scramble1), other control constructs (GFP expression vector, which served as non-specific vector control, or pSuper without any shRNA encoded), or mock (PEI only). See Figure 2E for location of sgRNA +T111. pSuper constructs were co-transfected with GFP-IRES-mCherry-2A-Puro expression constructs for selection purposes and experimental consistency (pSuper/GFP, pSuper-p53/GFP). Protein ladder (lane 2 from left) is not visible. See Figure S5 for uncropped immunoblots.

To summarize, our results demonstrate significant CRISPRi repression of *TP53* mRNA levels without use of a KRAB effector domain by targeting the TSS. Modest repression was also observed in four of five genes targeted at the TSS. Our study demonstrates that target site efficiencies can vary greatly. When one desires to utilize CRISPRi to repress genes in mammalian cells, one should evaluate a range of sgRNA target sites, including those near or at the transcriptional start site. However, the targeting of the transcriptional start site is not guaranteed to be effective for every gene and thus will need to be evaluated on a gene-by-gene basis. We propose that CRISPRi with dCas9 alone may prove useful for functional mapping of transcription factor

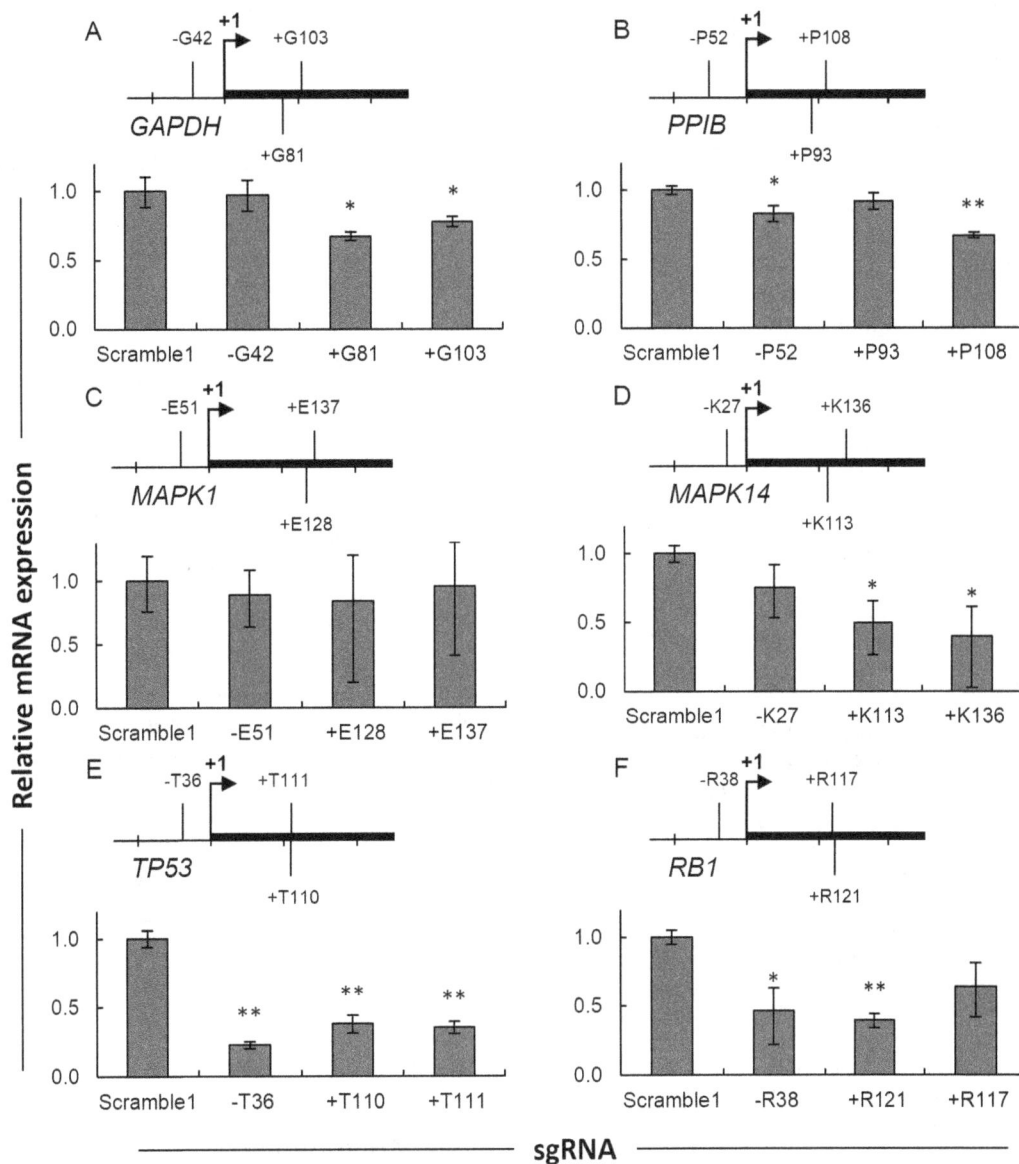

Figure 2. Evaluation of dCas9 targeting to the TSS of various genes. Diagram of sgRNA binding sites and plots of mRNA repression in HEK 293T by dCas9 and sgRNAs targeting (A) *GAPDH*, (B) *PPIB*, (C) *MAPK1*, (D) *MAPK14*, (E) *TP53*, and (F) *RB1*. Each sgRNA is numbered by the distance from the transcriptional start site. "−", upstream of +1; "+", downstream of +1; +1, transcriptional start site. Labeled sites above and below the transcribed region indicate sgRNAs targeting the template or non-template DNA strands, respectively. Relative expression of each mRNA in HEK 293T cells co-transfected with dCas9 and sgRNA constructs. After three days, cells transfected with dCas9 and indicated sgRNA constructs were sorted for BFP-positive cells, analyzed by qRT-PCR. Results were normalized, linearly rescaled, and calculated for mean fold change ($n = 3$)$\pm 95\%$ confidence interval, relative to Scramble1 negative control sgRNA. *$P < 0.02$, **$P < 0.002$ compared to Scramble1 sgRNA control by paired, one-sided t-test.

motifs and enhancer elements via transcription initiation blocking without the potential additional repressive effect of a KRAB domain.

Supporting Information

Figure S1   Evaluation of CRISPR-mediated repression using dCas9 and dCas9 fused to a KRAB repressor domain. (A) HEK 293T cells were co-transfected with dCas9 or dCas9 fused to KRAB domain (left) and sgRNA (right) expression plasmids. Both codon-optimized dCas9 and dCas9-KRAB were fused to three copies of nuclear localization signal (NLS) and were co-expressed with mCherry fluorescent protein. sgRNA plasmid

expresses mBFP and sgRNA off separate promoters. P_{CMV}, CMV promoter; 2A, ribosomal slippage site; $Puro^R$, puromycin resistance gene; IRES, internal ribosome entry site; mBFP, TagBFP fluorescent protein; $Blast^R$, blasticidin resistance gene; P_{U6}, U6 promoter. (B) Locations of sgRNA binding sites in the endogenous *TP53* locus. Each sgRNA is numbered by the distance (bp) from the transcriptional start site. "−", upstream of +1; "+", downstream of +1; "K", one thousand bp; +1, transcriptional start site. Labeled sites above and below the transcribed region indicate sgRNAs targeting the template or non-template DNA strands, respectively. (C–E) Relative expression of *TP53* mRNA in cells co-transfected with dCas9 and sgRNA constructs targeting (C) upstream, (D) downstream of the +1 site or

(E) in combinations of multiple sgRNA. After three days, cells co-transfected with indicated dCas9 and sgRNA constructs were analyzed by qRT-PCR. Data in (C, D) are fold change relative to Scramble1 or Scramble2 negative control sgRNA ± s.e. of three technical replicates. Data in (E) represents sorted cells and were normalized, linearly rescaled, and calculated for mean fold change ($n = 3$) ± 95% confidence interval, relative to Scramble1 negative control sgRNA. *$P < 0.01$ compared to non-targeted sgRNA control by paired, one-sided t-test. See also Figure 1B for additional targeted *TP53* sites.

Figure S2 CRISPR-mediated repression of GFP. Destabilized GFP (ds1GFP) expression cassette expressed using an (A) LTR (P_{LTR}) or (B) mutant EF-1α promoter (P_{EF-T05}) was retrovirally transduced into HEK 293 cells and selected with neomycin. LTR, retroviral long-terminal repeat; Neo^R, neomycin resistance gene; IRES, internal ribosome entry site. Cells were analyzed three days post-co-transfection with dCas9-KRAB and either GFP1 or GFP2 sgRNAs targeting the template strand. GFP fluorescence was measured via flow cytometry after gating for BFP positive cells. Values are arithmetic means of GFP fluorescence ± s.d. ($n = 3$) calculated from geometric means of each sample population and were normalized to dCas9:Scramble1 negative control.

Figure S3 Target sequence −T36 also leads to a reduction in *WRAP53α* isoform mRNA. (A) Locations of sgRNA binding sites in the endogenous *TP53* locus. Each sgRNA is numbered by the distance (bp) from the transcriptional start site. "−", upstream of +1; "+", downstream of +1; "K", one thousand bp; +1, transcriptional start site. Labeled sites above and below the transcribed region indicate sgRNAs targeting the template or non-template DNA strands, respectively. (B, C) Relative expression of *TP53*, *WRAP53α*, and all isoforms of *WRAP53* mRNA in cells co-transfected with (B) dCas9 and sgRNA constructs or (C) shRNA constructs targeting *TP53*. After three days, cells co-transfected with indicated constructs were sorted, analyzed by qRT-PCR. Data are fold change relative to (B) Scramble1 negative control sgRNA or (C) pSuper Control plasmid ± s.e. of three technical replicates.

Figure S4 Annealing temperature of sgRNA target sequence has minimal to no correlation with reduction in transcriptional expression. Minimal to no correlation was observed between transcriptional repression relative to Scramble1 control of all individual dCas9 knockdown experiments done in triplicate and the annealing temperature of the sgRNA target. Line represents linear regression of data.

Figure S5 Reduction of p53 protein in transfected HEK 293T cells. Uncropped immunoblot (see Figure 1E) containing 15 μg total protein/lane immunostained with (A) p53 antibody and (B) GAPDH antibody from HEK 293T cells transfected as indicated. Protein ladder (lane 2 from left) is not visible.

Table S1 Target sequences for sites within and flanking human *TP53*. The leading G nucleotide required for U6 promoter expression is in bold. The underlined following 19 to 24 nucleotides comprise the target sequence. The PAM site is in gray.

Table S2 Control target sequences. The leading G nucleotide required for U6 promoter expression is in bold. The underlined following 19 to 24 nucleotides comprise the target sequence. The PAM site is in gray.

Table S3 Target sequences within the promoter of various human genes of interest. The leading G nucleotide required for U6 promoter expression is in bold. The underlined following 19 to 24 nucleotides comprise the target sequence. The PAM site is in gray.

Table S4 Primer sequences used for quantitative RT-PCR.

Acknowledgments

Special thanks are owed to Jane Yang for helpful discussion and mentoring. We also thank the staff at the Stanford Shared FACS Facility for their help with cell sorting.

Author Contributions

Conceived and designed the experiments: IEBL JPF CLW. Performed the experiments: IEBL. Analyzed the data: IEBL. Contributed reagents/materials/analysis tools: IEBL JPF. Wrote the paper: IEBL JPF CLW.

References

1. Barrangou R, Fremaux C, Deveau H, Richards M, Boyaval P, et al. (2007) CRISPR Provides Acquired Resistance Against Viruses in Prokaryotes. Science 315: 1709–1712. doi:10.1126/science.1138140.

2. Wiedenheft B, Sternberg SH, Doudna JA (2012) RNA-guided genetic silencing systems in bacteria and archaea. Nature 482: 331–338. doi:10.1038/nature10886.

3. Marraffini LA, Sontheimer EJ (2010) CRISPR interference: RNA-directed adaptive immunity in bacteria and archaea. Nat Rev Genet 11: 181–190. doi:10.1038/nrg2749.

4. Deltcheva E, Chylinski K, Sharma CM, Gonzales K, Chao Y, et al. (2011) CRISPR RNA maturation by trans-encoded small RNA and host factor RNase III. Nature 471: 602–607. doi:10.1038/nature09886.

5. Jinek M, Chylinski K, Fonfara I, Hauer M, Doudna JA, et al. (2012) A Programmable Dual-RNA–Guided DNA Endonuclease in Adaptive Bacterial Immunity. Science 337: 816–821. doi:10.1126/science.1225829.

6. Mojica FJM, Díez-Villaseñor C, García-Martínez J, Almendros C (2009) Short motif sequences determine the targets of the prokaryotic CRISPR defence system. Microbiology 155: 733–740. doi:10.1099/mic.0.023960-0.

7. Shah SA, Erdmann S, Mojica FJM, Garrett RA (2013) Protospacer recognition motifs: Mixed identities and functional diversity. RNA Biol 10: 891–899. doi:10.4161/rna.23764.

8. Jiang W, Bikard D, Cox D, Zhang F, Marraffini LA (2013) RNA-guided editing of bacterial genomes using CRISPR-Cas systems. Nat Biotechnol 31: 233–239. doi:10.1038/nbt.2508.

9. Wang H, Yang H, Shivalila CS, Dawlaty MM, Cheng AW, et al. (2013) One-Step Generation of Mice Carrying Mutations in Multiple Genes by CRISPR/Cas-Mediated Genome Engineering. Cell 153: 910–918. doi:10.1016/j.cell.2013.04.025.

10. Mali P, Yang L, Esvelt KM, Aach J, Guell M, et al. (2013) RNA-Guided Human Genome Engineering via Cas9. Science 339: 823–826. doi:10.1126/science.1232033.

11. Jinek M, East A, Cheng A, Lin S, Ma E, et al. (2013) RNA-programmed genome editing in human cells. eLife 2: e00471. doi:10.7554/eLife.00471.

12. Cong L, Ran FA, Cox D, Lin S, Barretto R, et al. (2013) Multiplex Genome Engineering Using CRISPR/Cas Systems. Science 339: 819–823. doi:10.1126/science.1231143.

13. Hwang WY, Fu Y, Reyon D, Maeder ML, Tsai SQ, et al. (2013) Efficient genome editing in zebrafish using a CRISPR-Cas system. Nat Biotechnol 31: 227–229. doi:10.1038/nbt.2501.

14. Xu K, Ren C, Liu Z, Zhang T, Zhang T, et al. (2014) Efficient genome engineering in eukaryotes using Cas9 from Streptococcus thermophilus. Cell

Mol Life Sci. Available: http://link.springer.com/10.1007/s00018-014-1679-z. Accessed 2014 August 18.

15. Qi LS, Larson MH, Gilbert LA, Doudna JA, Weissman JS, et al. (2013) Repurposing CRISPR as an RNA-Guided Platform for Sequence-Specific Control of Gene Expression. Cell 152: 1173–1183. doi:10.1016/j.cell.2013.02.022.

16. Gilbert LA, Larson MH, Morsut L, Liu Z, Brar GA, et al. (2013) CRISPR-Mediated Modular RNA-Guided Regulation of Transcription in Eukaryotes. Cell 154: 442–451. doi:10.1016/j.cell.2013.06.044.

17. Bikard D, Jiang W, Samai P, Hochschild A, Zhang F, et al. (2013) Programmable repression and activation of bacterial gene expression using an engineered CRISPR-Cas system. Nucleic Acids Res 41: 7429–7437. doi:10.1093/nar/gkt520.

18. Kearns NA, Genga RMJ, Enuameh MS, Garber M, Wolfe SA, et al. (2014) Cas9 effector-mediated regulation of transcription and differentiation in human pluripotent stem cells. Development 141: 219–223. doi:10.1242/dev.103341.

19. Maeder ML, Linder SJ, Cascio VM, Fu Y, Ho QH, et al. (2013) CRISPR RNA-guided activation of endogenous human genes. Nat Methods 10: 977–979. doi:10.1038/nmeth.2598.

20. Cheng AW, Wang H, Yang H, Shi L, Katz Y, et al. (2013) Multiplexed activation of endogenous genes by CRISPR-on, an RNA-guided transcriptional activator system. Cell Res 23: 1163–1171. doi:10.1038/cr.2013.122.

21. Perez-Pinera P, Kocak DD, Vockley CM, Adler AF, Kabadi AM, et al. (2013) RNA-guided gene activation by CRISPR-Cas9-based transcription factors. Nat Methods 10: 973–976. doi:10.1038/nmeth.2600.

22. Chen B, Gilbert LA, Cimini BA, Schnitzbauer J, Zhang W, et al. (2013) Dynamic Imaging of Genomic Loci in Living Human Cells by an Optimized CRISPR/Cas System. Cell 155: 1479–1491. doi:10.1016/j.cell.2013.12.001.

23. Larson MH, Gilbert LA, Wang X, Lim WA, Weissman JS, et al. (2013) CRISPR interference (CRISPRi) for sequence-specific control of gene expression. Nat Protoc 8: 2180–2196. doi:10.1038/nprot.2013.132.

24. Farzadfard F, Perli SD, Lu TK (2013) Tunable and Multifunctional Eukaryotic Transcription Factors Based on CRISPR/Cas. ACS Synth Biol 2: 604–613. doi:10.1021/sb400081r.

25. Bieging KT, Mello SS, Attardi LD (2014) Unravelling mechanisms of p53-mediated tumour suppression. Nat Rev Cancer 14: 359–370. doi:10.1038/nrc3711.

26. Brummelkamp TR, Bernards R, Agami R (2002) A system for stable expression of short interfering RNAs in mammalian cells. Science 296: 550–553. doi:10.1126/science.1068999.

27. Hao D-L, Liu C-M, Dong W-J, Gong H, Wu X-S, et al. (2005) Knockdown of Human p53 Gene Expression in 293-T Cells by Retroviral Vector-mediated Short Hairpin RNA. Acta Biochim Biophys Sin 37: 779–783. doi:10.1111/j.1745-7270.2005.00107.x.

28. Suzuki Y, Yamashita R, Nakai K, Sugano S (2002) DBTSS: DataBase of human Transcriptional Start Sites and full-length cDNAs. Nucleic Acids Res 30: 328–331. doi:10.1093/nar/30.1.328.

29. Yamashita R, Sugano S, Suzuki Y, Nakai K (2012) DBTSS: DataBase of Transcriptional Start Sites progress report in 2012. Nucleic Acids Res 40: D150–D154. doi:10.1093/nar/gkr1005.

30. Tsuchihara K, Suzuki Y, Wakaguri H, Irie T, Tanimoto K, et al. (2009) Massive transcriptional start site analysis of human genes in hypoxia cells. Nucleic Acids Res 37: 2249–2263. doi:10.1093/nar/gkp066.

31. Ferreira JP, Overton KW, Wang CL (2013) Tuning gene expression with synthetic upstream open reading frames. Proc Natl Acad Sci 110: 11284–11289. doi:10.1073/pnas.1305590110.

32. Graham FL, Smiley J, Russell WC, Nairn R (1977) Characteristics of a Human Cell Line Transformed by DNA from Human Adenovirus Type 5. J Gen Virol 36: 59–72. doi:10.1099/0022-1317-36-1-59.

33. Stewart N, Bacchetti S (1991) Expression of SV40 large T antigen, but not small t antigen, is required for the induction of chromosomal aberrations in transformed human cells. Virology 180: 49–57.

34. Ferreira JP, Peacock RWS, Lawhorn IEB, Wang CL (2011) Modulating ectopic gene expression levels by using retroviral vectors equipped with synthetic promoters. Syst Synth Biol 5: 131–138. doi:10.1007/s11693-011-9089-0.

35. Willems E, Leyns L, Vandesompele J (2008) Standardization of real-time PCR gene expression data from independent biological replicates. Anal Biochem 379: 127–129. doi:10.1016/j.ab.2008.04.036.

36. Groner AC, Meylan S, Ciuffi A, Zangger N, Ambrosini G, et al. (2010) KRAB-zinc finger proteins and KAP1 can mediate long-range transcriptional repression through heterochromatin spreading. PLoS Genet 6: e1000869. doi:10.1371/journal.pgen.1000869.

37. Hagmann M, Georgiev O, Schaffner W (1997) The VP16 paradox: herpes simplex virus VP16 contains a long-range activation domain but within the natural multiprotein complex activates only from promoter-proximal positions. J Virol 71: 5952–5962.

38. Mahmoudi S, Henriksson S, Corcoran M, Méndez-Vidal C, Wiman KG, et al. (2009) Wrap53, a Natural p53 Antisense Transcript Required for p53 Induction upon DNA Damage. Mol Cell 33: 462–471. doi:10.1016/j.molcel.2009.01.028.

39. Polson A, Reisman D (2014) The bidirectional p53–Wrap53β promoter is controlled by common cis- and trans-regulatory elements. Gene 538: 138–149. doi:10.1016/j.gene.2013.12.046.

40. Saldaña-Meyer R, Recillas-Targa F (2011) Transcriptional and epigenetic regulation of the p53 tumor suppressor gene. Epigenetics 6: 1068–1077. doi:10.4161/epi.6.9.16683.

41. Selby CP, Drapkin R, Reinberg D, Sancar A (1997) RNA polymerase II stalled at a thymine dimer: footprint and effect on excision repair. Nucleic Acids Res 25: 787–793. doi:10.1093/nar/25.4.787.

42. Gao X, Tsang JCH, Gaba F, Wu D, Lu L, et al. (2014) Comparison of TALE designer transcription factors and the CRISPR/dCas9 in regulation of gene expression by targeting enhancers. Nucleic Acids Res: gku836. doi:10.1093/nar/gku836.

43. Wu X, Kriz AJ, Sharp PA (2014) Target specificity of the CRISPR-Cas9 system. Quant Biol. Available: http://link.springer.com/10.1007/s40484-014-0030-x. Accessed 2014 October 9.

44. Wu X, Scott DA, Kriz AJ, Chiu AC, Hsu PD, et al. (2014) Genome-wide binding of the CRISPR endonuclease Cas9 in mammalian cells. Nat Biotechnol 32: 670–676. doi:10.1038/nbt.2889.

45. Fu Y, Foden JA, Khayter C, Maeder ML, Reyon D, et al. (2013) High-frequency off-target mutagenesis induced by CRISPR-Cas nucleases in human cells. Nat Biotechnol 31: 822–826. doi:10.1038/nbt.2623.

46. Pattanayak V, Lin S, Guilinger JP, Ma E, Doudna JA, et al. (2013) High-throughput profiling of off-target DNA cleavage reveals RNA-programmed Cas9 nuclease specificity. Nat Biotechnol 31: 839–843. doi:10.1038/nbt.2673.

47. Cho SW, Kim S, Kim Y, Kweon J, Kim HS, et al. (2014) Analysis of off-target effects of CRISPR/Cas-derived RNA-guided endonucleases and nickases. Genome Res 24: 132–141. doi:10.1101/gr.162339.113.

48. Kuscu C, Arslan S, Singh R, Thorpe J, Adli M (2014) Genome-wide analysis reveals characteristics of off-target sites bound by the Cas9 endonuclease. Nat Biotechnol 32: 677–683. doi:10.1038/nbt.2916.

25

Thermostable Artificial Enzyme Isolated by *In Vitro* Selection

Aleardo Morelli[⍵], **John Haugner**[⍵], **Burckhard Seelig***

Department of Biochemistry, Molecular Biology, and Biophysics, University of Minnesota, Minneapolis, Minnesota, United States of America, & BioTechnology Institute, University of Minnesota, St. Paul, Minnesota, United States of America

Abstract

Artificial enzymes hold the potential to catalyze valuable reactions not observed in nature. One approach to build artificial enzymes introduces mutations into an existing protein scaffold to enable a new catalytic activity. This process commonly results in a simultaneous reduction of protein stability as an undesired side effect. While protein stability can be increased through techniques like directed evolution, care needs to be taken that added stability, conversely, does not sacrifice the desired activity of the enzyme. Ideally, enzymatic activity and protein stability are engineered simultaneously to ensure that stable enzymes with the desired catalytic properties are isolated. Here, we present the use of the *in vitro* selection technique mRNA display to isolate enzymes with improved stability and activity in a single step. Starting with a library of artificial RNA ligase enzymes that were previously isolated at ambient temperature and were therefore mostly mesophilic, we selected for thermostable active enzyme variants by performing the selection step at 65°C. The most efficient enzyme, ligase 10C, was not only active at 65°C, but was also an order of magnitude more active at room temperature compared to related enzymes previously isolated at ambient temperature. Concurrently, the melting temperature of ligase 10C increased by 35 degrees compared to these related enzymes. While low stability and solubility of the previously selected enzymes prevented a structural characterization, the improved properties of the heat-stable ligase 10C finally allowed us to solve the three-dimensional structure by NMR. This artificial enzyme adopted an entirely novel fold that has not been seen in nature, which was published elsewhere. These results highlight the versatility of the *in vitro* selection technique mRNA display as a powerful method for the isolation of thermostable novel enzymes.

Editor: Isabelle Andre, University of Toulouse - Laboratoire d'Ingénierie des Systèmes Biologiques et des Procédés, France

Funding: This work was supported in part by the US National Aeronautics and Space Administration (NASA) Agreement no. NNX09AH70A through the NASA Astrobiology Institute-Ames Research Center (to A.M. and B.S.), the US National Institutes of Health (NIH) (T32 GM08347 to J.C.H.) and by a grant from the Biocatalysis Initiative of the BioTechnology Institute at the University of Minnesota and the Minnesota Medical Foundation (to B.S.). The funders had no role in study design, data collection and analysis, decision to publish, or preparation of the manuscript.

Competing Interests: The authors have declared that no competing interests exist.

* Email: seelig@umn.edu

⍵ These authors contributed equally to this work.

Introduction

Protein stability is often a limiting factor for the application, engineering and structural studies of proteins. Low protein stability can result in aggregation, susceptibility to protease degradation and poor yields in the expression of soluble protein, thereby complicating the study and use of these proteins. For commercial applications, proteins commonly need to be particularly stable to increase their tolerance to process conditions like high temperatures or organic solvents [1]. Furthermore, proteins with low stability are less tolerant to mutations thereby limiting further engineering because even slightly destabilizing mutations can lead to unfolding. This can create situations where mutations that would improve enzyme activity in a protein engineering project appear ineffective because the enzyme was not stable enough to remain folded [2]. Conversely, improved thermal stability correlates with mutational robustness and evolvability [3].

Methods to increase the thermodynamic stability of proteins include rational design, consensus-based design, directed evolu-

tion, and commonly some combination of these approaches [4]. Rational design introduces mutations predicted to enable additional stabilizing interactions [5]. However, this approach requires extensive structural knowledge, substantial computing power and is technically challenging, which still limits the accessibility of this method. Consensus based-design utilizes phylogenetic information to determine which amino acids are preferred at certain positions [5]. This method can be used to reconstruct thermostable ancestral proteins or, be combined with structural knowledge, which likely further improves the prediction of stabilizing mutations. However, these approaches are dependent on the quality of the constructed phylogenetic tree, which is non-trivial to accurately assemble. Directed evolution is a combinatorial approach that introduces mutations at random and then screens for desired properties such as improved activity or stability [6–8]. High throughput screens are often performed *in vivo*, utilizing colorimetric [9] or fluorescent [10] reporters to measure levels of soluble expression as readout for stability, or *in vitro* using protease resistance and phage display [11,12]. Protein variants are also commonly assayed directly for thermostability and activity as purified proteins, but these methods have a relatively low throughput. Numerous examples have recently been discussed in excellent review articles [4,13]. As mutations are introduced randomly, the chance of success increases with the number of mutants sampled. This favors high throughput methods which can sample millions to trillions of mutants [14,15]. Individual methods aimed to generate more stable protein variants can also be combined for best results as was demonstrated by consensus design that used the sequence output of a library selection [16].

We previously reported the *in vitro* selection of *de novo* RNA ligase enzymes that catalyze a reaction not observed in nature [17]. These artificial enzymes ligate RNA with a 5'-triphosphate to the 3'-hydroxyl of second RNA forming a native 5'-3' linkage and releasing pyrophosphate. These artificial ligases are zinc dependent metalloenzymes of about 10 kDa. Several enzymes resulting from this *in vitro* selection experiment were analyzed in more detail. All examined enzymes were soluble when expressed as fusion proteins with maltose-binding protein (MBP), but most enzymes were poorly soluble when expressed on their own. NMR HSQC spectroscopy of the most soluble clone, ligase #6, revealed that a significant portion of the protein was well-folded, yet the overall resolution of the data was insufficient to solve the three-dimensional structure [17]. To overcome this issue, we again utilized *in vitro* selection. We modified the conditions of our original procedure and continued the selection to isolate ligase variants with improved stability in order to facilitate structural and mechanistic studies of these artificial enzymes.

Here, we describe in detail the *in vitro* selection of RNA ligases with increased stability. For this directed evolution experiment we utilized the mRNA display technology, an *in vitro* display method, which covalently links each protein to its encoding mRNA [18,19]. Using this technology, up to 10^{13} unique proteins can be sampled in a single experiment, which is orders of magnitude more than most other selection strategies [14]. To isolate enzymes with increased thermodynamic stability, we modified parts of the selection procedure and performed the ligation step at 65°C. For the selection reported here, we used the output library from our previous selection at room temperature [17] as starting material without further diversification. We hypothesized that enzymes, which are active at elevated temperature, will have a more stable protein fold that in turn will facilitate structural characterization. We also hoped that the increased structural stability would correspond to increased solubility and expression *in vivo*. After several rounds of selection, representative ligase clones were sequenced and tested for soluble expression in *E. coli*. The soluble and most active ligase 10C was characterized further and its activity and stability was compared to two closely related sequences from the previous selection at room temperature. The experiments revealed that ligase 10C is both more stable and more active than either of these ligases. We recently described the three-dimensional structure of ligase 10C solved by NMR, revealing a novel fold that has not been observed in nature and lacks secondary structural elements like α-helices or β-strands [20]. Furthermore, we reported a detailed analysis of the substrate specificity of ligase 10C showing that this enzyme can facilitate the selective isolation and sequencing of any RNA with a 5'-triphosphate [21].

This manuscript is the first report of an mRNA display selection at high temperature. These results demonstrate the efficacy of mRNA display for isolating thermostable enzymes as stability and activity are selected simultaneously in a high throughput experiment.

Materials and Methods

Preparation of Oligonucleotides

^{32}P-labeled PPP-substrate-23 used in the original selection at 23°C (5'-PPP–GGAGACUCUUU) and PPP-substrate-65 for the selection at 65°C (5'-PPP–GGAGAUUCACUAGCUGGUUU) were prepared through T7 transcription as reported previously [17,22]. The HO-substrate-23 (5'-CUAACGUUCGC), HO-substrate-65 (5'-UCACACUGUCUAACGUUCGC) and HO-substrate-65-Bio (5'-(PC)-UCACACUGUCUAACGUUCGC, (PC) represents PC biotin phosphoramidite from Glen Research, Sterling, VA) were purchased from Dharmacon (Lafayette, CO) and prepared according to the manufacturer's protocol. The DNA splint (5'-GAGTCTCCGCGAACGT) complementary to the substrates-23, and RNA splint (5'- AAACCAGCUAGUGAAU-CUCCGCGAACGUUAGACAGUGUGA) complementary to the substrates-65, were purchased from Integrated DNA Technologies (Coralville, IA). The reverse transcription primer (HEG$_4$-RT) was produced by ligating the PPP-substrate-65 to BS75P-HEG$_4$ in the presence of BS76 as template using T4 DNA ligase [23] and purified by denaturing PAGE. All oligonucleotides were dissolved in ultra-pure water and concentrations determined by UV absorbance.

Selection of RNA Ligases at 65°C

The mRNA display selection was performed as previously published [17], with the following exceptions. Primers BS99 and BS24RXR2 were used to amplify the DNA by PCR. Primer BS99 replaces the N-terminal FLAG affinity tag that was used in the previous selection at room temperature [17] with the E-tag. Accordingly, both FLAG affinity purification steps in the previous protocol were substituted by E-tag affinity purifications. For the first E-tag purification, the mRNA-displayed proteins eluted from the oligo(dT)cellulose were mixed with binding buffer (same as Flag binding buffer [17]) and then incubated for 30 min at 4°C with rotation with 25 μL Anti-E affinity gel (from Anti E-tag affinity column, GE healthcare Biosciences; prewashed with E clean buffer (100 mM glycine, pH 3.0, 0.05% Tween-20) and binding buffer). The Anti-E tag affinity gel was then washed with binding buffer and eluted with binding buffer containing two equivalents of E-peptide (Bachem, Osteocalcin (7–19, human); one equivalent of E-peptide saturates the antigen sites of the antibody resin) for 3 min at 4°C. The second E-tag purification was performed in a similar fashion using 50 μL Anti-E affinity gel and 6 equivalents of E-peptide to elute. The elution from the second E-tag affinity purification was incubated with the HO-substrate-65-Bio and the RNA splint in presence of 2 mM MgCl$_2$ and 100 μM ZnCl$_2$ for 1 hour at 65°C in selection rounds 1, 2, 3 and 5. In round 4, the sample was divided into two aliquots, one of which was incubated for 1 h, and the other aliquot was incubated for 5 min. The reaction was quenched and purified on streptavidin beads as described previously [17], and the photocleaved DNA was amplified by PCR and used as input for the following round. For the starting material in round 5, the photocleaved DNA from round 4 was used that resulted from the 5 min incubation.

Expression & Purification of RNA Ligases

RNA ligases were expressed and purified as previously described [20].

Screening for Ligase Activity by Gel-Shift Assay

5 μM ^{32}P-labeled PPP-substrate-65, 6 μM RNA splint, 7 μM HO-substrate-65, 20 mM HEPES pH 7.5, 100 mM NaCl, 100 μM ZnCl$_2$ and 1.7 μM enzyme (purified by Ni-NTA affinity

chromatography [20]) were combined and incubated for 16 hours at 23°C and 65°C. Reactions were stopped by the addition of EDTA to a final concentration of 10 mM. Immediately following, the RNA was denatured for 40 min at 65°C in 7.5% formaldehyde, 58% formamide and 11.6 mM MOPS pH 7.0. Samples were separated by 20% denaturing PAGE gel containing 2% formaldehyde. The gel was analyzed using the GE Healthcare (Amersham Bioscience) Phosphorimager and ImageQuant software (Amersham Bioscience). The amount of radiation in both the substrate and product bands was measured and the percentage of ligation was determined by dividing the intensity of the product band by the sum of the product and substrate bands.

Determination of Observed Rate Constants (k_{obs})

5 μM enzyme (purified by Ni-NTA affinity and size exclusion chromatography [20]) was incubated with 10 μM ^{32}P-labeled PPP-substrate-23, 15 μM DNA splint, 20 μM HO-substrate-23 and ligation was monitored for up to 2 hours at 23°C. Reactions were quenched with two volumes of 20 mM EDTA in 8 M urea after 0, 15, 30, 60 and 120 minutes, heated to 95°C for 4 min and separated by 20% denaturing PAGE gel. The gel was analyzed using the GE Healthcare (Amersham Bioscience) Phosphorimager and ImageQuant software (Amersham Bioscience). The rate constant (k_{obs}) was calculated by determining the slope of the linear fit of percentage of ligation over time and correcting for enzyme concentration by multiplying with the ratio of PPP-substrate to enzyme (2 = 10 uM/5 uM) resulting in a value with the unit h^{-1}. The reported values are an average of 3 independent replicates ± the standard deviation. Total conversion was <10% for all cases.

Circular Dichroism and Thermal Denaturation

Ligase enzymes (purified by Ni-NTA affinity and size exclusion chromatography [20]) were concentrated to 50 μM and dialyzed against CD buffer (150 mM NaCl, 2 mM HEPES, 0.5 mM 2-mercaptoethanol, 100 μM $ZnCl_2$). Circular dichroism spectra and thermal denaturation curves were recorded on a JASCO J-815 spectropolarimeter at 30 μM or 50 μM protein, respectively. The following parameters were used for both measurements: 1.5 nm band width, 2 seconds response time, standard sensitivity, 10 accumulations. The ellipticity at 222 nm was monitored to determine thermal denaturation curves over a temperature range from 5 to 91°C with a ramp rate of 1°C/min and a temperature pitch of 2°C.

Results

Setup of Selection Procedure

Sequence analysis of the artificial RNA ligase enzymes that resulted from the final round of the previous in vitro selection performed at 23°C [17] revealed substantial sequence diversity. The DNA encoding those diverse ligases was used as the starting library for the selection at 65°C described in this paper without introducing further sequence diversity. The RNA ligation reaction catalyzed by the previously selected enzymes was dependent on a complementary splint oligonucleotide that base-pairs to the two substrate RNAs [17] (Figure 1). During the selection at 23°C, this splint base-paired to eight nucleotides of each substrate (Figure 1B). In order to ensure stable base-pairing during a splinted ligation at elevated temperatures, a longer splint was chosen to extend the region complementary to each substrate to twenty nucleotides (Figure 1C). The 40-nucleotide-long splint resulted in a melting temperature of 76°C and 69°C with the PPP-substrate and the HO-substrate, respectively (Figure S1).

To enable the selection of active enzymes, the PPP-substrate was linked to the mRNA-displayed proteins via the reverse transcription (RT) primer that initiates the cDNA synthesis (Figure 1A). This linkage resulted in a high local concentration of substrate in the vicinity of each protein. In order to reduce this local concentration and thereby favor the selection of enzymes with an increased substrate affinity, we lengthened the RT primer by an additional eighteen non-complementary nucleotides and four flexible hexaethylene glycol linker units (HEG_4, Figure 1C). The hexaethylene glycol linker simply acted as a long unstructured tether to increase the average distance between protein and substrate. The use of the longer RT primer in combination with the splint of 40 nucleotides (nt) in length (Figure 1C) resulted in a ligase activity of about 50% compared to a ligation using the shorter RT primer and the 16 nt splint (Figure 1B).

We then evaluated the ligase activity of the starting library at increasing temperatures in order to determine a temperature at which the majority of the library members are inactive. Using the 40 nt splint and the HEG_4-RT primer, at 65°C no ligation was detectable (<10%), whereas at 60°C the ligation activity was about half of the activity measured at 23°C. Therefore, we decided to carry out the selection for higher stability at 65°C.

During the previous selection for ligases, 57% of the isolated enzymes had acquired a second FLAG binding sequence (DYKXXD) in addition to the FLAG binding sequence that was part of the N-terminal constant region. This was likely a result of a selection bias caused by two FLAG affinity purification steps per round of selection. In order to counteract this FLAG purification bias, we changed the selection protocol to using the E-tag affinity purification instead. Therefore, we replaced the FLAG tag coding sequence in the N-terminal constant region of the library with an E-tag sequence by PCR. The ligation activity was unaffected by the change of tags.

In Vitro Selection at 65°C

To enrich for RNA ligase enzyme with increased thermostability, we performed a total of six rounds of selection and amplification (Figure 1A). After reverse transcription, the mRNA-displayed proteins were incubated with the HO-substrate-65 and the RNA splint for 60 min and/or 5 min. The percentage of cDNA that was immobilized on streptavidin beads after each round of selection is shown in Figure 2. In the case of the 60 minute incubation, the percentage of immobilized cDNA increased steadily over the course of the selection, from 0.61% after round 1 to 6.6% after round 6. In order to increase the selection pressure by favoring enzymes with faster ligation rates, in round 4, we incubated a second aliquot of the mRNA-displayed proteins for only 5 min yielding 0.66% immobilized cDNA. This cDNA was used as input for following round, but no increase in the amount of immobilized cDNA after 5 min incubation was observed in round 5 (amount decreased to 0.41%). Therefore, we performed the sixth and final round of selection, again with 60 min incubation. The resulting DNA was cloned and sequenced for further analysis.

Sequence Analysis and Expression of Selected Ligases

The sequence alignment of 32 clones from the sixth round of selection at 65°C revealed two protein families (Figure S2). One representative clone from each family was cloned and expressed in E. coli to examine soluble expression (Figure S3). While both clones expressed well, ligase 10C was highly soluble whereas ligase 10H was largely insoluble. Furthermore, native Ni-NTA affinity purification of ligase 10H yielded no soluble protein (data not shown) and, therefore, ligase 10H was not characterized further.

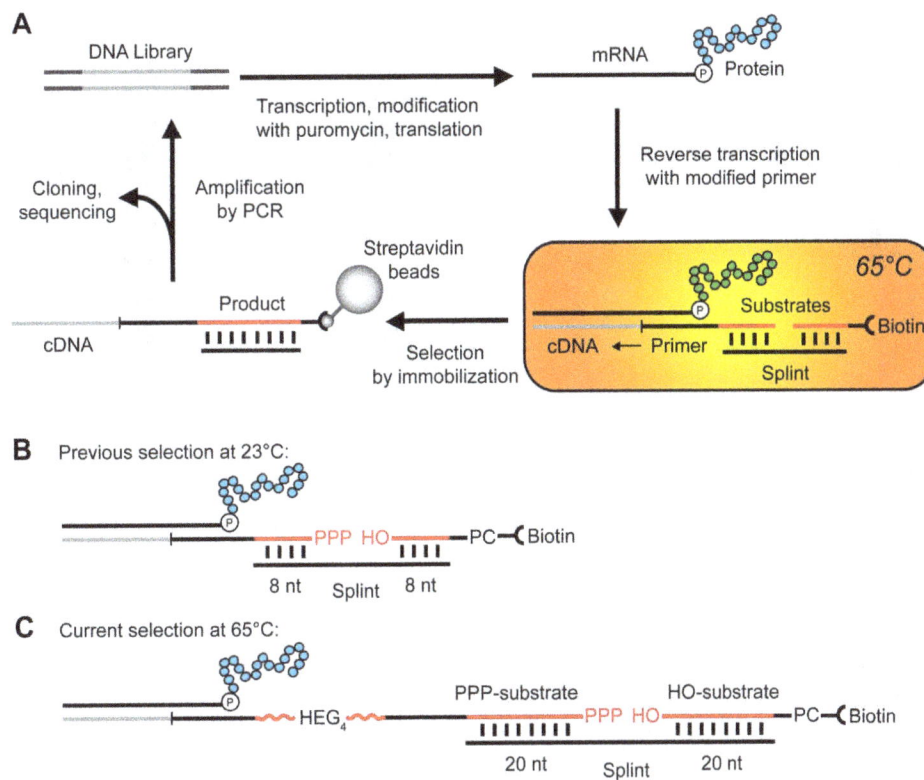

Figure 1. *In vitro* selection of artificial ligase enzymes with increased stability. (A) Schematic of the isolation of ligase enzymes. The DNA library encodes the library of proteins that resulted from the original selection of ligase enzymes at 23°C [17,22]. The DNA is transcribed into RNA, modified with puromycin at the 3'-end and translated *in vitro* yielding a library of mRNA-displayed proteins [22]. Reverse transcription with a primer containing one RNA substrate shown in red results in a complex of protein, mRNA, cDNA and substrate. This complex is incubated at 65°C with the second RNA substrate (red) and the complementary splint as highlighted in the orange box. The cDNA of ligases active at this temperature is immobilized on streptavidin beads and amplified for subsequent rounds of selection, or identified by cloning and sequencing. **(B)** Detailed view of ligation reaction substrates in complex with the mRNA-displayed protein. The two strands of RNA in red, the 5'-triphosphate RNA (PPP-substrate) and 3'-hydroxyl RNA (HO-substrate), are joined in a template-dependent ligation reaction. The PPP-substrate is part of the reverse transcription primer. The photocleavable site (PC) is used to release the cDNA that encodes active enzymes from streptavidin immobilization by irradiation at 365 nm. The splint acts as template of the ligation and base pairs with 8 nucleotides of each RNA substrate during the previously published selection at 23°C [17,22], and with **(C)** 20 nucleotides of each substrate during the current selection at 65°C. HEG4 represents the linker of four hexaethylene glycol units (red wavy line).

The sequence of ligase 10C shared similarities to ligases #6 and #7 from the original selection with #7 being more similar (Figure 3). All three ligases were almost identical in sequence in the formerly randomized region 2, and all three shared the deletion of 17 amino acids following region 1. Ligases 10C and #7 also shared the sequence in region 1, but 10C contained a second deletion of 13 amino acids near the C-terminus. This C-terminal deletion was also found in other clones from the selection at 23°C [17], but these proteins were poorly soluble when expressed without an maltose-binding protein fusion and therefore unsuited for a direct comparison.

Activity of Ligase Enzymes

To compare the enzymatic activity of ligase 10C to ligases #6 and #7, we assayed the three enzymes at 23°C and 65°C (Figure 4). Ligase 10C was the only enzyme active at 65°C. In comparison, ligases #6 and #7 were active at room temperature as expected, but had no measurable activity at 65°C. In addition to its activity at 65°C, ligase 10C was also active at room temperature. To compare the activity of the three enzymes more accurately, we measured the k_{obs} for each ligase at 23°C. At a subsaturating substrate concentration of 10 µM, ligase 10C had a

k_{obs} of 0.165 ± 0.015 h^{-1} while ligases #6 and #7 had k_{obs} of 0.0174 ± 0.0066 h^{-1} and k_{obs} of 0.0207 ± 0.0045 h^{-1}, respectively (Table S1). This represents an 8 to 10-fold increased activity of ligase 10C compared to ligases #6 and #7 even at 23°C. While the main goal of the selection was to isolate an enzyme with greater thermostability, as an added benefit, the most stable enzyme also featured an improved catalytic rate at room temperature.

Characterization of Thermal Stability by Circular Dichroism (CD)

In order to assess if the unique enzymatic activity of ligase 10C at 65°C was correlated to increased structural stability, we measured thermal denaturation curves of all three ligases by circular dichroism. In preparation of the thermal unfolding experiment, we measured the CD spectra of the three enzymes (Figure S4). All three spectra exhibited two minima of negative ellipticity at 205 nm and between 220 and 225 nm, respectively. While those minima suggested α-helical secondary structural content [24], the 205 nm minimum was substantially more negative than the second minimum, which differs from purely alpha helical proteins that have similar absolute values for both

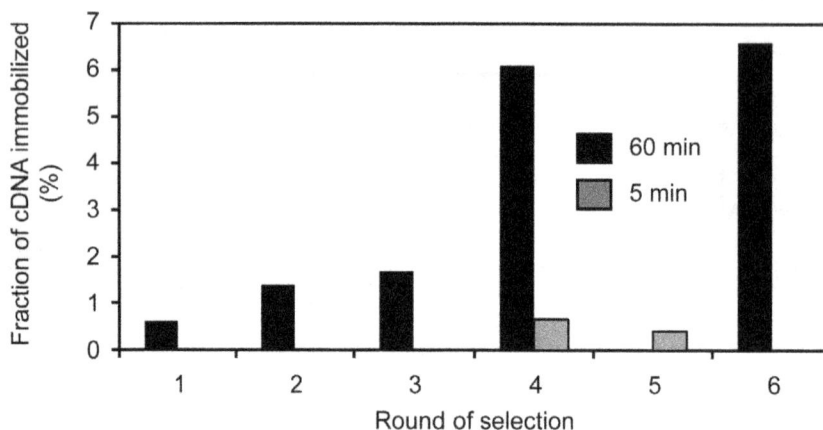

Figure 2. Progress of selection for ligases at 65°C. The fraction of ^{32}P-labelled cDNA that bound to streptavidin agarose after each round of selection is shown. The reaction time was either 60 min or 5 min as indicated by black or gray bars, respectively.

minima. In fact, the three-dimensional structure of ligase 10C recently solved by NMR revealed that α-helices and β-strand regions are essentially absent in ligase 10C [20]. Nevertheless, we used the strong negative ellipticity of all three ligases at 222 nm to monitor thermal unfolding of the proteins over a temperature range from 5 to 91°C. We found all three enzymes to give the characteristic single sigmoidal transition corresponding to a two-state unfolding reaction (Figure 5). As determined from the curves, the enzymes showed very different melting temperatures. Ligase 10C had the highest melting temperature ($T_m = 72°C$), which was 35 degrees higher than the T_m of ligase #6 (37°C), and 24 degrees higher than the T_m of ligase #7 (48°C). The high melting temperature of 72°C for ligase 10C was in agreement with its retained enzymatic activity at 65°C as the enzyme had not undergone unfolding yet. In contrast, ligases #6 and #7 were fully denatured at 65°C, and, therefore, their complete lack of enzymatic activity at 65°C could be explained by their unfolding.

Discussion

We isolated a thermostable artificial RNA ligase enzyme by *in vitro* selection at 65°C of a library of artificial ligases that were originally generated at 23°C. The isolated ligase 10C was more thermostable and more active than the two most closely sequence-related ligases #6 and #7 identified during the selection at 23°C. Ligase 10C had a melting temperature (T_M) of 72°C corresponding to a stability increase of 24 degrees compared to #7, and 35 degrees compared to ligase #6. Previously reported T_M improvements through protein engineering are commonly between 2 to 15 degrees [5]. The T_M increase by 35 degrees reported here favorably compares with those rare examples of 'record-setting

stabilizations' [4,25–27]. While the ligases #6 and #7 had no measurable enzymatic activity at 65°C, ligase 10C ligated RNA at 65°C with an activity that was similar to its activity at 23°C. Furthermore, the activity of ligase 10C at 23°C was about an order of magnitude higher than the activity of the ligases #6 and #7 at the same temperature.

The increased thermostability of ligase 10C was likely due to additional intramolecular contacts within the protein compared to the mesophilic ligases #6 and #7. In contrast to these enzymes isolated at 23°C, the properties of ligase variant 10C were suitable to solve its three-dimensional solution structure by NMR [20]. The structure featured a small, well-folded core coordinated by two Zn^{2+}-ions. In addition, the folding core also contained a highly dynamic internal loop and was framed by unstructured termini. In order to discuss a potential correlation between differences in primary sequence and altered thermal stability, we mapped sequence differences between ligase #7 and 10C onto the structure of 10C (Figure 6). We chose ligase #7 for comparison because despite the high sequence similarity it showed a large difference in thermostability. All sequence differences between these two ligases were found in or near the structured region responsible for zinc coordination. We previously demonstrated by NMR that residues Ile68 and His69 near the C-terminus of ligase 10C made long range NOE contacts with several residues at the N-terminus (Lys17, His18, Ala27 and Glu28) [20]. Notably, His18 was one of the zinc coordinating residues in ligase 10C [20] and mutating this position to Ala resulted in a drastically reduced solubility of the protein. In ligase #7, the residue corresponding to Ile68 was a methionine. In addition, ligase #7 contained an additional 13 amino acids located between the residues corresponding to Ile68 and His69 in ligase 10C, which likely moved

Figure 3. Sequence alignment of the library used as input for the original ligase selection with ligases #6, #7 [17] and 10C that were selected at 23°C and at 65°C, respectively [41,42]. The amino acids in regions 1 and 2 of the original library (on top) were randomized prior to the selection at 23°C and are shown as "x" [43]. Dashes symbolize amino acids that are identical to the starting library. A period highlighted in gray represents a deletion. The underlined N-terminal amino acids of the library and ligase 10C represent a Flag epitope tag and an E epitope tag, respectively.

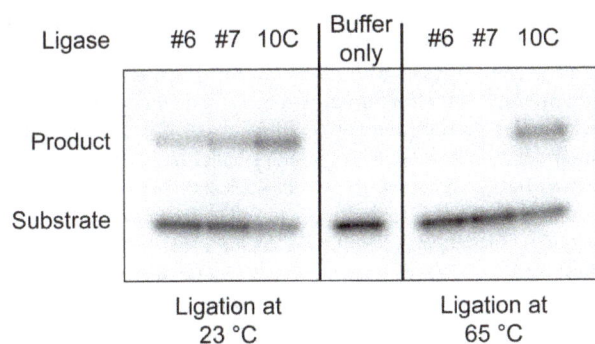

Figure 4. Activity of ligase enzymes assayed at different temperatures. Ligases #6 and #7 had been selected previously at 23°C [17,22] and ligase 10C was selected at 65°C. In this assay, the ^{32}P-labeled PPP-substrate-65, HO-substrate-65 and 40 nt splint were incubated with the individual enzymes for 16 h and the activity was monitored by a gel-shift assay.

His69 and prevented its contacts with Lys17, His18 at the N-terminus. Presumably, all these mutations could compromise the intramolecular interactions in these positions reported for ligase 10C and decrease the stability of ligase #7 at high temperature. Ligase 10C also differed from ligase #7 in two additional positions (Ser54 and Asp65) which may further influence protein stability. A direct comparison of the overall flexibility of ligase 10C and the two mesophilic ligases would require solving also the structures of ligases #6 and #7 by NMR. This would be beyond the scope of this paper and preliminary experiments suggested that ligase #6 is not amenable to detailed NMR studies.

The in vitro selection at 65°C not only yielded the family A of related sequences that included ligase 10C (Figure S2), but also a second family B represented by ligase 10H which could not be expressed solubly in E. coli. During the original selection at 23°C, we noted that of the seven ligases characterized, only #6 and #7 were soluble without being expressed as a MBP fusion. While ligase 10C was closely related to ligases #6 and #7, ligase 10H is most similar to ligase #1, which also did not express solubly. Isolating proteins like ligase 10H and ligase #1 is not surprising because mRNA display uses a eukaryotic in vitro translation system and therefore soluble expression in E. coli was never directly selected for. Additionally, the covalently linked RNA increases protein solubility which can also contribute to this result.

In general, this solubilizing effect is a favorable feature of mRNA display because it allows identifying proteins that might be lost during other selection techniques due to poor solubility. It is possible that ligase 10H could be solubilized through MBP fusion like ligase #1, but such a modification would have complicated subsequent structural studies.

Considering the high melting temperature of 72°C for ligase 10C, it is particularly surprising to discover the lack of secondary structural motifs like α-helices or β-strands combined with highly dynamic regions [20]. The structure of this artificial enzyme does appear to match with any known protein folds. While it is increasingly appreciated that catalytic activity of enzymes can require conformational flexibility [28–30], thermal stability is usually associated with tight packing and rigidity. Generally, thermophilic enzymes possess well packed hydrophobic cores [31], few exposed surface loops [32] and additional stabilizing interactions such as salt bridges [33] and a high number of hydrogen bonds [34]. These features lead to an increased rigidity that, while favoring stability at higher temperature, often appears to decrease activity at lower temperature. This observation has been interpreted to mean that stability, dynamics and catalysis are a tradeoff, but this common notion has recently been called into question [35]. The structure of the ligase 10C [20] combines a high flexibility and the absence of a packed hydrophobic core with thermostability, and is equally active at 65°C and at ambient temperature. The structure of this de novo enzyme challenges the common view of how enzymes are supposed to look – a view that is biased by proteins amenable to crystallization.

The high degree of disorder and flexibility present in ligase 10C might be a feature that favors its evolvability. For example, the presence of disordered regions and loosely packed structures found in viral proteins, structural characteristics similar to those found in ligase 10C, may allow for increased evolvability because each mutation, due to a lower amino acid interconnectivity, would lead to a slower loss in stability, compared to the more packed structures of thermophilic enzymes [36]. Similarly, ligase 10C might also be highly evolvable because of its flexible structure and disordered regions. Yet, this artificial enzyme was generated de novo and, unlike biological proteins, has not been shaped by billions of years of evolution. As its structure and function has just come into existence, ligase 10C could be considered a model protein for primordial enzymes. For these reasons, properties of this enzyme like its evolutionary potential will be interesting to

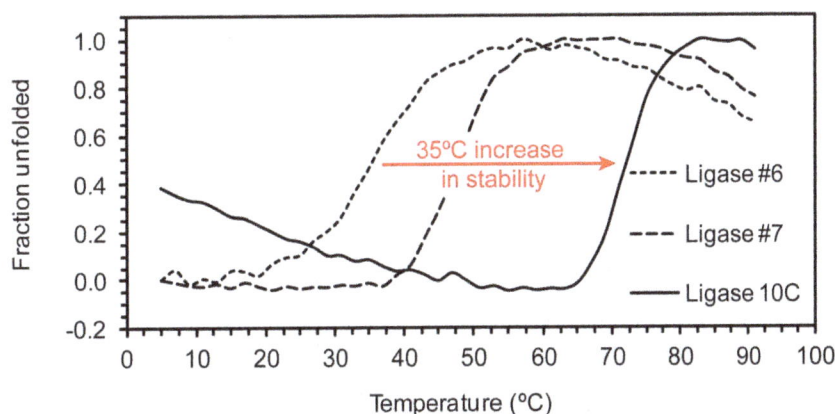

Figure 5. Thermal unfolding curves of ligases #6, #7 and 10C. Thermal unfolding was monitored by circular dichroism at 222 nm. For each measurement 10 accumulations were acquired.

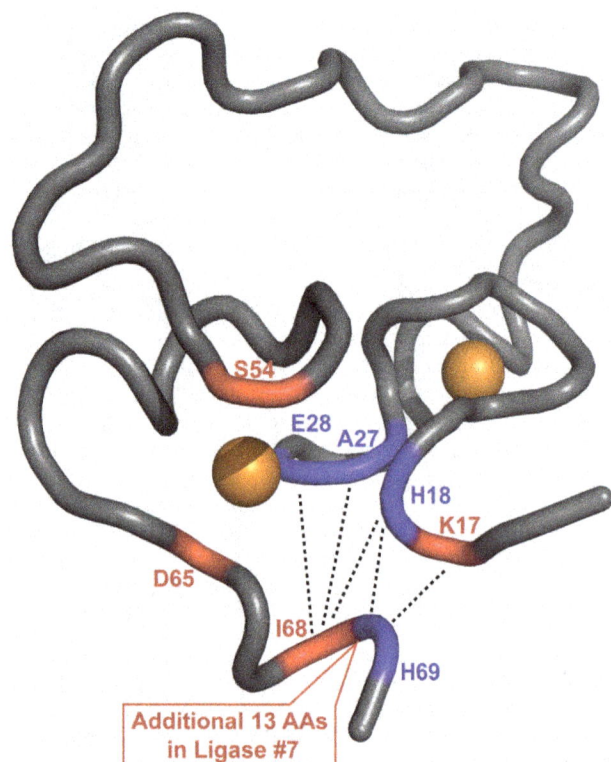

Figure 6. Sequence differences between ligase #7 and ligase 10C mapped onto the NMR structure of ligase 10C [20]. Mutations are shown in red. Residues potentially perturbed by the mutations are labeled in blue and long range NOEs are shown as dashed black lines. The two coordinated zinc ions as depicted as orange spheres and the residue numbers refer to ligase 10C. The unstructured termini of ligase 10C were omitted for clarity.

enables to search through large libraries of up to 10^{13} protein variants. This feature is beneficial because the chance of finding a desired activity increases with the number of variants interrogated. Furthermore, the *in vitro* format of this method allows selecting for activity under a wide range of conditions, which is similar to the common approach of screening much smaller libraries of purified proteins, but in contrast to *in vivo* selection strategies where maintenance of cell viability limits the experimental possibilities. Previous reports on mRNA display include the improvement of folding and stability of proteins by selecting for resistance to protease degradation [38], or by selecting in the presence of increasing amounts of the denaturant guanidine hydrochloride [39,40]. Interestingly, in parallel to our successful selection for RNA ligases at elevated temperature, we also attempted a similar selection in presence of guanidine hydrochloride, but no enrichment was observed even after six rounds (data not shown). Nevertheless, to our knowledge the work presented here is the first description of an mRNA display selection at elevated temperatures yielding thermostable proteins. The *in vitro* format of mRNA display should facilitate other selections at a variety of pH, temperatures, ionic strength, or in the presence of co-solvents, inhibitors or other chemicals. Such experiments will help to study the coevolution of protein stability and activity, and also has the potential to produce proteins that are more stable in industrial or biomedical applications.

Supporting Information

Figure S1 Thermal denaturation of substrate and splint oligonucleotides used in the selection and activity assays at 65°C. The first derivative of the melting curve for the 40 nt splint in the presence of both PPP-substrate-65 and HO-substrate-65 RNA oligonucleotides is presented. The concentration of each oligonucleotide was 0.5 µM in a buffer containing 70 mM KCl, 100 µM $ZnCl_2$, 5 mM 2-mercaptoethanol and 20 mM HEPES at pH 7.5.

Figure S2 Clones identified from round 6 of the *in vitro* selection at 65°C. Two protein families (A, B) were identified and a representative clone from each family was chosen for further characterization (ligase 10C and ligase 10H, shown in red).

Figure S3 Protein expression in *E. coli* of representative ligases selected at 65°C. A Coomassie-stained SDS-PAGE gel shows samples of whole cells pre- and post-induction and the insoluble and soluble fractions after cell lysis and centrifugation. Red boxes in the lane 'Soluble fraction' indicate the presence or absence of soluble ligases 10C and 10H, respectively.

Figure S4 Circular dichroism spectra of ligases #6, #7 and 10C at 25°C.

Table S1 Data for determining k_{obs}.

Table S2 Sequences of ligases 10C and 10H selected at 65°C; and ligases #6 and #7 selected previously at 23°C for comparison.

Table S3 Sequences of oligonucleotides.

study, however comparisons to natural proteins might be challenging.

The starting library for this selection at elevated temperature was a mixture of protein variants that was the final output of the previously described selection for artificial ligases at 23°C [17]. No further genetic diversity had been introduced. Sequencing of the starting library showed a diverse mixture of unrelated sequences and sequence families. Ligase 10C had not been observed during the sequencing of 49 individual clones and was only sufficiently enriched and detected after the subsequent selection at 65°C. It is conceivable that future mutagenesis and directed evolution of ligase 10C using the same selection strategy will further improve thermal stability and activity. These studies will help us understand the evolutionary potential of this artificial enzyme and also yield improved catalysts for a variety of applications [21].

The discovery of this thermostable enzyme and its unusual structure emphasizes the value of directed evolution approaches that do not require a detailed understanding of protein structure-function relationships, but instead randomly sample sequence space for functional proteins. In contrast, it would have been impossible to construct this particular artificial enzyme by rational design despite recent advances in rational protein engineering. In the current project, we employed the *in vitro* selection technique mRNA display [18,19]. This method uses product formation as the sole selection criterion and is independent of the mechanism of the catalyzed reaction. The technique has several advantages over other selection strategies [37]. The mRNA display technology

Acknowledgments

We thank R. J. Kazlauskas and D. Morrone for carefully reading the manuscript. We gratefully acknowledge generous support by J. W. Szostak in the beginning of the project.

Author Contributions

Conceived and designed the experiments: AM JCH BS. Performed the experiments: AM JCH BS. Analyzed the data: AM JCH BS. Wrote the paper: AM JCH BS.

References

1. Bornscheuer UT, Huisman GW, Kazlauskas RJ, Lutz S, Moore JC, et al. (2012) Engineering the third wave of biocatalysis. Nature 485: 185–194.
2. Tokuriki N, Tawfik DS (2009) Stability effects of mutations and protein evolvability. Curr Opin Struct Biol 19: 596–604.
3. Bloom JD, Labthavikul ST, Otey CR, Arnold FH (2006) Protein stability promotes evolvability. Proc Natl Acad Sci USA 103: 5869–5874.
4. Bommarius AS, Paye MF (2013) Stabilizing biocatalysts. Chem Soc Rev 42: 6534–6565.
5. Wijma HJ, Floor RJ, Janssen DB (2013) Structure- and sequence-analysis inspired engineering of proteins for enhanced thermostability. Curr Opin Struct Biol 23: 588–594.
6. Romero PA, Arnold FH (2009) Exploring protein fitness landscapes by directed evolution. Nat Rev Mol Cell Biol 10: 866–876.
7. Eijsink VGH, Gåseidnes S, Borchert TV, van den Burg B (2005) Directed evolution of enzyme stability. Biomol Eng 22: 21–30.
8. Lane MD, Seelig B (2014) Directed evolution of novel proteins. Curr Opin Chem Biol 22: 129–126.
9. Wigley WC, Stidham RD, Smith NM, Hunt JF, Thomas PJ (2001) Protein solubility and folding monitored in vivo by structural complementation of a genetic marker protein. Nat Biotechnol 19: 131–136.
10. Waldo GS, Standish BM, Berendzen J, Terwilliger TC (1999) Rapid protein-folding assay using green fluorescent protein. Nat Biotechnol 17: 691–695.
11. Martin A, Sieber V, Schmid FX (2001) In vitro selection of highly stabilized protein variants with optimized surface. J Mol Biol 309: 717–726.
12. Sieber V, Pluckthun A, Schmid FX (1998) Selecting proteins with improved stability by a phage-based method. Nat Biotechnol 16: 955–960.
13. Socha RD, Tokuriki N (2013) Modulating protein stability – directed evolution strategies for improved protein function. FEBS J 280: 5582–5595.
14. Golynskiy MV, Haugner III JC, Morelli A, Morrone D, Seelig B (2013) In vitro evolution of enzymes. Methods Mol Biol 978: 73–92.
15. Schmid FX (2011) Lessons about protein stability from in vitro selections. Chembiochem 12: 1501–1507.
16. Jäckel C, Bloom JD, Kast P, Arnold FH, Hilvert D (2010) Consensus protein design without phylogenetic bias. J Mol Biol 399: 541–546.
17. Seelig B, Szostak JW (2007) Selection and evolution of enzymes from a partially randomized non-catalytic scaffold. Nature 448: 828–831.
18. Roberts RW, Szostak JW (1997) RNA-peptide fusions for the in vitro selection of peptides and proteins. Proc Natl Acad Sci USA 94: 12297–12302.
19. Nemoto N, MiyamotoSato E, Husimi Y, Yanagawa H (1997) In vitro virus: Bonding of mRNA bearing puromycin at the 3′-terminal end to the C-terminal end of its encoded protein on the ribosome in vitro. FEBS Lett 414: 405–408.
20. Chao F-A, Morelli A, Haugner JC, III, Churchfield L, Hagmann LN, et al. (2013) Structure and dynamics of a primordial catalytic fold generated by in vitro evolution. Nat Chem Biol 9: 81–83.
21. Haugner III JC, Seelig B (2013) Universal labeling of 5′-triphosphate RNAs by artificial RNA ligase enzyme with broad substrate specificity. Chem Commun 49: 7322–7324.
22. Seelig B (2011) mRNA display for the selection and evolution of enzymes from in vitro-translated protein libraries. Nat Protocols 6: 540–552.
23. Moore MJ, Sharp PA (1992) Site-specific modification of pre-mRNA: the 2′-hydroxyl groups at the splice sites. Science 256: 992–997.
24. Greenfield NJ (2006) Using circular dichroism collected as a function of temperature to determine the thermodynamics of protein unfolding and binding interactions. Nat Protoc 1: 2527–2535.
25. Diaz JE, Lin C-S, Kunishiro K, Feld BK, Avrantinis SK, et al. (2011) Computational design and selections for an engineered, thermostable terpene synthase. Protein Sci 20: 1597–1606.
26. Reetz MT, Soni P, Acevedo JP, Sanchis J (2009) Creation of an amino acid network of structurally coupled residues in the directed evolution of a thermostable enzyme. Angew Chem Int Ed Engl 48: 8268–8272.
27. Palackal N, Brennan Y, Callen WN, Dupree P, Frey G, et al. (2004) An evolutionary route to xylanase process fitness. Protein Sci 13: 494–503.
28. Henzler-Wildman K, Kern D (2007) Dynamic personalities of proteins. Nature 450: 964–972.
29. Nashine VC, Hammes-Schiffer S, Benkovic SJ (2010) Coupled motions in enzyme catalysis. Curr Opin Chem Biol 14: 644–651.
30. Ramanathan A, Agarwal PK (2011) Evolutionarily conserved linkage between enzyme fold, flexibility, and catalysis. PLoS Biol 9.
31. Auerbach G, Ostendorp R, Prade L, Korndorfer I, Dams T, et al. (1998) Lactate dehydrogenase from the hyperthermophilic bacterium Thermotoga maritima: the crystal structure at 2.1 Å resolution reveals strategies for intrinsic protein stabilization. Structure 6: 769–781.
32. Russell RJ, Gerike U, Danson MJ, Hough DW, Taylor GL (1998) Structural adaptations of the cold-active citrate synthase from an Antarctic bacterium. Structure 6: 351–361.
33. Arnold FH, Wintrode PL, Miyazaki K, Gershenson A (2001) How enzymes adapt: lessons from directed evolution. Trends Biochem Sci 26: 100–106.
34. Macedo-Ribeiro S, Darimont B, Sterner R, Huber R (1996) Small structural changes account for the high thermostability of 1[4Fe-4S] ferredoxin from the hyperthermophilic bacterium Thermotoga maritima. Structure 4: 1291–1301.
35. Elias M, Wieczorek G, Rosenne S, Tawfik DS (2014) The universality of enzymatic rate-temperature dependency. Trends Biochem Sci 39: 1–7.
36. Tokuriki N, Oldfield CJ, Uversky VN, Berezovsky IN, Tawfik DS (2009) Do viral proteins possess unique biophysical features? Trends Biochem Sci 34: 53–59.
37. Golynskiy MV, Seelig B (2010) De novo enzymes: from computational design to mRNA display. Trends Biotechnol 28: 340–345.
38. Golynskiy MV, Haugner JC, Seelig B (2013) Highly diverse protein library based on the ubiquitous (β/α)₈ enzyme fold yields well-structured proteins through in vitro folding selection. ChemBioChem 14: 1553–1563.
39. Chaput JC, Szostak JW (2004) Evolutionary optimization of a nonbiological ATP binding protein for improved folding stability. Chem Biol 11: 865–874.
40. Smith MD, Rosenow MA, Wang MT, Allen JP, Szostak JW, et al. (2007) Structural insights into the evolution of a non-biological protein: importance of surface residues in protein fold optimization. PLoS ONE 2.
41. Hall TA (1999) BioEdit: a user-friendly biological sequence alignment editor and analysis program for Windows 95/98/NT. Nucleic Acids Symp Ser 41: 95–98.
42. Thompson JD, Higgins DG, Gibson TJ (1994) CLUSTAL-W - improving the sensitivity of progressive multiple sequence alignment through sequence weighting, position-specific gap penalties and weight matrix choice. Nucleic Acids Res 22: 4673–4680.
43. Cho GS, Szostak JW (2006) Directed evolution of ATP binding proteins from a zinc finger domain by using mRNA display. Chem Biol 13: 139–147.

In Vitro Evolution and Affinity-Maturation with Coliphage Qβ Display

Claudia Skamel[1], Stephen G. Aller[2], Alain Bopda Waffo[3]*

1 Campus Technologies Freiburg (CTF) GmbH, Agency for Technology Transfer at the University and University Medical Center Freiburg, Freiburg, Germany, **2** Department of Pharmacology and Toxicology and Center for Structural Biology, University of Alabama at Birmingham, Birmingham, Alabama, United States of America, **3** Department of Biological Sciences, Alabama State University, Montgomery, Alabama, United States of America

Abstract

The *Escherichia coli* bacteriophage, Qβ (Coliphage Qβ), offers a favorable alternative to M13 for *in vitro* evolution of displayed peptides and proteins due to high mutagenesis rates in Qβ RNA replication that better simulate the affinity maturation processes of the immune response. We describe a benchtop *in vitro* evolution system using Qβ display of the VP1 G-H loop peptide of foot-and-mouth disease virus (FMDV). DNA encoding the G-H loop was fused to the A1 minor coat protein of Qβ resulting in a replication-competent hybrid phage that efficiently displayed the FMDV peptide. The surface-localized FMDV VP1 G-H loop cross-reacted with the anti-FMDV monoclonal antibody (mAb) SD6 and was found to decorate the corners of the Qβ icosahedral shell by electron microscopy. Evolution of Qβ-displayed peptides, starting from fully degenerate coding sequences corresponding to the immunodominant region of VP1, allowed rapid *in vitro* affinity maturation to SD6 mAb. Qβ selected under evolutionary pressure revealed a non-canonical, but essential epitope for mAb SD6 recognition consisting of an Arg-Gly tandem pair. Finally, the selected hybrid phages induced polyclonal antibodies in guinea pigs with good affinity to both FMDV and hybrid Qβ-G-H loop, validating the requirement of the tandem pair epitope. Qβ-display emerges as a novel framework for rapid *in vitro* evolution with affinity-maturation to molecular targets.

Editor: Mark Isalan, Imperial College London, United Kingdom

Funding: The authors acknowledge the Deutsche Forschungsgemeinschaft for grant # DFG-BI 521/2-3 and the CNBR of ASU for grant # NSF-CREST (HRD-1241701). The funders had no role in study design, data collection and analysis, decision to publish, or preparation of the manuscript.

Competing Interests: Dr. Claudia Skamel, a co-author is currently working with a commercial company (Campus Technologies Freiburg GmbH).

* Email: abopdawaffo@alasu.edu

Introduction

Following its discovery by George Smith in the early 1980's, phage display technologies have been built predominantly from DNA phage platforms, particularly that of M13 [1–5]. M13 is DNA-filamentous bacteriophage with a genome size of 6.4 kb [6] and have very low mutation rates that limit their use in *in vitro* evolution processes. On the contrary, RNA-based replication systems possess attractive features, including high mutation rates, high population size and short replication times, that can be exploited for rapid *in vitro* evolution [7]. Additionally, RNA-replication systems lack recombination processes that can further complicate DNA-based replication systems and technologies. Early efforts to generate recombinant RNA had limited success due to limitations in technology and RNA instability. However, with the improvement of recombinant DNA technology, and the existence of reverse transcription techniques, the generation of recombinant RNA is now straightforward. Recent advancements have led to the generation and cloning of Qβ cDNA into several stable plasmids that are able to liberate phage upon bacterial transformation [8]. The cDNA of Qβ coliphage RNA has become amenable for use in displaying random peptide libraries *in vitro* followed by *in vivo* translation and phage production.

Qβ belongs to the family of *Leviviridae* and is found throughout the world in bacteria isolates associated with sewage [9]. Of the four groups of RNA coliphages, the genome and proteins of Qβ phages have been the most extensively characterized [10]. Some representatives of these groups are: group I (f2, MS2, R17, fr) group II (GA) group III (Qβ) and group IV (SP) [11]. In this report we present a framework of peptide display and affinity maturation using Qβ phage and the integrin receptor of Foot-and-Mouth-Disease-Virus (FMDV) as a proof-of-concept for acquiring binders to a highly infectious agent with many different serotypes. FMDV, the causative agent of the most economically important infectious diseases in farm animals, has seven serotypes (O, A, C, SAT$_1$, SAT$_2$, SAT$_3$ and Asia 1, [12]). The varied nature of the serotypes compromises the ability to control this disease using present vaccination strategies. Furthermore, the instability of currently available vaccines leaves farmers with no practical option but to slaughter, emphasizing the urgent need for new vaccines [13]. FMDV is a single stranded positive-sense RNA virus of ~8 kilobases (kb). FMDV particles consist of four major polypeptides, three outer capsid proteins (VP1, VP2 and VP3) and a fourth smaller capsid protein (VP4). The G-H loop of VP1 is of particular interest due to its major antigenic site at the carboxyl terminal [14–16].

Both FMDV and Qβ have icosahedral shells of 30 nm and 25 nm in diameter, respectively [17,18]. The Qβ genome is ~4.2 kb surrounded by a shell of 180 coat protein molecules

[17,18]. Of these proteins, A2, A1 (known as readthrough) and the replicase are encoded by the phage genome and are important for the formation of infectious phage [19]. Due to its copy number and position [20], we hypothesize that A1 can be utilized for phage display. Phage display, previously called phage exposition, consists of an insertion of a foreign DNA fragment into the minor structural phage A1 gene to create a fusion protein, which is then incorporated into a virion that retains its infectivity and exposes the foreign peptides in an accessible form at the surface [1].

We constructed hybrid phages bearing FMDV VP1 G-H loop C-terminus that efficiently binds monoclonal antibodies directed against the antigenic loop. Furthermore, display of randomized peptides allowed *in vitro* Qβ phage selection, evolution and convergence on a displayed peptide containing a tandem amino acid sequence required for anti-FMDV monoclonal antibody recognition. The specificity, productivity, affinity and efficiency of the hybrid phage were characterized. Additionally, our data provides an insight into FMDV antigen motif representing candidates for development of vaccines for livestock.

Materials and Methods

Reagents

All media for bacteria culture and phages were purchased from Fisher Scientific (Pittsburgh, PA). Restriction enzymes and *T4 DNA ligase* were purchased from New England BioLabs (Ipswich, MA). Unless otherwise indicated, chemical reagents (ie. RbCl and $CaCl_2$) were purchased from Sigma-Aldrich (St. Louis, MO).

Microorganisms

Escherichia coli MC1016 (Invitrogen, Grand Island, NY) was used to grow and maintain plasmids. *E. coli* HB101 was used to grow and maintain pBRT7Qβ, pQβ8 plasmids and all their recombinant derivatives. Three different indicator bacteria were used for phage production and titration: K12 (*E. coli* ATCC 23725), HfrH (*E. coli* ATCC23631) and Q13 (*E. coli* ATCC 29079) purchased from ATCC. The *E. coli* bacteriophage Q-β *ATCC 23631-B1* was used as a positive wild type (wt) control in experiments.

Antibodies

The FMDV VP1 G-H loop specific antibody, SD6, was obtained from Professor Esteban Domingo's laboratory from the Department of Virology and Microbiology of the University of Madrid, Spain. Anti-green fluorescent protein (GFP) polyclonal antibody was from Biofuture Group in Goettingen, Germany. Anti-protein tHisF and HisJ polyclonal antibodies were obtained from Professor Hans-Joachim Fritz'laboratory from the Institute for Microbiology and Genetics of the University of Goettingen, Germany.

Hybrid phage construction

Plasmids pBRT7Qβ and pQβ8 were obtained from Professor Weber [21] and from Professor Kaesberg groups [8] respectively. These plasmids pBRT7Qβ having 7489 bp (from 1 to 7489 when restricted with *Sma*I [21]) and pQβ8 having 7393 bp (from 1 to 7393 when restricted with *Sma*I endonuclease [8]) were used for this work since they both contain the entire cDNA of the Qβ phage with different orientation. For the cloning procedure into the pBRT7Qβ plasmid, *Afl*II and *Nsi*I restriction sites were used. All the primers used to amplify foreign functional protein genes were flanked with *Afl*II or Bpu10I (forward) and *Esp*I or *Nsi*I (reverse) restriction sites (Table 1). These primers were designed to confer some important features to the foreign gene after cloning: to

maintain the reading frame of the vector and to maintain the important secondary structure of phage RNA for replication transcription, translation, regulation and assembly. The *Esp*I site is absent at the end sequence of the A1 protein gene of the pBRT7Qβ plasmid. To introduce this site, DNA fragments were transiently cloned into some intermediate plasmids. After PCR, the foreign gene insert was cloned into the pCR2.1 Topo vector. The pCR2.1 vector is a linearized vector ready for direct ligation of unmodified, unpurified PCR products. This vector has a single overhanging "T" which facilitates the cloning of a *Taq* and *Phusion* amplified fragment. This vector does not contain the *Afl*II and *Esp*I restriction sites and the PCR products are cloned between two *Eco*RI restriction sites. The recombinant pCR2.1 plasmid enables amplification with higher fidelity and sequencing of the PCR fragments for further cloning.

The correct insert was later cloned into a pUC-cassette vector, which allows the introduction of new restriction sites like *Esp*I. The pUC-cassette is a recombinant pUC18, containing the C-terminal of the cDNA of the A1 protein gene (from 2129–2402), which introduces the *Esp*I restriction site prior to cloning, into pBRT7Qβ. This part of the A1 gene is cloned between the *Hind*III and *Kpn*I restriction sites. Modifications aiming to add or subtract part of the foreign gene fragment to be cloned into the pBRT7Qβ plasmid were performed on the recombinant pUC18. When these manipulations were successfully performed in small size plasmids, the foreign protein gene inserts were cloned back into the pBRT7Qβ/pQβ8 plasmid as presented in Fig. 1. The recombinant pBRT7Qβ was used to generate RNA phage displaying the exogenous functional peptide. To further explore this new display technology, other functional proteins with different specific motifs that are larger than, but related to the FMDV GH-loop in the structure were also studied. These functional proteins were: the green fluorescent protein (GFP), the imidazole glycerol phosphate subunit of the synthase thermostable subunit (tHisF) and the periplasmic histidine-binding protein (HisJ).

Deletion of A1 protein construction in pBRT7Qβ and pQβ8

To insert larger DNA fragments at the end of the A1 gene, we deleted the last 162 nucleotides of this gene keeping the interregional A1 gene and the replicase gene. These plasmids are called pBRT7QβΔA1 or pQβ8ΔA1 derived from pBRT7Qβ or pQβ8 respectively. To construct recombinant QβΔA1 plasmid, either a short portion of the cDNA of Qβ of about 420 bp was amplified with PCR or a gene part of 162 bp (between *Afl*II and *Nsi*I restriction sites) was removed and replaced by a short adaptor gene sequence. In the case of PCR, the forward primer (Table 1) used was flanked by *Bpu*10I and the reverse primer was flanked by the tag and the *Nhe*I and *Nsi*I sequences. *Nhe*I was added to monitor the cloning process. The PCR product was cloned into the pBRT7Qβ or pQβ8 plasmids using *Bpu*10I and *Nsi*I. The adaptor oligos were annealed and ligated into the *Esp*I enzyme restriction site.

Strategy for phage production

Positive clones were transformed into *E. coli* HB 101 after sequencing using the method of Taniguchi and collaborators [22]. The supernatants of overnight clones were checked by agar overlay method for the presence of phages. The phages were amplified using an indicator bacteria cell, *E. coli* HfrH, Q13, 1101 or K12. Fresh overnight cultures (on standard nutrient agar I plate) were amplified at 37°C to reach an OD_{600} of 0.6–0.8 (after 2–3 h) and inoculated with phage suspension at a multiplicity of 3.

Table 1. Oligonucleotides.

Name	Sequence	Functions
ABW1	atgcatttcatccttagGCTAGCttactacgacttaagatagatgaattgttcgatgttaccg	For A1 deletion with NheI and NsiI restriction site reverse
ABW2	cagctgaacccagcgtatTGAacgttgctcattgccggtggtggctc	For A1 deletion forward before Bpu10I site used
CB191	TTACACCGCCAGTGCACGCGCGGGGATCTTGCTCACCTAACGACGAC	FMD-loop adaptor
CB192	TAAGTCGTCGTTAGGTGAGCAAGATCCCCGCGTGCACTGGCGGTG	Complementary of CB191
CB193	TTACACCGCCAGNGCANNNNNNNNNNNNNNNTCACCTAACGACGA	Randomize-FMD-loop adaptor
CB194	CTGGCGGTG	Complementary of CB193 first
CB195	TAAGTCGTCGTTAGGTG	Complementary of CB193 second
CB197	ACCTTCAACCTCAATTCTTGTGTTC	For sequencing Qβ cDNA for reverse from G 2410
CB198	TGCGTGATCAGAAGTATGATATTCG	For sequencing Qβ cDNA of end of A1 gene and insert forward from T 2083

The infected cells were incubated at 37°C by shaking (150 rpm) for 5 h. After this incubation period, the phage titer was checked and a second round of amplification was performed to scale up the phage titer according to the same procedure. At the end of amplification, indicator cells were allowed to complete the lysis by adding few drops of chloroform to the culture suspension.

Agar overlays for spot test

This test was done according to Adam [23]. A bacteria culture was grown to log phase (OD_{600} of 0.6–0.8). A volume of 100 µl of this culture was added to 3 ml of YT-Top-agar and the mixture was poured on the surface of nutrient agar plates. The plates were left to solidify at 37°C for a few minutes. Thereafter, 4–7 µl of the phage suspension was dropped on the solidified plates. The plates were incubated at 37°C for 24 h and examined for lysis of the E. coli lawn where the droplet of phage suspension was placed.

Measurement of phage yield and plaque quality

Indicator bacteria (100 µl) were infected with 100 µl of the appropriate serial dilution of phage-containing tryptone glucose yeast (TGY) solution. After 10 minutes of incubation at room temperature, 3 ml of soft-agar (TGY with 0.6% agar) was added and the mixture was poured onto plain agar plates. The plates were allowed to solidify and incubate for 16 h at 37°C. The plaque count was done following the method of Pace & Spiegelman [24]. We observed both quality and size of plaques.

Reverse transcription (RT) PCR from plaques and/or purified phages

Phages from the clear zone of specific plaques were extracted by excising the soft media to a tube, adding 10–15 µl of H_2O and centrifuged at 3000 rpm for 5 min to remove the media. 10 µl of the supernatant was used for RT. For the purified phage, following RNA extraction, 2–5 µg of RNA was incubated at 99.6°C with 50 pmol of reverse primers for 2 min in a total volume of 11 µl (filled with RNase-free water). This was followed by incubation on ice to allow the annealing of primers to the template. To the mixture, the following components were added: 10 µl 5×RT-buffer, 2 µl $MgSO_4$ (25 mM), 1 µl 10 mM dNTP mix, 1.5 µl AMV-RT (5 U/µl) and 24.5 µl of H_2O RNase-free.

To allow reverse transcription, the mixture was incubated at 42°C for 1 h followed by AMV denaturation at 94°C for 2 min. For the PCR reaction, 2 µl of the reverse transcriptase was used with the following protocol. The cycling protocol consisted of 25 cycles of three temperatures: 94°C, 30 s (strand denaturation), 50–57°C, 1 min (primer annealing), 68°C, 2 min (primer extension), followed by a final extension at 68°C for 7 min.

Selection with biopanning

The antibodies were adsorbed to Xenobind™ microtiter plates (Dunn in Asbach, Germany) for biopanning. The middle wells of the plates were covered with 150 µl of the antibody solution (2.5 µg/ml in carbonate buffer: 15 mM Na_2CO_3 and 35 mM $NaHCO_3$ pH 9.6) and incubated at room temperature for overnight. To cover the surface of the wells, a solution of 5% bovine serum albumin (BSA) in the antibody solution was added and then incubated for 1 hour at room temperature. The excess of unbound BSA was removed by washing 3 times with wash buffer (137 mM NaCl, 2.7 mM KCl, 8.3 mM Na_2HPO_4 $2H_2O$, 1.5 mM NaH_2PO_4 at pH 7.2 and 0.05% Triton X-100). A phage solution of 150 µl was then added to experimental wells of the plate and the plate was incubated at room temperature for 4 h and washed twice with wash buffer and 2 additional times with phage buffer. The experimental wells were then covered with 200 µl of E coli Q13 or HfrH culture (grown to OD_{600} of 0.7) and incubated at 37°C for 20 min. The bacteria culture from experimental wells was transferred into tubes as aliquots under sterile conditions after incubation. One aliquot was plated for phage titration and the rest incubated at 37°C overnight. For the next round of biopanning, 150 µl of the previous overnight bacteria and panning were used as phage solution. To further characterize phages from rounds of panning, 50 µl of phages from each round were used to extract RNA. This RNA was subjected to RT-PCR and sequencing reactions.

Phage purification and analysis

Phage was collected using polyethylene glycol (PEG) precipitation as described in [25] with minor modifications (using PEG_{8000}). Phage suspension (cell debris and phage) was incubated with 8% PEG and 0.5 M NaCl (final concentration, respectively) overnight at 4°C. The phages were pelleted by centrifugation at 3000 rpm for 20 min at 4°C (Sorvall GSA). The pellet was resuspended in phage buffer (10 mM Tris HCl pH 7.5, 1 mM MgCl2, 100 mM NaCl, 10 mg/l gelatine with 1/5 of the volume phage suspension after amplification) at 4°C for 20 min and pelleted by centrifugation at 10000 rpm for 20 min at 4°C (Sorvall RC5B, SS34). The procedure was repeated. After overnight incubation the phages were collected by centrifugation and the pellet was suspended in a

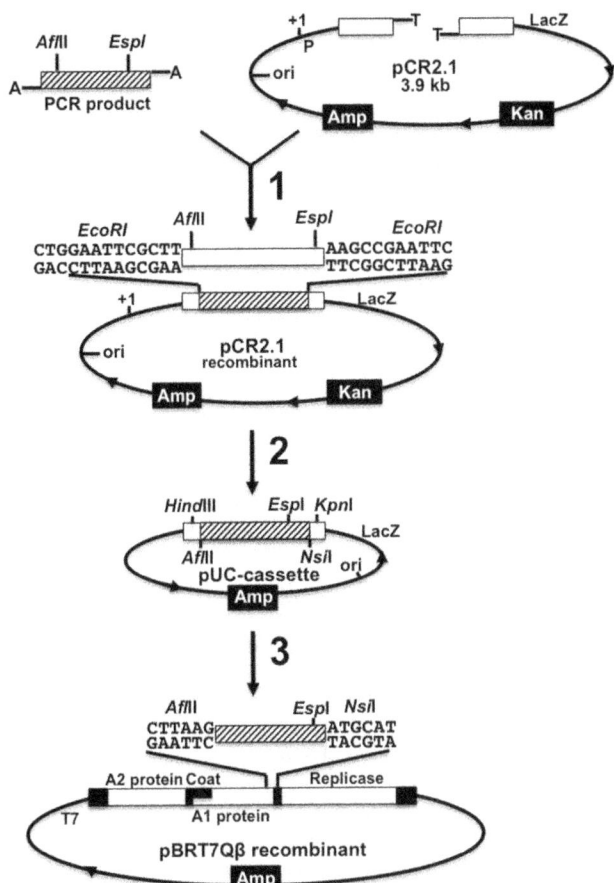

Figure 1. Schematic representation of the RNA phage display vector construction. General cloning procedure from PCR fragments to pBRT7Qβ with transient cloning in the pUC18-cassette working plasmid. Step 1: cloning of PCR fragment into pCR2.1 vector; Step 2: cloning of the foreign gene from PCR into the pUC-cassette (with *Nsi*I) using *Afl*II and *Esp*I sites; Step 3: Cloning of the foreign gene into pBRT7Qβ using *Afl*II and *Nsi*I. P: promoter; *Amp*: ampicillinase gene; *Kan*: kanamycin resistance gene; ori: origin of replication.

small amount of phage buffer (50 μl) without gelatine. The suspension was centrifuged at 15000 rpm for 20 min at 37°C (Sorvall RC5B, SS34) and the supernatant containing phages were collected and subjected to DEAE sepharose or CsCl-gradient purification. For long-term storage the phage phase was stored in 50% glycerol at −80°C.

Another phage amplification procedure was done based on Gschwender and Hofschneider [26]. In this procedure at the log phase infected cells were incubated in high Mg^{2+} (200 mM) to inhibit cell lysis after infection. Phages were extracted from bacterial sedimentation after lysis were induced with 50 mM EDTA on ice. The suspension was adjusted to pH 9.5 by the addition of 1 M NaOH under vigorous stirring. The cellular debris was removed by low-speed centrifugation.

Immuno-precipitation of hybrid phages against respective displayed peptide-antibody

Agarose double diffusion of Ouchterlony and Nilsson [27] was used to test the presence of foreign protein on phage surface with modifications. 1% agarose gel solution in assay buffer (50 mM Tris-HCl pH 7.5; 0.1 M NaCl) was poured into 10 cm petri dish, and allowed to solidify. Six wells were punched at equal distance

from the center well. To each well 50 μl of the appropriate concentration of phages or antibody was added and incubated for 24 h at room temperature.

Electron microscopy

A carbon-coated Formvar grid was filled with 5 μl of purified phage solution diluted to the titer of 10^9 plaque-forming units per ml (p.f.u/ml). The solution was left on the grid for a short while and then a few drops of aqueous uranyl acetate were added. Slides were then observed under a JEOL 1200EX electron microscope.

Results

Tolerance of Qβ A1 gene to manipulation

Initially, two variants of the pBRT7Qβ plasmid were constructed: pBRT7QβESPI and pBRT7QβNOTI. In these plasmids, additional nucleotides were added to the 3′end of A1 gene to introduce multiple cloning sites (Fig. 1). For pBRT7QβESPI, 6 nucleotides were added to introduce an *Esp*I site, and 9 nucleotides were added to pBRT7QβNOTI to introduce a *Not* I site. We tested if these extensions allowed proper DNA packing and the production of infectious-competent phage. Indeed, we show that 3 different gene fusions with A1 placed in front of the natural opal and ochre stop codons (TGA and TAA), produced phage plaques in bacterial lawns (Fig. 2). These results suggest that the 3′- end of A1 can accept minor extensions without disturbing the function of phage infectivity. We next explored the lengths of extensions and their effect on infectivity. Various DNA lengths (15–850 bp) were successfully fused with the A1 gene (Figs. 2 & 3), but only recombinant plasmids containing foreign inserted DNA with lengths between 15–300 bp produced phage plaques. These results show that the length of the inserted DNA is critically important for this novel system. Next we tested whether the 3′ end of the A1 gene is critical and important. To accomplish this, we constructed the plasmid pBRT7QβΔA1, in which non-essential sequences of the cDNA of the Qβ genome were deleted from the 3′-terminus of the A1 protein gene. Specifically, we deleted a 162 bp part of the 3′ terminus of the A1 gene (between nucleotides 2271 and 2333) and replaced it with a short adaptor gene sequence of 33 bp leaving the original intercistronic region between A1 gene and the replicase gene intact. Interestingly, these recombinant plasmids with 3′ truncations of A1, still produced phage plaques. However, further deletion of the A1 protein gene beyond nucleotides 2271 at 5′ end or 2333 at 3′ end abolished phage production. Furthermore, we tested whether the orientation of Qβ cDNA within the plasmid is critical. We created identical constructs using pBRT7Qβ and pQβ8, both of which contain the entire cDNA of phage Qβ albeit in opposite orientations. These plasmids yielded phages with similar titers to the wt, suggesting that the orientation of phage cDNA does not influence the phage production. Positive recombinant pBRT7Qβ or pQβ8 plasmids were identified via restriction enzyme (Fig. 3) prior to sequencing and transforming into Qβ for characterizing the display of foreign peptides and proteins.

Phage production and resulting titers

To produce wild type and recombinant phages, all plasmid vectors and variant constructs were transformed into *E. coli* HB101. *E. coli* HB101 bacteria were selected because they lack the pili appendage (F$^-$) necessary for Qβ absorption and infection. This insures exclusive usage of high-fidelity DNA polymerase-mediated replication of Qβ genes and prevents premature evolutionary events. Similar phage titers for both wt and recombinant phages (∼10^8 to 10^9 p.f.u/ml) were obtained with

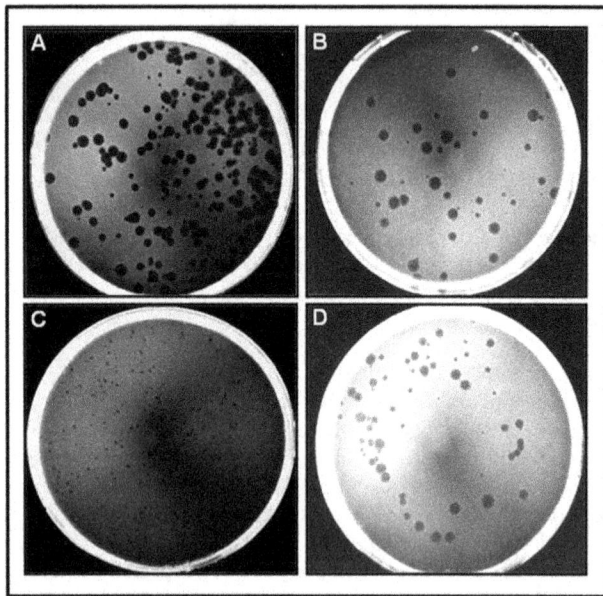

Figure 2. Morphology of wild type vs. hybrid Qβ phage plaques. Panel A) wild type Qβ phages; Panel B) Qβ-FMDV VP1 G-H loop phages; Panel C) Qβ-GFP rescued phages from *E. coli* SURE (expression host with F⁺) over-expressing A1-GFP protein infected with wild type Qβ. Panel D) QβΔA1 phages. All at very low multiplicity of infection (MOI), and all plates are exactly 1 day (24 hours) old when photographed.

plaque sizes ranging between 1 mm and 3 mm in diameter in both wt and variants. The plate of phage harboring the 3′-truncation of A1 minor coat protein gene was dominated by smaller (1 mm and 70%) than larger (3 mm and 30%) sized plaques as shown in Fig. 2. However, some minor differences in plaque size were also observed. We interpret these differences as due to either the nature of the quasispecies within the phage population and/or effects associated with the insert size (Fig. 2). Next, we analyzed the phages with different sized plaques using RT-PCR and wt Qβ was used as a standard (Fig. 4). Finally, all cDNAs were sequenced to confirm the presence of the appropriate foreign gene within the hybrid phage. Results show that sequences encoding foot-and-mouth disease virus (FMDV), HisJ and HisF, appended onto Qβ-A1 allow assembly of plaque-forming phage particles containing the gene fusions.

Efficient Qβ-FMDV phage library display and biopanning

Rapid fitness gains are the main goal of molecular evolution and are directly proportional to the population size and the selection pressure. To mimic this process *in vitro*, we synthesized a randomized VP1 G-H loop library (YTAXA**XXXX**XHLTT) that corresponded to the immunodominant region of VP1 including the canonical RGD epitope (YTASA**RGD**LAHLTT) using three oligonucleotides CB193, CB194 and CB195 as depicted in (Fig. 5). We then cloned the randomized library into Qβ plasmids (pBRT7Qβ and derivatives) and used monoclonal antibody (mAb) SD6 as the constant selective target in a biopanning assay. Additionally, the original sequence of the VP1 G-H loop of FMDV serotype C clone C-S8c1 was cloned into pBRT7Qβ using the annealed oligonucleotides: CB191 and CB192 (Fig. 5) as a control. Recombinant plasmid (pBRT7Qβ-FMDV), derived from the previous vector harboring the VP1 G-H loop fused with A1 within the phage cDNA was used to transform *E. coli* HB101.

Only 50–55% of these clones produced phage plaques (not shown). Positive VP1 G-H loop clones were validated through RT-PCR and sequencing. We further confirmed the presence of the G-H loop of the hybrid phages through dot blot using mAb SD6 (Fig. 6). We further randomized the G-H loop to form a library, which was directly ligated with A1 of cDNA of Qβ. The Qβ-FMDV phages were found to produce clear plaques in all wt Qβ natural hosts namely: *E. coli* K12, Q13, and HfrH. As before, we recapitulated our data in HB101, showing that, as with wt phage, Qβ-FMDV phage can be propagated in the other *E. coli* strains. There were no significant differences in the yield of phage particles between Qβ-FMDV and Qβ. Finally, these Qβ-FMDV phages were amplified in Q13 cells (chosen among other Qβ hosts) and a high titer (10⁹ p.f.u./ml) was obtained and purified by ultracentrifugation on CsCl gradient for guinea pig immunization.

Non-canonical FMDV epitope

To gain fitness, the synthesized library of the G-H loop was selected using a modified biopanning protocol [1,2,28] with mAb SD6. This modified protocol selects and amplifies phage while avoiding acidic elution of phages selected (Fig. 7). We reasoned that removal of an acidic elution step would enhance phage viability and the overall efficiency of *in vitro* evolution. Additionally, media containing the indicator bacteria *E. coli* Q13 grown to the log phase was added directly to the plate to further enhance survival of hybrid phages. After each round, an aliquot of phage was used for RT-PCR and the resulting DNA sequence was compared to the wild type sequence (Fig. 8). The sequence comparison of the randomized VP1 G-H loop after six rounds of biopanning revealed a shift of mAb SD6 binding motif from *Arg-Gly-Asp* to *Xxx-Arg-Gly*. Due to the preservation of the *Arg* and *Gly* in all three rounds of biopanning, we conclude that this tandem pair is essential for mAb SD6 binding. The third amino acid was substituted without disturbing the binding capacity of the peptide to the mAb SD6. Over 80% of glycine exposed by the phage was in contact with the mAb SD6. This clearly shows that arginine and glycine were not just only together in the antibody binding motif but were representing optimized amino acid from a randomized pool.

Immunization characterization of Qβ-FMDV phages

To assess the immunogenicity of the hybrid phages, guinea pigs were immunized with purified Qβ-FMDV phages. The serum obtained after immunization tested positive for FMDV antibody (Professor Esteban Domingo, personal communication). We validated this finding with a qualitative Ouchterlony assay (27) using serum obtained from immunization with Qβ-FMDV phages against the same phages on one hand, then Qβ wt. In this assay, antibody and antigen solution are placed in nearby wells cut out of a thin layer of agarose and allowed to diffuse toward each other forming a visible line of precipitation where they meet. The lines produced by the two adjacent wells containing Qβ-FMDV and Qβ wt join together in a pattern of partial identity (Fig. 9A). More specifically, at least two epitopes on Qβ-FMDV were recognized by the serum antibody. A similar result was obtained with immunoglobulin purified from the serum with protein A affinity column (results not shown). The fractionated Igs from the column did not react with phage displaying the C-terminal deletion of A1 (pQβ8ΔA1; Fig. 9C). Furthermore, the thickness of the precipitation line in the double diffusion was reduced with the reduction in the serum amount shifting towards the phages wells (Fig. 9B). The second line of precipitation close to the phage wells was reduced with half of the phages titer (Fig. 9C). The additional line of precipitation on Ouchterlony assay plate showed the presence

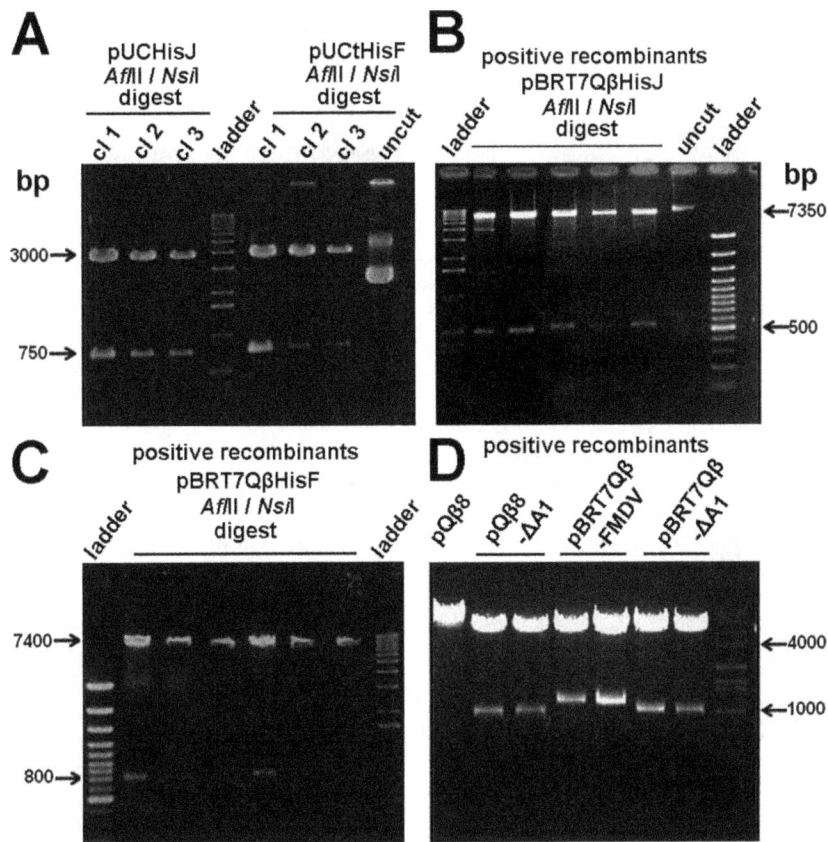

Figure 3. Agarose gel electrophoresis of the RNA display system vector construction. Panel A) Lanes 1–3: positive recombinant pUCHisJ plasmid clone (cl) restricted with AflII and NsiI; Lanes 5–7: positive pUCtHisF and Lane 8: negative clone. Panel B) Lanes 2–6: positive recombinants pBRT7QβHisJ restricted with AflII and NsiI; Lane 7: negative clone. Panel C) Lanes 2–7: positive recombinants pBRT7QβtHisF restricted with AflII and NsiI. Panel D): Lane 1: pQβ8 negative control; Lanes 2 and 3: positive recombinants pQβ8ΔA1; Lanes 4 and 5: positive recombinants pBRT7Qβ-FMDV; Lanes 6 and 7: positive recombinants pBRT7QβΔA1 all restricted with NheI. Lanes "ladder" were loaded with the 100 bp or 1 kb DNA ladder.

of an epitope on Qβ-FMDV that is absent on both wt Qβ and pQβ8ΔA1. We conclude that this additional line of precipitation can only be the VP1 G-H loop exposed on the exterior surface of Qβ. This result illustrated and clarified the heterogeneity and specificity of the serum obtained from immunized guinea pigs.

Finally, Qβ-FMDV presence and display was validated using negative stain electron microscopy that shows the presence of antigen-antibody interaction. The wt phage particle was found with a clear zone around its particles (Fig. 10A), while Qβ-FMDV phages displaying the G-H loop of FMDV showed dots decorating its surface by mAb SD6 (Fig. 10C). We theorize that, these dots

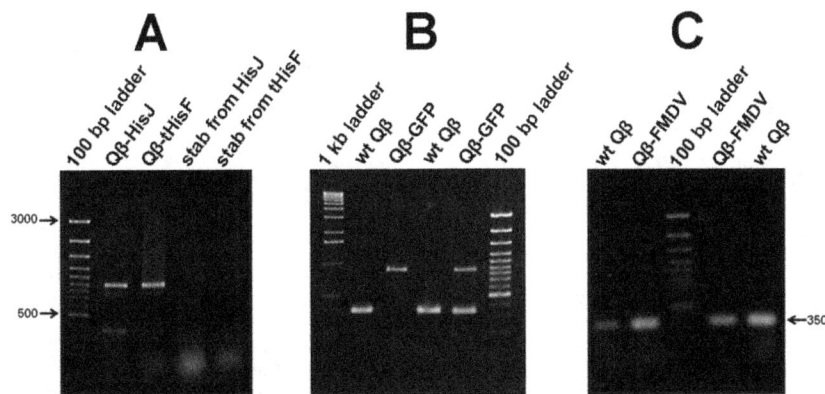

Figure 4. RT-PCR of RNA purified from Qβ-phage plaques. Panel A) Lane 2: Qβ-HisJ; lane 3: Qβ-tHisF; lane 4: soft agar stab from HisJ plate; lane 4: soft agar stab from tHisF. Panel B) Lanes 2 and 4: wild type Qβ; Lanes 3 and 5: Qβ-GFP. Panel C) Lanes 2 and 4: Qβ-FMDV; Lanes 1 and 5: wild type Qβ (positive control). The 100 bp and 1 kb DNA ladder were used.

A

CB 191

5' - TTACACCGCCAGTGCACGCGGGGATCTTGCTCACCTAACGACGAC___ - 3'

3' -   GTGGCGGTCACGTGCGCCCCTAGAACGAGTGGATTGCTGCTGAAT - 5'

                                                            CB 192

B

CB 193

5' - TTACACCGCCAGNGCNNNNNNNNNNNNNNNNTCACCTAACGACGAC___ - 3'

3' -   GTGGCGGTC   GTGGATTGCTGCTGAAT - 5'

CB 194   **CB 195**

Figure 5. Design and schematic of FMDV VP1 G-H loop (serotype C-S8cl) oligonucleotide sequences. Panel A) Original sequence used for cloning (with *Esp*I site at both ends) into Qβ and production of phages Qβ-FMDV for guinea pig immunization. Panel B) Randomized sequence synthesis (with ends similar to Panel A) for library generation and phage population production and selection against mAb SD6.

represent the position of A1 fused with the G-H loop of FMDV on the exterior surface of phage Qβ (arrow). The combined results of animal immunization, EM and serological assays, suggested that, on the exterior surface of Qβ phages, there can be exposed epitopes which may be used to induce the production of specific antibody.

Discussion

Current phage display technologies have been exclusively designed using DNA-based phage platforms such as M13 [1–5,28]. Use of such systems for library screening purposes requires highly diverse starting libraries encoding the displayed proteins since in vitro evolution is difficult, due to the relatively high-fidelity of proof-reading bacterial DNA polymerases. The work of Drake [29] showed a surprisingly consistent overall genomic mutation rate for DNA-based replication in plants, yeast, bacteria and bacteriophage. When the mutation rate per base was multiplied by the genome size, the mutation rate per genome for these DNA-based replicating organisms fell within an astonishingly narrow range (0.0022–0.0046; Table 2). Genomic replication error rates for RNA-based replicating MNV11RNA and bacteriophage Qβ were 7× and ~500× greater, respectively, revealing a high degree of inherent genomic evolution potential. Compared to M13-based mutation rates, Qβ has a 420× greater mutation rate per base, indicating a major advantage in utilization of Qβ for in vitro evolution. Moreover, relatively harsh acidic elution procedures common to DNA phage systems add another obstacle in

developing a practical system for *in vitro* evolution. The RNA phage Qβ possesses key features that can be exploited for *in vitro* evolutionary display that overcomes both obstacles [30]. We describe here a framework for Qβ display that allows a robust *in vitro* evolution at the lab benchtop.

Developing Qβ phage for peptide display included four steps. Firstly, the DNA sequence coding for the displayed peptides or proteins was fused to the end of the A1 minor coat protein gene in the DNA plasmids utilized. A1 is essential for infection, but specific roles in the Qβ virion cycle have not yet been elucidated. Our results indicated that the C-terminus of A1 (nucleotides 2271 to 2333), plays only a minor role in function and is consistent with a previous report describing the importance of the N-terminus [31]. Secondly, recombinant plasmids were sequenced and transformed in bacteria cells to produce a population of phages. Thirdly, hybrid

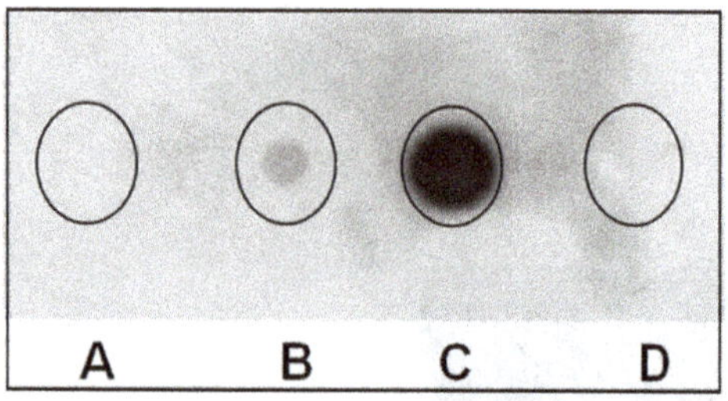

Figure 6. Dot blotting analysis of the FMDV VP1 G-H loop displayed on Qβ phages with SD6 mAb. Spot A: Supernatant of the Qβ wild-type culture infection (negative control); Spot B: Qβ-G-H loop phages from the supernatant of culture after 5 h of infection; Spot C: Qβ-G-H loop phages from the supernatant after overnight infection (higher concentration); Spot D: the phage buffer only as a negative control. Phages from spots A, B and C were purified by PEG8000/NaCl precipitation and CsCl gradient. The same pattern was obtained with the spotted corresponding crud lysat.

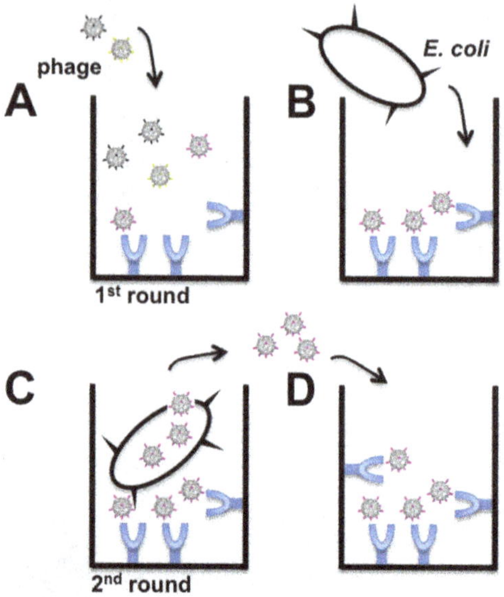

Figure 7. Schematic of biopanning assay with Qβ phage derivatives without the usual acidic elution. A) A population of phages displaying the library of interest (here randomized VP1 G-H loop) was added to the well of a plate pre-coated with the desired target (in this case, mAb SD6 covalently immobilized with F_c region). B) Indicator *E. coli* are added to the well after phages having low-affinity to the target are removed. C) High-affinity phage bound to target can infect *E. coli* Q13 by adsorbing and injecting its RNA via the F^+ pilus. D) Phages newly obtained after indicator *E. coli* infection were transferred to new wells containing the immobilized target for the next round of biopanning.

AGTGCACGCGGGGATCTTGCT Ser Ala Arg Gly Asp Leu Ala	FMDV strain S8C
AGnGCAnnnnnnnnnnnnnnnT	Randomized sequence
AGgGCAng t nggngnn t nn t T	after transformation
AGgCARnTASGGTGcMSATGT	1 biopan round
AGaGCRnTASKGKSYCCaYGT	2 biopan rounds
AGAGCaRTASgGGG t CCRYGT	3 biopan rounds
Arg Ala V/I R/G Gly Pro R/C	3 biopan rounds

Figure 8. Sequences comparison of the randomized FMDV G-H loop displayed after three rounds of Biopanning. The first line up is the original loop motif *Arg-Gly-Asp* of VP1 G-H loop of FMDV strain S8C. On the left hand side under the original sequence are the sequences obtained after a round of biopanning. The low case letters show the different between the sequences. The last line is the expression of the evolution of the original motif sequence on the 3rd round shown the maintanance of *Arg-Gly* and change of the last amino acid.

phages were subjected to several rounds of selection and amplification to optimize interactions with the target. Fourthly, selected phages were analyzed for the presence of the correct recombinant RNA and the ability to display the appropriate protein. We examined the limits of the sizes of displayed proteins by attempting display of HisJ (726 bp), HisF (753 bp) and GFP (714 bp). Unfortunately, all three proved too large for Qβ, as evidenced by poor infectivity and/or plaque formation, even when fused to full-length A1 or A1 3′ truncation. The transformation efficiency of these constructs dropped with the increasing size of the insert but could be somewhat improved by using the rubidium chloride method with heat shock at 43.5°C, in contrast to conventional methods [32]. The Qβ phage remains functional (able to absorb and infect) with up to 60 nucleotides inserted into the A1 gene. The phage can also function with 162 nucleotides removed from the 3′ end of the A1 gene. These observations together give the A1 gene a total loading capacity of ~222 nucleotides (74 amino acids protein). When more than 300 nucleotides of foreign gene were inserted, no viable, stable, infective replicable phages were obtained. Although spontaneous Qβ particles have been previously obtained with the A1 protein extended to 195 amino acids [33], neither replication competency

nor recombinant RNA-genomic packaging limits were determined. An enormous drop in phage viability was found using MS2 phage to display five amino acids (Ala-Ser-Ile-Ser-Ile) on the exterior surface [34] revealing potential limitations on the secondary structure of the recombinant RNA on its influence on the replication, regulation and assembly to form viable phage [35].

Successful transformation and subsequent amplification was achieved with the A1 3′ truncation plasmid and the 14 amino acid G-H loop of FMDV plasmid. FMDV is a particular threat for animal livestock worldwide and the large number of serotypes and diversity of strains make the development of a universal vaccine very challenging. Through its RNA replication system, FMDV has a very high mutation rate which allows the virus to escape drug suppression [7]. Considering the highly mutable character of RNA viruses, we reasoned that a vaccine system that can also exploit this feature of RNA viruses would be highly valuable. The displayed G-H loop was found to occupy the corners of the icosahedron of Qβ as visualized by negative-stain electron microscopy, which corresponds to its natural positions in the FMDV structure [36]. The recombinant A1-G-H loop was found to decorate the phage at 12 copies per virion, indicating that the wt A1 protein must also have 12 copies per Qβ phage. Prior to these results, the exact copy number of the A1 protein was not well known but was estimated to be between 3–7% of total phage protein [20,37,38]. An octa-peptide of β-tubulin motif was also found to decorate the 12 corners of the Qβ icosahedron (unpublished data). This is in contrast to the commonly used phage M13 where any fusion to the minor coat protein gene, pIII, would display the foreign peptide on only one side of the phage [39].

We next exploited the high error-rate of the RNA-dependent RNA polymerase of Qβ to test the feasibility of using this system for *in vitro* evolution. Transforming a completely degenerate DNA that appended seven randomly encoded amino acids to the A1 protein, with a total theoretical diversity of 2×10^{14}, has resulted in a six amino acid peptide library with an actual diversity of 2×10^{8}. In only six rounds of biopanning and selective pressure in the presence of immobilized SD6 antibody, we isolated the Qβ phage that contained a conserved tandem pair, *(Arg-Gly)*, that is essential for mAb SD6 recognition. The main motif of the G-H loop was known to be Arg-Gly-Asp that recognizes and binds to mAb SD6 [14,15]. Upon randomization and selection, the Arg-Gly motif was found to be enough for binding to the same mAb SD6, amongst the spectrum of variants generated. Panning with Qβ-FMDV phages has a double advantage of binding to antibody (selection)

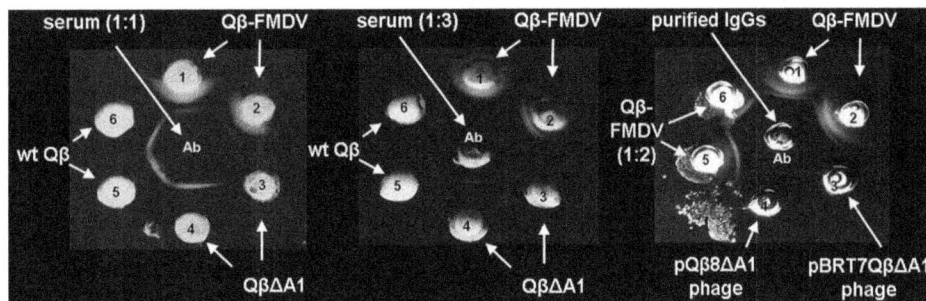

Figure 9. Ouchterlony double diffusion assay. A) Wells 1 and 2 represent Qβ-FMDV phages; wells 3 and 4 represent QβΔA1 phages; wells 5 and 6 represent wild-type Qβ; center well contains polyclonal serum from immunized guinea pig (labeled "Ab"). B) Same as panel A but with 1/3 of the serum concentration. C) Wells 1 and 2 Qβ-FMDV are the same as panel A; wells 5 and 6 contain half the phage titer of wells 1 and 2; well 3 represents phages from pBRT7QβΔA1 and well 4 represents phages from pQβ8ΔA1; center well contains IgGs purified from serum panel A and B (labeled "Ab"). The line of precipitation is visible as a white haze forming a half-circle around some of the wells in the experiments.

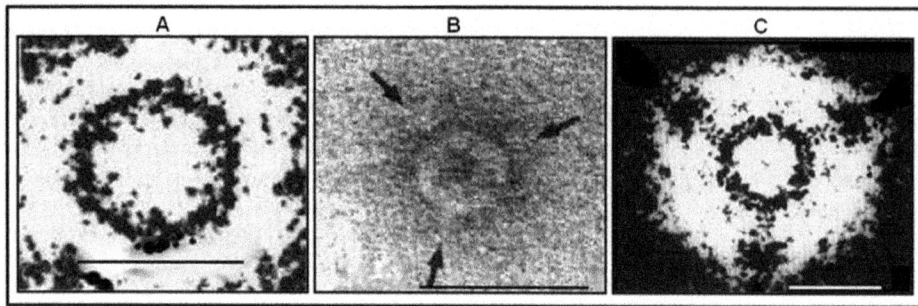

Figure 10. Field light micrograph of modified Qβ. A) Unlabelled Qβ virion: the original particle projection obtained with conventional transmission electron microscopy of a negatively stained sample was treated by Marham rotation 3 times 120\grad\intervals and printed at reversed contrast; Magnification Bar: 50 nm. B) Negatively stained Qβ, modified at A1 gene products by additional of FMDV G-H loop motif decorating the corners by IgG of mAb SD6 against VP1 G-H loop motif at 120\grad\intervals (arrows); Magnification Bar 25 nm. C) Phage particles projection depicted in B was treated by Makham rotation and printed at reversed contrast; arrows showing antibody against integrin motif attached to the corners of the virus particle; Magnification 25 nm.

and being amplified without conventional acidic elution of the phage. The phages were selected by the A1 protein extension which does not hinder adsorption via the pilus, allowing the phage to inject RNA. To mimic the natural infection of Qβ during panning, prevents acidic elution and neutralization steps previously needed before further enrichment and amplification [40–42]. This newly developed panning method can be exploited in new antibody selection from a pool of mRNA since it is very similar to the natural one without intervention of rough chemicals like in the case of most DNA phage display technologies.

Serum from guinea pigs immunized with Qβ-FMDV exhibited cross reactivity with intact FMDV as well as the hybrid Qβ phage displaying the G-H loop in a qualitative Ouchterlony assay. The fact that Qβ exist as quasi species with very high mutation rate [43–48] is a double advantage for Qβ-FMDV hybrid phages which contain a pool of antigens (quasispecies with mutant spectra) important for immunization and potential vaccine. A pool of different antigen FMDV G-H loop strains can be exposed on Qβ surface and used as vaccine.

In conclusion, we have developed a peptide library display system using the Qβ RNA-coliphage that efficiently mimics evolutionary adaptation and affinity maturation. A randomized G-H loop of the FMDV VP1 protein was exposed on the exterior surface of Qβ and selected against the G-H loop mAb SD6. Guinea pig serum immunized with the hybrid phages (Qβ-FMDV) contained immunoglobulin specific to FMDV and the Qβ-FMDV hybrid phages. These hybrid phages could principally serve as good candidates for FMDV vaccine development. Robust viability and infectivity was achieved with a C-terminal A1 deletion that maintained 12 copies per virion. Current size limitations of display are ~74 amino acid peptide/protein domain since larger domains (e.g. GFP) could not be displayed. With further optimization of A1-appended sequences, Qβ display of larger protein domains may eventually prove possible. The Qβ *in vitro* evolution platform we describe here may be readily adapted for the development of nanotechnology including novel biosensors, therapeutics, immunization reagents or crystallization scaffolds.

Table 2. Table adapted from Drake [29].

Species name	Genome size (bases)	Target	Mutation rate (per base)[a]	Mutation rate (per genome)
N. crassa	4.19×10^7	ad-3AB, mtr	7.2×10^{-11}	0.0030
S. cerevisiae	1.38×10^7	URA3, CAN1	2.2×10^{-10}	0.0031
E. coli	4.70×10^6	lacI, hisGDCBHAFE	4.6×10^{-10}	0.0022
bacteriophage M13	6.41×10^3	lacZα	7.2×10^{-7}	0.0046
bacteriophage λ	4.85×10^4	cI	7.7×10^{-8}	0.0038
bacteriophage T2	1.60×10^5	rII	2.7×10^{-8}	0.0043
bacteriophage T4	1.66×10^5	rII	2.0×10^{-8}	0.0033
MNV11RNA[b]	86	itself	3×10^{-4}	0.026
bacteriophage Qβ[b]	4.2×10^3	replicase	3×10^{-4}	1.9
Taq polymerase	n/a	n/a	2×10^{-5}	n/a

[a]In the cases where multiple targets were measured, the average is presented.
[b]Taken from Domingo [48].

Acknowledgments and Dedication

This paper is dedicated to the memory of Prof. Dr. Christof K Biebricher of the Max-Planck-Institute of Biophysical Chemistry of Gottingen, Germany, who was an initial sponsor of this research project. Our thanks and sincere appreciation to Professor Esteban Domingo, for the SD6 mAb, neutralization assay, and the serum production against Qβ-FMDV. The authors are grateful to Professor Weber for pBRT7Qβ and to Professor Alexander Chetverin for pQβ7 and pQβ8 plasmids.

Author Contributions

Conceived and designed the experiments: ABW CS. Performed the experiments: ABW CS. Analyzed the data: ABW CS SGA. Contributed reagents/materials/analysis tools: ABW SGA. Wrote the paper: ABW SGA.

References

1. Smith GP (1985) Filamentous fusion phage: novel expression vectors that display cloned antigens on the virion surface. Science 234: 211–212.
2. Smith GP, Scott JK (1993) Libraries of peptides and proteins displayed on filamentous phage. Meth Enzymol 217: 228–257.
3. Smith GP, Petrenko VA (1997) Phage display. Chem Rev 97: 391–410
4. Petrenko VA, Smith GP (2000) Phages from landscape libraries as substitute antibodies. Protein Eng 13: 589–592.
5. Rakonjac J, Bennett NJ, Spagnuolo J, Gagic D, Russel M (2011) Filamentous bacteriophage: biology, phage display and nanotechnology application. Curr Issues Mol Biol 13: 51–76.
6. Marvin DA (1998) Filamentous phage structure, infection and assembly. Curr Opin Struct Biol 8: 150–158.
7. Domingo E, Holland JJ (1997) RNA virus mutations and fitness for survival. Annu Rev Microbiol 51: 151–178.
8. Shaklee PN, Miglietta JJ, Palmenberg AC, Kaesberg P (1988) Infectious positive- and negative-strand transcript RNAs from bacteriophage Qβ cDNA clones. Virology 163: 209–213.
9. Furuse K, Osawa K, Kawashiro J, Tanaka R, Ozawa A et al. (1983) Bacteriophage distribution in human faeces: continuous survey of healthy subjects and patients with internal and leukaemic diseases. J Gen Virol 64: 2039–2043.
10. Furuse K (1987) Distribution of coliphage in the general environment: general considerations. In Phage Ecology, pp. 87–124. Edited by S. M. . Goyal, C. P. Gerba & G. . Bitton. New York: Wiley.
11. Bollback JP, Huelsenbeck JP (2001) Single-stranded RNA bacteriophage (family Leviviridae). J Mol Evol 52: 117–128.
12. Brown F (1999) Foot-and-mouth disease and beyond: vaccine design, past, present and future. Arch Virol 15: 179–188.
13. Kahn S, Geale DW, Kitching PR, Bouffard A, Allard DG et al. (2002) Vaccination against foot-and-mouth disease: the implications for Canada. Can Vet J 43: 349–354.
14. Logan D, Abu-Ghazaleh R, Blakemor W, Curry S, Jackson T (1993) Structure of a major immunogenic site on foot-and-mouth disease virus. Nature 362: 566–568.
15. Domingo E, Verdaguer N, Ochoa WF, Ruiz-Jarabo CM, Sevilla N et al. (1999) Biochemical and structural studies with neutralizing antibodies raised against foot-and-mouth disease virus. Virus Res 62: 169–175.
16. Verdaguer N, Sevilla N, Valero ML, Stuart D, Brocchi E et al. (1998) A similar pattern of interaction for different antibodies with a major antigenic site of foot and mouth disease virus: implications for intratypic antigenic variation. J Virol 72: 739–748.
17. Brown F, Cartwright B (1961) Dissociation of foot-and-mouth disease virus into its nucleic acid and protein components. Nature 192: 1163–1164.
18. Blumenthal T, Carmichael GC (1979) RNA replication: function and structure of Qβ replicase. Annu Rev Biochm 48: 525–548.
19. Weber H, Konigsberg W (1975) Proteins of RNA phages. In RNA Phages, pp 51–84. Edited by N. D. . Zinder. Cold Spring Harbor, New York: Cold Spring Harbor Laboratory.
20. Hofstetter H, Monstein HJ, Weissmann C (1974) The read-through protein A₁ is essential for the formation of viable Qβ particles. Biochim Biophs Acta 374: 238–251.
21. Barrera I, Schuppli D, Sogo JM, Weber H (1993) Different mechanisms of recognition of bacteriophage Qβ plus and minus-strand RNAs by Qβ replicase. J Mol Biol 232: 512–521.
22. Taniguchi T, Palmiri M, Weissmann C (1978) Qβ DNA-containing hybrid plasmids giving rise to Qβ phage formation in the bacteria host. Nature 274: 223–228.
23. Adam MH (1959) In: Bacteriophages, pp. 473–490. Inter-Science Publishers: New York.
24. Pace NR, Spiegelman S (1966) In vitro synthesis of an infectious mutant RNA with a normal RNA replicase. Science 153: 64–67.
25. Yamamoto KR, Alberts BM, Benzinger R, Lawhorne L, Treiber G (1970) Rapid bacteriophage sedimentation in the presence of polyethylene glycol and its application to large-scale virus purification. Virology 40: 734–744.
26. Gschwender HH, Hofschneider PH (1969) Lysis inhibition of φX174-, M12 and Q β-infected Escherichia coli bacteria by magnesium ions. Biochim Biophys Acta 190: 454–459.
27. Ouchterlony O, Nilsson LA (1978) Immunodiffusion and immunoelectrophoresis. In: Handbook of experimental immunity. Edited by, Weir, D. M., 3rd Edition, Oxford: Blackwell Scientific Publication.
28. Azzazy HME, Highsmith Jr EW (2002) Phage display technology: clinical applications and recent innovations. Clininical Biochemistry 35: 425–445.
29. Drake JW (1991) A constant rate of spontaneous mutation in DNA-based microbes. Proceedings of the National Academy of Sciences 88(16): 7160–7164.
30. Ferrer-Orta C, Arias A, Escarmis C, Verdaguer N (2006) A comparison of viral RNA-dependent RNA polymerases. Curr Opin Struct Biol 16: 27–34.
31. Vasiljeva I, Kolzlovska T, Cielens I, Strelnnikova A, Kazaks A et al. (1998) Mosaic Qβ coats as a new presentation model. FEBS Letters 431: 7–11.
32. Kerri M, Titball RW (1996) Transformation of Burkholderia pseudomallei by electroporation. Anal Biochem 242: 73–76.
33. Kozlovska TM, Cielens I, Vasiljeva I, Strelnikova A, Kazaks A et al. (1996) RNA phage Qβ coat protein as a carrier for foreign epitopes. Intervirology 39(1–2): 9–15.
34. Van Meerten D, Olsthoorn RCL, van Duin J, Verhaert RM D (2001) Peptide display on live MS2 phage: restriction at the RNA genome level. J Gen Virol 82: 1797–1805.
35. Beekwilder MJ, Nieuwenhuizen R, van Duin J (1995) Secondary structure model for the last two domains of single-stranded RNA phage Qβ. J Mol Biol 247: 903–917.
36. Long D, Abu-Ghazaleh R, Blakemore W, Curry S, Jackson T et al. (1993) Structure of a major immunogenic site on foot-and-mouth disease virus. Nature 362: 566–568.
37. Rumnieks J, Kaspars T (2011) Crystal structure of the read-through domain from bacteriophage Qβ A1 protein. Protein Science 20: 1707–1712.
38. Weiner AM, Weber K (1971) Natural read-through at the UGA termination signal of Qβ coat protein cistron. Nat New Biol 234: 206–209.
39. van Rooy I, Hennink WE, Storm G, Schiffelers RM, Mastrobattista E (2012) Attaching the phage display-selected GLA peptide to liposomes: factors influencing target binding. Eur J Pharm Sci 45: 330–335.
40. Jenkins GM, Rambaut A, Pybus OG, Holmes E (2002) Rates of molecular evolution in RNA viruses: a quantitative phylogenetic analysis. J Mol Evol 54: 156–165.
41. Scott JK, Smith GP (1990) Searching for peptides ligands with an epitope library. Science 349: 386–390.
42. Beer M, Liu C-Q (2012) Panning of a phage display library against a synthetic capsule for peptide ligands that bind to the native capsule of Bacillus anthracis. Plos One 7: e45472. doi:10.1371/journal.pone.0045472
43. Eigen M (1971) Selforganization of matter and evolution of biological macromolecules. Naturwissenschaften 58: 465–523.
44. Eigen M, Schuster P (1977) The hypercycle – a principal of natural selforganisation. Naturwissenschaften 64: 541–565.
45. Eigen M, McCaskill J, Schuster P (1989) The molecular quasispecies. In Prigogine, I. & Rice, S. A. Edited by, Adv Chem Phys 75: 149–263 John Wiley & Sons, Inc.
46. Steinhauer D, Domingo E, Holland JJ (1992) Lack of evidence for proofreading mechanisms associated with an RNA virus polymarase. Gene 122: 281–288.
47. Schuster P, Stadler PF (1994) Landscapes: complex optimization problems and biopolymer structures. Comput Chem 18: 295–324.
48. Domingo E, Biebricher CK, Eigen M, Holland JJ (2001) Quasispecies and RNA virus evolution: principles and consequences (p. 173). Austin: Landes Bioscience.

27

Proteome Folding Kinetics Is Limited by Protein Halflife

Taisong Zou[1], Nickolas Williams[2], S. Banu Ozkan[1], Kingshuk Ghosh[2]*

1 Center for Biological Physics, Department of Physics, Arizona State University, Tempe, Arizona, United States of America, **2** Department of Physics and Astronomy, University of Denver, Denver, Colorado, United States of America

Abstract

How heterogeneous are proteome folding timescales and what physical principles, if any, dictate its limits? We answer this by predicting copy number weighted folding speed distribution – using the native topology – for E.coli and Yeast proteome. E.coli and Yeast proteomes yield very similar distributions with average folding times of 100 milliseconds and 170 milliseconds, respectively. The topology-based folding time distribution is well described by a diffusion-drift mutation model on a flat-fitness landscape in free energy barrier between two boundaries: i) the lowest barrier height determined by the upper limit of folding speed and ii) the highest barrier height governed by the lower speed limit of folding. While the fastest time scale of the distribution is near the experimentally measured speed limit of 1 microsecond (typical of barrier-less folders), we find the slowest folding time to be around seconds (≈ 8 seconds for Yeast distribution), approximately an order of magnitude less than the fastest halflife (approximately 2 minutes) in the Yeast proteome. This separation of timescale implies even the fastest degrading protein will have moderately high (96%) probability of folding before degradation. The overall agreement with the flat-fitness landscape model further hints that proteome folding times did not undergo additional major selection pressures – to make proteins fold faster – other than the primary requirement to "sufficiently beat the clock" against its lifetime. Direct comparison between the predicted folding time and experimentally measured halflife further shows 99% of the proteome have a folding time less than their corresponding lifetime. These two findings together suggest that proteome folding kinetics may be bounded by protein halflife.

Editor: Emanuele Paci, University of Leeds, United Kingdom

Funding: KG acknowledges support from NSF (award number 1149992), and TZ and SBO acknowledge ASU-CLAS funding. The funders had no role in study design, data collection and analysis, decision to publish, or preparation of the manuscript.

Competing Interests: The authors have declared that no competing interests exist.

* Email: kghosh@du.edu

Introduction

Diverse pool of protein sequences give rise to an astonishing degree of heterogeneity in the biophysical properties across the proteome. This raises a fundamental question: how heterogeneous is the proteome? Recent work showed biophysical properties have broad distributions across the proteome and their consequences at the phenotypic level [1–5]. While sequence variation alone would lead to such diverse biophysical properties, there are other features of the cellular environment – for example protein abundance, role of chaperones, co-translational folding – that can further influence these distributions. Protein copy number – although neglected in the earlier calculations of distributions – in particular can play a crucial role due to a possible correlation with biophysical properties such as folding stability [6]. It has been well established that highly abundant proteins are slowly mutating [7,8]. The reason behind this negative correlation is believed to be the selection pressure against cytotoxicity of misfolded proteins arising due to lower stability. Rules of protein biophysics has been used to quantitatively establish the relation between abundance and stability [6,8]. On the other hand, it is believed that there may be a possible correlation between stability and folding speed [9–11]. Thus, it is tempting to hypothesize that protein abundance and folding speed may be related as well. A natural question arises – how does protein abundance alter, if at all, the folding time distribution? Without *a priori* knowledge of the effect of protein abundance on the folding time distribution, it is imperative that any attempt to predict the folding time distribution of a proteome should consider the effect of abundance as well.

Learning about the extent of heterogeneity in biophysical properties across the proteome in itself is a fundamental question – leading further inquires on the details of the distribution. For example in case of folding time distribution, what are the lower and upper speed limits? What physical principle dictates these limits? What is the peak value, if any, of the distribution? Is there a limiting behavior due to competition with other time scales such as diffusion, protein synthesis, degradation? If kinetic stability [12] – introducing higher barrier height while keeping the same value for the free energy difference between the folded and the unfolded state – is a strategy cells use to minimize exposure to unfolded states to avoid lethal effects of aggregation or degradation [13], do we expect proteomes to be biased towards higher folding times? And if so, how do these timescales compare with protein halflife, in other words is the proteome folding timescale still able to beat the degradation clock with an increased barrier height? While outpacing degradation appears to be important, are there any other selection pressures that may have influenced proteome folding kinetics? Furthermore, how do these distributions vary across different kingdoms of life – for example between Escherichia coli (E.coli) and Yeast – or is there an universality in the shape of the distribution? In this article, we attempt to determine proteome folding kinetics distribution and address some of these fundamental questions.

Materials and Methods

Determining the folding speed of a protein

Plaxco, Simons, Baker [14] made the observation that relative contact order (CO), a metric based on the native topology of the protein, correlates well with the folding speed measured *in vitro*. CO is defined as the average residue separation – normalized by the chain length – of atomic contacts present in the native structure of the protein [14]. Since the pioneering work of Plaxco, Simons, Baker there have been numerous efforts to understand its implication [15] and establish the role of other native-centric metric [16–21] and their relative performances to predict the folding speed of proteins using native structure [18,20,21]. One such effort has shown absolute contact order (ACO) – defined as the product of CO and the chain length – predicts folding speeds more accurately than CO for bigger set of proteins [16]. In a nutshell, all these different metrics provide a prescription to predict the folding speed of a protein with the knowledge of the native structure alone. We utilize this powerful idea to predict the folding time distribution for proteins in the proteome for which the exact (or highly homologous) native structures are known. Recent work by Rustad and Ghosh [21] has provided a first principle explanation – employing polymer physics arguments – for the observed correlation between absolute contact order (ACO) [16] and folding speed. Furthermore, within a perturbative scheme, the work has proposed an extension of the metric (ACO) that captures the effect of different loop topologies [21]. This new metric, minor variation of ACO, provides slight improvement over ACO when benchmarked against the largest set (116 proteins) of *in vitro* folding speed data. We use this new modified metric, instead of ACO, to predict the folding speed from the native structure of the protein. For a given protein, we predict folding speeds for different domains, assuming each domain folds independently. Since the domain with the slowest folding speed is rate limiting, we use the folding speed of the slowest folding domain to be the folding speed of the protein.

Curating the fraction of proteome that have both the structure and abundance data available

In order to predict folding speed, as described above, we need the information about the native structures of proteins in the proteome. We collect proteins from the Yeast and E.coli proteome for which the structures of proteins are available. For the Yeast proteome we use domain assignment from Yeast resource center (YRC) database [22]. Next we perform a BLAST search of the corresponding sequences to identify the best possible match for their structures. We list only those proteins that simultaneously satisfy a minimum of 80% sequence coverage and 50% identity match. In order to predict copy number weighted folding time distribution, we gather proteins for which both the structure and abundance information are available. We cross reference the curated list of proteins with available structure, described above, against the integrated list from PaxDB database [23]. The integrated list is the most comprehensive list of protein abundance values. We choose this list to ensure maximum coverage of proteins from the proteome. This method yields a total of 755

Yeast proteins. For E.coli proteome, we follow a similar approach but use the dataset collected by O'Brien *et al.* [24]. The original dataset reported in O'Brien *et al.* categorizes proteins (and their domains) based on a single abundance scale. We cross reference the combined list against the integrated list of abundance from PaxDb [23] yielding a total of 848 E.coli proteins. In summary, our datasets (Table S1 and S2) provide the largest fraction of proteomes (in E.coli and Yeast) for which both the abundance and structural informations are now available.

Results and Discussion

Folding time distribution is heterogeneous

Copy number weighted folding speed (lnk_f, k_f being the folding speed) distributions in E.coli and Yeast show a broad range of folding speeds, from several microseconds^{-1} to minutes^{-1} (Figure 1). The fastest folding time is in the neighborhood of microseconds. This is consistent with studies on ultrafast folding proteins defining the speed limit of protein folding [21,25,26]. It is interesting to note the lower speed limit is of the order of seconds to minutes, in proximity to the scale of halflives of short-lived proteins [27]. The implication of this observation will be discussed in detail in the section below. The average folding time (τ_f) for copy number weighted distribution is calculated as

$$\ln \tau_f \approx -\langle \ln k_f \rangle = -\frac{\sum_i \ln k_{fi} N_i}{\sum_i N_i} \qquad (1)$$

where, k_{fi} and N_i are the folding speed and the copy number, respectively, of the i th protein. Average folding time without accounting for differential protein abundance levels can be obtained by simply setting $N_i = 1$. For E.coli, we find the average is approximately 100 milliseconds for copy number weighted distribution. The average remains almost unaltered when the distribution is not weighted by the protein expression level (i.e. setting $N_i = 1$, distribution not shown here). The average folding time for Yeast proteome is 170 milliseconds and 60 milliseconds for copy number weighted and unweighted distributions, respectively.

Recent work – grounded in the hypothesis of global selection against toxic effect of misfolding explaining observed correlation between abundance and evolution rate [8] – predicts highly abundant proteins are more stable [6]. Given this link between stability-abundance and *possible* interdependence between stability and folding kinetics [9–11], it is natural to expect a possible relation between abundance and folding speed as well. However, based on the results stated above, we do not see any noticeable effect of abundance on folding kinetics in E.coli. A possible explanation, among many other alternative ones, could be that the proteome can not afford to under-express slow folding proteins due to functional reasons. Furthermore, we notice a marginal slowing down of the proteome folding speed in Yeast upon weighting by protein abundance. Given the inherent uncertainties in predicting folding speed from native topology, a three-fold slowing down of the proteome is probably a very weak effect. However, if slowing down of the proteome due to copy number weighting is indeed beyond uncertainty, it may imply slow folding proteins are over-expressed for strong functional reasons despite the threat of misfolding. It may also imply the proteome is equipped with mechanisms such as chaperone-assisted folding, complex chaperone-substrate network [28] to mitigate possible deleterious effects of misfolding due to lower folding speed. As will be seen in later sections, three fold lowering of the speed around 60 millisecond timescale still allows proteins enough time to fold

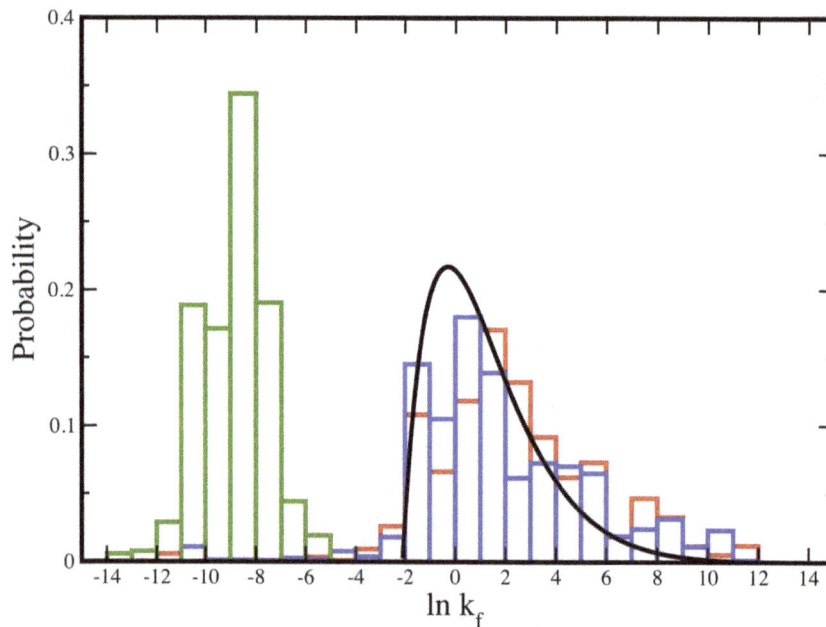

Figure 1. Folding speed (lnk_f) distribution – calculated using native topology – of E.coli (in red) and Yeast (in blue) weighted by protein copy number. The distribution of average lifetime for proteins in Yeast [27] is shown in green. The predicted folding time distribution using a diffusion-drift model (equation 5) with the boundary condition of the maximum folding time of 8 seconds is shown in black. Maximum folding time of 8 seconds was determined by best fitting Yeast distribution.

before degradation. It is interesting to note folding speed distributions in E.coli and Yeast – baring minor variations mentioned above – are very similar, indicating a universal behavior in the folding kinetics.

One caveat of our analysis is that the folding speed is predicted using models that have been benchmarked against *in vitro* folding data. However recent work, although limited, does not show significant differences between folding times measured *in vivo* and *in vitro* [29]. It is also important to note major conclusions remain the same if other metric such as ACO is used to predict the folding speed.

Diffusion-drift model of mutations on a flat-fitness landscape explains the predicted distribution of folding speed

Apart from minor differences in details, the overall shape and the range of the distributions for E.coli and Yeast are roughly similar. The universal distribution (Figure 1) of the folding speed, irrespective of the details of the species, is well explained by a diffusion-drift model of mutations altering folding free energy barrier ($\Delta G^\dagger$). Shakhnovich *et al.* [1] used a similar model to describe a universal distribution of stability (ΔG). Due to close analogy between the two models, we briefly describe the stability model first. Further details of the model can be found in the work of Shakhnovich *et al.* [1]. Their model uses diffusion - arising from random mutations - with a drift to explain the stability distribution $P(\Delta G)$. The model also imposes two boundary conditions $P(\Delta G_{min}) = P(\Delta G_{max}) = 0$ at the maximum (ΔG_{max}) and minimum (ΔG_{min}) values of allowed stability. These two constraints can be explained as follows (Figure 2A): from design perspective, it is impossible to make proteins indefinitely stable, thus imposing an upper limit on the stability, hence $P(\Delta G_{max}) = 0$. The boundary condition on the lower limit of stability, on the other hand, arises

from the requirement of minimal stability to avoid misfolding that can be lethal to the phenotype of the organism. The model assumes a flat-fitness landscape for all values of stability greater than the minimum, i.e. $\Delta G > \Delta G_{min}$. The fitness is severely compromised if stability drops below the threshold i.e. $\Delta G < \Delta G_{min}$, imposing the constraint $P(\Delta G_{min}) = 0$. Thus, the fitness landscape is 'step-like' near the threshold (see Figure 2A). The time evolution of the probability distribution of stability in this mutational model with the flat 'step-like' landscape is given by [1]

$$\frac{\partial P}{\partial t} = cP - mh\frac{\partial P}{\partial \Delta G} + \frac{m}{2}(h^2 + D)\frac{\partial^2 P}{\partial(\Delta G)^2}; \tag{2}$$

$$p(\Delta G_{max}) = p(\Delta G_{min}) = 0$$

where, c is a constant related to the birth rate of the population, m is the mutation rate per gene (or protein), h and D are the average and variance, respectively, of the distribution of stability changes upon mutation. Formally, $h = \langle \Delta\Delta G \rangle$ and $h^2 + D = \langle (\Delta\Delta G)^2 \rangle$, where $\langle ... \rangle$ denotes the average over all possible mutations and $\Delta\Delta G = \Delta G_{mutant} - \Delta G_{wt}$. The second derivative in equation 2 describes diffusion, while drift is captured by the first derivative (in the right hand side of the equation). Using the long-time limit solution $P(\Delta G, t) = \exp(\lambda t)P(\Delta G)$ [1], we require the steady state solution to be the eigenfunction of the differential equation

$$-mh\frac{\partial P}{\partial \Delta G} + \frac{m}{2}(h^2 + D)\frac{\partial^2 P}{\partial(\Delta G)^2} \tag{3}$$

subject to the boundary conditions. Thus, the steady state solution – within a normalization constant A – is given by

$$P(\Delta G) = A \exp\left(\frac{h\Delta G}{h^2 + D}\right) \sin\left(\pi \frac{\Delta G - \Delta G_{min}}{\Delta G_{max} - \Delta G_{min}}\right) \quad (4)$$

Noticing one-to-one relation between folding speed (k_f) and barrier height ($\Delta G^\dagger$), we employ similar idea to model the distribution of barrier height to ultimately predict the folding speed distribution. We use the same diffusion-drift model where mutations alter the free energy barrier of folding instead of folding stability. Analogous to the stability model, we impose two boundary conditions, $P(\Delta G_{min}^\dagger) = P(\Delta G_{max}^\dagger) = 0$, at the two extremities of the free energy barrier, $\Delta G_{min}^\dagger$ and $\Delta G_{max}^\dagger$ (see Figure 2B). On one hand it is simply impossible to make proteins that fold faster than the speed limit of folding, setting the lower limit of the barrier $\Delta G_{min}^\dagger$. On the other hand, extremely slow folding proteins – if not folded at birth – even if highly stable will not be able to fold in time before degradation. Stated differently, for functional reasons, proteins would require to fold before their lifetime (inside the cell) expires. Also, slow folding proteins would be a potential hazard due to unfolded-state induced aggregation propensity. This sets a selection pressure against slow folding proteins with extremely high barriers ($\Delta G_{max}^\dagger$). Similar to the stability model, we assume a flat-fitness landscape for $\Delta G^\dagger < \Delta G_{max}^\dagger$, with a severe drop in fitness for $\Delta G^\dagger > \Delta G_{max}^\dagger$ (Figure 2B). In reality, fitness can gradually decrease around the threshold value of $\Delta G_{max}^\dagger$. However, in order to keep the calculation simple and analogous to the work of Shakhnovich et al., we make the simplifying assumption of a 'step-like' fitness function. Thus the model assumes all proteins are subjected to a single global constraint of lifetime implying a single value of $\Delta G_{max}^\dagger$. Noticing the exact analogy between the model for stability and the barrier height, the predicted distribution for the free energy barrier can be easily obtained by replacing the stability (ΔG) by the barrier height $\Delta G^\dagger$ in equation 4. Thus,

$$P(\Delta G^\dagger) = A \exp\left(\frac{h\Delta G^\dagger}{h^2 + D}\right) \sin\left(\pi \frac{\Delta G^\dagger - \Delta G_{min}^\dagger}{\Delta G_{max}^\dagger - \Delta G_{min}^\dagger}\right) \quad (5)$$

where, A is a normalization constant, $h = \langle \Delta\Delta G^\dagger \rangle$, $h^2 + D = \langle(\Delta\Delta G^\dagger)^2\rangle$; $\Delta\Delta G^\dagger = \Delta G_{mutant}^\dagger - \Delta G_{wt}^\dagger$, and $\langle...\rangle$ denotes the average over all possible mutations of barrier height. Three parameters of the model h, D, and $\Delta G_{min}^\dagger$, can be estimated from the literature. From the dataset of 858 mutations across 24 different proteins [30], we find $h = 0.6(k_b T)$ and $h^2 + D = 1.12(k_b T)^2$; k_b is the Boltzmann constant and T is the room temperature.

The lower limit of the barrier is assumed to be zero, ($\Delta G_{min}^\dagger = 0$), consistent with barrier-less folding proteins that define the speed limit of folding [25,26].

Now we focus on the determination of $\Delta G_{max}^\dagger$. We hypothesize the lower speed limit i.e. the maximum folding time ($t_{f,max}$) – setting the upper limit of folding barrier ($\Delta G_{max}^\dagger$) – has to be less than the protein halflife ($t_{1/2}$). Experimentally reported halflife measures the time scale over which the copy number of a given protein, upon inhibition of synthesis, decreases by half [27]. This timescale does not distinguish between unfolded or folded state degradation, instead simply provides an estimate of the lifetime of a protein inside a cell. Based on this definition of halflife, it is natural to expect that proteins would be required to fold in a timescale lower than their halflife. Assuming lifetime distribution to be Poisson, average lifetime (t_l) and halflife ($t_{1/2}$) are related $t_l = t_{1/2}/\ln 2$. If the average folding time of a given protein is t_f, the probability of folding before degradation (P_{fbd}) is

$$P_{fbd} = \frac{1}{1 + t_f/t_l}. \quad (6)$$

Clearly, if $t_f >> t_l$ most of the proteins will be degraded before folding. At the other extreme if $t_l >> t_f$, almost all of the proteins will be folded before degradation. It is also important to note, even if $t_l \approx t_f$, nearly 50% of the proteins will be degraded before folding which is not very efficient either. Thus we do not assume the boundary condition due to the maximum folding time to be exactly equal to the average lifetime of the fastest degrading protein. Instead, we fit topology-based folding speed distribution to determine the maximum allowed folding time for the diffusion-drift model. We find the best fit value of $\Delta G_{max}^\dagger$ to be $16k_b T$, yielding the maximum folding time $t_{f,max} \approx 8$ seconds (for Yeast distribution). In the above we used the speed-barrier height relation $k_f = k_0 \exp(-\Delta G^\dagger/k_b T)$ and $k_0 \approx 1\,\text{microsecond}^{-1}$. The numerical value of k_0 is consistent with several estimates of folding speed limit [21,25,26,31,32].

Figure 1 shows the best fit distribution is in reasonable agreement with the Yeast distribution. The implication of this is threefold: i) the diffusion-drift model provides an independent test of our topology-based model prediction for the distribution of folding kinetics; ii) $t_f = t_{f,max} = 8$ seconds and $t_l = 2/.69 = 3$ min

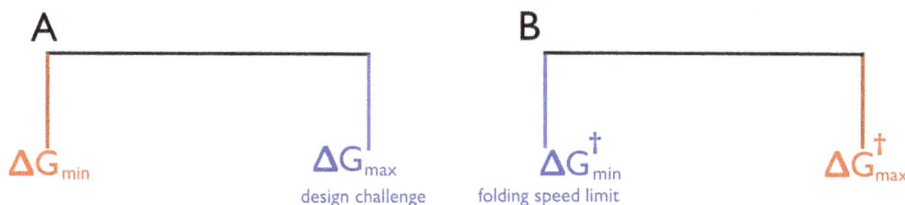

Figure 2. A) Accessible range in stability (ΔG increasing towards right) is shown between blue and red lines. Black line shows the flat-fitness landscape for all values of stability greater than the minimum; i.e. $\Delta G > \Delta G_{min}$, with the red line showing the drop in fitness when stability is lower than the minimum due to cytotoxic effects from aggregation/misfolding. Blue line shows the upper limit of stability (ΔG_{max}) due to design challenge. B) Accessible range in the folding free energy barrier height ($\Delta G^\dagger$ increasing to the right) between blue and red lines. Black line shows the flat-fitness landscape for all values of barrier heights less than the maximum allowed i.e. $\Delta G^\dagger < \Delta G_{max}^\dagger$, with the red line showing the compromised fitness when the barrier height is greater than the maximum leading to slow folding proteins, prone to aggregation and degradation. Blue line shows it is not possible to create proteins faster than the speed limit of folding set by barrier-less folders.

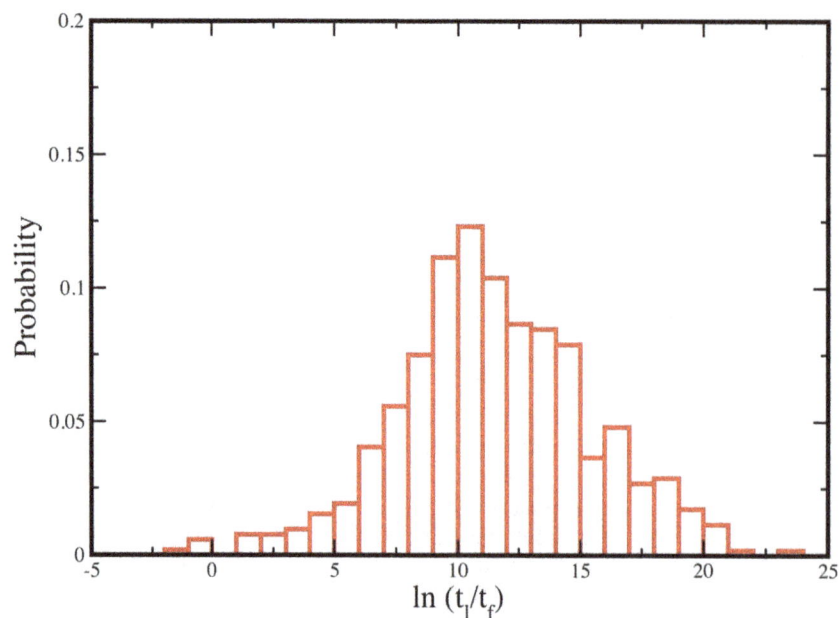

Figure 3. Distribution of the ratio of protein lifetime and protein folding time.

(for the fastest degrading protein in Yeast) argues even the fastest degrading protein in Yeast has roughly 96% probability of folding before the expiration of its lifetime. This supports the hypothesis that the slowest folding processes may be constrained by protein lifetime allowing sufficient chance for proteins to fold before degradation; iii) the assumption of flat-fitness landscape is reasonable. This implies proteome folding kinetics is not subjected to any major selection criteria to make it faster other than the primary requirement of staying sufficiently below the maximum allowed timescale set by protein halflife. However, it can not be ruled out that there are other secondary pressures to alter folding kinetics that can further improve the agreement between the diffusion-drift and topology-based model of folding kinetics. We have also fitted E.coli speed distribution with the diffusion-drift model, yielding $t_{f,max} = 2$ seconds (data not shown). However we do not provide details since a corresponding comparison with lifetime is not possible due to lack of lifetime information for E.coli proteome.

Diffusion-drift mutation model makes further prediction on the upper limit of the number of mutations per portion of the genome encoding essential genes per replication. As mentioned above, long time limit solution is given by $P(\Delta G^\dagger, t) = \exp(\lambda t) P(\Delta G^\dagger)$. In order for the population to survive, we require $\lambda \geq 0$. This requirement sets an upper limit on the number of mutations per portion of the genome encoding essential genes per replication. This limit can be obtained in terms of h, D, $\Delta G^\dagger_{max} - \Delta G^\dagger_{min}$ (see equation 8 from [1] for details). Using the values for the parameters noted above, our estimate for the upper limit is ≈ 5.5. This is indeed close to $5.7 (\approx 6)$ predicted by Shakhnovich et al. from the consideration of the stability distribution and matches well with experiments [1].

Proteome folding time is lower than the lifetime

The analysis above provides indirect support to the hypothesis that proteome lifetime may limit folding kinetics. We further test this hypothesis by directly plotting the distribution of average lifetime (t_l converted from experimentally measured halflife) values

[27] for Yeast proteome (Figure 1 in green). It is evident that the folding time and lifetime distributions are well separated. However, we also notice slight overlap between the two time scales at the boundary. This observation, at first, may indicate existence of some proteins for which the folding time may be higher than the lifetime, implying a possible contradiction to our hypothesis that protein folding is faster than degradation. In order to further test the validity of our hypothesis, we directly compare these measured lifetime values [27] and predicted folding times for each individual proteins. We select proteins from our list – used to predict the folding time in the Yeast proteome – for which lifetimes are known [27]. We compute the ratio of the lifetime and folding time for each protein in our dataset (Table S3). Figure 3 shows the distribution of the ratios of these two time scales. We find less than 1% of the proteome (4 out of 520 proteins in our list) has a folding time higher than their lifetime. The overwhelming number of proteins with a lower folding time than their lifetime, further supports the hypothesis that the lower limit of protein folding speed is indeed bounded by protein lifetime.

Although 1% is a minor fraction, one can further reason these possible exceptions. First, chaperones can play an important role to facilitate folding [5,33–35]. Chaperones can favorably alter the ratio of lifetime and folding time to help proteins escape the selection against degradation. Second, it is possible that the kinetics of the slowest folding domains are altered due to possible interdependence between multiple domains [36], an aspect not included in our model. Third, it should also be noted that the reported halflife in the work of O'Shea et al. [27] has an inherent uncertainty of a factor of two. In order to determine if any of the reasons mentioned above may be responsible, we further studied in detail the four proteins (corresponding open reading frames of YER070W, YFL041W, YJL200C and YLR304C) for which the predicted folding time is higher than the lifetime. We find three of these proteins (YFL041W, YJL200C and YLR304C) have folding time within twice their average lifetime, within the measurement uncertainty [27]. The only protein that has significantly higher folding time (fourfold higher than the lifetime) is YER070W with 80% probability of degradation before folding. However, it is interesting to note that this protein is also one of the highly abundant (top 5%) protein in the Yeast proteome [23]. The high abundance is likely due to its important biological function of facilitating synthesis of DNA. Furthermore, high abundance may offset the effect of slow folding ensuring enough copies (in absolute numbers) of the protein are present inside the cell despite the low probability of folding before degradation. Moreover, this protein has eighteen chaperone interaction partners as reported in ChaperoneDB database [28]. While the exact role of such unusually high number of chaperones to folding speed is not known at this time, it may be possible that some specific chaperones from this list or the entire chaperone network – in concert – facilitate folding of this protein in reasonable time scale to lower the burden of degradation.

Conclusions

In summary, we predict the folding time distributions for E.coli and Yeast proteome weighted by protein expression levels. We make four key observations. First, we notice E.coli and Yeast have broad distributions of folding speed with roughly similar features and ranges of the distribution. Second, the underlying distribution is reasonably explained by an independent model of diffusion-drift of mutations in free energy barrier on a "flat-fitness landscape" with two boundary conditions. While the boundary at the upper speed limit (minimum folding time) is determined by barrierless folding proteins, we find the maximum folding time to be $t_{f,max} \approx 8$ seconds (for Yeast proteome). Comparing this with the average lifetime of the fastest degrading protein ($t_l = 3$ min), we find even the fastest degrading protein in Yeast has roughly 96% probability of folding before the expiration of its lifetime. This supports the hypothesis that the slowest folding time may be bounded by protein lifetime allowing sufficient chance for proteins to fold before degradation. Third, direct comparison between measured lifetime and predicted folding time shows 99% of the proteome has a folding time less than the corresponding lifetime. Finally, the reasonable agreement between the topology-based speed distribution and the diffusion-drift model on "flat-fitness

landscape" further justifies the assumption of flat-fitness landscape. This implies the primary selection pressure for proteome folding kinetics is perhaps to outrun degradation only.

Supporting Information

Table S1 Dataset of folding time and abundance for E.coli proteome. First column reports protein name as reported in O'Brien et al. [24]; second column reports $\ln k_f$ where k_f is the folding speed (in the units of s^{-1}) for the slowest folding domain; third column reports abundance value (in ppm) from PaxDB Integrated list [23].

Table S2 Dataset of folding time and abundance for Yeast proteome. First column reports Open Reading Frame as reported in YRC [22]; second column reports $\ln k_f$ where k_f is the folding speed for the slowest folding domain in the units of s^{-1}; third column reports abundance value (in ppm) from PaxDB Integrated list [23].

Table S3 Dataset of folding time and halflife for Yeast proteome. First column reports Open Reading Frame as reported in YRC [22]; second column reports halflife (in minutes) from O'Shea et al. [27]; third column reports $\ln k_f$ where k_f is the folding speed for the slowest folding domain in the units of s^{-1}.

Acknowledgments

TZ and SBA acknowledge XSede for CPU time. We dedicate the paper in the memory of Nickolas Williams.

Author Contributions

Conceived and designed the experiments: SBO KG. Performed the experiments: TZ NW KG. Analyzed the data: TZ SBO KG. Contributed reagents/materials/analysis tools: TZ. Wrote the paper: SBO KG.

References

1. Zeldovich K, Chen P, Shakhnovich E (2007) Protein stability imposes limits on organism complexity and speed of molecular evolution. Proc Natl Acad Sci 104: 16152–16157.
2. Ghosh K, Dill K (2010) Cellular proteomes have broad distributions of protein stability. Biophys J 99: 3996–4002.
3. Sawle L, Ghosh K (2011) How do thermophilic proteins and proteomes withstand high temperature? Biophys J 101: 217–227.
4. Dill K, Ghosh K, Schmit J (2011) Physical limits of cells and proteomes. Biophys J 108: 17876.
5. Rollins G, Dill K (2014) General mechanism of two-state protein folding kinetics. J Am Chem Soc 136: 11420–11427.
6. Serohijos A, Lee S, Shakhnovich E (2013) Highly abundant proteins favor more stable 3d structures in yeast. Biophys J 104: L1–3.
7. Drummond D, Bloom J, Adami C, Wilke C, Arnond F (2005) Why highly expressed proteins evolve slowly. Proc Natl Acad Sci 102: 14338–14343.
8. Serohijos A, Rimas Z, Shakhnovich E (2012) Protein biophysics explains why highly abundant proteins evolve slowly. Cell Reports 2: 249–256.
9. Clarke J, Cota E, Fowler S, Hamill S (1999) Folding studies of immunoglobulin-like beta-sandwich proteins suggest that they share a common folding pathway. Structure 7: 1145–1153.
10. Dinner A, Karplus M (2001) The roles of stability and contact order in determining protein folding rates. Nature Structural Biology 8: 21–22.
11. Wang T, Zhu Y, Gai F (2004) Folding of a three-helix bundle at the folding speed limit. J Phys Chem B 108: 3694–3697.
12. Baker D, Agard D (1994) Kinetics versus thermodynamics in protein folding. Biochemistry 33: 7505–7509.
13. Braselmann E, Chaney J, Clark P (2013) Folding the proteome. Trends in Biochemical Sciences 38: 337–344.
14. Plaxco K, Simons K, Baker D (1998) Contact order, transition state placement and the refolding rates of single domain proteins. J Mol Biol 277: 985.
15. Chan H (1998) Protein folding: Matching speed and locality. Nature 392: 761–763.
16. Ivankov D, Garbuzynskiy S, Alm E, Plaxco K, Baker D, et al. (2003) Contact order revisited: Influence of protein size on the folding rate. Protein Sci 12: 2057–2062.
17. Gromiha M, Selvaraj S (2001) Comparison between long-range interactions and contact order in determining the folding rate of two-state proteins: application of long-range order to folding rate prediction. J Mol Biol 310: 27–32.
18. Ouyang Z, Liang J (2008) Predicting protein folding rates from geometric contact and amino acid sequence. Protein Sci 17: 1256.
19. De Sancho D, Munoz V (2011) Integrated prediction of protein folding and unfolding rates from only size and structural class. Phys Chem Chem Phys 13: 17030–17043.
20. Zou T, Ozkan S (2011) Local and non-local native topologies reveal the underlying folding landscape of proteins. Physical Biology 8: 066011.
21. Rustad M, Ghosh K (2012) Why and how does native topology dictate the folding speed of a protein? J Chem Phys 137: 205104.
22. Drew K, Winters P, Butterfoss G, Berstis V, Uplinger K, et al. (2011) The proteome folding project: proteome-scale prediction of structure and function. Genome Res 21: 1981–1994.
23. Wang M, Weiss M, Simonovic M, Haertinger G, Schrimpf S, et al. (2012) Paxdb, a database of protein abundance averages across all three domains of life. Mol Cell Proteomics 11: 492–500.
24. Ciryam P, Morimoto R, Vendruscolo M, Dobson C, O'Brien E (2013) In vivo translation rates can substantially delay the cotranslational folding of the e. coli cytosolic proteome. Proc Natl Acad Sci 110: E132–140.
25. Hagen S, Hofrichter J, Szabo A, Eaton W (1996) Diffusion-limited contact formation in unfolded cytochrome c: estimating the maximum rate of protein folding. Prot Natl Acad Sci 93: 11615–17.
26. Ghosh K, Ozkan S, Dill K (2007) The ultimate speed limit to protein folding is conformational searching. J Am Chem Soc 129: 11920–11927.

27. Belle A, Tanay A, Bitincka L, Shamir R, O'Shea E (2006) Quantification of protein half-lives in the budding yeast proteome. Proc Natl Acad Sci 103: 13004–13009.

28. Gong Y, Kakihara Y, Krogan N, Greenblatt J, Emili A, et al. (2009) An atlas of chaperoneprotein interactions in saccharomyces cerevisiae: implications to protein folding pathways in the cell. Molecular Systems Biology 5: 275.

29. Guo M, Xu Y, Gruebele M (2012) Temperature dependence of protein folding kinetics in living cells. Proc Natl Acad Sci 109: 17863–17867.

30. Naganathan A, Munoz V (2010) Insights into protein folding mechanisms from large scale analysis of mutational effects. Proc Natl Acad Sci 107: 8611–8616.

31. Yang W, Gruebele M (2003) Folding at the speed limit. Nature 423: 193–197.

32. Changbong H, Thirumalai D (2012) Chain length determines the folding rates of RNA. Biophys J 102: L11–L13.

33. Mashaghi A, Kramer G, Bechtluft P, Zachmann-Brand B, Driessen A, et al. (2013) Reshaping of the conformational search of a protein by the chaperone trigger factor. Nature 500: 98–101.

34. Brinker A, Pfeifer G, Kerner M, Naylor D, Hartl F, et al. (2001) Dual function of protein confinement in chaperonin-assisted protein folding. Cell 107: 223–233.

35. Cuyle J, Texter F, Ashcroft A, Masselos D, Robinson C, et al. (1999) Groel accelerates the refolding of hen lysozyme without changing its folding mechanism. Nature 6: 683–690.

36. Batey S, Clarke J (2006) Apparent cooperativity in the folding of multidomain proteins depends on the relative rates of folding of the constituent domains. Proc Natl Acad Sci 103: 18113–8.

28

Bioinformatic Analysis Reveals Genome Size Reduction and the Emergence of Tyrosine Phosphorylation Site in the Movement Protein of New World Bipartite Begomoviruses

Eric S. Ho[1]*, Joan Kuchie[2], Siobain Duffy[3]

1 Department of Biology, Lafayette College, Easton, Pennsylvania, United States of America, **2** New Jersey City University, Jersey City, New Jersey, United States of America, **3** Department of Ecology, Evolution and Natural Resources, Rutgers University, New Brunswick, New Jersey, United States of America

Abstract

Begomovirus (genus Begomovirus, family *Geminiviridae*) infection is devastating to a wide variety of agricultural crops including tomato, squash, and cassava. Thus, understanding the replication and adaptation of begomoviruses has important translational value in alleviating substantial economic loss, particularly in developing countries. The bipartite genome of begomoviruses prevalent in the New World and their counterparts in the Old World share a high degree of genome homology except for a partially overlapping reading frame encoding the pre-coat protein (PCP, or AV2). PCP contributes to the essential functions of intercellular movement and suppression of host RNA silencing, but it is only present in the Old World viruses. In this study, we analyzed a set of non-redundant bipartite begomovirus genomes originating from the Old World (N = 28) and the New World (N = 65). Our bioinformatic analysis suggests ~120 nucleotides were deleted from PCP's proximal promoter region that may have contributed to its loss in the New World viruses. Consequently, genomes of the New World viruses are smaller than the Old World counterparts, possibly compensating for the loss of the intercellular movement functions of PCP. Additionally, we detected substantial purifying selection on a portion of the New World DNA-B movement protein (MP, or BC1). Further analysis of the New World MP gene revealed the emergence of a putative tyrosine phosphorylation site, which likely explains the increased purifying selection in that region. These findings provide important information about the strategies adopted by bipartite begomoviruses in adapting to new environment and suggest future *in planta* experiments.

Editor: Darren P. Martin, Institute of Infectious Disease and Molecular Medicine, South Africa

Funding: ESH is supported by NIH K12 GM093854-01. JK is supported by New Jersey City University and the RiSE program at Rutgers. SD is supported by NSF DEB 1026095 and BMGF/DFID OPP1052391. The funders had no role in study design, data collection and analysis, decision to publish, or preparation of the manuscript.

Competing Interests: The authors have declared that no competing interests exist.

* Email: hoe@lafayette.edu

Introduction

Begomoviruses (genus Begomovirus, family *Geminiviridae*) are single-stranded DNA viruses of dicots with small genomes - one or two circular segments of ~2.5–2.9 K nucleotides (nts). Begomoviruses are transmitted by the whitefly *Bemisia tabaci* [1,2] and their damaging infections pose a severe threat to commercial and subsistence production of key crops worldwide, including tomato, squash, cassava and bean [3]. Understanding the molecular biology and adaptation of begomoviruses to novel hosts has an important socioeconomic impact as they are emerging problems in developing countries [3]. The vast majority of begomovirus sequences also exhibit a classic biogeographic pattern: they fall into clades of New World (the Americas, and Caribbean) and Old World (rest of the world) viruses, with New World viruses thought to be derived from those in the Old World [4–6]. Bipartite begomoviruses, which have two similarly-sized, ambisense genomic segments termed DNA-A and DNA-B, are found worldwide,

with monopartite begomoviruses largely restricted to the Old World [7]. The DNA-A segment contains five or six genes, including the capsid protein (CP, also known as AV1), the replication-associated protein (REP, also known as AC1), a transcriptional activator (TrAP, also known as AC2), a replication enhancer (REn, also known as AC3) that overlaps with both the REP and TrAP genes and a virulence factor (AC4) that overlaps the reading frame within REP. The DNA-B segment contains two non-overlapping genes: the nuclear shuttle protein (NSP, also known as BV1), and the movement protein (MP, also known as BC1).

Old and New World bipartite begomoviruses share a high degree of homology, with the largest exception being the gene for the pre-coat protein (PCP, also known as AV2), which partially overlaps the CP gene and is only present in Old World viruses [8]. PCP and the monopartite V2 has been shown to localize at the cell periphery and is thought to act as a "movement protein" by increasing the size exclusion limit of the plasmodesmata [9,10].

They also suppresses RNA silencing by binding to the host's SGS3 protein [11]. V2 is thought to be the key movement protein in monopartite Old World viruses, but two genes on the DNA-B segment (NSP and MP) also contribute to systemic infection of plants by bipartite begomoviruses [12]. Virulent New World begomoviruses must rely on their other seven proteins to cope with the loss of PCP, and this is frequently invoked as the reason the DNA-B segment is required for infectivity of the overwhelming majority of New World begomoviruses [13]. Despite this assumption, the selective pressures imposed by the loss of PCP on the remaining New World viral genes have not been examined.

In this report we have compared the genome size, degree of variability and purifying selection of the viral genes between the Old and New World. Results indicate a loss of 100 nts in PCP's promoter region, stronger purifying selection on the two DNA-B genes in the New World, and the emergence of a putative tyrosine phosphorylation site in the New World MP. Studies with RNA plant viruses have shown that phosphorylation of MP regulates their localization and may account for cell-to-cell movement [14]. We speculate that the reduction in viral cell-to-cell movement caused by the loss of the PCP in the New World begomoviruses may be compensated by systematic genome size reduction and/or the gain of additional phosphorylation activity in the MP.

Materials and Methods

Compilation of bipartite begomovirus genomes

Genomes of begomovirus were downloaded from the June-2012 release of the viral genome database hosted in NCBI (ftp://ftp.ncbi.nih.gov/refseq/release/viral/). Only genomes containing the distinct, invariant nonamer "TAATATT|AC" were included in this study (the vertical bar represents the cleavage site). The pairing of DNA-A and DNA-B genomes, and the classification of genomes into Old and New Worlds were done semi-automatically according to the information stated in NCBI's RefSeq records [15] and ICTV report [16]. To ease sequence comparison, the beginning of the cleavage site "AC" was adopted as reference position 1 and the original genomic coordinates stated in NCBI's RefSeq records of the begomoviruses were adjusted accordingly. 33 and 83 Old and New World bipartite begomoviruses were collected, respectively, before further redundancy checking.

Identification of common regions

The DNA-A and DNA-B genomes of a bipartite begomovirus share a 200- to 250-nt long highly identical segment (>85%), namely the common region (CR), in which the invariant nonamer "TAATATT|AC" resides near to the middle of it. To determine the 5′ and 3′ termini of the CR, a pair of segments consisting of 250 nts upstream and downstream flanking regions of the invariant nonamer from DNA-A and DNA-B was aligned. Based on the alignment, the longest stretch of highly identical (at least 20 nts long and 80% identity) segment flanking the invariant nonamer was taken as the CR.

Identification of non-redundant genomes and ORFs

We clustered DNA-As together if their CRs shared >80% similarity. If more than one species was found in a cluster, only one species was retained arbitrarily for further analysis. A Peruvian begomovirus, Tomato leaf deformation virus (ToLDeV), was confirmed to be the first New World monopartite begomovirus in 2013 [7], but this was after our dataset had been finalized. ToLDeV does not appear to have a PCP gene. As a result, 26 out of 33 (85%) and 65 out of 83 (78%) non-redundant Old World and New World bipartite begomovirus genomes were included in this

analysis (Table 1). The full list of bipartite begomovirus genomes used and their sizes can be found in the Table S1 in File S1. Additionally, ORFs specified in RefSeq records were verified. We required the stated coding sequences or ORFs be translated exactly to the protein sequences specified in the RefSeq records. Genes failed to meet this requirement were excluded from this study (Table 1). Genomes and viral protein sequences used in this study can be downloaded as Data S1 or through this web link: http://sites.lafayette.edu/hoe/files/2014/01/bipartite_seqs_eh_jk_sd.tar_.gz.

dN/dS calculation

dN/dS represents the log ratio of the rate of non-synonymous substitutions to the rate of synonymous (silent) substitutions. A negative, zero, or positive dN/dS value indicates purifying (negative), neutral, or positive selection, respectively. Protein sequences were aligned by T_COFFEE [17] using default parameters. Protein alignments were converted to codon alignments using pal2nal v14 [18]. The codon alignments were submitted to the tool SLAC [19] hosted in the Datamonkey web server http://www.datamonkey.org/ [20] for site dN/dS calculation. Substitution models were selected by iterating the likelihood ratio tests between nested and non-nested models. This procedure is implemented in Datamonkey web server and detailed discussion of the procedure can be found in [21]. The results calculated by SLAC were downloaded in CSV format for analysis.

Pairwise protein sequence alignment

As dynamic programming approach to local pairwise sequence alignment produces the optimal alignment for a given scoring scheme. We used the percentage of identity calculated by an implementation of such approach i.e. Smith-Waterman water program [22], to determine the diversity of each viral protein for either Old or New World regions. BLOSUM62 score matrix was used and gap opening penalty and gap extension penalty were 10 and 0.5, respectively.

D-statistic of the Kolmogorov-Smirnov test

In order to ascertain the statistical significance of the difference between two non-Gaussian, cumulative distributions of protein sequence similarities and dN/dS values, we quantified the difference using the D-statistic of the two-sample Kolmogorov-Smirnov (KS) test. In both worlds, the viral protein AC4 exhibited the highest diversity. Thus, AC4 was chosen as the reference for two-sample D-statistic calculation. D-statistics were computed using the R function `ks.test()` [23]. All the values of D-statistic calculated showed significant differences between the two worlds with p-value in the range of 10^{-16}.

Scanning of functional sites in the movement protein

We developed a Python script (available upon request) to scan for functional sites in protein sequences using the BioPython scanProsite package [24], where the option for skipping of high probability of occurrence was turned off. In addition, our script used the bootstrap approach to compute the p-value of hits through these steps: 1. Obtain the list of functional sites detected in the input sequences through scanProsite, 2. Scramble input sequences, 3. Scan for functional sites in scrambled sequences, 4. Register the list of functional sites found in scrambled sequences, 5. Repeat steps 2 to 4 100 times (a user-defined parameter), 6. Estimate the p-value of a functional site by dividing the occurrence of the functional site in scrambled sequences by the occurrence of the same site in the original input sequences.

Table 1. Number of bipartite begomovirus genomes and proteins included in this study.

	Old World (28)	New World (65)
Coat protein (CP/AV1)	28	65
Pre-coat protein (PCP/AV2)	23	0
Replication-associated protein (REP/AC1)	28	65
Transcription activator protein (TrAP/AC2)	27	63
Replication enhancer (REn/AC3)	27	64
AC4	26	42
Nuclear shuttle protein (NSP/BC1)	28	65
Movement protein (MP/BV1)	28	64

The bracketed numbers in the column head is the number of genomes included in this report. Note that not all genes from included genomes were automatically accepted in this study because we found some of the ORFs documented in NCBI RefSeq database showed discrepancies with the associated protein sequences.

Results and Discussion

New World begomoviruses have smaller segments

We discovered that the genome size of DNA-A in the New World is on average 121 nts shorter than their counterparts in the Old World (Figure 1C). Intriguingly, though no apparent gene loss event was reported previously in the New World DNA-B, their genomes (mean size is 2,589 nts, standard deviation or s.d. 43) are also on average 113 nts smaller than the Old World DNA-B (mean is 2,702 nts and s.d. 55) as shown in Figure 1C. This commensurate genome size reduction does not seem to be coincidental as bipartite genome segments (DNA-A and DNA-B) in the New World begomoviruses show a higher correlation in size ($R = 0.91$, p-value $< 2.2 \times 10^{-16}$) than those in the Old World ($R = 0.74$, p-value $< 8.1 \times 10^{-6}$). Besides, the genome size differences between DNA-A and DNA-B concurred this point as we found smaller and less variable differences between DNA-A and DNA-B in the New World (mean 37 nts, s.d. 18) than the Old World (mean 45 nts, s.d. 37). Regardless of the geographical factor, this result may suggest size codependency of the bipartite genomes, which is still largely unknown. Our findings are unlikely confounded by biased samples as viral genes AC4 (pink) and REP (blue) exhibit similar spectra of sequence diversity between the two worlds (Figure 2). We further investigated whether or not deletions are localized at a particular region and how it may explain the loss of PCP in the New World begomoviruses.

Deletions are localized at PCP's promoter region

We compared dinucleotide profiles in the 400-nt upstream, homologous region of all viral genes using a 60-nt sliding window between the two worlds (see Materials and Methods, and Figure S1A–G in File S1). We found that dinucleotides were better than single nucleotides in insulating the profiles from random nucleotide fluctuation. If short (<5 nts) insertions or deletions are scattered, dinucleotide profiles between the two worlds should exhibit similar patterns; otherwise we should see a direct shift between the two profiles. Among dinucleotide profiles of all genes, only C+G profiles, i.e. CC, CG, GC, and CG, of the CP gene were found to differ between the two worlds in which the region with high concentration of C+G in the New World was shifted ~100 nts closer to the start of the ORF (Figure 1A). The elevated C+G content is chiefly due to the stem of the highly conserved hairpin structure found in all begomoviruses in which the loop region contains the invariant "TAATATT|AC" nonamer ("|" represents the cleavage site during complementary strand synthesis). Corroborating results were found when we examined the

distance between the cleavage site "AC" and the start of the CP genes in both worlds (Figure 1B) where the New World's CP gene is on average 100 nts closer to the cleavage site "AC" than those in the Old World. This accounts for much of the 121 nts shorter average genome size of DNA-A of the New World viruses compared to those of the Old World (Figure 1C). In New World begomoviruses the distance from the cleavage site to CP is highly correlated with DNA-A size ($R = 0.93$), but to a lesser extent in the Old World ($R = 0.72$). Additionally, the C+G content in the non-overlapping region of PCP (from −164 to 0 in Figure 1A) remains at similar level between the two worlds. Therefore our analysis indicates one or more deletions totaling more than 100 nts were mainly localized in the proximal promoter region of PCP, not in other genomic regions, and that these deletions may have led to PCP inactivation in the New World begomoviruses.

Currently, little is known about the effect of genome size on cell-to-cell transport through plasmodesma but studies have shown that the plasmodesmata impose a size limit [25,26]. Effective shuttling of viral genomes between cells without passing through the cell wall is critical for maintaining infectivity of plant viruses as small viruses do not encode enzymes to breakdown the cell wall, which other phytopathogens such as fungi employ [27]. This finding suggests genome size reduction may be one of the evolutionary paths selected for in the New World begomoviruses in order to maintain virulence despite the loss of cell-to-cell movement conferred by PCP.

New World NSP and MP are under enhanced purifying selection

The compact begomovirus genome encodes only a small number of highly overlapping genes in ambisense, most known to have multiple functions during infection. The lost functions of the PCP gene are likely compensated by remaining genes in New World begomoviruses. Therefore, we took a comparative approach to identify the presence of purifying selection in the New World viral proteins. We measured the within-world diversity of each gene by pairwise protein sequence alignment. A high sequence similarity indicates strong conservation pressures on the genes. Figure 2 shows the cumulative distributions of pairwise identity (%id) of seven or eight viral proteins from the two worlds. In the Old World (Figure 2A), AC4 exhibits the highest variability followed by the two DNA-B proteins NSP and MP, then REn, TrAP, REP and finally CP. The CP is known to be the most conserved of all begomovirus proteins, and under the greatest amount of purifying selection [12,28]. All distributions were tested

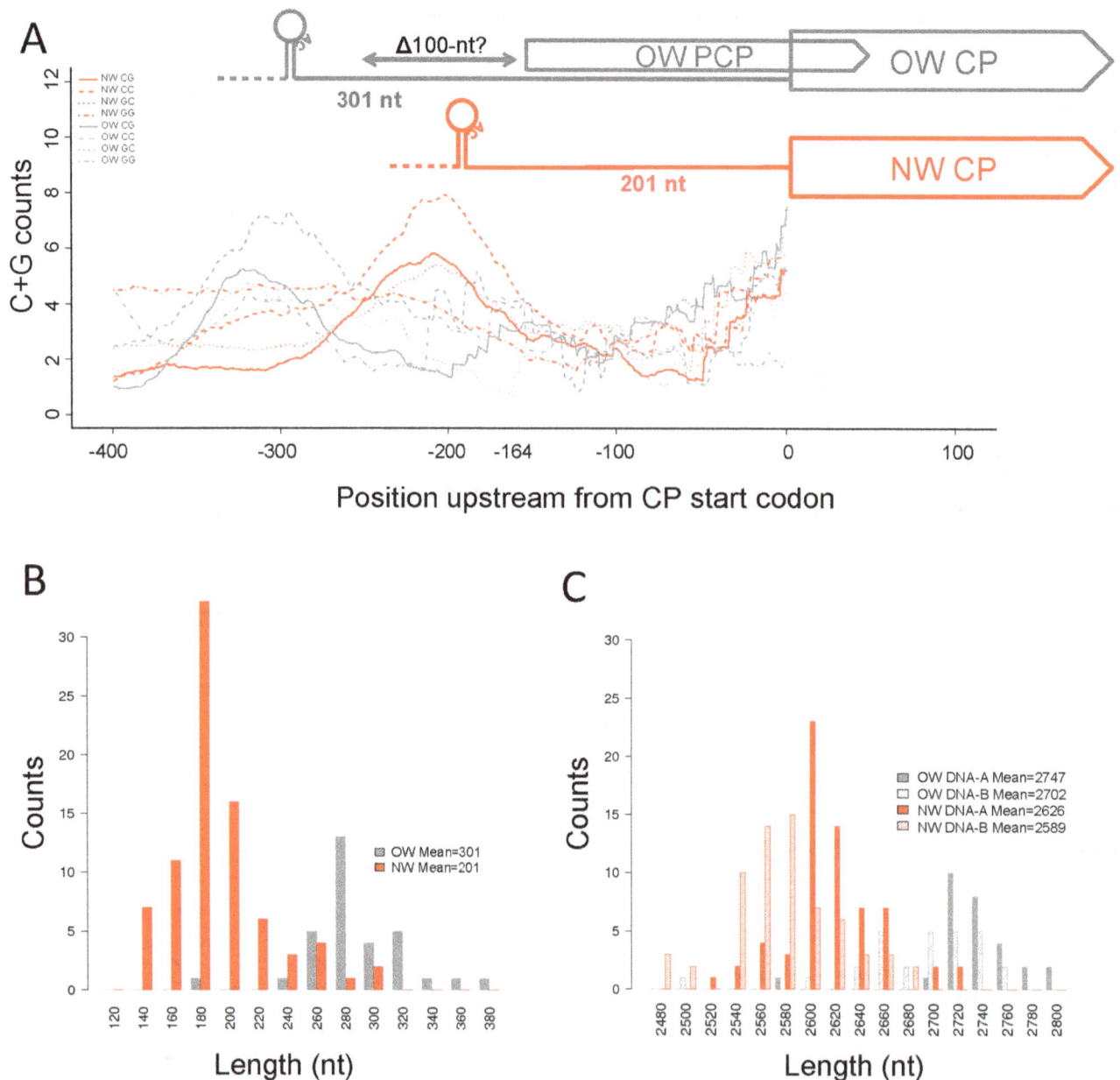

Figure 1. The loss of 100 nts from the promoter region of the New World PCP. A) C+G profiles of the homologous regions upstream from the CP from Old and New World bipartite begomoviruses. All positions labeled in the gene structure diagram are average values. Each plot represents the average number of CC, CG, GC, or GG in a 60-nt window. B) Distributions of the distance between the cleavage site "AC" of the invariant nonamer and the beginning of CP gene. C) Distribution of genome size.

for statistical significance (p-value $\sim 10^{-26}$) according to the two-sample Kolmogorov-Smirnov test with AC4 as the reference protein. Results from the New World viral proteins show much lower levels of diversity except for AC4 and REP. Such results are consistent with the presumed more recent origin of New World begomoviruses [4], but show a different pattern in protein variability (Figure 2B). The seven proteins are still bounded by AC4 as the most variable protein, and CP being the most conserved. The most striking difference is in the reduced variability of the New World MP (black plot in Figure 2B), which has become the most conserved protein after CP. Additionally, the New World NSP (gray plot in Figure 2B) is also found to show significant reduction in variability, comparable to the essential

replication protein REP (blue plot in Figure 2B). It appears that both DNA-B genes are under stronger selective pressure in the absence of PCP. These results suggest that the less genomically compressed DNA-B genomic segment was more able to accommodate new or enhanced functions than the more constrained DNA-A segment, which already has several overlapping open reading frames.

To further confirm this point, we sought evidence for adaptive evolution at the nucleotide level to corroborate these protein sequence analyses. Adaptive evolution is measured by the log ratio of the rate of non-synonymous substitutions versus synonymous substitutions (dN/dS). If the rate of non-synonymous substitution is lower than synonymous substitution, dN/dS will yield a negative

A

Old World

B

New World

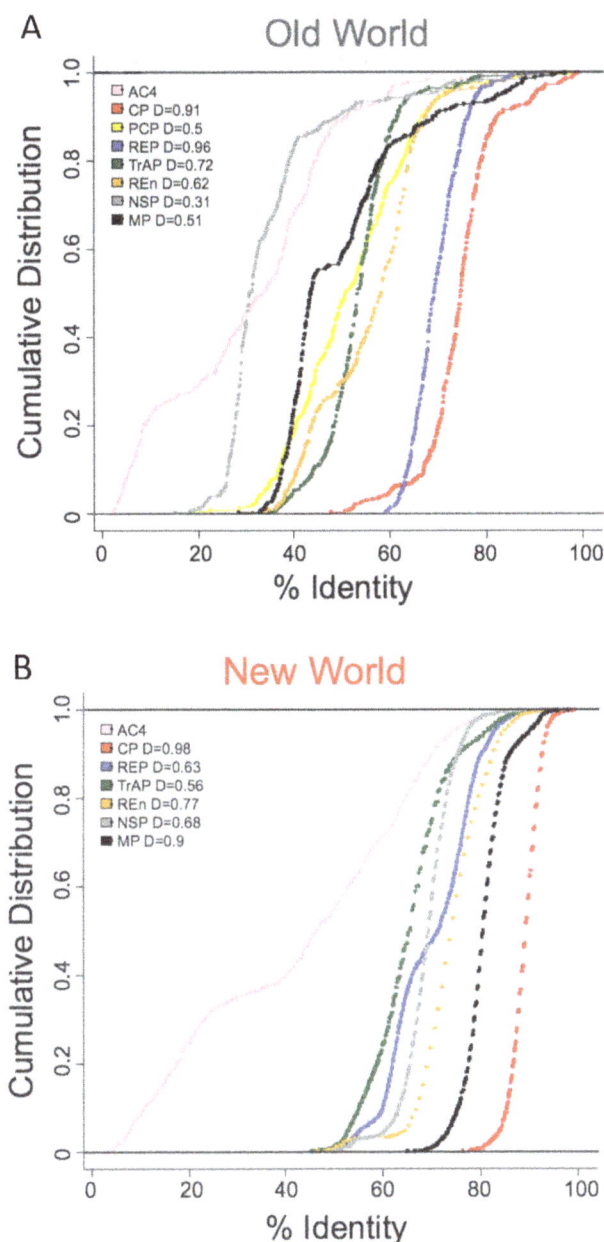

Figure 2. Protein sequences variability by gene. A) Cumulative distributions of percentage of identity (%id) of viral proteins from the Old World bipartite begomoviruses. The D values printed beside the protein name in the legend represent the magnitude of deviation of the plot from the AC4's curve and it was determined by two-sample Kolmogorov test. Larger the D value, the great is the deviation from AC4. B) Proteins from the New World.

value, indicating amino acid substitution is unfavorable. Conversely, a positive dN/dS value indicates amino acid substitution is permissible, suggesting the protein is under positive or adaptive selection. Aligned protein sequences were converted to corresponding codon alignments before dN/dS calculation using the Single Likelihood Ancestor Counting (SLAC) method from the Datamonkey website [20,29]. Figure 3A–B show the cumulative dN/dS ratios for MP and NSP. The New World MP shows the biggest deviation (D = 0.54, p-value = 0) from the Old World counterpart and nearly all dN/dS values fall in the negative

region, reconfirming elevated purifying selection in the New World MP. But we did not see this in other viral genes (Figure S2 in File S1). We further explored whether or not purified residues in the New World MP constitute to any functional motif(s).

The emergence of tyrosine phosphorylation site in the New World MP

In order to uncover the specific nucleotides subjected to elevated purifying selection in the New World MP, we compared the functional sites of MP in both worlds. Among all functional sites discovered, a putative tyrosine phosphorylation site [RK]-x(2,3)-[DE]-x(2,3)-Y (PROSITE ID: PS00007; the notation means the site starts with either R or K, followed by any 2 to 3 residues, and then a D or E residue, followed by any 2 to 3 residues, and ends with Y) shows the greatest difference between the two worlds: 62/65 New World MP sequences were found to carry the tyrosine phosphorylation site compared to only 1/28 from the Old World.

We also evaluated the likelihood for the emergence of this site by comparing the putative tyrosine phosphorylation site in the New World MP with the homologous region in the Old World MP. We aligned the eight-residue site and the corresponding codons. The amino acid consensus of the Old World (Figure 3C) shows only two residue substitutions are needed to transform the functionally indeterminate eight-residue site in the Old World MP to the tyrosine phosphorylation site found in the New World MP. From the codon perspective, four nucleotide substitutions from the first and fifth codons are sufficient to transform the site (Figure 3D).

Conclusions

Our analysis strongly suggests one or more deletions of 100 nts in the promoter region of the New World PCP, which may be linked to the inactivation of PCP. The resultant shrunken genome may have had an advantage in cell-to-cell movement through plasmodesmata. Furthermore, our genome size analysis unraveled putative size codependency of the bipartite genomes. As the New World begomoviruses are presumably originated from the Old World counterparts recently [4], the conspicuous correlation between the New World DNA-A and DNA-B could be alluded to bottleneck effect. However, the more diverse Old World begomoviruses still maintain a high level of correlation (R = 0.74, p-value $<8.1 \times 10^{-6}$) between segment size of their bipartite genomes. DNA-B's functional sequences – the common region ($\sim$200 nts), ORFs of NSP ($\sim$800 nts) and MP ($\sim$900 nts) and their promoter regions ($\sim$200 nts) – occupy $\sim$2,100 nts of the 2,700-nt genome on average, leaving $\sim$600 nts (22% of the genome) available for size reduction. According to our data (Table S1 in File S1), the mean, median and maximum difference between the bipartite genomes of the Old World viruses are only 45, 40, and 170 nts, respectively, which are far smaller than the 600 nts permissible range without interrupting the genomic structure of the viruses. This result is surprising as begomoviruses are fast mutating [30] and recombining [31,32] ssDNA viruses, indicating the presence of unknown constraints that limit the variance in size between segments in bipartite genomes.

Our prediction aligns with findings in closely related monopartite begomoviruses. PCP has been reported previously to perform some MP functions, such as intracellular movement and cell periphery localization [9,25,26]. Additionally, in-vitro phosphorylation activity was reported in MP of Abutilon mosaic virus [33]. Our thorough bioinformatic comparison of geographically separated begomovirus species has produced a candidate region for detailed wet lab analysis. If the tyrosine phosphorylation site is

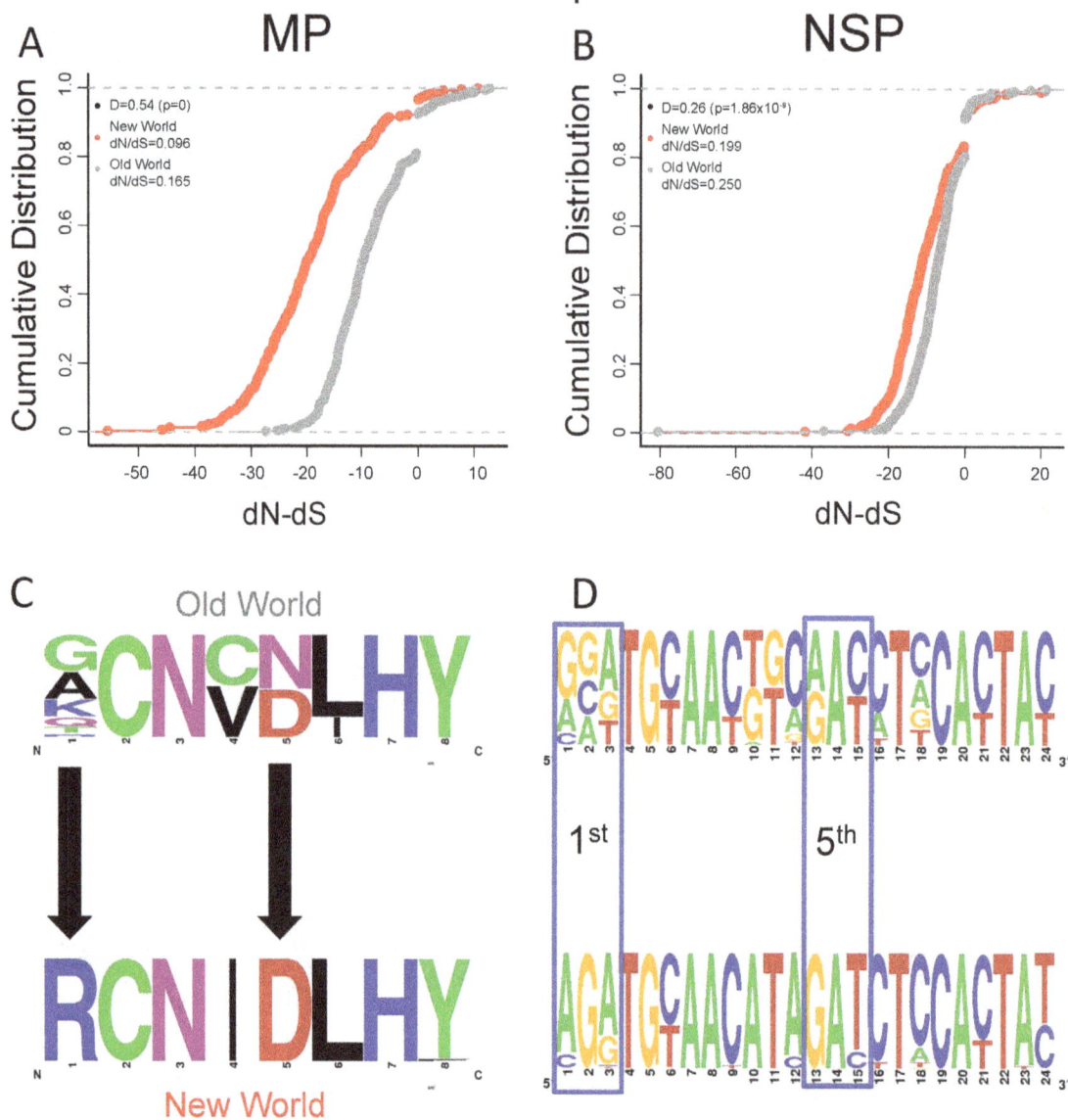

Figure 3. Cumulative distributions of site dN/dS values of the viral proteins. D-value, with p-value, represents the deviation of the New World curve from the Old World curve. D-value was calculated by the two-sample Kolmogorov test. Averaged site dN/dS is displayed on the top left, which reflects the overall selection pressure on the protein. A) MP. B) NSP. C) The emergence of putative tyrosine phosphorylation site in the New World MP. Consensus sequence pictures were created using Weblogo [34]. Searching is based on PROSITE database [35]. PROSITE ID of the tyrosine phosphorylation site is PS00007 where its consensus is [RK]-x(2,3)-[DE]-x(2,3)-Y. D) Codon alignment of the homologous region of the site.

critical to infectivity of New World begomoviruses, it will be a novel target for sequence-specific, anti-viral strategies.

Supporting Information

File S1 Contains the following files: **Table S1:** Selected bipartite begomoviruses and their genome size. **Figure S1:** Dinucleotide profiles in 400-nt upstream regions. Window size is 60 nts. Y-axis denotes the average occurrences of the specified dinucleotide in the 60-nt window. Plots of dinucleotides AA, AC, ..., TG, TT are arranged from top left to bottom right. **Figure S2:** Cumulative dN/dS values by gene.

Data S1 Genomes and protein sequences used in this study.

Author Contributions

Conceived and designed the experiments: ESH JK SD. Performed the experiments: ESH JK. Analyzed the data: ESH SD. Contributed reagents/materials/analysis tools: ESH JK. Contributed to the writing of the manuscript: ESH SD.

References

1. Nault LR (1997) Arthropod transmission of plant viruses: A new synthesis. Annals of the Entomological Society of America 90: 521–541.
2. Zhang WM, Fu HB, Wang WH, Piao CS, Tao YL, et al. (2014) Rapid Spread of a Recently Introduced Virus (Tomato Yellow Leaf Curl Virus) and Its Vector Bemisia tabaci (Hemiptera: Aleyrodidae) in Liaoning Province, China. Journal of Economic Entomology 107: 98–104.
3. Seal SE, Jeger MJ, Van den Bosch F (2006) Begomovirus evolution and disease management. Plant Virus Epidemiology. 297–316.
4. Rybicki EP (1994) A phylogenetic and evolutionary justification for 3 genera of geminiviridae. Archives of Virology 139: 49–77.
5. Xu XZ, Liu QP, Fan LJ, Cui XF, Zhou XP (2008) Analysis of synonymous codon usage and evolution of begomoviruses. Journal of Zhejiang University Science B 9: 667–674.
6. Rojas MR, Hagen C, Lucas WJ, Gilbertson RL (2005) Exploiting chinks in the plant's armor: evolution and emergence of geminiviruses. Annu Rev Phytopathol 43: 361–394.
7. Melgarejo TA, Kon T, Rojas MR, Paz-Carrasco L, Zerbini FM, et al. (2013) Characterization of a new world monopartite begomovirus causing leaf curl disease of tomato in Ecuador and Peru reveals a new direction in geminivirus evolution. J Virol 87: 5397–5413.
8. Ha C, Coombs S, Revill P, Harding R, Vu M, et al. (2008) Molecular characterization of begomoviruses and DNA satellites from Vietnam: additional evidence that the New World geminiviruses were present in the Old World prior to continental separation. The Journal of general virology 89: 312–326.
9. Rothenstein D, Krenz B, Selchow O, Jeske H (2007) Tissue and cell tropism of Indian cassava mosaic virus (ICMV) and its AV2 (precoat) gene product. Virology 359: 137–145.
10. Poornima Priyadarshini CG, Ambika MV, Tippeswamy R, Savithri HS (2011) Functional characterization of coat protein and V2 involved in cell to cell movement of Cotton leaf curl Kokhran virus-Dabawali. PLoS ONE 6: e26929.
11. Glick E, Zrachya A, Levy Y, Mett A, Gidoni D, et al. (2008) Interaction with host SGS3 is required for suppression of RNA silencing by tomato yellow leaf curl virus V2 protein. Proc Natl Acad Sci U S A 105: 157–161.
12. Padidam M, Beachy RN, Fauquet CM (1995) Classification and identification of geminiviruses using sequence comparisons. Journal of General Virology 76: 249–263.
13. Briddon RW, Patil BL, Bagewadi B, Nawaz-ul-Rehman MS, Fauquet CM (2010) Distinct evolutionary histories of the DNA-A and DNA-B components of bipartite begomoviruses. BMC Evol Biol 10: 97.
14. Modena NA, Zelada AM, Conte F, Mentaberry A (2008) Phosphorylation of the TGBp1 movement protein of Potato virus X by a Nicotiana tabacum CK2-like activity. Virus Res 137: 16–23.
15. Pruitt KD, Brown GR, Hiatt SM, Thibaud-Nissen F, Astashyn A, et al. (2014) RefSeq: an update on mammalian reference sequences. Nucleic Acids Res 42: D756–763.
16. King AMQ, Adams MJ, Carstens EB, Lefkowitz EJ, editors (2012) Virus taxonomy: classification and nomenclature of viruses: Ninth Report of the International Committee on Taxonomy of Viruses. San Diego: Elsevier.
17. Notredame C, Higgins DG, Heringa J (2000) T-Coffee: A novel method for fast and accurate multiple sequence alignment. J Mol Biol 302: 205–217.
18. Suyama M, Torrents D, Bork P (2006) PAL2NAL: robust conversion of protein sequence alignments into the corresponding codon alignments. Nucleic Acids Res 34: W609–612.
19. Kosakovsky Pond SL, Frost SDW (2005) Datamonkey: rapid detection of selective pressure on individual sites of codon alignments. Bioinformatics 21: 2531–2533.
20. Delport W, Poon AF, Frost SD, Kosakovsky Pond SL (2010) Datamonkey 2010: a suite of phylogenetic analysis tools for evolutionary biology. Bioinformatics 26: 2455–2457.
21. Posada D, Buckley TR (2004) Model selection and model averaging in phylogenetics: advantages of akaike information criterion and bayesian approaches over likelihood ratio tests. Syst Biol 53: 793–808.
22. Rice P, Longden I, Bleasby A (2000) EMBOSS: the European Molecular Biology Open Software Suite. Trends Genet 16: 276–277.
23. Team RDC (2012) R: a languate and environment for statistical computing. Vienna: R Foundation for Statistical Computing.
24. Cock PJ, Antao T, Chang JT, Chapman BA, Cox CJ, et al. (2009) Biopython: freely available Python tools for computational molecular biology and bioinformatics. Bioinformatics 25: 1422–1423.
25. Gilbertson RL, Sudarshana M, Jiang H, Rojas MR, Lucas WJ (2003) Limitations on geminivirus genome size imposed by plasmodesmata and virus-encoded movement protein: insights into DNA trafficking. Plant Cell 15: 2578–2591.
26. Rojas MR, Jiang H, Salati R, Xoconostle-Cazares B, Sudarshana MR, et al. (2001) Functional analysis of proteins involved in movement of the monopartite begomovirus, Tomato yellow leaf curl virus. Virology 291: 110–125.
27. Tonukari NJ, Scott-Craig JS, Walton JD (2000) The Cochliobolus carbonum SNF1 gene is required for cell wall-degrading enzyme expression and virulence on maize. Plant Cell 12: 237–248.
28. Duffy S, Holmes EC (2009) Validation of high rates of nucleotide substitution in geminiviruses: phylogenetic evidence from East African cassava mosaic viruses. J Gen Virol 90: 1539–1547.
29. Kosakovsky Pond SL, Frost SDW (2005) Not so different after all: a comparison of methods for detecting amino acid sites under selection. Mol Biol Evol 22: 1208–1222.
30. Duffy S, Shackelton LA, Holmes EC (2008) Rates of evolutionary change in viruses: patterns and determinants. Nat Rev Genet 9: 267–276.
31. Monjane AL, Pande D, Lakay F, Shepherd DN, van der Walt E, et al. (2012) Adaptive evolution by recombination is not associated with increased mutation rates in Maize streak virus. BMC Evol Biol 12: 252.
32. Rocha CS, Castillo-Urquiza GP, Lima AT, Silva FN, Xavier CA, et al. (2013) Brazilian begomovirus populations are highly recombinant, rapidly evolving, and segregated based on geographical location. J Virol 87: 5784–5799.
33. Kleinow T, Holeiter G, Nischang M, Stein M, Karayavuz M, et al. (2008) Post-translational modifications of Abutilon mosaic virus movement protein (BC1) in fission yeast. Virus Res 131: 86–94.
34. Crooks GE, Hon G, Chandonia JM, Brenner SE (2004) WebLogo: a sequence logo generator. Genome research 14: 1188–1190.
35. Sigrist CJ, de Castro E, Cerutti L, Cuche BA, Hulo N, et al. (2013) New and continuing developments at PROSITE. Nucleic Acids Res 41: D344–347.

PERMISSIONS

LIST OF CONTRIBUTORS

John Daniel DeBord and Francisco A. Fernandez-Lima
Department of Chemistry and Biochemistry, Florida International University, Miami, Florida, United States of America

Donald F. Smith and Ron M. A. Heeren
FOM Institute AMOLF, Science Park 104, Amsterdam, The Netherlands

Christopher R. Anderton and Ljiljana Paša-Tolić
Environmental Molecular Sciences Laboratory, Pacific Northwest National Laboratory, Richland, Washington, United States of America,

Richard H. Gomer
Department of Biology, Texas A&M University, College Station, Texas, United States of America

Mei-Lin Wu, Zhao-Yu Jiang and Hao Cheng
State Key Laboratory of Tropical Oceanography, South China Sea Institute of Oceanology, Chinese Academy of Sciences, Guangzhou, China

Fu-Lin Sun, You-Shao Wang and Cui-Ci Sun
State Key Laboratory of Tropical Oceanography, South China Sea Institute of Oceanology, Chinese Academy of Sciences, Guangzhou, China
Daya Bay Marine Biology Research Station, South China Sea Institute of Oceanology, Chinese Academy of Sciences, Shenzhen, China

Xiao-Lin Li, Shubha P. Kale, Harris McFerrin, Madhusoodanan Mottamal, Xin Yao, Fengkun Du, Baihan Gu, Kim Hoang, Yen H. Nguyen, Nichelle Taylor, Chelsea R. Stephens and Qian-Jin Zhang
Department of Biology, Xavier University of Louisiana, New Orleans, Louisiana, United States of America

Marjolein Sluijter, Elien M. Doorduijn and Thorbald van Hall
Clinical Oncology, K1-P, Leiden University Medical Center, Leiden, the Netherlands

Yong-Yu Liu
Department of Basic Pharmaceutical Sciences, University of Louisiana at Monroe, Monroe, Louisiana, United States of America

Yan Li
Department of Biology, Xavier University of Louisiana, New Orleans, Louisiana, United States of America

College of Chemistry & Environmental Science, Hebei University, Hebei Province, Baoding, China

Meghan May
Department of Biomedical Sciences, College of Osteopathic Medicine, University of New England, Biddeford, Maine, United States of America

Dylan W. Dunne
Department of Biological Sciences, Jess and Mildred Fisher College of Science and Mathematics, Towson University, Towson, Maryland, United States of America

Daniel R. Brown
Department of Infectious Diseases and Pathology, College of Veterinary Medicine, University of Florida, Gainesville, Florida, United States of America

Janine T. Bossé, Yanwen Li and Paul R. Langford
Section of Paediatrics, Imperial College London, St Mary's Campus, London, United Kingdom

Denise M. Soares-Bazzolli
Section of Paediatrics, Imperial College London, St Mary's Campus, London, United Kingdom
Laboratório de Genética Molecular de Micro-organismos, Departamento de Microbiologia - DMB – BIOAGRO, Universidade Federal de Viçosa – Viçosa, Brazil

Brendan W. Wren
Department of Pathogen Molecular Biology, London School of Hygiene and Tropical Medicine, London, United Kingdom

Alexander W. Tucker and Duncan J. Maskell
Department of Veterinary Medicine, University of Cambridge, Cambridge, United Kingdom

Andrew N. Rycroft
Department of Pathology and Pathogen Biology, The Royal Veterinary College, North Mymms, Hatfield, United Kingdom

Martin Axelsson and Francesco Gentili
Department of Wildlife, Fish, and Environmental Studies, Swedish University of Agricultural Sciences, Umeå, Sweden

Ryan O. Emerson
Adaptive Biotechnologies Corporation, Seattle, Washington, United States of America

Iwona M. Konieczna and Joseph R. Leventhal
Department of Surgery, Comprehensive Transplant center, Northwestern University, Chicago, Illinois, United States of America

James M. Mathew
Department of Surgery, Comprehensive Transplant center, Northwestern University, Chicago, Illinois, United States of America
Department of Microbiology-Immunology, Northwestern University, Chicago, Illinois, United States of America

Harlan S. Robins
Public Health Sciences Division, Fred Hutchinson Cancer Research Center, Seattle, Washington, United States of America

Adrian De Stefano and Florencia Karlanian
Laboratory of Animal Biotechnology, Faculty of Agriculture, University of Buenos Aires, Buenos Aires, Argentina

Andrés Gambini, Daniel Felipe Salamone and Romina Jimena Bevacqua
Laboratory of Animal Biotechnology, Faculty of Agriculture, University of Buenos Aires, Buenos Aires, Argentina
National Institute of Scientific and Technological Research, Buenos Aires, Argentina

Manuela Mandl and Michael Hristov
Institute for Cardiovascular Prevention, Ludwig-Maximilians-University (LMU), Munich, Germany

Susanne Schmitz
Institute for Cardiovascular Prevention, Ludwig-Maximilians-University (LMU), Munich, Germany
Center of Allergy & Environment (ZAUM), Technical University of Munich, Munich, Germany

Christian Weber
Institute for Cardiovascular Prevention, Ludwig-Maximilians-University (LMU), Munich, Germany
Munich Heart Alliance, Munich, Germany

Geon A Kim, Hyun Ju Oh, Min Jung Kim, Young Kwang Jo, Jin Choi, Jung Eun Park, Eun Jung Park, Goo Jang and Byeong Chun Lee
Department of Theriogenology & Biotechnology, College of Veterinary Medicine, Seoul National University, Seoul, Republic of Korea

Sang Hyun Lim and Sung Keun Kang
Central Research Institutes, Kstem cell, Seoul, Republic of Korea

Byung Il Yoon
Laboratory of Histology and Molecular Pathogenesis, College of Veterinary Medicine, Kangwon National University, Chuncheon, Gangwon-do, Republic of Korea

Kyle H. Elliott
Department of Biological Sciences, University of Manitoba, Winnipeg, Canada

Mikaela Davis
Department of Biological Sciences, Simon Fraser University, Burnaby, Canada

John E. Elliott
Science & Technology Branch, Environment Canada, Delta, Canada

David F. Burke and Derek J. Smith
Department of Zoology, University of Cambridge, Cambridge, United Kingdom

Karen Staines, Lawrence G. Hunt, John R. Young and Colin Butter
The Pirbright Institute, Compton, United Kingdom

Kristen Anderson and Morgan S. Pratchett
ARC Centre of Excellence for Coral Reef Studies, James Cook University, Townsville, Australia

Chiara Pisapia
ARC Centre of Excellence for Coral Reef Studies, James Cook University, Townsville, Australia
AIMS@JCU Australian Institute of Marine Science, School of Marine Biology, James Cook University, Townsville, Australia

Ailsa H. C. McLean and H. Charles J. Godfray
Department of Zoology, University of Oxford, Oxford, United Kingdom

Bo Tian
Key Laboratory of Tropical Plant Resource and Sustainable Use, Xishuangbanna Tropical Botanical Garden, Chinese Academy of Sciences, Kunming, People's Republic of China

Guowei Zheng and Weiqi Li
Key Laboratory for Plant Diversity and Biogeography of East Asia, Kunming Institute of Botany, Chinese Academy of Sciences, Kunming, Yunnan, People's Republic of China

Plant Germplasm and Genomics Center, Germplasm Bank of Wild Species, Kunming Institute of Botany, Chinese Academy of Sciences, Kunming, Yunnan, People's Republic of China

Amornrat Chaiyasen and Saisamorn Lumyong
Department of Biology, Faculty of Science, Chiang Mai University, Chiang Mai, Thailand

J. Peter W. Young
Department of Biology, University of York, York, United Kingdom

Neung Teaumroong
Schoool of Biotechnology, Institute of Agricultural Technology, Suranaree University of Technology, Nakhon Ratchasima, Thailand

Paiboolya Gavinlertvatana
Thai Orchid Labs Co. Ltd., Khannayao, Bangkok, Thailand

Akihiro Itoh, Atsushi Uchiyama and Junji Sagara
Department of Biomedical Laboratory Sciences, Health Sciences, Shinshu University, Matsumoto, Japan

Shunichiro Taniguchi
Department of Molecular Oncology, Medical Sciences, Shinshu University Graduate School of Medicine, Matsumoto, Japan

Tingming Liang, Chen Yang, Ping Li and Chang Liu
Jiangsu Key Laboratory for Molecular and Medical Biotechnology, College of Life Science, Nanjing Normal University, Nanjing, 210023, China

Li Guo
Department of Epidemiology and Biostatistics, School of Public Health, Nanjing Medical University, Nanjing, 211166, China

Mario Núñez, Carmen Sánchez-Jiménez, José Alcalde and José M. Izquierdo
Centro de Biología Molecular 'Severo Ochoa', Consejo Superior de Investigaciones Científicas, Universidad Autónoma de Madrid (CSIC/UAM), Madrid, Spain

Bernhard M. Fuchs
Department of Molecular Ecology, Max Planck Institute for Marine Microbiology, Bremen, Germany

Stefan Thiele
Department of Molecular Ecology, Max Planck Institute for Marine Microbiology, Bremen, Germany
Stazione Zoologica Anton Dohrn, Naples, Italy

Christian Wolf and Katja Metfies
Department of Polar Biological Oceanography, Division of Bioscience, Alfred Wegener Institute - Helmholtz Centre for Polar and Marine Research, Bremerhaven, Germany

Isabelle Katharina Schulz
Department of Polar Biological Oceanography, Division of Bioscience, Alfred Wegener Institute - Helmholtz Centre for Polar and Marine Research, Bremerhaven, Germany
Bremen International Graduate School for Marine Sciences (GLOMAR), MARUM - Center for Marine Environmental Sciences, University of Bremen, Bremen, Germany
Red Sea Research Center, King Abdullah University of Science and Technology (KAUST), Thuwal, Kingdom of Saudi-Arabia

Philipp Assmy
Center for Ice, Climate and Ecosystems (ICE), Norwegian Polar Institute, Fram Centre, Tromsø, Norway

Silvija Miosic, Jana Thill, Malvina Milosevic, Christian Gosch, Sabrina Pober, Karl Stich and Heidi Halbwirth
Vienna University of Technology, Institute of Chemical Engineering, Vienna, Austria

Christian Molitor and Annette Rompel
Institut für Biophysikalische Chemie, Fakultät für Chemie, Universität Wien, Vienna, Austria

Shaghef Ejaz
Bahauddin Zakariya University, Department of Horticulture, Multan, Pakistan

Edoardo Trotta
Institute of Translational Pharmacology, Consiglio Nazionale delle Ricerche (CNR), Roma, Italy

Ingrid E. B. Lawhorn, Joshua P. Ferreira and Clifford L. Wang
Department of Chemical Engineering, Stanford University, Stanford, California, United States of America

Aleardo Morelli., John Haugner and Burckhard Seelig
Department of Biochemistry, Molecular Biology, and Biophysics, University of Minnesota, Minneapolis, Minnesota, United States of America, & BioTechnology Institute, University of Minnesota, St. Paul, Minnesota, United States of America

Claudia Skamel
Campus Technologies Freiburg (CTF) GmbH, Agency for Technology Transfer at the University and University Medical Center Freiburg, Freiburg, Germany

Stephen G. Aller
Department of Pharmacology and Toxicology and Center for Structural Biology, University of Alabama at Birmingham, Birmingham, Alabama, United States of America

Alain Bopda Waffo
Department of Biological Sciences, Alabama State University, Montgomery, Alabama, United States of America

Taisong Zou and S. Banu Ozkan
Center for Biological Physics, Department of Physics, Arizona State University, Tempe, Arizona, United States of America

Kingshuk Ghosh and Nickolas Williams
Department of Physics and Astronomy, University of Denver, Denver, Colorado, United States of America

Eric S. Ho
Department of Biology, Lafayette College, Easton, Pennsylvania, United States of America

Joan Kuchie
New Jersey City University, Jersey City, New Jersey, United States of America

Siobain Duffy
Department of Ecology, Evolution and Natural Resources, Rutgers University, New Brunswick, New Jersey, United States of America

Index

www.ingramcontent.com/pod-product-compliance
Lightning Source LLC
Chambersburg PA
CBHW080525200326
41458CB00012B/4333